Selenium in pig nutrition and health

Selenium
in pig nutrition and health

edited by:

Peter F. Surai

Wageningen Academic Publishers

EAN: 9789086863594
e-EAN: 9789086869121
ISBN: 978-90-8686-359-4
e-ISBN: 978-90-8686-912-1
DOI: 10.3920/978-90-8686-912-1

First published, 2021

Wageningen Academic Publishers
P.O. Box 220
6700 AE Wageningen
The Netherlands
www.WageningenAcademic.com
copyright@WageningenAcademic.com

Dedication

To my wife Helen, my daughter Katie, my son Anton,
my grandsons Oscar, Arthur and Henry
and my granddaughter Aiste
who supported me in my 40-year research career.

About the editor

Dr Peter Surai started his studies at Kharkov University, Ukraine, where he obtained his PhD and DSc in biochemistry studying effects of antioxidants on poultry. Later he became Professor of Human Physiology. In 1994 he moved to Scotland to continue his antioxidant related research in poultry and in 2000 he was promoted to a full Professor of Nutritional Biochemistry at the Scottish Agricultural College. Recently he was awarded Honorary Professorships in 7 universities in various countries, including UK, Hungary, Bulgaria, Russia and Ukraine. In 2010 he was elected to the Russian Academy of Sciences as a foreign member. He has more than 850 research publications, including 360 papers in peer-reviewed journals, 15 books and 46 chapters in various books. In 1999 he received the prestigious John Logie Baird Award for Innovation for the development of 'super-eggs' and, in 2000, The World's Poultry Science Association Award for Research in recognition of an outstanding contribution to the development of the poultry industry. In 2017 he became a member of the team at the Moscow State Academy of Veterinary Medicine and Biotechnology named after K.I. Skryabin to conduct a research under a mega-grant of the Government of Russian Federation (Contract No. 14.W03.31.0013). He successfully transferred the vitagene concept of stress adaptation from medical sciences into poultry and animal sciences. For the last 20 years he has been lecturing all over the world visiting more than 70 countries.

Table of contents

Abbreviations

AA	ascorbic acid
Ab	antibody
ADFI	average daily feed intake
ADG	average daily gain
AEC	abdominal exudate cells
AFB1	aflatoxin B1
AGE	advanced glycation end products
AI	artificial insemination
AMP	antimicrobial peptides
AO	antioxidant
AOA	antioxidant activity
AOPP	advanced oxidation protein products
APC	antigen-presenting cell
APP	acute phase protein
APR	acute phase response
AR	antioxidant response
ARE	antioxidant response element
AST	aspartate aminotransferase
ATF	activating transcription factor
ATP	adenosine triphosphate
BAP	biological antioxidant potential
BCR	B cell receptor
BD	basal diet
BW	body weight
C/EBP	CCAAT-enhancer-binding protein
CAT	catalase
CD	cluster of differentiation
ConA	concanavalin A
CoQ	coenzyme Q
COX-2	cyclooxygenase-2
CTL	cytotoxic T cell
CVB3	Coxsackie virus B3
Cys	cysteine
DAMP	damage-associated molecular patterns
DC	dendritic cells
DHA	docosahexaenoic acid
DHT	dihydrotestosterone
Dio	iodothyronine deiodinase
DON	deoxynivalenol
DTH	delayed-type hypersensitivity
EC-SOD	extracellular superoxide dismutase
ED	exudative diathesis
eEFSec	Sec-specific elongation factor
EFX	enrofloxacin

ELAM-1	endothelial-leukocyte adhesion molecule 1
ER	endoplasmic reticulum
ERAD	ER-associated degradation
ERK	extracellular signal-regulated kinase
ESI	electrospray ionisation
FCR	feed conversion ratio
FDA	US Food and Drug Administration
FOXO	forkhead box transcription factor
GI	gastrointestinal
GOT	glutamic oxaloacetic transaminase
GPx	glutathione peroxidase
GR	glutathione reductase
GSH	reduced glutathione
GSH-Px	glutathione peroxidase
GSSG	oxidised glutathione
GST	glutathione S-transferase
H/L	ratio heterophil to lymphocyte ratio
H_2O_2	hydrogen peroxide
HI	hemaglutination inhibition
HIV	human immunodeficiency virus
HMSeBA	selenomethionine hydroxyanalogue, 2-hydroxy-4-methylselenobutanoic acid
HO-1	haeme oxygenase-1
HPLC	high-performance liquid chromatography
HS	heat stress
HSP	heat shock proteins
IBD	infectious bursal disease
ICAM-1	intercellular adhesion molecule 1
ICP	inductively coupled plasma
ID	iodothyronine deiodinase
IELs	intraepithelial lymphocytes
IFN	interferon
Ig	immunoglobulin
IL-1	interleukin 1
IL-2R	interleukin 2 receptor
IL-6	interleukin 6
iNOS	inducible nitric oxide synthase
LA	linoleic acid
LITAF	lipopolysaccharide-induced TNF-α factor
LOO*	hydroperoxyl radical
LOOH	lipid hydroperoxide
LOX	lipoxygenase
LPL	lamina propria lymphocytes
LPS	lipopolysaccharide
LXA4	lipoxin A4
mAbs	monoclonal antibodies
MAP	microangiopathy

MAPK	mitogen-activated protein kinase
MD	Marek's disease
MDA	malondialdehyde
MeSeCys	methyl selenocysteine
Met	methionine
Met-SO	methionine sulfone
MHC	major histocompatibility complex
MHD	Mulberry heart disease
MLTC	mixed lymphocyte/tumour cell cultures
MPO	myeloperoxidase
MS	mass spectrometry
Msr	methionine sulfoxide reductase
NADH	nicotinamide-adenine-dinucleotide
NADPH	Nicotinamide-adenine-dinucleotide phosphate
NETs	neutrophil extracellular traps
NF-κB	nuclear factor-kappa B
NK cells	natural killer cells
NKT cells	natural killer T cells
NLR	nucleotide-binding oligomerisation domain-like receptors
NMD	nutritional muscular dystrophy
NO	nitric oxide
Nrf2	NF-E2-related factor 2
OAS	oligoadenylate synthetase
OH-SeMet	hydroxy-selenomethionine
ONOO-	peroxynitrite
ORAC	oxygen radical absorbance capacity
OS	oxidative stress
PAMP	pathogen-associated molecular patterns
PBMC	peripheral blood mononuclear cell
PCB	polychlorinated biphenyls
PCV2	porcine circovirus type 2
pDC	plasmacytoid dendritic cell
PGE2	prostaglandin E2
pGSH-Px	plasma glutathione peroxidase
PHA	phytohemagglutin
PH-GPx	phospholipid hydroperoxide glutathione peroxidase
PH-GSH-Px	phospholipid glutathione peroxidase
pHu	ultimate pH
Pig-MAP	pig major acute phase protein
PMN	polymorphonuclear leukocytes
PON	paraoxonase
PPAR	peroxisome proliferator-activated receptor
PRR	pattern recognition receptors
Prx	peroxiredoxin
PSE	pale, soft, exudative
PSS	porcine stress syndrome

PTGE	prostaglandin E
PUFA	polyunsaturated fatty acid
PWM	pokeweed mitogen
RDA	recommended daily allowance
RNS	reactive nitrogen species
RONS	reactive oxygen and nitrogen species
ROS	reactive oxygen species
SAA	sulphur amino acids
SBP2	SECIS-binding protein 2
SeAlb	selenoalbumin
Sec	selenocysteine
SECIS	selenocysteine insertion sequence
SeCys	selenocysteine
SeHLan	selenohomolanthionine
SelN	selenoprotein N
SelP	selenoprotein P
SelP-L	long-form selenoprotein P
SeMet	selenomethionine
SeP	selenoprotein P
SM	silymarin
SO	Selisseo, OH-SeMet
SOD	superoxide dismutase
SP	selenoproteins
SPS	selenophosphate synthetase-2
SRBC	sheep red blood cells
SS	sodium selenite
SY	selenium enriched yeast
T	testosterone
T3	triiodothyronine
T4	thyroxine
T-AOC	total antioxidant capacity
TBARS	thiobarbituric acid reactive substances
TCR	T cell receptor
TEAC	Trolox equivalent antioxidant capacity
Th cells	T helper cells
TLR	Toll-like receptors
TNF-α	tumour necrosis factor alpha
TRAP	total radical antioxidant potential
Treg	regulatory T cell
Trx	thioredoxin
TrxR	thioredoxin reductase
TSH	thyroid-stimulating hormone
UGTR	UDP-glucose:glycoprotein glucosyltransferase
VCAM-1	vascular cell adhesion molecule 1
WHC	water holding capacity
WNT	wingless/integrated
XANES	X-ray absorption near-edge structure

Chapter 1
Antioxidant defence systems in animal health

P.F. Surai[1-7]
[1]*Vitagene and Health Research Centre, Bristol, BS4 2RS, United Kingdom;* [2]*Department of Hygiene and Poultry Sciences, Moscow State Academy of Veterinary Medicine and Biotechnology Named After K.I. Skryabin, 109472 Moscow, Russia;* [3]*Department of Microbiology and Biochemistry, Faculty of Veterinary Medicine, Trakia University, 6000 Stara Zagora, Bulgaria;* [4]*Department of Animal Nutrition, Faculty of Agricultural and Environmental Sciences, Szent Istvan University, 2103 Gödöllo, Hungary;* [5]*Saint-Petersburg State Academy of Veterinary Medicine, 196084 St. Petersburg, Russia;* [6]*Sumy National Agrarian University, 40000 Sumy, Ukraine;* [7]*Odessa National Academy of Food Technologies, 65039 Odessa, Ukraine; psurai@feedfood.co.uk*

Abstract

Redox (antioxidant-prooxidant) balance in the cell is an important determinant of various physiological functions. Indeed, oxidative stress occurs when this balance is disturbed due to overproduction of free radicals or compromised antioxidant defences. Free radical overproduction and oxidative stress are considered as a pathobiochemical mechanism involved in the initiation or progression phase of various diseases in farm animals, including pigs. Dietary antioxidants may be especially important in protecting against the development of the oxidative stress. The mechanisms by which natural antioxidants act at the molecular and cellular level include roles in gene expression and regulation, apoptosis, and signal transduction and antioxidants are involved in fundamental metabolic and homeostatic processes. Indeed, free radicals are now considered to take part in signal transduction in the cell and at least two well-defined transcription factors, NF-κB and Nrf2 have been identified to be regulated by the intracellular redox state. The regulation of gene expression by oxidants, antioxidants, and redox state has emerged as a novel subdiscipline in molecular biology that has promising implications for the feed industry and animal production. However, there are still many gaps in our knowledge of the basic molecular mechanisms of oxidative damage and antioxidant defences. Selenium in the form of SeCys is incorporated into 25 selenoproteins and more than half selenoproteins are involved in antioxidant defences and redox homeostasis maintenance being the 'chief executive' of the antioxidant defence network.

Keywords: antioxidants, selenium, oxidative stress, redox balance, pigs

DOI 10.3920/978-90-8686-912-1_1 ,

1.1 Introduction

It is well known that oxygen is an essential element for eukaryotes, and they rely on oxygen for the efficient production of energy. However, energy production in mitochondria is not 100% efficient and some electrons can escape from the electron transport chain forming free radicals. For the last three decades free radical research has generated valuable information for deeper understanding not only detrimental, but also beneficial role of free radicals in cell signalling, stress adaptation and other physiological processes. In fact, the benefit or harm of free radicals ultimately depend on the level of their production and efficiency of antioxidant defence.

1.2 Free radicals and reactive oxygen and nitrogen species

Free radicals are atoms or molecules containing one or more unpaired electrons. Free radicals are highly unstable and reactive and are capable of damaging biologically relevant molecules such as DNA, proteins, lipids or carbohydrates.

The animal body is under constant attack from free radicals, formed as a natural consequence of the body's normal metabolic activity and as part of the immune system's strategy for destroying invading microorganisms. The internal and external sources of free radicals are shown in Table 1.1. Collective terms reactive oxygen species (ROS) and reactive nitrogen species (RNS) have been introduced (Halliwell and Gutteridge, 2015) including not only the oxygen or nitrogen radicals, but also some non-radical reactive derivatives of oxygen and nitrogen (Table 1.2).

Superoxide (O_2^{-*}) is the main free radical produced in biological systems during normal respiration in mitochondria and by autoxidation reactions with half-life at 37 °C in the range of 1×10^{-6} second. Superoxide can inactivate some enzymes due to formation of unstable complexes with transition metals of enzyme prosthetic groups,

Table 1.1. Internal and external sources of free radicals (adapted from Surai, 2006).

Internally generated	Factors promoted reactive oxygen species formation
Mitochondria (ETC)	Cigarette smoke
Phagocytes (NADPH-oxidase)	Radiation
Xanthine oxidase	UV light
Reactions with Fe^{2+} or Cu^{+}	Pollution
Arachidonate pathways	Certain drugs
Peroxisomes	Chemical reagents
Inflammation	Industrial solvents
Biomolecule oxidation (adrenaline, dopamine, tetrahydrofolates, etc.)	High level of ammonia
	Mycotoxins

Table 1.2. Reactive oxygen and nitrogen species (adapted from Surai, 2006).

Radicals	Non-radicals
Alkoxyl, RO*	Hydrogen peroxide, H_2O_2
Hydroperoxyl, HOO*	Hypochlorous acid, HOCl
Hydroxyl, *OH	Ozone, O_3
Peroxyl, ROO*	Singlet oxygen, $^1O_2$
Superoxide, O_2*	Peroxynitrite, $ONOO^-$
Nitric oxide, NO*	Nitroxyl anion, NO^-
Nitrogen dioxide, NO_2*	Nitrous acid, HNO_2

followed by oxidative self-destruction of the active site (Chaudiere and Ferrari-Iliou, 1999). Depending on condition, superoxide can act as oxidising or a reducing agent. It is necessary to mention that superoxide, by itself, is not extremely dangerous and does not rapidly cross lipid membrane bilayer (Kruidenier and Verspaget, 2002). However, superoxide is a precursor of other, more powerful ROS. For example it reacts with nitric oxide with a formation of peroxynitrite ($ONOO^-$), a strong oxidant, which leads to formation of reactive intermediates due to spontaneous decomposition (Kontos, 2001; Mruk *et al.*, 2002). In fact ONOO- was shown to damage a wide variety of biomolecules, including proteins (via nitration of tyrosine or tryptophan residues or oxidation of methionine or selenocysteine residues), DNA and lipids (Groves, 1999). Superoxide can also participate in the production of more powerful radicals by donating an electron, and thereby reducing Fe^{3+} and Cu^{2+} to Fe^{2+} and Cu^+, as follows:

$$O_2^{\cdot -} + Fe^{3+}/Cu^{2+} \longrightarrow Fe^{2+}/Cu^+ + O_2$$

Further reactions of Fe^{2+} and Cu^+ with H_2O_2 are a source of the hydroxyl radical (*OH) in the Fenton reaction:

$$H_2O_2 + Fe^{2+}/Cu^+ \longrightarrow {}^*OH + OH^- + Fe^{3+}/Cu^{2+}$$

The sum of reaction of superoxide radical with transition metals and transition metals with hydrogen peroxide is known as the Haber-Weiss reaction. It is necessary to underline that superoxide radical is a 'double-edged sword'. It is beneficial when produced by activated polymorphonuclear leukocytes and other phagocytes as an essential component of their bactericidal activities but in excess it may result in tissue damage associated with inflammation.

Hydroxyl radical is the most reactive species with an estimated half-life of only about 10^{-9} second. It can damage any biological molecule it touches, however, its diffusion capability is restricted to only about two molecular diameters before reacting (Yu, 1994). Therefore, in most cases damaging effect of hydroxyl radical is restricted to the site of its formation. In general, hydroxyl radical can be generated in human/animal body as a result of radiation exposure from natural sources (radon gas, cosmic radiation) and from man-made sources (electromagnetic radiation and radionuclide contamination). In fact in many cases hydroxyl radical is a trigger of chain reaction in lipid peroxidation.

Therefore, ROS/RNS (Table 1.2) are constantly produced *in vivo* in the course of the physiological metabolism in tissues. It is generally accepted that the electron-transport chain in the mitochondria is responsible for major part of superoxide production in the body (Halliwell and Gutteridge, 2015). Mitochondrial electron transport system consumes more than 85% of all oxygen used by the cell and, because the efficiency of electron transport is not 100%, about 1-3% of electrons escape from the chain and the univalent reduction of molecular oxygen results in superoxide anion formation (Chow *et al.*, 1999; Halliwell, 1994; Singal *et al.*, 1998). About 10^{12} O_2 molecules processed by each rat cell daily and if the leakage of partially reduced oxygen molecules is about 2%, this will yield about 2×10^{10} molecules of ROS per cell per day (Chance *et al.*, 1979). An interesting calculation has been made by Halliwell (1994), showing that in the human body about 1.72 kg/year of superoxide radical is produced. In stress condition it would be substantially increased. Clearly, these calculations showed that free radical production in the body is substantial and many thousand biological molecules can be easily damaged if not protected. The activation of macrophages in stress conditions is another important source of free radical generation. Immune cells produce ROS/RNS and use them as an important weapon to destroy pathogens (Kettle and Winterbourn, 1997; Schwarz, 1996).

The most important effect of free radicals on the cellular metabolism is due to their participation in lipid peroxidation reactions (Surai, 2006). The first step of this process is called the initiation phase, during which carbon-centred free radicals are produced from a precursor molecule, for example a polyunsaturated fatty acid (PUFA):

$$LH \xrightarrow{\text{Initiator}} L^{*}$$

The initiator in this reaction could be the hydroxyl radical, radiation or some other events or compounds. In presence of oxygen these radicals (L*) react with oxygen producing peroxyl radicals starting the next stage of lipid peroxidation called the propagation phase:

$$L^{*} + O_2 \longrightarrow LOO^{*}$$

At this stage a relatively unreactive carbon-centred radical (L*) is converted to a highly reactive peroxyl radical. A resulted peroxyl radical can attack any available peroxidazable material producing hydroperoxide (LOOH) and new carbon-centred radical (L*):

$$LOO^{*} + LH \longrightarrow LOOH + L^{*}$$

Therefore, lipid peroxidation is a chain reaction and potentially large number of cycles of peroxidation could cause substantial damage to cells. In membranes the peroxidazable material is represented by PUFAs. It is generally accepted that PUFA susceptibility to peroxidation is proportional to the amount of double bounds in the molecules. In fact, docosahexaenoic acid (DHA, 22:6n-3) and arachidonic acid (AA, 20:4n-6) are among major substrates of the peroxidation in the membrane. It is necessary to underline that the same PUFAs are responsible for maintenance of physiologically important membrane properties including fluidity and permeability. As a result of lipid peroxidation within the biological membranes their structure and functions are compromised. Proteins and DNA are also important targets for ROS.

It has been shown that the DNA in each cell of a rat is hit by about 100,000 free radicals a day and each cell sustains as many as 10,000 potentially mutagenic (if not repaired) lesions per day arising from endogenous sources of DNA damage (Ames, 2003; Ames and Gold, 1997; Diplock, 1994; Helbock *et al.*, 1998). Therefore, some oxidative lesions escape repair and the steady state level of oxidative lesions increased with age, and an old rat has accumulated about 66,000 oxidative DNA lesions per cell (Ames, 2003). Oxidation, methylation, deamination and depurination are four endogenous processes leading to significant DNA damage with oxidation to be most significant one and approximately 20 types of oxidatively altered DNA molecules have been identified. The chemistry of attack by ROS on DNA is very complex and lesions in chromatin include damage to bases, sugar lesions, single strand-breaks, basic lesions and DNA-nucleoprotein cross-links (Diplock, 1994).

The complex structure of proteins and a variety of oxidisable functional groups of the amino acids make them susceptible to oxidative damage. In fact, the accumulation of oxidised proteins has been implicated in the aging process and in other age-related pathologies. A range of oxidised proteins and amino acids has been characterised in biological systems (Dean *et al.*, 1997; Kehrer, 2000). In general, the accumulation of oxidised proteins depends on the balance between antioxidants, prooxidants and removal/repair mechanisms. Oxidation of proteins leads to the formation of reversible disulphide bridges. More severe protein oxidation causes a formation of chemically modified derivatives e.g. Schiff's base (Tirosh and Reznick, 2000). Nitric oxide, hydroxyl radical, alkoxyl and peroxyl radicals as well as carbon-centred radicals, hydrogen peroxide, aldehydes or other products of lipid peroxidation can attack protein molecules. Usually oxidative modification of proteins occurs by two different mechanisms: a site-specific formation of ROS via redox-active transition metals and non-metal-dependent ROS-induced oxidation of amino acids (Tirosh and Reznick, 2000). The modification of a protein occurs by either a direct oxidation of a specific amino acid in the protein molecule or cleavage of the protein backbone. In both cases biological activity of the modified proteins would be compromised. The degree of protein damage depends on many different factors (Grune *et al.*, 1997):

- the nature and relative location of the oxidant or free radical source;
- nature and structure of protein;
- the proximity of ROS to protein target;
- the nature and concentrations of available antioxidants.

Free radicals are implicated in the initiation or progression phase of various diseases, including cardiovascular disease, some forms of cancer, cataracts, age-related macular degeneration, rheumatoid arthritis and a variety of neuro-degenerative diseases (Hogg, 1998; McCord, 2000). In poultry production, overproduction of free radicals and oxidative stress are considered to be related to various type of stresses, including, nutritional, technological, environmental and internal stresses (Surai and Fisinin, 2016a,b; Surai *et al.*, 2019b). In general, it is widely believed that most human and animal/poultry diseases at different stages of their development are associated with free radical production and metabolism (Surai, 2002, 2006, 2018). Normally, there is a delicate balance between the amount of free radicals generated in the body and the antioxidants to protect against them (Figure 1.1). For the majority of organisms on Earth, life without oxygen is impossible, animals, plants and many micro-organisms relying on oxygen for the efficient production of energy. However, they pay a high price for pleasure of living in an oxygenated atmosphere since high oxygen concentration in the atmosphere is potentially toxic for living organisms (Figure 1.2).

1.3 Three levels of antioxidant defence

During evolution living organisms have developed specific antioxidant protective mechanisms to deal with ROS and RNS (Halliwell and Gutteridge, 2015). Therefore, it is only the presence of natural antioxidants in living organisms which enable them to survive in an oxygen-rich environment (Halliwell, 1994, 2012). These mechanisms are described by the general term 'antioxidant system'. It is diverse and responsible for the protection of cells from the actions of free radicals. This system includes (Surai *et al.*, 2019b):

- natural fat-soluble antioxidants (vitamin E, carotenoids, ubiquinones, etc.);
- water-soluble antioxidants (ascorbic acid, uric acid, carnitine, taurine, etc.);
- antioxidant enzymes: glutathione peroxidase (GPx), catalase (CAT) and superoxide dismutase (SOD);
- thiol redox system consisting of the glutathione system (glutathione/glutathione reductase/glutaredoxin/GPx and a thioredoxin system (thioredoxin/thioredoxin peroxidase/thioredoxin reductase).

The antioxidant capacity of a compound is determined by multiple factors, including the chemical reactivity toward free radicals; fate of antioxidant-derived radicals; interaction with other antioxidants; concentration, distribution, mobility, and metabolism at the micro-environment (Niki, 2014, 2016). The protective antioxidant compounds are located in organelles, subcellular compartments or the extracellular space enabling maximum cellular protection to occur. Thus, antioxidant systems of the living cell includes three major levels of defence (Niki, 1996; Surai, 1999, 2002, 2006, 2018, 2020a,b):

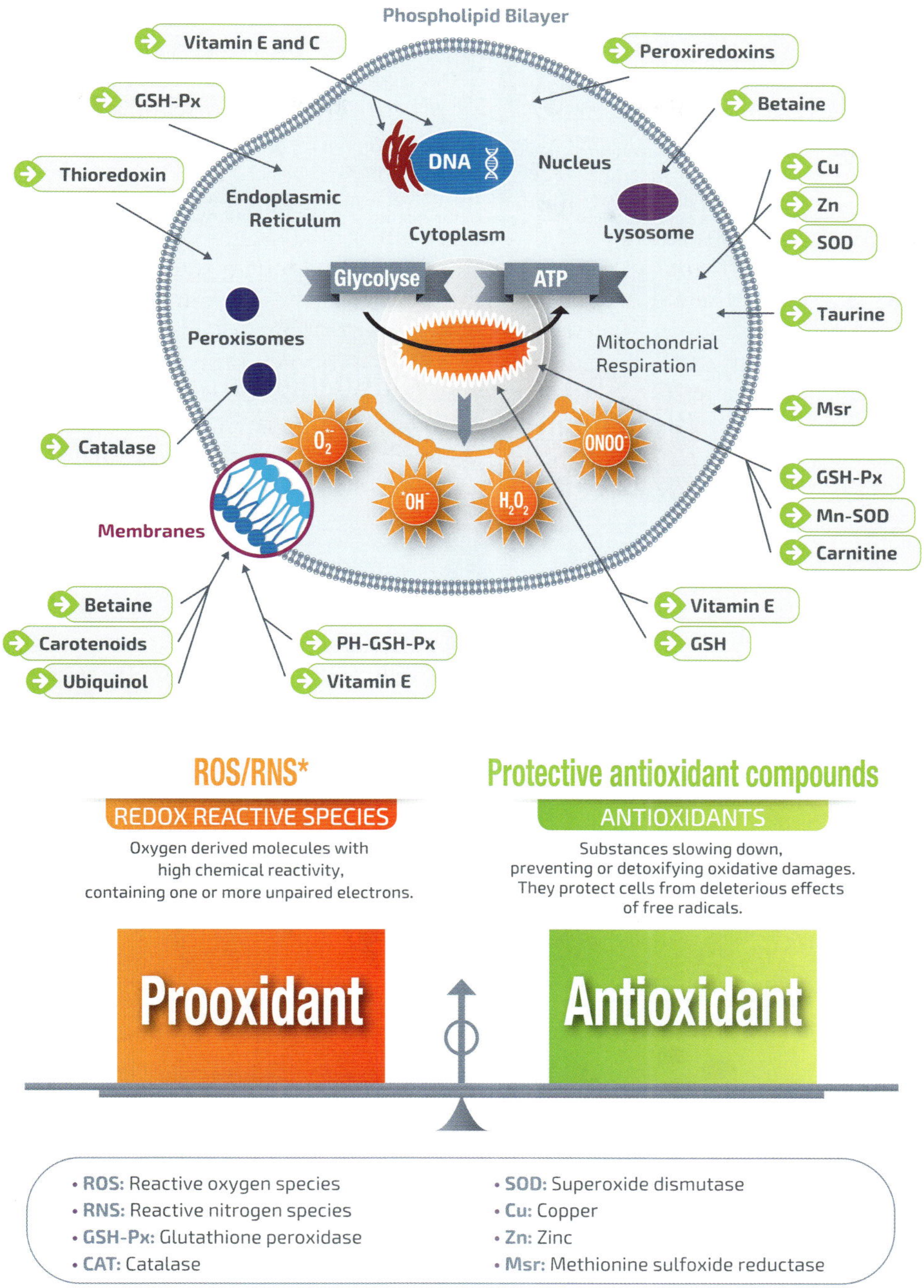

Figure 1.1 The right balance to maintain cellular functions.

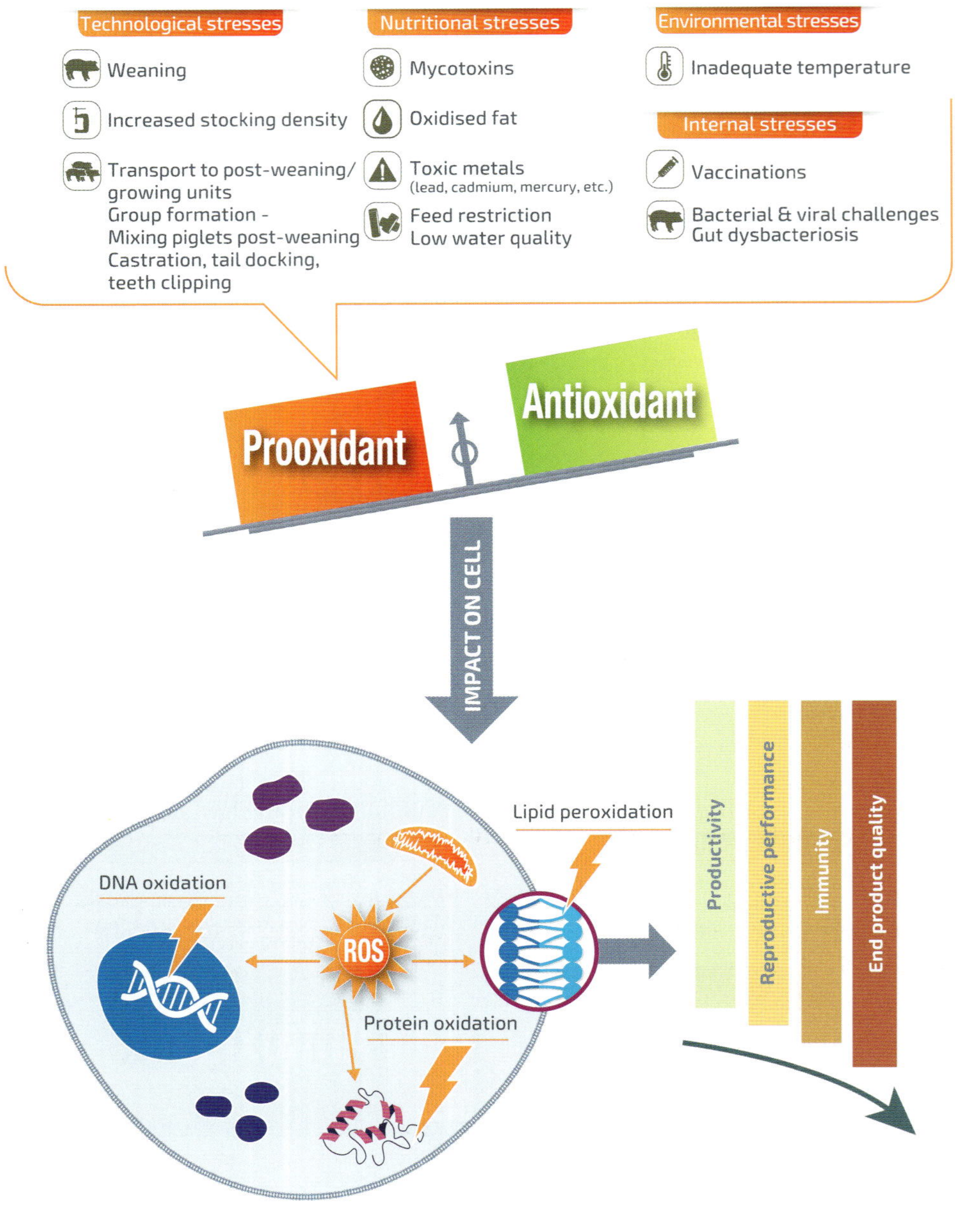

Figure 1.2. External stresses impact cellular functions and animal productivity.

The first level of defence is responsible for prevention of free radical formation by removing precursors of free radicals or by inactivating catalysts and consists of three antioxidant enzymes namely SOD, GPx and CAT plus metal-binding proteins (Figure 1.3 and Figure 1.4).

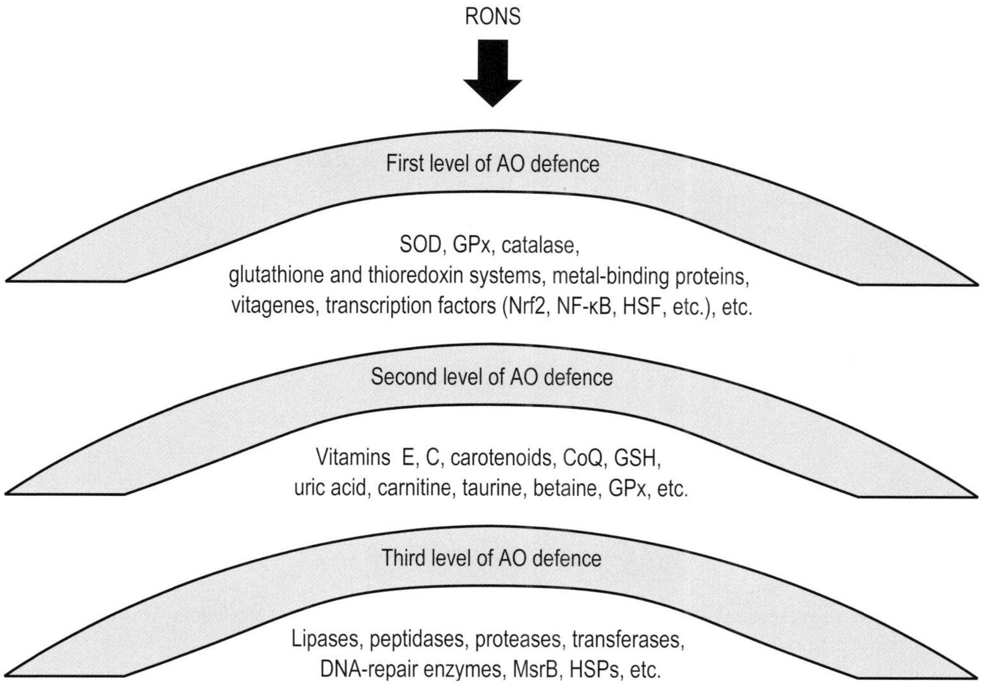

Figure 1.3 Three lines of antioxidant defence in animal cells (adapted from Surai, 1999, 2020).

1.4 Antioxidant enzymes in biological systems

Since the superoxide radical is the main free radical produced in physiological conditions in the cell (Halliwell, 1994) superoxide dismutase (EC 1.15.1.1) is considered to be the main element of the first level of antioxidant defence in the cell (Surai, 1999, 2016, 2020b). This enzyme dismutates the superoxide radical in the following reaction:

$$2O_2^{*} + 2H^{+} \xrightarrow{\text{SOD}} H_2O_2 + O_2$$

On one hand, SOD is the main cellular antioxidant mechanism dealing with overproduction of free radicals in stress conditions. On the other hand, in biological systems SODs are important source of H_2O_2, main signalling molecule participating in stress adaptation (Surai, 2020a,b; Surai *et al.,* 2019b).

The hydrogen peroxide formed by SOD action can be detoxified by GPx or CAT which reduce it to water as follows:

$$H_2O_2 + 2GSH \xrightarrow{\text{GPx}} GSSG + 2H_2O$$

$$2H_2O_2 \xrightarrow{\text{Catalase}} 2H_2O + O_2$$

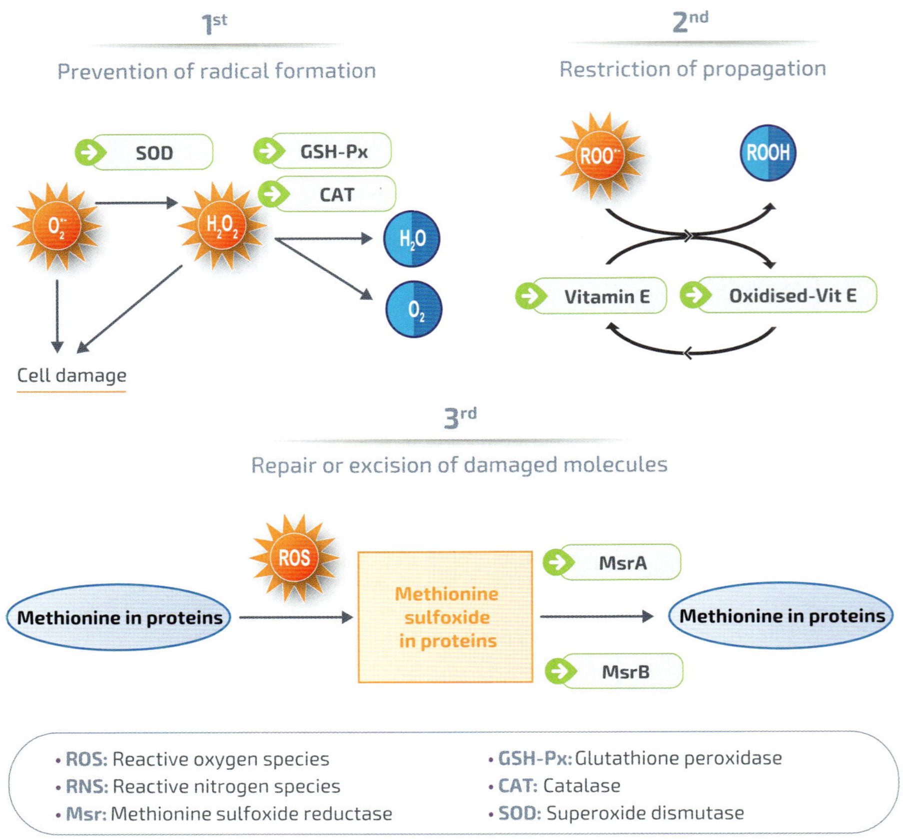

Figure 1.4. The antioxidant system. 3 levels of defences.

Catalase (EC 1.11.1.6) is a tetrameric enzyme consisting of four identical subunits of 60 kDa containing a single ferriprotoporphyrin group per subunit. It plays an important role in the acquisition of tolerance to oxidative stress in the adaptive response of cells (Mates *et al.,* 1999). In mammalian cells, NADPH is bound to catalase protecting it from inactivation by H_2O_2 (Chaudiere and Ferrari-Illiou, 1999). Since GPx has a much higher affinity for H_2O_2 than CAT (Jones *et al.*, 1981) and wider distribution in the cell (catalase is located mainly in peroxisomes), the H_2O_2 removal from the cell is very much dependent on GPx.

Recently it has been shown that thioredoxin peroxidases are also capable of directly reducing hydrogen peroxide (Nordberg and Arner, 2001). It is interesting that the levels of antioxidant enzymes are regulated by gene expression as well as by post-translational modifications (Fujii and Taniguchi, 1999; Surai, 2020a).

Hydrogen peroxide can also be detoxified by peroxiredoxins (Prx):

$$\text{Prx (reduced)} + H_2O_2 \longrightarrow \text{Prx (oxidised)} + 2H_2O$$

The oxidised form of Prx is inactive, requiring the donation of electrons from reduced thioredoxin (Trx) to restore its catalytic activity:

$$\text{Prx (oxidised)} + \text{Trx (reduced)} \longrightarrow \text{Prx (reduced)} + \text{Trx (oxidised)}$$

Furthermore, Se-dependent enzyme TR reduce oxidised Trx to its reduced active form:

$$\text{Trx (oxidised)} \xrightarrow{\text{TR}} \text{Trx (reduced)}$$

Mammalian cells express 6 Prx isoforms, including Prx3 and Prx5 in the mitochondria. Prxs function by undergoing oxidation by H_2O_2 at an active site cysteine and then subsequent reduction by thioredoxin, thioredoxin reductase, and NADPH. There are eight GPXs, which are oxidised by H_2O_2 and reduced by glutathione (GSH) and catalase found in peroxisomes (Sena and Chandel, 2012; Surai, 2020a).

Transition metal ions also accelerate the decomposition of lipid hydroperoxides into cytotoxic products such as aldehydes, alkoxyl radicals and peroxyl radicals:

$$\text{LOOH} + Fe^{2+} \longrightarrow \text{LO}^{*} + Fe^{3+} + OH^{-}$$

$$\text{LOOH} + Fe^{3+} \longrightarrow \text{LOO}^{*} + Fe^{2+} + H^{+}$$

Therefore, metal-binding proteins (transferrin, lactoferrin, haptoglobin, hemopexin, metallothionein, ceruloplasmin, ferritin, albumin, myoglobin, etc.) also belong to the first level of defence. It is necessary to take into account that iron and copper are powerful promoters of free radical reactions and therefore their availability in 'catalytic' forms is carefully regulated *in vivo* (Halliwell and Gutteridge, 2015). Organisms have evolved to keep transition metal ions safely sequestered in storage or transport proteins. In this way the metal-binding proteins prevent formation of hydroxyl radical by preventing them from participation in radical reactions. For example, transferrin binds the iron (about 0.1% of the total body reserves), transports it in the plasma pool and attaches it to the transferrin receptor. The important point is that iron associated with transferrin will not catalyse free radical reaction. Ferritin is considered to be involved in iron storage (about 30% of total body reserves) within the cytosol in various tissues including liver and spleen. Major part of iron in the body (55-60%) is associated with haemoglobin within red cells and about 10% with myoglobin in muscles (Galey, 1997). A range of other iron-containing proteins (mainly enzymes) can be found in the body including NADH dehydrogenase, cytochrome P450, ribonucleotide reductase, proline hydroxylase, tyrosine hydroxylase, peroxidases, catalase, cyclooxygenase, aconitase, succinate dehydrogenase, etc. (Galey, 1997). Despite an importance of iron in various biochemical reactions, iron can be extremely dangerous when not carefully handled by proteins. In fact, in many stress conditions a release of free iron from its normal sites and its participation in Fenton chemistry mediate damages to cells. For example, superoxide radical can release iron from ferritin and H_2O_2 degrades the heme of haemoglobin to liberate iron ions (Halliwell, 1987).

Ceruloplasmin is another major protein mediating free radical metabolism being a copper-binding protein. Under physiological conditions it binds six or seven copper ions per molecule preventing their participation in free radical generation. About 5% of human plasma copper is bound to albumin or to amino acids and the rest is bound to ceruloplasmin. Furthermore, ceruloplasmin possesses antioxidant properties itself being able to scavenge superoxide radical (Yu, 1994). Therefore, it is now quite clear that metal sequestration is an important part of extracellular antioxidant defence (Surai, 2020a).

1.5 Chain-breaking antioxidants and integrated antioxidant defence mechanisms

Unfortunately, this first level of antioxidant defence in the cell is not sufficient to completely prevent free radical formation and some radicals do escape through the preventive first level of antioxidant safety screen initiating lipid peroxidation and causing damage to DNA and proteins. Therefore, the second level of defence consists of chain-breaking antioxidants – vitamin E, ubiquinol, carotenoids, ascorbic acid, taurine, carnitine, uric acid and some other antioxidants. Glutathione and thioredoxin systems also have a substantial role in the second level of antioxidant defence. Chain-breaking antioxidants inhibit peroxidation by keeping the chain length of the propagation reaction as small as possible. Therefore, they prevent the propagation step of lipid peroxidation by scavenging peroxyl radical intermediates in the chain reaction:

$$\text{LOO}^* + \text{Toc} \longrightarrow \text{Toc}^* + \text{LOOH}$$

Where LOO* is lipid peroxyl radical, Toc is tocopherol, Toc* is tocopheroxyl radical and LOOH is lipid hydroperoxide. The antioxidant defence, therefore, includes several options (Figure 1.5; Surai *et al.*, 2019b).

Indeed, antioxidant defence mechanisms include various options, including decrease local oxygen concentration; decrease activity of pro-oxidant enzymes (carnitine, silymarin); improve efficiency of electron transport chain in the mitochondria and decreasing electron leakage leading to superoxide production (carnitine); induction of various transcription factors (e.g. NF-E2-related factor 2 [Nrf2], nuclear factor-kB [NF-κB] and others) and ARE-related synthesis of AO enzymes (SOD, GPx, CAT, glutathione reductase [GR], glutathione S-transferase [GST], etc.); binding metal ions (metal-binding proteins) and metal chelating; decomposition of peroxides by converting them to non-radical, nontoxic products (Se-GPx); chain breaking by scavenging intermediate radicals such as peroxyl and alkoxyl radicals (vitamins E, C, GSH, uric acid, carnitine, ubiquinol, bilirubin, etc.); repair and removal of damaged molecules (methionine sulfoxide reductase, DNA-repair enzymes, chaperons, etc.); redox-signalling and vitagene activation with synthesis and increased expression of protective molecules (GSH, thioredoxins, SOD, heat shock proteins [HSPs], sirtuins, etc.); antioxidant recycling mechanisms, including vitamin E recycling; apoptosis activation and removal terminally damaged cells and restriction of mutagenesis.

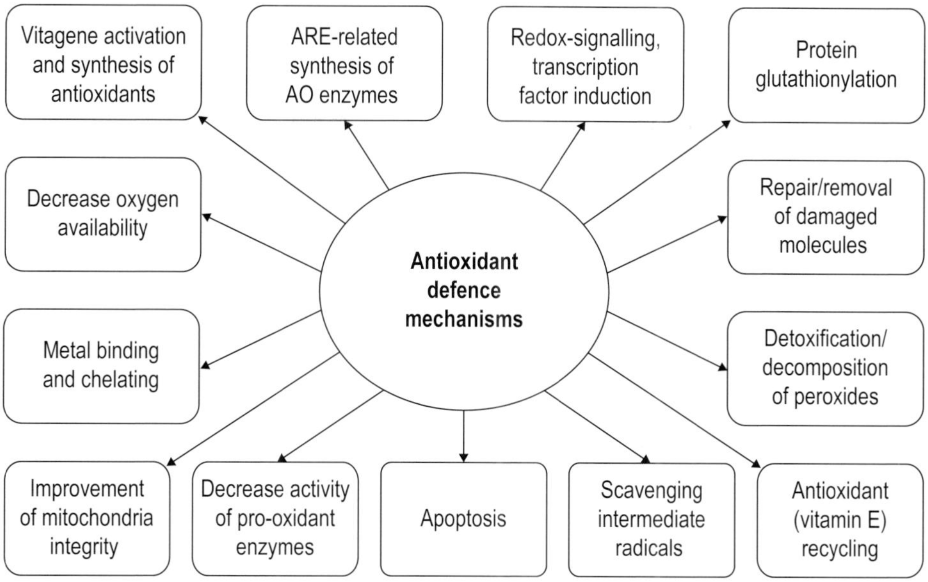

Figure 1.5. Antioxidant defence mechanisms (adapted from Surai, 2020a; Surai *et al.*, 2019b).

1.6 Antioxidant-prooxidant (redox) balance in the body

In the body a delicate critical balance exists between antioxidant defence and repair systems and free radical generation (Figure 1.6).

In physiological conditions the right and left parts of the so-called 'balances' are in equilibrium i.e. free radical generation is neutralised by the antioxidant system. Exogenous factors are among the most important elements, which increase an efficiency of the antioxidant system of the organism. Natural and synthetic antioxidants in the feed as well as optimal levels of Mn, Cu, Zn and Se help to maintain the efficient levels of endogenous antioxidants in the tissues. Optimal diet composition allows the antioxidants of the food to be efficiently absorbed and metabolised. Optimal temperature, humidity and other environmental conditions are also required for the effective protection against free radical production. The prevention of different diseases by antibiotics and other drugs is an integral part of the efficient antioxidant defence as well. Different stress conditions are associated with overproduction of free radicals and cause oxidative stress i.e. a disturbance in the prooxidant-antioxidant balance leading to potential tissue damage (Jaeschke, 1995). Stress conditions can be generally divided into four main categories, including nutritional, environmental, technological and internal stresses (Surai *et al.*, 2019b; Table 1.3).

The most important part is nutritional stress conditions including high levels of PUFAs, deficiencies of vitamin E, Se, Zn or Mn, Fe-overload, hypervitaminosis A and presence of different mycotoxins and other toxic compounds, low water quality, usage of coccidiostats or other drugs via feed or water. A second group of stress factors

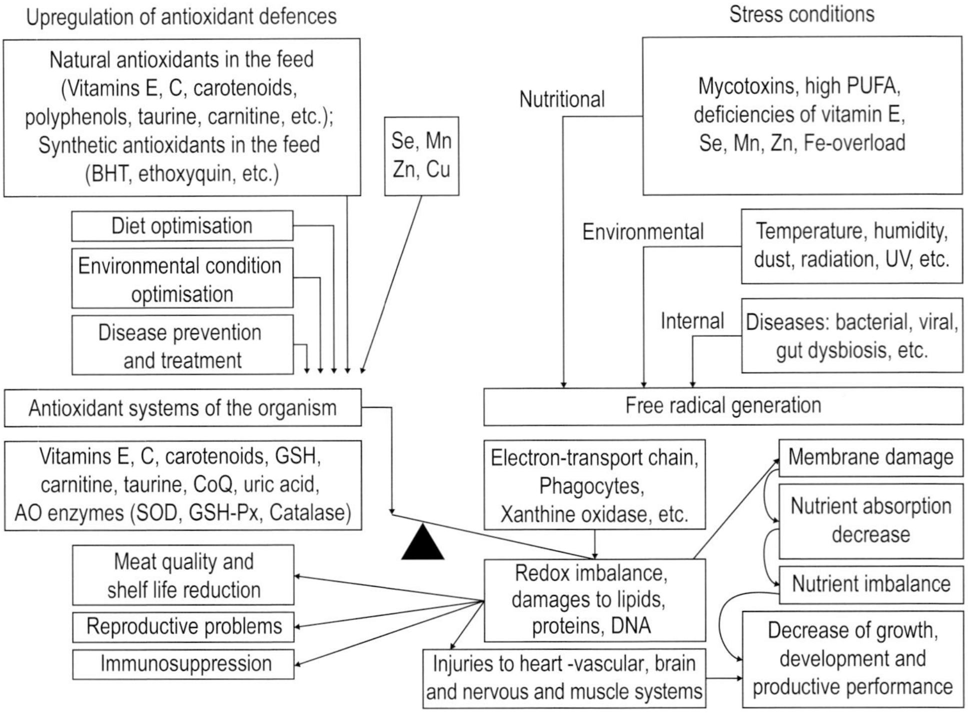

Figure 1.6. Antioxidant-prooxidant balance in the organism (adapted from Surai, 1999).

Table 1.3. Stresses in pig production (adapted from Surai *et al.,* 2019b).

Technological stresses
Weaning
Increased stocking density
Weighing, grading, group formation, etc.
Environmental stresses
Inadequate temperature
Inadequate ventilation and increased dust
Inadequate lightning
Nutritional stresses
Mycotoxins
Oxidised fat
Toxic metals (lead, cadmium, mercury, etc.)
Imbalance of minerals (Se, Zn, Cu, etc.) and other nutrients
Low water quality
Usage of various drugs via feed or water
Internal stresses
Vaccinations
Microbial or virus challenges
Gut dysbacteriosis

includes environmental conditions: inadequate temperature, humidity or lightning, inadequate ventilation and increased dust, etc. A third group of stresses includes technological stresses, such as chick placement, increased stocking density, weighing, grading, group formation, catching, transferring to breeder houses, prolonged egg storage, egg transportation, inadequate egg storage conditions, incorrect incubation regimes, etc. A fourth group of stress factors includes internal stress factors such as vaccination, microbial or viral challenges, gut dysbacteriosis, pipping and hatching (Surai, 2019a,b; Surai and Fisinin, 2016b). In the commercial conditions of pig production, there are four main groups of stressors: physical, chemical, biological and psychological (Puppel *et al.*, 2015; Surai, 2018). Indeed, fluctuations in ambient temperature, mechanical injuries, irregularities in habits and or routines, nutrient disbalance, mycotoxins contamination of feed, microbial and viral infections, etc., are important stress-factors contributing to oxidative stress which can potentially decrease productive and reproductive performance and compromise animal health.

Redox biology is an extremely quickly developing branch of biological sciences. It is proven that redox homeostasis in the cell/tissue/body is a major regulator of many physiological processes and biochemical pathways associated with animal health, growth, development, reproduction and is of paramount importance in stress adaptation.

This topic has been extensively characterised and interested readers are invited to the related chapters of the recent monographs (Surai, 2018, 2020a; Surai *et al.*, 2018).

1.7 Biomarkers of oxidative stress in animals

In general, there is a range of various biomarkers used for detection of OS and their advantages and limitations are summarised in Table 1.4.

It is necessary to emphasise that there is no single test that would give an ultimate answer to what the optimal oxidative status of an animal should be, and therefore a combination of various tests is preferable. For example, the ratio of pro-oxidant to antioxidant capacity, known as the oxidative status index (OSi), was suggested as a tool to assess the redox status and possible oxidative stress in dairy cattle (Abuelo *et al.*, 2013). However, it is important to measure macromolecule damage (e.g. lipid and/or protein oxidation products) that can occur due to free radical overproduction to make a conclusion about oxidative stress in farm animals (Putman *et al.*, 2018; Surai, 2018).

Table 1.4. Advantages and disadvantages of various biomarkers as indicators of oxidative stress in animals (adapted from Surai *et al.*, 2019a).[1]

Biomarker	Advantages	Disadvantages
MDA	Sensitive and reproducible	Non-specific product of lipid peroxidation
TBARS	Rapid, popular, easy, and economical	Non-specific, non-reproducible, no quantitative relationship with lipid peroxidation
F2-Isoprostane	Specific, reproducible, sensitive	Expensive, auto-oxidation of samples, sample derivatisation is required
ORAC	Sensitive and covers a wide variety of antioxidants	Requires spectrofluorometer; AAPH, a free radical source is sensitive to temperature, low reactivity of fluorescein toward ROO* radicals
FRAP	Inexpensive, reagents are simple to prepare, results are highly reproducible, and the procedure is straightforward and speedy	The reaction is non-specific, and the result of the test depends on the reaction time.
TEAC	Extremely fast and simple	Results vary with sample dilution; antioxidant used may interact with solvent molecules; specificity varies
TRAP	Gives an idea of the rate of free radical formation	Antioxidant employed may not trap all types of free radicals
ROMs	Extremely fast, simple; can be performed directly in whole blood, inflammatory fluids, cell extracts and respiratory condensate	Inhibited by sodium azide, lack of reference values
RONS	Fast, commercial kits are available	Lack of reference values
BAP	Fast, simple and covers a wide variety of antioxidants	Can be performed only in plasma and serum samples; hyperlipaemic samples can underestimate results
AOPPs	Novel markers of protein oxidation, quickly developing, mediators of pro-inflammatory response	Lack of reference values
Protein carbonylation	Easy to perform	Lack of reference values
AO enzymes (SOD, GPx, catalase, etc.)	Common, widely used tests, commercial kits are available	Difficulties with results interpretation, since some enzymes are stress-inducible
Plasma total thiols	Important part of the Redox system, commercial kits are available	Very sensitive to oxidation during sample preparation and storage
Non-enzymatic antioxidants: glutathione, α-tocopherol, β-carotene, uric acid, etc.	Common, widely used tests.	Individually reflect only a small proportion of the antioxidant defence potential
HSP	Important elements of antistress protection	Difficult to perform, difficulties with results interpretation, since HSP are stress-inducible

[1] AAPH = 2,2′-azobis(2-amidinopropane) dihydrochloride; AGE = advanced glycation end products; AO = antioxidant; AOPPs = advanced oxidation protein products; BAP = biological antioxidant potential; FRAP = ferric reducing ability of plasma; GPx = glutathione peroxidase; HSP = heat shock proteins; MDA = malondialdehyde; ORAC = oxygen radical absorbance capacity; ROM = reactive oxygen metabolite; RONS = reactive oxygen and nitrogen species; SOD = superoxide dismutase; TBARS = thiobarbituric acid reactive substances; TEAC = Trolox equivalent antioxidant capacity; TRAP = total radical antioxidant potential.

1.8 Vitagene concept development

Since at the molecular level most stresses are associated with overproduction of free radicals and oxidative stress, the development of the effective antioxidant solutions to decrease negative consequences of commercially-relevant stresses is an important task for pig scientists. One of such approaches is based on possibilities of modulation of vitagenes, a family of genes responsible for animal/poultry adaptation to stress. The term 'vitagene' was introduced in 1998 by Rattan who wrote 'Our survival and the physical quality of life depends upon an efficient functioning of various maintenance and repair processes. This complex network of the so-called longevity assurance processes is composed of several genes, which may be called *vitagenes*' (Figure 1.7).

Later vitagene concept has been further developed in medical sciences by Professor Calabrese and colleagues in 2004-2016. In accordance with Calabrese *et al.* (2007, 2009, 2014) the term vitagenes refers to a group of genes that are strictly involved in preserving cellular homeostasis during stress conditions and the vitagene family includes heat shock proteins (HSPs), including heme oxygenase-1 (HSP32, HO-1) and HSP70, the thioredoxin system and sirtuins. The list of potential candidates to vitagene family was extended. In particular, SOD, a major inducible enzyme of the first level of antioxidant defence, has been included into the vitagene family (Surai, 2016; Surai and Fisinin, 2016c). The products of the mentioned genes actively operate in detecting and controlling diverse forms of stress and cell injuries.

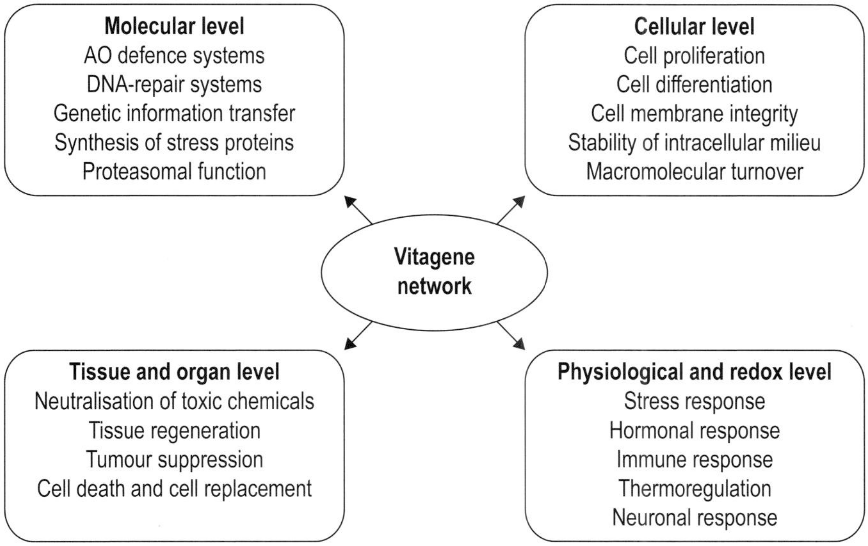

Figure 1.7. Major components of the vitagene network (adapted from Calabrese *et al.*, 2007; Rattan, 1998; Surai *et al.*, 2019b).

The cooperative mechanisms of the vitagene network are reviewed in recently published aforementioned comprehensive reviews with a major conclusion indicating an essential regulatory role of the vitagene network in cell and whole organism adaptation to various stresses. Indeed, cellular stress response is mediated via the regulation of pro-survival pathways and vitagene activation with the following synthesis of a range of protective antioxidant molecules is the central event in such an adaptation. The vitagene concept helped in developing effective strategies to fight oxidative stress in various human diseases, including neurodegenerative disorders, neuroprotection, aging and longevity, dermatology, osteoporosis and Alzheimer pathology, and other free radical-related diseases (for review and references there see Surai and Fisinin, 2016c,d). Indeed, HSPs, including heme oxygenase-1 (HO-1) and HSP70, are responsible for protein homeostasis in stress conditions of poultry/animal production (Surai and Kochish, 2017), while the thioredoxin system is the major player in maintaining redox status of the cell involved in protein and DNA synthesis and repair as well as in regulation of expression of many important genes (Surai, 2019a,b; Surai and Fisinin, 2016c,d). Furthermore, sirtuins regulate the biological functions of various molecules post-translationally by removing acetyl groups from protein substrates ranging from histones to transcription factors and orchestrate cellular stress response by maintenance of genome integrity and protein stability. Finally, SODs belong to the first level of antioxidant defence preventing lipid and protein oxidation at the very early stages (Surai, 2016, 2020b,c). All the aforementioned vitagenes operate in concert building a reliable system of stress detection and adequate response and are considered to be key elements in stress adaptation.

1.9 Conclusions

Redox (antioxidant-prooxidant) balance in the cell is an important determinant of various physiological functions. Indeed, oxidative stress occurs when this balance is disturbed due to overproduction of free radicals or compromised antioxidant defences. Free radical overproduction and oxidative stress are considered as a pathobiochemical mechanism involved in the initiation or progression phase of various diseases in farm animals. In fact, in animal production free radial generation and lipid peroxidation are responsible for the development of various diseases as well as for the decrease of animal productivity and product quality. Dietary antioxidants may be especially important in protecting against the development of the oxidative stress.

Free radicals are now considered to take part in signal transduction in the cell and at least two well-defined transcription factors, NF-κB and Nrf2 have been identified to be regulated by the intracellular redox state. The regulation of gene expression by oxidants, antioxidants, and redox state has emerged as a novel subdiscipline in molecular biology that has promising implications for the feed industry and animal production. Thus, the redox state of the cell, which reflects antioxidant/prooxidant balance, can be considered as an important element of gene regulation. It is important to mention that selenium in the form of SeCys is incorporated into

25 selenoproteins (for more details see Chapter 2; Zhao *et al.*, 2021) and more than half selenoproteins are involved in antioxidant defences and redox homeostasis maintenance (Surai, 2018).

Therefore, the effect of antioxidants on animal health is much deeper than one could expect several years ago. The mechanisms by which natural antioxidants act at the molecular and cellular level include roles in gene expression and regulation, apoptosis, and signal transduction and antioxidants are involved in fundamental metabolic and homeostatic processes. However, there are still many gaps in our knowledge of the basic molecular mechanisms of oxidative damage and antioxidant defences.

Indeed, a range of selenoproteins is involved in a regulation of key elements of antioxidant defences. By maintaining optimal activities of such antioxidant enzymes as GPx, TrxR and MsrB as well as directly participating in regulation of DNA-repair enzymes Se belongs to all three major levels of antioxidant defences. Furthermore, Se also interacts with other antioxidants such as vitamin E, ascorbic acid, glutathione and some others. That is why I would call selenium 'chief executive of antioxidant defences' (Figure 1.8).

In the next chapters information will be presented proving that Se is key to effective antioxidant defences and to maintaining animal and human health. Indeed 'Diet cures more than lancet'.

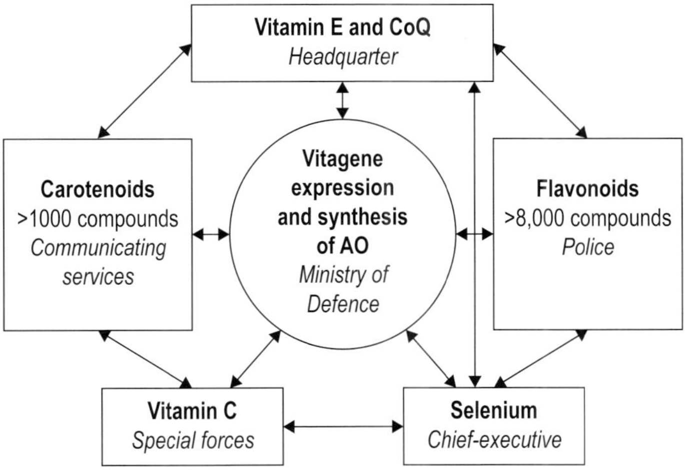

Figure 1.8. Antioxidant defence network in the cell/tissue/body (adapted from Surai, 2018).

Acknowledgements

Peter F. Surai is supported by a grant of the Government of Russian Federation (Contract No. 14.W03.31.0013).

References

Abuelo, A., Hernández, J., Benedito, J.L. and Castillo, C., 2013. Oxidative stress index (OSi) as a new tool to assess redox status in dairy cattle during the transition period. Animal 7: 1374-1378.

Ames, B.N. and Gold, L.S., 1997. The causes and prevention of cancer: gaining perspective. Environmental Health Perspectives 105 (Suppl 4): 865-873.

Ames, B.N., 2003. An enthusiasm for metabolism. Journal of Biological Chemistry 278: 4369-4380.

Calabrese, V., Cornelius, C., Mancuso, C., Barone, E., Calafato, S., Bates, T., Rizzarelli, E. and Kostova, A.T., 2009. Vitagenes, dietary antioxidants and neuroprotection in neurodegenerative diseases. Frontiers in Bioscience 14: 376-397.

Calabrese, V., Guagliano, E., Sapienza, M., Panebianco, M., Calafato, S., Puleo, E., Pennisi, G., Mancuso, C., Butterfield, D.A. and Stella, A.G., 2007. Redox regulation of cellular stress response in aging and neurodegenerative disorders: role of vitagenes. Neurochemical Research 32: 757-773.

Calabrese, V., Scapagnini, G., Davinelli, S., Koverech, G., Koverech, A., De Pasquale, C., Salinaro, A.T., Scuto, M., Calabrese, E.J. and Genazzani, A.R., 2014. Sex hormonal regulation and hormesis in aging and longevity: role of vitagenes. Journal of Cell Communication and Signaling 8: 369-384.

Chance, B., Sies, H. and A. Boveries, A., 1979. Hydroperoxide metabolism in mammalian organs. Physiological Reviews 59: 527-605.

Chaudiere, J. and Ferrari-Iliou, R., 1999. Intracellular antioxidants: from chemical to biochemical mechanisms. Food and Chemical Toxicology 37: 949-962.

Chow, C.K., Ibrahim, W., Wei, Z. and Chan, A.C., 1999. Vitamin E regulates mitochondrial hydrogen peroxide generation. Free Radical Biology and Medicine 27: 580-587.

Dean, R.T., Fu, S., Stocker, R. and Davies, M.J., 1997. Biochemistry and pathology of radical-mediated protein oxidation. Biochemical Journal 324: 1-18.

Diplock, A.T., 1994. Antioxidants and disease prevention. In: Baum, H. (ed.) Molecular aspects of medicine. Volume. 15, Pergamon Press, Oxford, UK, pp. 295-376.

Fujii, J. and Taniguchi, N., 1999. Down regulation of superoxide dismutases and glutathione peroxidase by reactive oxygen and nitrogen species. Free Radical Research 31: 301-308.

Galey, J-B., 1997. Potential use of iron chelators against oxidative damage. In: Sies, H. (ed.) Antioxidants in disease mechanisms and therapy. Academic Press, San Diego, CA, USA, pp. 167-203.

Groves, J.T., 1999. Peroxynitrite: reactive, invasive and enigmatic. Current Opinion in Chemical Biology 3: 226-235.

Grune, T., Reinheckel, T. and Davies, K.J., 1997. Degradation of oxidized proteins in mammalian cells. FASEB Journal 11: 526-534.

Halliwell, B., 1987. Oxidants and human disease: some new concepts. FASEB Journal 1: 358-364.

Halliwell, B., 1994. Free radicals and antioxidants: a personal view. Nutrition Reviews 52: 253-265.

Halliwell, B., 2012. Free radicals and antioxidants: updating a personal view. Nutrition Reviews 70: 257-265.

Halliwell, B. and Gutteridge, J.M.C., 2015. Free radicals in biology and medicine. Oxford University Press, Oxford, UK.

Helbock, H.J., Beckman,K.B., Shigenaga, M.K., Walter, P.B., Woodall, A.A., Yeo, H.C. and Ames, B.N., 1998. DNA oxidation matters: the HPLC-electrochemical detection assay of 8-oxo-deoxyguanosine and 8-oxo-guanine. Proceedings of the National Academy of Science of the USA 95: 288-293.
Hogg, N., 1998. Free radicals in disease. Seminars in Reproductive Endocrinology 16: 241-248.
Jaeschke, H., 1995. Mechanisms of oxidant stress-induced acute tissue injury. Experimental Biology and Medicine 209: 104-111.
Jones, D.P., Eklow, L., Thor, H. and Orrenius, S., 1981. Metabolism of hydrogen peroxide in isolated hepatocytes: relative contributions of catalase and glutathione peroxidase in decomposition of endogenously generated H_2O_2. Archives of Biochemistry and Biophysics 210: 505-516.
Kehrer, J.P., 2000. The Harber-Weiss reaction and mechanism of toxicity. Toxicology 149: 43-50.
Kettle, A.J. and Winterbourn, C.C., 1997. Myeloperoxidase: a key regulator of neutrophil oxidant production. Redox Report 3: 3-15.
Kontos, H.A., 2001. Oxygen radicals in cerebral ischemia: the 2001 Willis lecture. Stroke 32: 2712-2716.
Kruidenier, L. and Verspaget, H.W., 2002. Review article: oxidative stress as a pathogenic factor in inflammatory bowel disease – radicals or ridiculous? Alimentary Pharmacology and Therapeutics 16: 1997-2015.
Mates, J.M. and Sanchez-Jimenez, F., 1999. Antioxidant enzymes and their implications in pathophysiologic processes. Frontiers in Bioscience 4: D339-D345.
McCord, J.M., 2000. The evolution of free radicals and oxidative stress. The American Journal of Medicine 108: 652-659.
Mruk, D.D., Silvestrini, B., Mo, M.Y. and Cheng, C.Y., 2002. Antioxidant superoxide dismutase – a review: its function, regulation in the testis, and role in male fertility. Contraception 65: 305-311.
Niki, E., 1996. α-tocopherol. In: Cadenas, E. and Packer, L. (eds) Handbook of antioxidants. Marcel Dekker, New York, NY, USA, pp. 3-25.
Niki, E., 2014. Antioxidants: basic principles, emerging concepts, and problems. Biomedical Journal 37: 106-111.
Niki, E., 2016. Oxidative stress and antioxidants: distress or eustress? Archives of Biochemistry and Biophysics 595: 19-24.
Nordberg, J. and Arner, E.S., 2001. Reactive oxygen species, antioxidants, and the mammalian thioredoxin system. Free Radical Biology and Medicine 31: 1287-1312.
Puppel, K., Kapusta, A. and Kuczyńska, B., 2015. The etiology of oxidative stress in the various species of animals, a review. Journal of the Science of Food and Agriculture 95: 2179-2184.
Putman, A.K., Brown, J.L., Gandy, J.C., Wisnieski, L. and Sordillo, L.M., 2018. Changes in biomarkers of nutrient metabolism, inflammation, and oxidative stress in dairy cows during the transition into the early dry period. Journal of Dairy Science 101: 9350-9359.
Rattan, S.I., 1998. The nature of gerontogenes and vitagenes. Antiaging effects of repeated heat shock on human fibroblasts. Annals of the New York Academy of Science 854: 54-60.
Schwarz, K.B., 1996. Oxidative stress during viral infection: a review. Free Radical Biology and Medicine 21: 641-649.
Sena, L.A. and Chandel, N.S., 2012. Physiological roles of mitochondrial reactive oxygen species. Molecular Cell 48: 158-167.
Singal, P.K., Khaper, N., Palace, V. and Kumar, D., 1998. The role of oxidative stress in the genesis of hart disease. Cardiovascular Research 40: 426-432.
Surai, P.F., 1999. Vitamin E in avian reproduction. Poultry and Avian Biology Review 10: 1-60.
Surai, P.F., 2002. Natural antioxidants in avian nutrition and reproduction. Nottingham University Press, Nottingham, UK.

Surai, P.F., 2006. Selenium in nutrition and health. Nottingham University Press, Nottingham, UK.

Surai, P.F., 2016. Antioxidant systems in poultry biology: superoxide dismutase. Animal Nutrition 1: 8.

Surai, P.F., 2018. Selenium in poultry nutrition and health. Wageningen Academic Publishers, Wageningen, the Netherland, 430 pp.

Surai, P.F., 2019a. Vitagenes in poultry production: stresses and antioxidants. Suchasne Ptakhivnitstvo (Modern Poultry Science, Ukraine) 9-10: 24-27.

Surai, P.F., 2019b. Vitagenes in poultry production: adaptation to commercially relevant stresses. Suchasne Ptakhivnitstvo (Modern Poultry Science, Ukraine) 7-8: 28-32.

Surai, P.F., 2020a. Vitagenes in avian biology and poultry health. Wageningen Academic Publishers, Wageningen, the Netherlands.

Surai, P.F., 2020b. Antioxidants in poultry nutrition and reproduction: an update. Antioxidants 9, 2: 105.

Surai, P.F., 2020c. Superoxide dismutase as a new entrant into the vitagene family in animals/poultry. EC Nutrition 15(3): 01-03.

Surai P.F. and Kochish I.I., 2017. Antioxidant systems and vitagenes in poultry biology: heat shock proteins. In: Asea, A.A.A. and Kaur, P. (eds) Heat shock proteins in veterinary. Springer, Cham, Switzerland, pp. 123-177.

Surai, P.F. and Fisinin, V.I., 2016a. Vitagenes in poultry production. Part 1. Technological and environmental stresses. World's Poultry Science Journal 72: 721-733.

Surai, P.F. and Fisinin, V.I., 2016b. Vitagenes in poultry production. Part 2. Nutritional and internal stresses. World's Poultry Science Journal 72: 761-772.

Surai, P.F. and Fisinin, V.I., 2016c. Vitagenes in poultry production. Part 3. Vitagene concept development. World's Poultry Science Journal 72: 793-804.

Surai, P.F. and Fisinin, V.I., 2016d. Antioxidant system regulation: from vitamins to vitagenes. In: Watson, R.R. and De Meester, F. (eds) Handbook of cholesterol. Wageningen Academic Publishers, Wageningen, the Netherlands, pp. 451-481.

Surai, P.F., Kochish, I.I., Fisinin, V.I. and Juniper, D.T., 2019a. Revisiting oxidative stress and the use of organic selenium in dairy cow nutrition. Animals 9: 7.

Surai, P.F., Kochish, I.I., Fisinin, V.I. and Kidd, M.T., 2019b. Antioxidant defence systems and oxidative stress in poultry biology: an update. Antioxidants 8: 7.

Surai, P.F., Kochish, I.I., Fisinin, V.I., Grozina, A.A. and Shatskikh, E.V., 2018. Molecular mechanisms of gut health support in poultry: role of microbiota. Monograph. Agricultural Technologies, Moscow, Russia.

Tirosh, O. and Reznick. A.Z., 2000. Chemical bases and biological relevance of protein oxidation. In: Sen, C.K., Packer, L. and Hanninen, O.O.P. (eds) Handbook of oxidants and antioxidants in exercise. Elsevier, Amsterdam, the Netherlands, pp. 89-114.

Yu, B.P., 1994. Cellular defences against damage from reactive oxygen species. Physiological Reviews 74: 139-162.

Zhao, H., Zhao, Z. and Lei, X.G., 2021. Function and regulation of porcine selenogenome and selenoproteome. Chapter 2. In: Surai, P.F. (ed.) Selenium in pig nutrition and health. Wageningen Academic Publishers, Wageningen, the Netherlands, pp. 39-59. https://doi.org/10.3920/978-90-8686-912-1_2

Chapter 2
Function and regulation of porcine selenogenome and selenoproteome

H. Zhao[1], Z. Zhao[2] and X.G. Lei[3]*
[1]Animal Nutrition Institute, Sichuan Agricultural University, Chengdu 611130, Sichuan, China, P.R.; [2]Department of Surgery, Weill Cornell Medicine, New York, NY 10065, USA; [3]Department of Animal Science, Cornell University, Ithaca, NY 14853, USA; xl20@cornell.edu

Abstract

Nutritional roles and metabolic impacts of trace element Se are largely mediated by Se-dependent proteins. There are 25 porcine selenoprotein genes identified. These genes are highly conserved and share a close evolutionary relationship with those of humans and rodents. Selenoproteins are widely recognised as 'antioxidant' by their verified and(or) presumed functions in free radical scavenging and redox control. Revealing their role and mechanism in regulating glucose homeostasis and energy metabolism signifies a major breakthrough of Se biology. Nearly all porcine selenoprotein genes are responsive to dietary Se deficiency, which is similar to the rodent response pattern. In contrast, expression of porcine selenogenome displays variable responses to moderately high dietary Se intakes, and is more sensitive to high dietary Se intakes than that of rodents. Functional expressions of porcine selenogenome and selenoproteome are also regulated or affected by dietary fat and serine and other factors including heat stress, inflammation, and immunomodulators. Comparatively, more porcine selenoprotein genes (23 vs 15) are responsive to high dietary fat intakes than those of rodents. Dietary serine affects expression of 5 porcine selenoprotein genes. Expression of 60-70% porcine selenoprotein genes in a number of tissues is altered by heat stress and inflammation challenge. Overall, this chapter offers a comprehensive review and critical analysis of dietary Se requirements by pigs, selenoprotein biosynthesis, and evolution, function, and regulation of porcine selenogenome and selenoproteome. Comparative Se biology between pigs and rodents or humans and physiological implications of recent progress are also highlighted.

Keywords: gene expression, pig, selenium, selenogenome, selenoproteome, tissue

2.1 Introduction

Dietary Se requirements by different types of pigs are 0.15-0.30 mg/kg of diet (NRC, 2012). These requirements were established mainly based on saturation of Se-dependent glutathione peroxidase (GPX) activities in blood and tissues of pigs under physiological states, but could be altered by age, metabolic stresses, and environmental conditions.

DOI 10.3920/978-90-8686-912-1_2, © Wageningen Academic Publishers 2021

Newborn and weanling pigs require greater dietary Se intakes than grower or finisher pigs because of limited placental transfer of Se, low status of vitamin E, ascorbic acid, and other antioxidants, and high susceptibility to reactive oxygen species (ROS)-induced oxidative damages due to the rapid growth (Surai, 2016). These challenges might be partially relieved by supplementing sow with Se-enriched yeast (Jia, 2007) and selenomethionine (SeMet) (Falka *et al.*, 2020; Zhan *et al.*, 2011). Growing and finishing pigs can ingest adequate Se from the commercial feed formulated based on the NRC and FDA standards, and feeding organic Se forms instead of inorganic forms may help enrich Se deposition in pork and improve the meat quality (Silva *et al.*, 2019). Sows and boars also require high dietary Se intakes, and likely benefit from organic Se supplementation (Mahan, 2000; Mahan and Peters, 2004; Surai and Fisinin, 2016).

2.1.1 Biosynthesis of selenoproteins

It is widely accepted that Se plays its metabolic role mainly through incorporation into the 24/25 Se-dependent proteins (selenoproteins) (Table 2.1) in the chemical moiety of selenocysteine (Sec). Sec was named as the 21st amino acid and is encoded by UGA, a terminal codon in normal translation (Hatfield and Portugal, 1970). Sec is structurally similar to cysteine (Cys) and is derived from a serine (Ser) precursor (Nauser *et al.*, 2006). Unlike any other known amino acids in eukaryotes, Sec biosynthesis requires a special tRNA designated as $tRNA^{[Ser]Sec}$, which is identified in 1970 (Maenpaa and Bernfied., 1970). A selenoprotein encoding gene contains a stem-loop structure in the 3′ untranslated region (3′ UTR), an element named the Sec insertion sequence element or SECIS element (Berry *et al.*, 1991). All selenoprotein encoding genes have two characteristic features: (1) UGA codon; and (2) SECIS element (Kryukov *et al.*, 2003). The incorporation of Sec into selenoproteins requires complex synthetic machinery (for review see Hatfield *et al.*, 2016) and is dependent on cis-acting factor (SECIS element) within the 3-untranslated region of selenoprotein mRNA and trans-acting factors involved in bringing a selenocysteyl-tRNA (Sec-$tRNA^{[Ser]Sec}$) to the ribosome. SECIS binding protein 2 (SBP2) is a factor required for Sec incorporation (Copeland *et al.*, 2000) and SBP2 itself is a selenoprotein. A Sec-specific elongation factor (eEFSec) is found to bind Sec-$tRNA^{[Ser]Sec}$ (Fagegaltier *et al.*, 2000; Tujebajeva *et al.*, 2000). The selenoprotein biosynthesis (Figure 2.1) consists of the following steps:

- Sec is synthesised from a serine (Ser) precursor and forms Sec-$tRNA^{[Ser]Sec}$.
- Sec-$tRNA^{[Ser]Sec}$ is delivered to the ribosome and recognises the UGA stop codon in the coding region of selenoprotein mRNA. This recognition depends on *cis*-acting factor SECIS element, *trans*-acting factor SBP2, Sec-specific elongation factor eEFSec, and other factors.
- Sec is co-translationally incorporated into peptides of selenoproteins.

Table 2.1. Porcine selenogenome and selenoproteome.[1,2]

Name	Accession #	Chromosome location	Gene length (bp)	Protein size (aa)	Gene function
DIO1	NC_010448.4	Chr 6 (158376744-158392745)	16,001	249	Convert the prohormone thyroxine (T4) to bioactive thyroid hormone (T3) to provide most of the circulating T3
DIO2	NC_010449.5	Chr 7 (103096414-103110749)	14,335	269	
DIO3	NC_010449.4	Chr 7 (130202342-130204449)	2,107	305	Catalyse T4 and T3 to 3,3',5'-triiodothyronine (RT3) and 3,3'-diiodothyronine (T2), respectively
GPX1	NC_010455.5	Chr 13 (31916269-31917337)	1,068	206	Catalyse the reduction of organic hydroperoxides and hydrogen peroxide (H_2O_2) by glutathione, and thereby protect cells against oxidative damage.
GPX2	NC_010449.5	Chr 7 (89007636-89011620)	3,984	190	
GPX3	NC_010458.4	Chr 16 (71980467-71989011)	8,544	225	
GPX4	NC_010444.4	Chr 2 (77321568-77323931)	2,363	197	This isozyme has a high preference for phospholipid hydroperoxides and protects cells against membrane lipid peroxidation and cell death
GPX6	NC_010449.5	Chr 7 (22256013-22267992)	11,979	220	The exact function of this gene is not known
MSRB1[3]	NC_010445.4	Chr 3 (40017161-40021736)	4,575	116	Protects cells from oxidative stress by reducing methionine-R-sulfoxides to methionine
SEPHS2	NC_010445.4	Chr 3 (17886965-17889128)	2,163	451	Catalyses the production of monoselenophosphate from selenide and ATP
SELENOF	NC_010446.5	Chr 4 (129258943-129287727)	28,784	162	Associated with UDP-glucose:glycoprotein glucosyltransferase (UGTR) (quality control of protein folding, ER stress)
SELENOH	NC_010444.4	Chr2 (13229296-13229879)	583	122	Oxidoreductase protects neurons, promotes mitochondrial biogenesis and function, and regulates redox status.
SELENOI	NC_010445.4	Chr 3 (112632985-112671127)	38,142	383	Produces phosphatidylethanolamine from diacylglycerol, regulates lipid metabolism, and protein folding (ER stress).

Table 2.1. Continued.

Name	Accession #	Chromosome location	Gene length (bp)	Protein size (aa)	Gene function
SELENOK	NC_010455.5	Chr 13 (35989461-35996824)	7,363	94	Regulates ER-associated degradation (ERAD) of misfolded, glycosylated proteins (ER stress)
SELENOM	NC_010456.5	Chr 14 (47902862-47905350)	2,488	142	May be involved in neurodegenerative disorders
SELENON	NC_010448.4	Chr 6 (83246934-83262748)	15,814	555	Protects cells from oxidative stress and regulates redox-related calcium homeostasis.
SELENOO	NC_010447.5	Chr 5 (467546-480836)	13,290	666	Thought to have redox activity
SELENOP	NC_010458.4	Chr 16 (27537885-27547863)	9,978	390	Extracellular antioxidant and transport Se
SELENOS	NC_010443.5	Chr 1 (139829002-139836923)	7,921	190	Involved in the degradation process of misfolded proteins in the ER
SELENOT	NC_010455.5	Chr 13 (90818409-90839242)	20,833	137	Protects dopaminergic neurons against oxidative stress and controls glucose homeostasis in beta-cells.
SELENOV	NC_010448.4	Chr 6 (48162428-48167438)	5,010	341	Regulates redox ?
SELENOW	NC_010448.4	Chr 6 (53470563-53474662)	4,099	97	Regulates muscle growth and differentiation and protects neurons from oxidative stress during neuronal development
TXNRD1	NC_010447.5	Chr 5 (80170334-80213889)	43,555	499	Reduces thioredoxins, and plays a key role in redox homoeostasis
TXNRD2	NC_010456.5	Chr 14 (51355731-51385096)	29,365	606	
TXNRD3	NC_010455.5	Chr 13 (73276249-73336782)	60,533	589	

[1] aa = amino acids; Accession # = accession numbers; ATP = adenosine triphosphate; bp = base pairs; chr = chromosome; ER = endoplasmic reticulum.

[2] Supported by database NCBI (Gene/Protein) (https://www.ncbi.nlm.nih.gov/gene/?term=) / (https://www.ncbi.nlm.nih.gov/protein/) and Genecards (https://www.genecards.org/).

[3] Known as *SELENX* or *SELENR*.

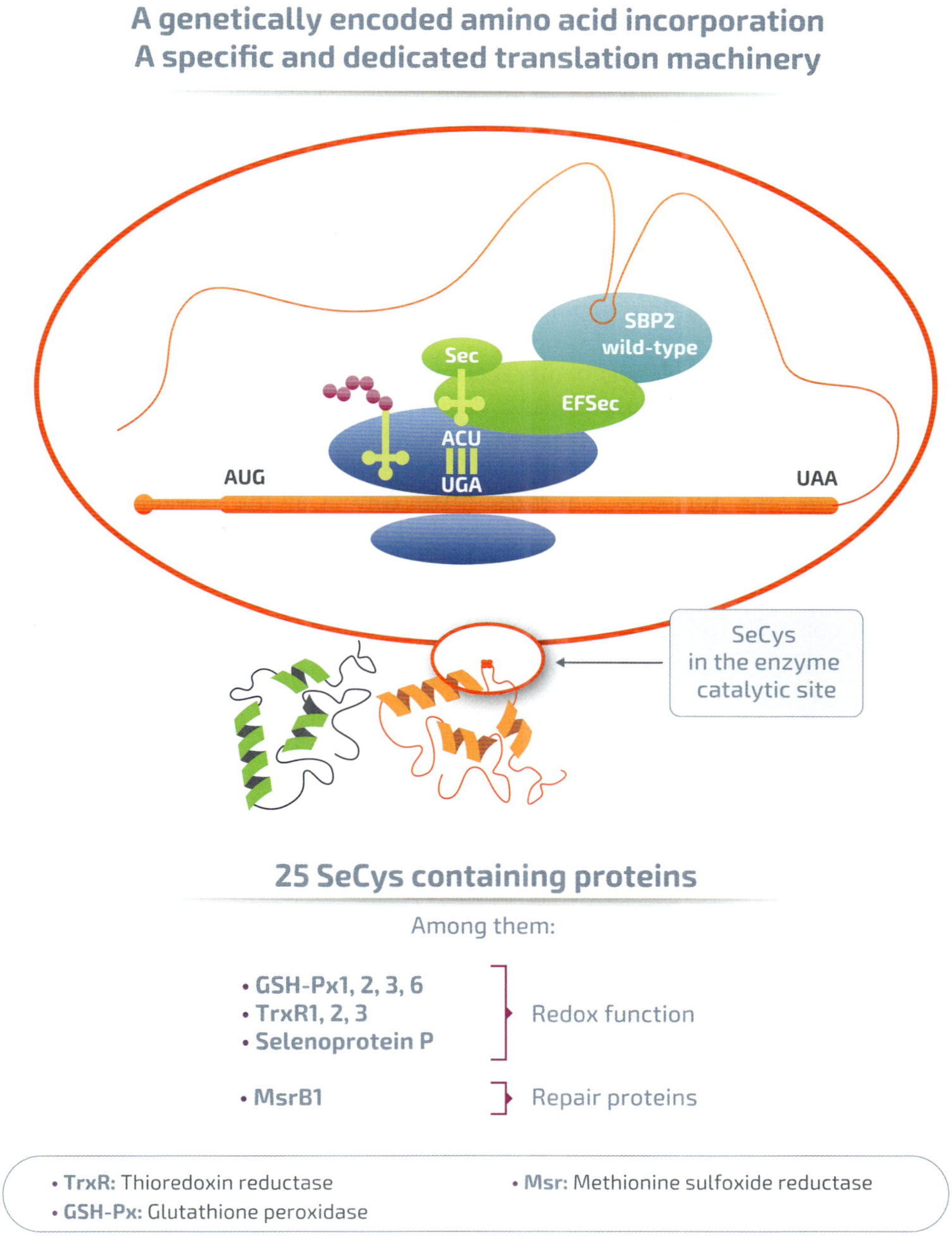

Figure 2.1. SelenoCysteine. The 21th amino acid.

Selenoenzymes, with Sec in their active sites, are of high redox potentials, and have greater catalytic efficiencies and faster chemical reaction rates with electrophiles than their Cys counterparts (Nauser *et al.*, 2006; Steinmann *et al.*, 2008). The selenoproteins are also more resistant to oxidative inhibition than the Cys-containing enzymes (Snider *et al.*, 2013). These redox characteristics help explain the oxidation/reduction roles in the electron transfer by selenoenzymes such as glutathione peroxidases (GPX) (Figure 2.2), iodothyronine deiodinases (DIO) and thioredoxin reductases (TR) (Figure 2.3) and sulphide oxidases.

2.1.2 Evolution and functions of porcine selenogenome and selenoproteome

Human, rodent and pig selenogenomes share a close evolutionary relationship and are highly conserved. In all three species, *SELENOH, SELENOT,* and *SELENOW* contain Sec and CXXU domain in the peptides (Groenen *et al.*, 2012). In contrast, these three proteins do not contain Sec (replaced by Cys) and the domain becomes CXXC motif in some other species (Chen *et al.*, 2018). Among all the selenoprotein genes, *SELENOT* is the most conserved one across the humans, rodents, and pigs (Chen *et al.*, 2018). Specifically, *DIO2, DIO3, GPX1, GPX2, GPX3, GPX6, SELENOI, SELENOO* and *SELENOP* are well conserved between the humans and rodents (Chen *et al.*, 2018). In contrast, *GPX4, MSRB1, SELENOF, SELENOK, SELENOM, SELENON,*

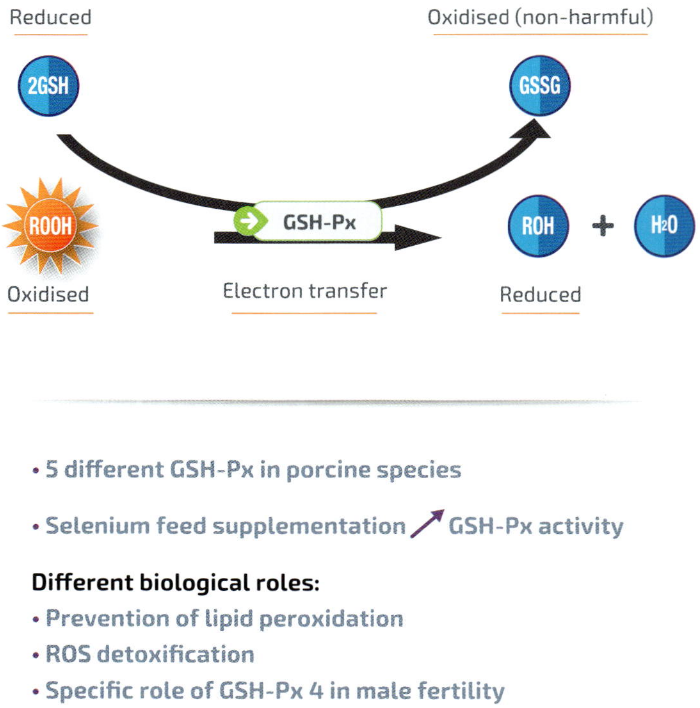

Figure 2.2. Glutathione peroxidase.

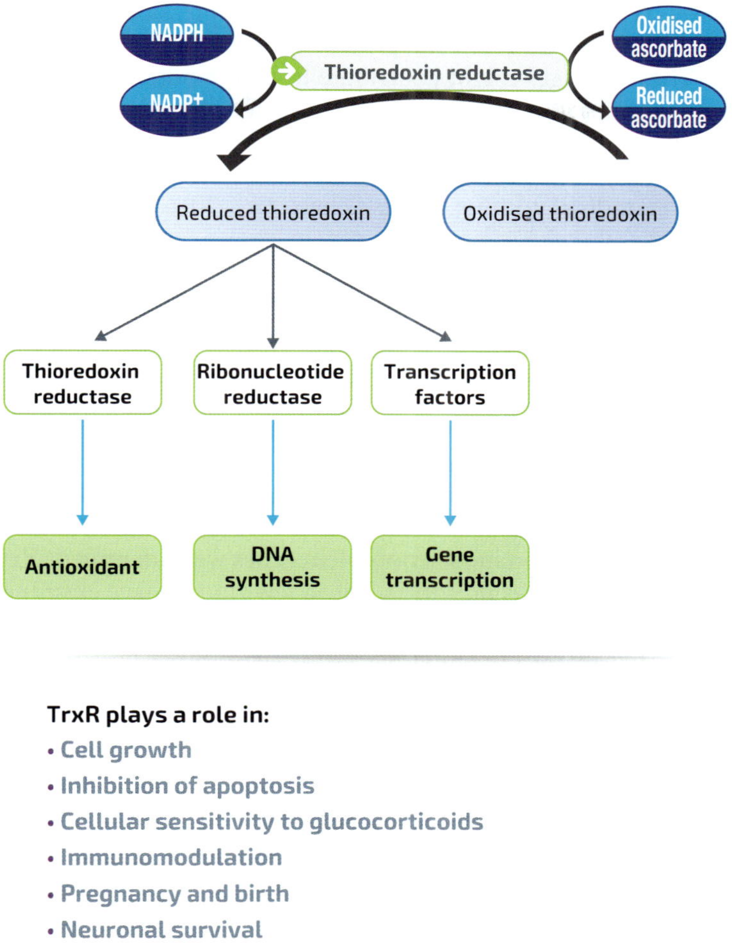

Figure 2.3. Thioredoxin reductase.

SELENOT, TXNRD1, and *TNXRD2* are better conserved between the rodents and pigs (Chen *et al.*, 2018). Meanwhile, *DIO1, SELENOH, SELENOS, SELENOV, SELENOW*, and *SEPHS2* are better conserved between the humans and pigs (Chen *et al.*, 2018). Notably, porcine *SEPHS2* has a high similarity to chicken *GPX4, SELENOW1*, and *TXNRD2* (Chen *et al.*, 2018). During evolution, selenoprotein Pb (SELENOPb) and selenophosphate synthetase 2a (SEPHS2a) lost Sec that was replaced by Cys (Chen *et al.*, 2018).

Recent studies have demonstrated novel roles of porcine selenoproteins in regulating insulin secretion and signalling and energy metabolism (Zhao *et al.*, 2016). Up-regulation of GPX1, GPX3, *SELENOH*, SELENOP and *SELENOV* and down-regulation of *DIO1* and *TXNRD1* in pig tissues were associated with decreased functional expression related to insulin signalling (INSR and AKT), glycolysis (glucokinase) and gluconeogenesis (PEPCK) and increased functional expression related to lipogenesis (FOXO1, SREBP1, ACC and FAS), and protein synthesis (mTOR and RPS6/S6) (Zhao *et al.,* 2016). These changes might contribute to high Se diet-led hyperinsulinemia, insulin resistance, elevated lipid, protein, saturated fatty acids (myristic, palmitic, stearic, and behenic acids), and aspartic acid, glutamine, and alanine, and decreased polyunsaturated fatty acids (dihomo-γ-linolenic, oleic, α-linolenic, and eicosatrienoic acids) in the liver and(or) adipose tissues (Zhao *et al.,* 2016).

2.2 Regulation of porcine selenogenome and selenoproteome

2.2.1 Regulation by dietary selenium

Body Se status affects tissue production of selenoproteins, not only via supplying Se for the formation of Sec in their peptides, but also via regulating their mRNA abundances (Sunde *et al.,* 2009). Dietary Se concentrations (mg Se/kg diet): deficiency (<0.02-0.03), adequacy (0.1-0.3), and excess (0.4-5), are often used to determine such regulations. Although 25 porcine selenoprotein genes were identified (Kryukov *et al.,* 2003; Sun *et al.,* 2017; Zhao *et al.*, 2015), only a few studies have examined collective responses of these genes to dietary Se concentrations ranging from deficiency to excess in pigs (Li *et al.*, 2011; Liu *et al.*, 2012, 2018; Lu *et al.,* 2019; Zhang *et al.*, 2020a,b; Zhao *et al.*, 2016; Zhou *et al.*, 2009).

Expressions of nearly all porcine selenoprotein genes, except for *GPX6,* were responsive to dietary Se intakes of pigs (Table 2.2). Among these responsive genes, *GPX1*, *SELENOH, SELENOP*, and *SELENOS* had their mRNA levels decreased in 6-8 tissues by dietary Se deficiency (<0.02 mg Se/kg diet) compared with the Se-adequate (0.3 mg/kg) diet. Meanwhile, dietary Se excess (3.0 mg Se/kg diet) caused diverse responses (no change or decrease) of *GPX1*, *SELENOP* and *SELENOS* (Liu *et al.*, 2012; Zhao *et al.*, 2016; Zhou *et al.,* 2009). SELENOH is a nucleus-localised protein, and its gene expression is highly sensitive to body Se status (Sunde *et al.,* 2009). Dietary Se deficiency or excess affected its mRNA levels in 8 tissues of pigs (Table 2.2). Overall, *GPX1*, *SELENOH, SELENOP*, and *SELENOS*, the four abundantly expressed selenoprotein genes, were readily responsive to dietary Se changes in pigs. These responses were similar to those in the chickens (Huang *et al.*, 2011; Xu *et al.*, 2017), mice (Sunde *et al.*, 2009), and rats (Barnes *et al.,* 2009).

The mRNA abundances of all selenoprotein genes but *DIO1*, *SELENOO*, *SELENOT,* and *SELENOV* in various tissues of pigs were decreased by dietary Se deficiency (<0.02-0.03 mg/kg), compared with dietary Se adequacy (0.3 mg/kg). Dietary Se deficiency caused diverse responses of *SEPHS2* in various tissues. In multiple porcine tissues including pituitary, spleen, and thyroid, both dietary Se deficiency and excess decreased the expression of *SELENOF* and *SELENOI* (Liu *et al.,* 2012). Compared with dietary Se adequacy, dietary Se deficiency decreased *GPX3* expression in the pituitary, testes (Liu *et al.,* 2012), spleen (Lu *et al.,* 2019), and peripheral blood lymphocytes (Liu *et al.,* 2018), and dietary Se excess up-regulated its expression in the liver and

Table 2.2. Regulation of selenogenome and selenoproteome by dietary selenium and fat intakes and other factors in pigs.[1]

Name	Location	<0.02 mg Se/kg[2]	0.4-3 mg Se/kg[2]	High fat diet	Se resource 0.25 mg/kg	Heat/immune stress	Serine
DIO1	Memb	L/Pi/Te/Th↔	M↓	Pa↓		Ic/S/Th↑;L↓	
DIO2	Memb[4]	S↓		Th↑		Ic/L/S/Th↑	
DIO3		Ly/S↓		Ly↓		Ic/L↑;S↓	
GPX1[3]	Cyto/Mito	H/K/[a]L/Ly/M/Pi/S/[a]Th↓	H/K/Th↔; [a]L/[a]M/[a]Th↑; Pi/S↓	Ly↑	L↑	Ic↓L↑	L/M↑
GPX2	Cyto	Ly/S↓		Ly↑		Ic↑;L↓S↓	L/M↑
GPX3	Secret	Ly/[a]Pl/Pi/Te↓S	[a]Pl↔; L/M↑	HT↑;Ly↓		Ic/S/Th↑	L↑;M↔
GPX4	Memb/Mito/Cyto	Ly↓	L↑	Ly↓		Ic↑	
GPX6	Secret[4]			M↑		Ic↑	
MSRB1	Cyto/Nu	Ly/S↓	S↓	A↓Ly↓		Ic/S↑;L↓	
SEPHS2	Cyto	L/Th↔; H↑; HT/Ly/Pi↓	H/L↔; HT/Pi/Th↓	Ly↓		Ic↑;L↓	
SELENOF[3]	ER	K/Pi/S/Th↓	Pi/S/Te/Th↓	L↑	L↑	Ic/S↑	
SELENOH[3]	Nu	K/L/Ly/Pi/S/Th↓	K/Th↔; L↑; S↓	Pa↓Ly↓	L↑	Ic↓	
SELENOI[3]	Memb[4]	Ly/Pi/S/Th↓	Pi/S/Th↓	A/HT/Pa/Pi↓	L↑	Ic↓;L/S↑	
SELENOK	ER/Memb[4]	Ly↓		Ly/Pa↓		S/Th↑	
SELENOM	ER/Gol[4]	Ly/S/Th↓	S↓	A/K↓; Pi/Ly↑	L/M↑	Ic↓	L/M↔
SELENON		Ly/S↓	M↑	Ly↓		Ic↔;L↓	L/M↔
SELENOO	Mito[4]	H/HT↑	H/HT↔	K↑		L↓S↓	
SELENOP[3]	Secret	K/L/Ly/M/Te/Th/S↓	H/K/L↔; M/Te/Th↑	Ly↓	L↑	Ic↔;L/S/Th↓	L/M↑
SELENOS[3]	ER/Memb	B/K/L/Pi/S/Th↓	B/H/K/L/M↔; Th↑	Ly↓	L↑	Ic/L/S/Th↑	
SELENOV	Un	HT/K/M↑; Th↓S	HT/K/M↔; Th↓	A/L↓; M↑			
SELENOW	Cyto	M/Ly/L↓	L↓;M↑	Pa↓Ly↓	L↑	Ic↑	
SELENOT	ER	H/L/M↑S;Ly↓	H/L/M↔	Ly↑	L↑	Ic/S/Th↑	
TXNRD1	Cyto/Nu	L/Ly↓; Te↔	M/Te↓; L↔↓	Th↑; HT/Pi↓	L↑	Ic/L/S/Th↑	
TXNRD2	Mito	S↓				Ic/L/S/Th↓	L/M↑
TXNRD3		S↓				S/Th↑	L/M↑

[1] Abbreviations and arrows: superscript a ([a]) = enzyme activity of the selenoprotein; A = adipose tissue; B = blood; Cyto = cytosol; ER = endoplasmic reticulum; Gol = Golgi apparatus; H = heart; HT = hypothalamus; K = kidney; L = liver; Ly = lymphocyte; M = muscle; Memb = membrane; Mito = mitochondria; Nu = nucleus; Pa = pancreas; Pl = plasma; Pi = pituitary; S = spleen; Se = selenium; Te = testis; Th = thyroid; Ic = intestinal cells; Un = unknown; ↑ = increase; ↓ = decrease; ↔ = no change or difference.

[2] Compared with 0.1-0.3 mg Se/kg diet.

[3] Changed in many tissues.

[4] Supported by the database of uniprot (http://www.uniprot.org/uniprot/).

muscle (Liu *et al.*, 2012). Dietary Se deficiency decreased expression of *SELENOW* in the liver, muscle (Li *et al*, 2011; Liu *et al.*, 2012), and peripheral blood lymphocytes (Liu *et al.*, 2018), whereas dietary Se excess down-regulated *SELENOW* in the liver (Li *et al.*, 2011; Liu *et al.*, 2012; Zhang *et al.*, 2020b) and up-regulated its expression in the muscle (Li *et al.*, 2011; Liu *et al.*, 2012). The increased expression of *SELENOW* in muscle was associated with enhanced water-holding capacity of meat (Li *et al.*, 2011) or improved meat quality (Zhang *et al.*, 2020b). SEPHS2 is an enzyme involved in Sec biosynthesis and its subsequent incorporation into selenoproteins. Dietary Se deficiency caused diverse responses of *SEPHS2* in various tissues of pigs (Table 2.2). However, mRNA levels of *DIO2, GPX4, SELENOK, TXNRD2* and *TXNRD3* in tissues or cells of pigs were somewhat resistant to dietary Se deficiency or excess (Liu *et al.*, 2012, 2018; Lu *et al.*, 2019).

Porcine *DIO1* transcripts were unaltered by dietary Se deficiency in the liver, pituitary, testis, and thyroid (Liu *et al.*, 2012), but decreased by high dietary Se intake in the muscle (Zhao *et al.*, 2016). Dietary Se deficiency down-regulated *SELENOO* in the heart and hypothalamus, while high dietary Se intake (3.0 mg/kg) showed no effect on its expression in those two tissues (Liu *et al.*, 2012). *SELENOT* expression was increased in the liver, heart, muscle (Liu *et al.*, 2012) and spleen (Lu *et al.*, 2019), but down-regulated in peripheral blood lymphocytes (Liu *et al.*, 2018) by dietary Se deficiency compared with dietary Se adequacy (Liu *et al.*, 2012; Zhou *et al.*, 2009). Expression of *SELENOV* was also affected by dietary Se intakes (Table 2.2) in 5 tissues of pigs. However, expression of *SELENOV* was readily detected only in testes of pigs (Liu *et al.*, 2012).

Collectively, dietary Se intakes exerted no simple or global mechanism in regulating transcripts of multiple selenoprotein genes across porcine tissues (Liu *et al.*, 2012). Instead, the regulation of porcine selenogenome was tissue-specific, hierarchy of selenoprotein-dependent, and time-variable (Liu *et al.*, 2012). Pigs fed Se deficiency (<0.02 mg/kg) or excess (3.0 mg/kg) diets for 8 weeks did not show changes in mRNA levels of any selenoprotein genes in thyroid or pituitary (Zhou *et al.*, 2009), while prolonged the trial for 16 weeks resulted in changes of mRNA levels of 8 selenoprotein genes in those tissues (Liu *et al.*, 2012). Furthermore, pigs seemed to have a lower number of selenoprotein genes affected by dietary Se deficiency (Huang *et al.*, 2011; Xu *et al.*, 2017) or Se excess (Huang *et al.*, 2016) than broilers. Their overall responses (Liu *et al.*, 2012) were similar to those of turkeys (Taylor and Sunde,2016; Taylor *et al.*, 2019), mice (Akahoshi *et al.*, 2019; Sunde *et al.*, 2009), and rats (Barnes *et al.*, 2009).

Due to the lack of reagents such as antibodies against porcine selenoproteins or protocols for assay their biochemical functions, there has been limited information on responses of selenoproteome or selenoenzyme activity to dietary Se intakes. Elevating dietary Se from 0.3 mg/kg to 3 mg/kg did not alter the mRNA or protein abundances of GPX1 in the heart, kidney, liver, muscle, and thyroid (Liu *et al.*, 2012; Zhao *et al.*, 2016; Zhou *et al.*, 2009), but enhanced GPX activities in the liver, muscle, and thyroid (Liu *et al.*, 2012; Zhou *et al.*, 2009). Compared with 0.17 mg Se/kg, 0.5 mg Se/kg enhanced GPX1 activity in the muscle (Pinto *et al.*, 2012). The Se excess diet did not alter protein levels of SELENOP or SELENOS in the heart, kidney, and liver, but elevated SELENOP in the muscle (Zhao *et al.*, 2016) and thyroid (Liu *et al.*, 2012), and SELENOS in the thyroid (Liu *et al.*, 2012). Plasma GPX3 activity of pigs was decreased by dietary Se deficiency, but remained unchanged at week 8 (Zhou *et al.*, 2009) or was elevated at week 16 after the treatment of the high Se diet (Liu *et al.*, 2012).

2.2.2 Regulation by dietary fat

One of the most important findings from the recent studies is the illustration of the regulation of selenogenome by dietary fat. Among the 25 porcine selenoprotein genes, mRNA expression of 23 genes (except for *TXNRD2* and *TXNRD3*) was altered by high dietary fat intakes (Table 2.2) (Liu *et al.*, 2018; Wang *et al.*, 2020; Zhao *et al.*, 2015). A high-fat diet increased expression of 12 selenoprotein genes in 6 tissues and decreased expression of 13 selenoprotein genes in 7 tissues of pigs (Zhao *et al.*, 2015). High fat diet also down-regulated mRNA levels of many selenoprotein genes, but elevated mRNA levels of *GPX1*, *GPX2*, *SELENOM*, and *SELENOT* in the pig peripheral blood lymphocytes (Liu *et al.*, 2018). While high dietary fat intake elevated the expressions of *GPX3* in the hypothalamus, *SELENOF* in the liver, *SELENOM* in the pituitary, *SELENOO* in the kidney, *SELENOV* in the muscle, and *TXNRD1* in the thyroid, it decreased the expressions of several selenoprotein genes in multiple endocrine organs including adipocyte tissue, hypothalamus, pancreas, and pituitary (Zhao *et al.*, 2015). Interesting, high fat diets exhibited little impact on expression of *GPX1*, *SELENOH*, *SELENOP*, and *SELENOS* that were highly sensitive to dietary Se changes. The high fat diet up-regulated the gene expressions of *DIO2* in the thyroid and *GPX6* in the muscle, down-regulated *SELENOK* expression in the pancreas and peripheral blood lymphocytes (Liu *et al.*, 2018; Zhao *et al.*, 2015), while expression of *GPX6* was not affected by dietary Se (Table 2.2).

Metabolically, the high fat intake-mediated changes of selenogenome expression were associated with the elevated serum malondialdehyde (MDA) (Wang *et al.*, 2020) and plasma TNFα and IL-6 (Zhao *et al.*, 2015), along with risk of obesity and chronic inflammation. Gene expression of cytosolic selenoproteins was readily affected by dietary Se and fat intakes. The same seemed to be true for those ER-resident selenoproteins (*SELENOI* and *SELENOM*). Because of the role of ER in lipid metabolism (Han and Kaufman, 2016), altering the ER-resident selenoproteins may impair protein folding (Labunskyy *et al.*, 2007; Shchedrina *et al.*, 2010) and partially explains the adverse metabolic effects of high dietary fat intakes (Wang and Kaufman, 2014).

2.2.3 Regulation by other factors

The expression of porcine selenogenome is affected by factors other than dietary Se or fat intakes (Table 2.2). The chemical forms of supplemental Se, either inorganic (sodium selenite) or organic (yeast Se, SeMet or its analogous), is one of these factors. A recent study fed finishing pigs a basal diet (contain 0.1 mg Se/kg) supplemented without or with three different sources of Se (SeNa, MeSeCys, or SeMet) at 0.25 mg Se/kg (Zhang *et al.*, 2020b). The organic Se supplementation (MeSeCys and SeMet) produced high mRNA levels of liver *SELENOF*, *SELENOH*, *SELENOI*, *SELENOM*, *SELENOT*, *SELENOW*, *SEPHS2*, and *TXNRD1* and muscle *SELENOW* than inorganic Se (SeNa). Between the two organic Se sources, SeMet was more efficient in increasing tissue Se deposition, but less efficient in enhancing the gene expression of selenoproteins than MeSeCys (Zhang *et al.*, 2020b).

Supplemental dietary serine elevated the mRNA levels of *GPX1*, *GPX2*, *SELENOI*, *TXNRD1* and *TXNRD2* in the skeletal muscle and liver of pigs (Long *et al.*, 2020). The serine supplementation also elevated activities of selenoenzymes in the liver and muscle, but had no effect on tissue Se contents of growing pigs. Environmental stress is another factor that affects selenoprotein gene expression. Cao *et al.* (2016) revealed that the expression of selenoprotein genes were affected by heat stress in porcine IPEC-J2 cells (porcine small intestinal epithelial). Heat stress elevated reactive oxygen species (ROS) levels in IPEC-J2 cells and caused inflammation (Cao *et al.*, 2016; Tang *et al.*, 2019a). The heat stress-induced cell injury was associated with the up-regulation of 6 inflammation-related genes and 14 selenoprotein genes (*DIO1*, *DIO2*, *GPX2*, *GPX3*, *GPX4*, *GPX6*, *MSRB1*, *SELENOF*, *SELENOK*, *SELENOS*, *SELENOT*, *SELENOW*, *SEPHS2* and *TXNRD1*) and down-regulation of 2 inflammation-related genes and 5 selenoprotein genes (*GPX1*, *SELENOH*, *SELENOI*, *SELENOM* and *TXNRD2*) (Tang *et al.*, 2019a). In cultured rodent C2C12 cells, heat stress up-regulated mRNA levels of 18, 11 and 8 selenoprotein genes after 4, 6 and 8 days of hyperthermia, and down-regulated only *DIO2* after 6 and 8 days of hyperthermia, respectively (Tang *et al.*, 2018). Administration of LPS to pigs altered mRNA levels of 14, 16, 10, and 6 selenoprotein genes in the liver, spleen, thymus, and lymph node of piglets, respectively (Sun *et al.*, 2017). Similar results were shown in mice (Tang *et al.*, 2019b).

2.3 Comparative regulations of porcine and rodent selenogenomes and selenoproteomes

2.3.1 Comparative regulations by dietary selenium deficiency

Rodents express 24 selenoprotein genes (Santesmasses *et al.*, 2020). The difference between rodent and porcine or human selenogenome is GPX6 which contains Cys in the rodents instead of Sec in the pigs or humans (Gladyshev *et al.*, 2016). Among the 24 rodent genes, 19 of them were regulated by dietary Se and fat intakes (Akahoshi *et al.*, 2019; Chen *et al.*, 2020; Han *et al.*, 2014; Labunskyy *et al.*, 2007; Mueller *et al.*, 2008; Novoselov *et al.*, 2010; Raines and Sunde, 2011; Sunde and Raines, 2011; Sunde *et al.*, 2018; Yang *et al.*, 2019; Yim *et al.*, 2019; Zeng *et al.*, 2012; Zhou *et al.*, 2017). Dietary Se deficiency exerted similar regulations of selenogenome and selenoproteome between pigs and rodents. While 5 rodent selenoprotein genes (*Gpx2, Selenon, Selenoo, Selenov*, and *Sephs2)* showed response to neither dietary fat nor Se changes, 13 of the remaining 19 selenoproteins (GPX1, GPX3, *Gpx4, Msrb1*, *Selenok, Selenoh, Selenom, Selenop, Selenos, Selenow, Txnrd2* and *Txnrd3*) were down-regulated by dietary Se deficiency in various tissues of both pigs and rodents (Table 2.2 and 2.3). Liver was the most sensitive tissue to dietary Se deficiency, displaying down-regulation of GPX1, GPX3, *Gpx4*, *Msrb1*, *Selenoh*, *Selenok*, *Selenom*, *Selenop*, *Selenow*, *Txnrd2*, and *Txnrd3* (Akahoshi *et al.*, 2019; Chen *et al.*, 2020; Novoselov *et al.*, 2010; Sunde and Raines, 2011; Sunde *et al.*, 2018; Yim *et al.*, 2019). In contrast, expression of *Selenok*, *Selenom*, *Txnrd2*, and *Txnrd3* was less affected by dietary Se deficiency than that of the other 7 selenoprotein genes in other tissues of rodents (Akahoshi *et al.*, 2019; Chen *et al.*, 2020; Novoselov *et al.*, 2010; Sunde and Raines, 2011; Sunde *et al.*, 2018; Yim *et al.*, 2019; Zhou *et al.*, 2017). Similarly low responsiveness of these four genes was also seen in pigs (Table 2.2).

Table 2.3. Regulations of selenogenome and selenoproteome by dietary Se and fat intakes in rodents.[1]

Name	Location	Se deficiency[2]	1 mg Se/kg[2]	3-5 mg Se/kg[2]	High fat diet
Responding to only dietary Se					
Dio1	Memb	L↓;A/K↔	A↓	A↔	
TXNRD2	Cyto	L↓;K/B↔	L↔	L/B↑	
Txnrd3	Mito	L↓	L↔	L↔	
GPX3	Secret	[a]Pl/L/O/K/P/S/Th↓;B/H/Lu/T/M↔	[a]Pl/L↑	[a]Pl/L/B/H/Lu/K/P/S/T/M/Th↔	
Selenot	ER	L↓			
Responding to both dietary Se and fat					
Dio2	ER		A/L↓		L↑
Selenof	ER		A↓;L↑		A/L↑
Selenok	ER	L↓;K↔	A↓		A↓
Selenom	ER	L↓;K↔	A/L↓		A↓;L↑
Selenos	ER	M↓; L↔		L↑; M↔	A↓;L↑
Gpx4	Cyto/Nu/Mito	L/O/K/P↓;B/H/Lu/T/S/M/Th↔	A↓;L/T↔	L/T/B/H/Lu/K/P/S/M/Th↔	A↓;L↑;T↔
Msrb1	Cyto/Nu	[a]L/B↓;P↔	A↓	B↓;L/P↔	L↑
Selenoh	Nu	L/P/M↓;K↔	A↓	L↑;P/M↔	A↓;L↑
Dio3	Memb		A↓		A/L↑
Selenoi	Memb[3]	L/M↔	A↓	L/M↔	A↓;L↑
Gpx6	Secret[3]		A↓		A↓;L↑
Selenop	Secret	P/M↔;L/T/K↓	A/L↑;T↔	L/T/P/M↔	A↑;T↓;L↔
Selenow	Cyto	L/P/M↓;K↔	L↓	L/P/M↔	M↓
[a]GPX1	Cyto/Mito	L/T/O/K/RBC/H/P/S/Th↓;B/Lu/M↔	L↓;T↑	L/T/B/H↑;Lu/K/P/S/M/Th↔	T↓;L/A↑
TXNRD1	Cyto/Nu	L/T/K↓;P↔;H↑	L/T↔	L/T↔	A↓;T/L↔

[1] Abbreviations and arrows: superscript a ([a]) = enzyme activity of the selenoprotein; A = adipose tissue; Cyto = cytosol; ER = endoplasmic reticulum; Memb = membrane; Mito = mitochondria; Nu = nucleus; Pl = Plasma; Se = selenium; ↑ = increase; ↓ = decrease; ↔ = no differences.

[2] Compared with 0.1-0.3 mg Se/kg diet.

[3] Supported by the database of uniprot (http://www.uniprot.org/uniprot/).

GPX1 and GPX activity were decreased by dietary Se deficiency in many tissues of pigs and rodents, with little changes in the brain, lung and muscle of rodents (Akahoshi *et al.*, 2019; Chen *et al.*, 2020; Novoselov *et al.*, 2010; Sunde and Raines, 2011; Sunde *et al.*, 2018; Yang *et al.*, 2019; Yim *et al.*, 2019). GPX3 protein was decreased in porcine and rodent plasma, liver, ovary, kidney, pancreas, spine and thyroid, but not in other 5 tissues (Akahoshi *et al.*, 2019; Sunde and Raines, 2011; Sunde *et al.*, 2018; Yang *et al.*, 2019). Whereas dietary Se deficiency suppressed *Gpx4* mRNA levels in the ovary, kidney and pancreas, but not brain, heart, lung, testicle, spleen, muscle and thyroid of rodents (Akahoshi *et al.*, 2019; Chen *et al.*, 2020; Sunde and Raines, 2011; Sunde *et al.*, 2018; Yang *et al.*, 2019), it decreased *GPX4* mRNA levels only in the porcine lymphocytes (Table 2.2). Responses of rodent *Selenow* and porcine *SELENOW to* dietary Se deficiency were similar (Chen *et al.*, 2020; Raines and Sunde, 2011; Sunde and Raines, 2011; Sunde *et al.*, 2018; Yim *et al.*, 2019; Zeng *et al.*, 2012). Among the 13 selenoprotein genes responding to dietary Se deficiency, *Selenos* was the only one not changed in the liver and down-regulated in muscle of rodents (Zeng *et al.*, 2012). Compared with the rodent genes, porcine *SELENOH, SELENOP* and *SELENOS* were more responsive (downregulated in more tissues) to dietary Se deficiency (Table 2.2).

There were differences in responses of selenoprotein genes related to hormone metabolism and ER-stress to dietary Se deficiency between pigs and rodents (Table 2.3). While rodent *Dio2, Dio3, Selenof* and *Selenoi* expression remained unchanged by dietary Se deficiency (Table 2.3), the porcine counterparts were downregulated in 4 tissues (Table 2.2). In contrast, porcine *DIO1* remained unchanged, but rodent *Dio1* was decreased in the Se-deficient liver (Akahoshi *et al.*, 2019; Hofstee *et al.*, 2020; Sunde and Raines, 2011; Sunde *et al.*, 2018). Porcine TXNRD1 was up-regulated in the heart and down-regulated in the liver, testicles and kidney by dietary Se deficiency, whereas the rodent protein remained unchanged in the pancreas (Akahoshi *et al.*, 2019; Chen *et al.*, 2020; Sunde *et al.*, 2018; Yim *et al.*, 2019; Zhou *et al.*, 2017). In addition, porcine *SELENOT* expression was elevated in many tissues (Table 2.2), while rodent *Selenot* expression was decreased in the liver (Akahoshi *et al.*, 2019; Sunde and Raines, 2011; Sunde *et al.*, 2018).

2.3.2 Comparative regulations by moderately high intakes of dietary selenium

In the adipose tissues of rodents, expression of GPX1, *Gpx4*, *Gpx6, Msrb1, Selenof, Selenoh*, *Selenoi*, *Selenok*, *Selenom*, and *Selenow* was down-regulated by elevating dietary Se concentration from 0.3 to 1.0 mg/kg (Zhao *et al.*, 2017, 2020). This dietary Se intake change did not affect the expressions of *Selenot*, TXNRD1, *Txnrd2* and *Txnrd3* (Akahoshi *et al.*, 2019; Chen *et al.*, 2020; Zhao *et al.*, 2017, 2020), but up-regulated the expression of GPX1, GPX3, SELENOP protein and(or) activity and *Selenof* mRNA level in the liver, testis, and plasma of mice (Chen *et al.*, 2020; Zhao *et al.*, 2017, 2020). Likewise, expression of *Selenos* failed to respond to moderately high intakes of dietary Se in assayed tissues of rodents (Chen *et al.*, 2020; Zhao *et al.*, 2017, 2020). In contrast, expression of porcine *GPX1*, *GPX3*, *DIO2*, *TXNRD3*, *SELENOH*, *SELENOK*, *SELENOM*, *SELENON*, *SELENOP*, *SELENOS*, *SELENOT* and

SELENOV was up-regulated by a moderately high intake of dietary Se (Lu *et al.*, 2019; Wang *et al.*, 2020).

2.3.3 Comparative regulations by dietary selenium excess

Many rodent selenoprotein genes were more responsive to high intakes of dietary fat than dietary Se excess (Akahoshi *et al.*, 2019; Chen *et al.*, 2020; Novoselov *et al.*, 2010; Raines and Sunde, 2011; Shin *et al.*, 2019; Sunde and Raines, 2011; Yim *et al.*, 2019; Zeng *et al.*, 2012; Zhao *et al.*, 2017; Zhou *et al.*, 2017). Among those genes, only GPX1, *Msrb1*, *Selenoh* and *Selenos* in mice were affected by dietary Se excess (Table 2.3). However, dietary Se excess suppressed expression of *DIO1, SELENOF, SELENOI*, and *SELENOM* and *TXNRD1,* and elevated GPX3, *GPX4* and SELENOP in several tissues of pigs (Table 2.2 and 2.3). Dietary Se excess exerted opposite regulation of *SELENOW* mRNA levels in the liver and muscle of pigs, but had no effect on the gene expression in mice (Akahoshi *et al.*, 2019). Dietary Se excess elevated rodent GPX1, *Selenoh, Selenos,* and TXNRD2 (italic for Selenoh and Selenos) in the liver (Akahoshi *et al.*, 2019; Chen *et al.*, 2020; Labunskyy *et al.*, 2007; Mueller *et al.*, 2008; Raines and Sunde, 2011; Sunde and Raines, 2011; Zeng *et al.*, 2012) and *SELENOS* in the thyroid of pigs. Porcine GPX1 and *SELENOH* in several tissues showed variable responses to dietary Se excess (Table 2.2). *Msrb1* was down-regulated in the brain of mice by dietary Se excess (Raines and Sunde, 2011; Yim *et al.*, 2019; Zeng *et al.*, 2012), and there was a similar decrease of the porcine gene in the spine of pigs (Table 2.2).

2.3.4 Comparative regulations by dietary fat

Unlike porcine selenogenome and selenoproteome, no single member of the rodent selenogenome or selenoproteome, except for the Se-independent *Gpx6*, showed responses solely to dietary fat (Zhao *et al.*, 2017, 2020). Overall, 15 rodent selenoprotein genes were regulated by both dietary Se and fat intakes (Table 2.3). Among these 15 proteins, 5 of them are located in endoplasmic reticulum (ER), namely DIO2, SELENOF, SELENOK, SELENOM, and SELENOS and 5 of them are located in cytoplasm namely GPX1, GPX4, MSRB1, SELENOW and TNXRD1 (www.uniprot.org). All 5 cytosolic selenoproteins except for SELENOW also exist in nucleus and mitochondria (www.uniprot.org).

Dietary fat exerted reversed or opposite regulations between different tissues of porcine *GPX3, SELENOM, SELENOV,* and *TXNRD1* and rodent GPX1, *Gpx4, Gpx6, Selenoh, Selenoi, Selenom, Selenop* and *Selenos* (Table 2.2 and 2.3). Among the remaining selenoprotein genes, 14 and 5 porcine genes were down- and up-regulated by high dietary fat diet, whereas 4 and 5 rodent selenoprotein genes were up- and down-regulated by high dietary fat diet (Table 2.3). Specifically, expressions of rodent *Dio2, Dio3*, GPX1, *Gpx4, Msrb1, Selenof, Selenoh, Selenoi,* and *Selenom* were up-regulated in the liver, and expressions of *Selenoh, Selenoi, Selenok, Selenom*, SELENOP, *Selenos* and *Selenow* were down-regulated in the adipose tissue, testis, and muscle, respectively, by high dietary fat intake (Chen *et al.*, 2020; Zhao *et al.*, 2017, 2020). In comparison, porcine *DIO1*, GPX1, GPX3, *GPX4, MSRB1, SELENOH,*

SELENOI, SELENOP, and *SELENOS* were down-regulated by high dietary fat intake, largely in the lymphocytes and to some extent in the pancreas, adipose tissue, hypothalamus and pituitary (Table 2.2). Porcine TXNRD1 showed variable responses to high dietary fat intakes in the hypothalamus-pituitary-thyroid axis (Table 2.2). However, 7 rodent selenoprotein genes (*Dio2, Selenof, Selenok, Selenom, Selenow, Txnrd2* and *Txnrd3*) showed similar responses to high dietary fat intakes to those of their porcine counterparts (Table 2.2 and 2.3). Likewise, neither rodent *Txnrd2* and *Txnrd3* nor their porcine counterparts were altered by high intakes of dietary fat in any of assayed tissues (Table 2.2 and 2.3).

2.4 Concluding remarks

Various types of pigs require different dietary Se intakes to meet needs for the biosynthesis and function of 25 selenoproteins. This chapter reviews key steps of selenoprotein biosynthesis, evolution of porcine selenogenome, and functions of porcine selenoproteome. Detailed coverage and analysis are given to regulations of porcine selenogenome and selenoproteome expression in various tissues of pigs by dietary Se deficiency, dietary Se excess, and high intakes of dietary fat. These regulations displayed no universal patterns across different selenoprotein genes or tissues, but rather tissue-, time-, and hierarchy of selenoprotein-dependent. Although dietary Se deficiency and high fat intake exerted similar impacts on expressions of selenogenomes and selenoproteomes in the rodent and porcine tissues, dietary Se excess produced fairly variable responses in that regard. In addition, chemical forms of dietary Se supplements, dietary serine concentrations, heat stress, and immunomodulators could also affect expressions of porcine selenogenome and selenoproteome. Future research should develop appropriate reagents and experimental protocols to reveal complete nutritional and environmental regulations of selenoprotein production and selenoenzyme activity and their metabolic impacts in pigs.

References

Akahoshi, N., Anan, Y., Hashimoto, Y., Tokoro, N., Mizuno, R., Hayashi, S., Yamamoto, S., Shimada, K.I., Kamata, S. and Ishii, I., 2019. Dietary selenium deficiency or selenomethionine excess drastically alters organ selenium contents without altering the expression of most selenoproteins in mice. The Journal of Nutritional Biochemistry 69: 120-129.

Barnes, K.M., Evenson. J.K., Raines, A.M. and Sunde, R.A., 2009. Transcript analysis of the selenoproteome indicates that dietary selenium requirements of rats based on elenium-regulated selenoprotein mRNA levels are uniformly less than those based on glutathione peroxidase activity. The Journal of Nutrition 139: 199-206.

Berry, M.J., Banu, L., Chen, Y., Mandel, S.J., Kieffer, J.D., Harney, J.W. and Larsen, P.R., 1991. Recognition of uga as a selenocysteine codon in type I deiodinase requires sequences in the 3′ untranslated region. Nature 353: 273-276.

Cao, L., Tang, J.Y, Li, Q., Xu, J.Y, Jia, G., Liu, G.M., Chen, X.L., Shang, H.Y., Cai, J.Y. and Zhao, H., 2016. Expression of selenoprotein genes is affected by heat stress in IPEC-J2 cells. Biological Trace Element Research 172: 354-360.

Chen, L.L., Huang, J.Q., Xiao, Y., Wu, Y.Y., Ren, F.Z. and Lei, X.G., 2020. Knockout of selenoprotein V affects regulation of selenoprotein expression by dietary selenium and fat intakes in mice. The Journal of Nutrition 150: 483-491.

Chen, X.D., Zhao, Z.P., Zhou, J.C. and Lei, X.G., 2018. Evolution, regulation, and function of porcine selenogenome. Free Radical Biology & Medicine 127: 116-123.

Copeland, P.R., Fletcher, J.E., Carlson, B.A., Hatfield, D.L. and Driscoll, D.M., 2000. A novel RNA binding protein, SBP2, is required for the translation of mammalian selenoprotein mRNAs. EMBO Journal 19: 306-314.

Fagegaltier, D., Hubert, N., Yamada, K., Mizutani, T., Carbon, P. and Krol, A., 2000. Characterization of mselb, a novel mammalian elongation factor for selenoprotein translation. EMBO Journal 19: 4796-4805.

Falka, M., Bernhoftb, A., Reinoso-Masetc E., Salbuc, B., Lebedc, P., Framstadd, T., Fuhrmanne, H. and Oropeza-Moea, M., 2020. Benefificial antioxidant status of piglets from sows fed selenomethionine compared with piglets from sows fed sodium selenite. Journal of Trace Elements in Medicine and Biology 58: 126439.

Gladyshev, V.N., Arner, E.S., Berry, M.J., Brigelius-Flohe, R., Bruford, E.A., Burk, R.F., Carlson, B.A., Castellano, S., Chavatte, L., Conrad, M., Copeland, P.R., Diamond, A.M., Driscoll, D.M., Ferreiro, A., Flohé, L., Green, F.R., Guigó, R., Handy, D.E., Hatfield, D.L., Hesketh, J., Hoffmann, P.R., Holmgren, A., Hondal, R.J., Howard, M.T., Huang, K., Kim, H.Y., Kim, I.Y., Köhrle, J., Krol, A., Kryukov, G.V., Lee, B.J., Lee, B.C., Lei, X.G., Liu, Q., Lescure, A., Lobanov, A.V., Loscalzo, J., Maiorino, M., Mariotti, M., Prabhu, K.S., Rayman, M.P., Rozovsky, S., Salinas, G., Schmidt, E.E., Schomburg, L., Schweizer, U., Simonović, M., Sunde, R.A., Tsuji, P.A., Tweedie, S., Ursini, F., Whanger, P.D. and Zhang, Y., 2016. Selenoprotein gene nomenclature. The Journal of Biological Chemistry 291: 24036-24040.

Groenen, M.A., Archibald, A.L., Uenishi, H., Tuggle, C.K., Takeuchi, Y., Rothschild, M.F., Rogel-Gaillard, C., Park, C., Milan, D., Megens, H.J., *et al.*, 2012. Analyses of pig genomes provide insight into porcine demography and evolution. Nature 491: 393-398. https://doi.org/10.1038/nature11622

Han, J and Kaufman, R.J., 2016. The role of ER stress in lipid metabolism and lipotoxicity. Journal of Lipid Research 57: 1329-1338.

Han, S.J., Lee, B.C., Yim, S.H., Gladyshev, V.N. and Lee, S.R., 2014. Characterization of mammalian selenoprotein o: a redox-active mitochondrial protein. PloS One 9: e95518.

Hatfield, D. and Portugal, F.H., 1970. Seryl-tRNA in mammalian tissues: chromatographic differences in brain and liver and a special response to the codon, UGA. Proceeding of the National Academy of Sciences of the USA 67: 1200-1206.

Hatfield, D.L., Schweizer, U., Tsui, P.A. and Dladyshev, V.N., 2016. Selenium: its molecular biology and role in human health. 4th edition. Springer, New York, NY, USA.

Hofstee, P., Cuffe, J.S.M. and Perkins, A.V., 2020. Analysis of selenoprotein expression in response to dietary selenium deficiency during pregnancy indicates tissue specific differential expression in mothers and sex specific changes in the fetus and offspring. International Journal of Molecular Science 21: 2210. https://doi.org/10.3390/ijms21062210

Huang, J.Q., Li, D.L., Zhao, H., Sun, L.H, Xia, X.J., Wang, K.N., Luo, X. and Lei, X.G., 2011.The selenium deficiency disease exudative diathesis in chicks is associated with downregulation of seven common selenoprotein genes in liver and muscle. The Journal of Nutrition 141: 1605-1610.

Huang, X.F., Tang, J.Y., Xu, J.Y., Jia, G., Liu, G.M., Chen, X.L., Shang, H.Y., Cai, J.Y. and Zhao, H., 2016. Supranutritional dietary selenium induced hyperinsulinmia and dyslipidemia via affected expression of selenoprotein genes and insulin signal-related genes in broiler. RSC Advances 6: 84990-84998.

Jia, J.Y., 2007. Effect of different source selenium on piglets of sows, tissue selenium deposition and blood biochemical parameters. Master Thesis. Sichuan Agricultural University, Sichuan, China, P.R.

Kryukov, G.V., Castellano, S., Novoselov, S.V, Lobanov, A.V., Zehtab, O., Guigo, R. and Gladyshev, V.N., 2003. Characterization of mammalian selenoproteomes. Science 300: 1439-1443.

Labunskyy, V.M., Hatfield, D.L. and Gladyshev, V.N., 2007. The Sep15 protein family: roles in disulfide bond formation and quality control in the endoplasmic reticulum. IUBMB Life 59: 1-5.

Li, J.G., Zhou, J.C., Zhao, H., Xia, X.J., Lei, X.G. and Wang, K.N., 2011. Enhanced water-holding capacity of meat was associated with increased Sepw1 gene expression in pigs fed selenium-enriched yeast. Meat Science 87: 95-100.

Liu, T., Yang, T., Pan, T., Liu, C. and Li, S., 2018. Effect of low-selenium/high-fat diet on pig peripheral blood lymphocytes: perspectives from selenoproteins, heat shock proteins, and cytokines. Biological Trace Element Research 183: 102-113.

Liu, Y., Zhao, H., Zhang, Q., Tang, J., Li, K., Xia, X.J., Wang, K.N., Li, K. and Lei, X.G., 2012. Prolonged dietary selenium deficiency or excess does not globally affect selenoprotein gene expression and/or protein production in various tissues of pigs. The Journal of Nutrition 142: 1410-1416.

Long, J., Liu, Y., Zhou, X. and Liu, H., 2020. Dietary serine supplementation regulates selenoprotein transcription and selenoenzyme activity in pigs. Biological Trace Element Research 199: 148-153. https://doi.org/10.1007/s12011-020-02117-8

Lu, Z., Wang, P., Teng, T., Shi, B., Shan, A. and Lei, X.G., 2019. Effects of dietary selenium deficiency or excess on selenoprotein gene expression in the spleen tissue of pig. Animals 9: 1122. https://doi.org/10.3390/ani9121122

Maenpaa, P.H. and Bernfied, M.R., 1970. A specific hepatic transfer RNA for phosphoserine. Proceedings of the National Academy of Science of the USA 67: 688-695.

Mahan, D.C., 2000. Effect of organic and inorganic selenium sources and levels on sow colostrum and milk selenium content. Journal of Animal Science 78: 100-105.

Mahan, D.C. and Peters, J.C., 2004. Long-term effects of dietary organic and inorganic selenium sources and levels on reproducing sows and their progeny. Journal of Animal Science 82: 1343-1358.

Mueller, A.S., Klomann, S.D., Wolf, N.M., Schneider, S., Schmidt, R., Spielmann, J., Stangl, G., Eder, K. and Pallauf, J., 2008. Redox regulation of protein tyrosine phosphatase 1B by manipulation of dietary selenium affects the triglyceride concentration in rat liver. The Journal of Nutrition 138: 2328-2336.

National Research Council (NRC), 2012 . Nutrient requirements of swine. 11th revised edition. National Academies Press, Washington, DC, USA.

Nauser, T., Dockheer, S., Kissner, R. and Koppenol, W.H., 2006. Catalysis of electron transfer by selenocysteine. Biochemistry 45: 6038-6043.

Novoselov, S.V., Kim, H.Y., Hua, D., Lee, B.C., Astle, C.M., Harrison, D.E., Friguet, B., Moustafa, M.E., Carlson, B.A., Hatfield, D.L. and Galdyshev, V.N., 2010. Regulation of selenoproteins and methionine sulfoxide reductases A and B1 by age, calorie restriction, and dietary selenium in mice. Antioxidants & Redox Signaling 12: 829-838.

Pinto, A., Juniper, D.T., Sanil, M., Morgan, L., Clark, L., Sies, H., Rayman, M.P. and Steinbrenner, H., 2012. Supranutritional selenium induces alterations in molecular targets related to energy metabolism in skeletal muscle and visceral adipose tissue of pigs. Journal of Inorganic Biochemistry 114: 47-54.

Raines, A.M., and Sunde, R.A., 2011. Selenium toxicity but not deficient or super-nutritional selenium status vastly alters the transcriptome in rodents. BMC Genomics 12: 26.

Santesmasses, D., Mariotti, M. and Gladyshev, V.N., 2020. Tolerance to selenoprotein loss differs between human and mouse. Molecular Biology and Evolution 37: 341-354.

Shchedrina, V.A., Zhang, Y., Labunskyy, V.M., Hatfield, D.L., and Gladyshev, V.N., 2010. Structure-function relations, physiological roles, and evolution of mammalian ER-resident selenoproteins. Antioxidants & Redox Signaling 12: 839-49.

Shin, M.G., Cha, H.N., Park, S., Kim, Y.W., Kim, J.Y. and Park, S.Y., 2019. Selenoprotein W deficiency does not affect oxidative stress and insulin sensitivity in the skeletal muscle of high-fat diet-fed obese mice. American Journal of Physiology Cell Physiology 317: C1172-C1182.

Silva, V.A., Bertechini, A.G., Clemente, A.H.S., De Freitas, L.F.V.B., Nogueira, B.R.F., De Oliveira, B.L. and Ramos, A.L.S., 2019. Different levels of selenomethionine on the meat quality and selenium deposition in tissue of finishing pigs. Journal of Animal Physiology and Animal Nutrition (Berl) 103: 1866-1874.

Snider, G.W., Ruggles, E., Khan, N. and Hondal, R.J., 2013. Selenocysteine confers resistance to inactivationby oxidation in thioredoxin reductase: comparison of selenium andsulfur enzymes. Biochemistry 52: 5472-5481.

Steinmann, D., Nauser, T., Beld, J., Tanner, M., Günther, D., Bounds, P.L. and Koppenol, W.H., 2008. Kinetics of tyrosyl radical reduction by selenocysteine. Biochemistry 47: 9602-9607.

Sun, L.H., Pi, D.A., Zhao, L., Wang, X.Y., Zhu, L.Y., Qi, D.S. and Liu, Y.L., 2017. Response of selenium and selenogenome in immune tissues to LPS-induced inflammatory reactions in pigs. Biological Trace Element Research 177: 90-96.

Sunde, R.A. and Raines, A.M., 2011. Selenium regulation of the selenoprotein and nonselenoprotein transcriptomes in rodents. Advances in Nutrition 2: 138-150.

Sunde, R.A., Raines, A.M., Barnes, K.M. and Evenson, J.K., 2009. Selenium status highly regulates selenoprotein mRNA levels for only a subset of the selenoproteins in the selenoproteome. Bioscience Reports 29: 329-38.

Sunde, R.A., Zemaitis, E.T., Blink, A.B. and Lawinger, J.A., 2018. Impact of glutathione peroxidase-1 (Gpx1) genotype on selenoenzyme and transcript expression when repleting selenium-deficient mice. Biological Trace Element Research 186: 174-184.

Surai, P.F., 2016. Antioxidant systems in poultry biology: superoxide dismutase. Animal Nutrition 1: 8. https://doi.org/10.21767/2572-5459.100008

Surai, P.F. and Fisinin, V.I., 2016. Selenium in sow nutrition. Animal Feed Science and Technology 211: 18-30.

Tang, J.Y., Cao, L., Jia, G., Liu, G.M., Chen, X.L., Tian, G., Cai, J.Y., Shang, H.Y. and Zhao, H., 2019a. The protective effect of selenium from heat stress-induced porcine small intestinal epithelial cell line (IPEC-J2) injury is associated with regulation expression of selenoproteins. British Journal of Nutrition 122: 1081-1090.

Tang, J.Y., He, A.H., Yan, H., Jia, G., Liu, G.M., Chen, X.L., Tian, G., Cai, J.Y., Shang, H.Y. and Zhao, H., 2018. Damage to the myogenic differentiation of C2C12 Cells by heat stress is associated with up-regulation of several selenoproteins. Scientific Reports 8: 10601-10609.

Tang, J.Y., Wang, L.Q., Jia, G., Liu, G.M., Chen, X.L., Tian, G., Cai, J.Y., Shang, H.Y. and Zhao, H., 2019b. The hydroxy-analogue of selenomethionine alleviated lipopolysaccharide-induced inflflammatory responses is associated with recover expression of several selenoprotein encoding genes in the spleens of kunming mice. RSC Advances 9: 40462-40470.

Taylor, R.M. and Sunde, R.A., 2016. Selenoprotein transcript level and enzyme activity as biomarkers for selenium status and selenium requirements of turkeys (*Meleagris gallopavo*). PLoS One. 11: e0151665.

Taylor, R.M., Bourget, V.G. and Sunde, R.A., 2019. High dietary inorganic selenium has minimal effects on turkeys and selenium status biomarkers. Poultry Science 98: 855-865.

Tujebajeva, R.M., Copeland, P.R., Xu, X.M., Carlson, B.A. and Berry, M.J., 2000. Decoding apparatus for eukaryot selenocysteine insertion. EMBO Reports 1: 158-163.

Wang, P., Lu, Z., He, M., Shi, B., Lei, X.G. and Shan, A.S., 2020. The effects of endoplasmic-reticulum-resident selenoproteins in a nonalcoholic fatty liver disease pig model induced by a high-fat diet. Nutrients 12: 692.

Wang, S. and Kaufman, R.J., 2014. How does protein misfolding in the endoplasmic reticulum affect lipid metabolism in the liver? Current Opinion in Lipidology 25: 125-132.

Xu, J.Y., Wang, L.Q., Tang, J.Y, Jia, G., Liu, G.M., Chen, X.L., Tian, G., Cai, J.Y., Shang, H.Y. and Zhao, H., 2017. Pancreatic atrophy caused by dietary selenium deficiency induced hypoinsulinemic hyperglycemia via global down-regulation of selenoprotein encoding genes in broiler. Plos One 12(8): e0182079. https://doi.org/10.1371/journal.pone.0182079

Yang, H., Qazi, I.H., Pan, B., Angel, C., Guo, S., Yang, J., Zhang, Y., Ming, Z., Zeng, C., Meng, Q., Han, H. and Zhou, G., 2019. Dietary selenium supplementation ameliorates female reproductive efficiency in aging mice. Antioxidants 8: 634.

Yim, S.H., Clish, C.B. and Gladyshev, V.N., 2019. Selenium deficiency is associated with pro-longevity mechanisms. Cell Reports 27: 2785-2797.

Zeng, M.S., Li, X., Liu, Y., Zhao, H., Zhou, J.C., Li, K., Huang, J.Q., Sun, L.H., Tang, J.Y., Xia, X.J., Wang, K.N. and Lei, X.G., 2012. A high-selenium diet induces insulin resistance in gestating rats and their offspring. Free Radical Biology & Medicine 52: 1335-1342.

Zhan, X., Qie, Y., Wang, M., Li, X. and Zhao, R.Q., 2011. Selenomethionine: an effective selenium source for sow to improve Se distribution, antioxidant status, and growth performance of pig offspring. Biological Trace Element Research,142: 481-49.

Zhang, K., Han, Y., Zhao, Q., Zhan, T., Li, Y., Sun, W., Li, S., Sun, D., Si, X., Yu, X., Qin, Y., Tang, C. and Zhang, J., 2020a. Targeted metabolomics analysis reveals that dietary supranutritional selenium regulates sugar and acylcarnitine metabolism homeostasis in pig liver. The Journal of Nutrition 150(4): 704-711. https://doi.org/10.1093/jn/nxz317

Zhang, K., Zhao, Q., Zhan, T., Han, Y., Tang, C. and Zhang, J., 2020b. Effect of different selenium sources on growth performance, tissue selenium content, meat quality, and selenoprotein gene expression in finishing pigs. Biological Trace Element Research 196: 463-471. https://doi.org/10.1007/s12011-019-01949-3

Zhao, H., Li, K., Tang, J.Y., Zhou, J.C., Wang, K.N., Xia, X.J. and Lei, X.G., 2015. Expressions of selenoprotein genes is affected by obesity pigs fed a high-fat diet. The Journal of Nutrition 145(7): 1394-401. https://doi.org/10.3945/jn.115.211318

Zhao, Z., Barcus, M., Kim, J., Lum, K.L., Mills, C. and Lei, X.G., 2016. High dietary selenium intake alters lipid metabolism and protein synthesis in liver and muscle of pigs. The Journal of Nutrition 146: 1625-1633.

Zhao, Z., Kim, J. and Lei, X.G., 2020. High dietary fat and selenium concentrations exert tissue- and glutathione peroxidase 1-dependent impacts on lipid metabolism of young-adult mice. The Journal of Nutrition 150(7): 1738-1748. https://doi.org/10.1093/jn/nxaa130

Zhao, Z., Zhou, J. and Lei, X.G., 2017. Differential regulation of transcriptional factors of selenoprotein genes by moderately high dietary concentrations of Se and fat in liver and adipose tissue of wildtype and gpx1$^{-/-}$ mice. The FASEB Journal 31: 459.6-459.6.

Zhou, J.C., Zhao, H., Li, J.G., Xia, X.J., Wang, K.N., Zhang, Y.J., Liu, Y., Zhao, Y. and Lei, X.G., 2009. Selenoprotein gene expression in thyroid and pituitary of young pigs is not affected by dietary selenium deficiency or excess. The Journal of Nutrition 139: 1061-1066.

Zhou, J.C., Zheng, S., Mo, J., Liang, X., Xu, Y., Zhang, H., Gong, C., Liu, X.L. and Lei, X.G., 2017. Dietary selenium deficiency or excess reduces sperm quality and testicular mRNA abundance of nuclear glutathione peroxidase 4 in rats. The Journal of Nutrition 147: 1947-1953.

Chapter 3

Selenium in feed: organic selenium concept development

P.F. Surai[1-7], I.I. Kochish[2] and V.I. Fisinin[8]*
[1]Vitagene and Health Research Centre, Bristol, BS4 2RS, United Kingdom; [2]Department of Hygiene and Poultry Sciences, Moscow State Academy of Veterinary Medicine and Biotechnology Named After K.I. Skryabin, 109472 Moscow, Russia; [3]Department of Microbiology and Biochemistry, Faculty of Veterinary Medicine, Trakia University, 6000 Stara Zagora, Bulgaria; [4]Department of Animal Nutrition, Faculty of Agricultural and Environmental Sciences, Szent Istvan University, 2103 Gödöllo, Hungary; [5]Saint-Petersburg State Academy of Veterinary Medicine, 196084 St. Petersburg, Russia; [6]Sumy National Agrarian University, 40000 Sumy, Ukraine; [7]Odessa National Academy of Food Technologies, 65039 Odessa, Ukraine; [8]All-Russian Institute of Poultry Husbandry, 141311 Sergiev Posad, Russia; psurai@feedfood.co.uk

Abstract

Selenium is an essential element in animal nutrition and its dietary supplementation in optimal form and amount is key for maintaining their health, productive and reproductive performance. It has been proven that organic form of Se (mainly SeMet) in the animal/pig diet has a range of advantages in comparison to traditional sodium selenite. In fact, organic Se concept considers SeMet as a storage form of Se in animal body. Since pigs are not able to synthesise SeMet its provision with a diet is a key element of a strategy to fight commercially-relevant stresses. Indeed, in stress conditions, when increased selenoprotein expression requires additional Se, its provision with feed usually decreased due to reduction in feed consumption. Therefore, Se reserves in the body (mainly in muscles) could help to maintain an effective antioxidant defence. Pig industry is looking for most effective sources of organic Se to be commercially used. Se-yeast received substantial attention and commercial applications as a reliable source of organic Se, however, active Se component of the yeast, SeMet, comprises only 50-70% of the total Se. From the production point of view it is difficult to guarantee certain percentage of SeMet in the Se-Yeast, since there is a range of factors (yeast strain, medium composition and selenite concentration in it, temperature, etc.) affecting the aforementioned parameter. Analytical difficulties in precise determination of SeMet amount further complicate the issue. Stability issues with pure SeMet in the pig diets restrict its wide usage. Indeed, the first generation of Se supplements for pigs includes selenite and selenate. The second generation of organic Se for pigs includes Se-Yeast, Se-Met and Zn-Se-Met products. Their advantages and disadvantages are discussed in the chapter. Indeed, a recently developed product (hydroxy-SeMet, OH-SeMet) can be considered as the next, third generation of Se supplement for pigs combining advantages of both – Se-Yeast (SeMet stability) and pure SeMet (high proportion of SeMet) and could be considered as a next step in Se application in pig nutrition.

Keywords: selenium, selenomethionine, OH-SeMet, feed, pigs

DOI 10.3920/978-90-8686-912-1_3,

3.1 Introduction

Selenomethionine (SeMet) was first studied as a possible cause of toxicity of seleniferous wheat in the 1950s and was later proven to be synthesised from inorganic selenium sources by various plants, including yeast, marine algae, Candida albicans, as well as by *Escherichia coli* and rumen bacteria (for review see Schrauzer, 2000, 2003; Schrauzer and Surai, 2009). Detailed analysis of the literature suggests that organic selenium in the form of various selenoamino acids is a natural form of selenium in animal and human diets and that the digestive system adapted to this nutrient form during evolution, thereby explaining why there are principal differences in assimilation and metabolism between organic and inorganic forms of selenium (Surai, 2002, 2006, 2018). Therefore, there is an inconsistency in common practise of selenium supplementation of pig diets. On one hand, naturally occurring organic selenium is represented by a mixture of selenoamino acids with SeMet comprising more than 50% of total selenium in major feed ingredients including grains, oil seeds, etc. On the other hand, until recently the supplemental form of selenium for pigs has been inorganic, either selenite or selenate.

3.2 Selenium in soils and plants

Selenium (Se) is known to be a chemical element with atomic number 34 and atomic weight 78.96 belonging to group VI of the periodic table of elements. This group also includes such non-metals as sulphur and oxygen (Figure 3.1).

In nature Se is known to exist in two chemical forms, organic and inorganic. Elemental Se can be reduced to the Se^{-2} oxidation state (selenide) or oxidised to the Se^{+4} (SO_3^{-2}, selenite) or Se^{+6} (SO_4^{-2}, selenate). In general, inorganic Se can be found in different minerals in the form of selenite, selenate and selenide as well as in the metallic (Se^0) form. In contrast, selenium in feed ingredients (forages, grains, oilseed meals, etc.) is an integral part of a range of Se-amino acids, including selenomethionine and selenocysteine, and exists in the Se^{-2} oxidation state. As a result, in nature pigs receive Se mainly in the form of selenomethionine (SeMet; Combs and Combs, 1986; Surai, 2006, 2018). Indeed, SeMet is considered a natural nutritional form of selenium for pigs (Table 3.1).

The selenium cycle in the food chain of pigs starts from soils and includes plant and animal sources ultimately dependent on its assimilation from the soil. Indeed, soils are the major source of Se for plants and therefore for pigs eating those plants and pigs consuming plant and animal-derived feed ingredients. It should be mentioned that Se concentration in soils varies significantly (Reilly, 2006). In fact, the Se content of most soils ranges between 0.1 and 2 mg/kg; and soil Se exists in various forms, including selenides, elemental Se, selenites, selenates and organic Se compounds (NRC, 1983) (Figure 3.2). High concentrations of Se are found mainly in sedimentary rocks and shales formed during the cretaceous period, at the same time lower concentrations of Se are characteristic for igneous (volcanic) rock, sandstone, granite and limestone

(Van Metre and Callan, 2001). Research conducted in China indicated that soils developed under tropic and subtropic conditions (laterite, yellow soil and red soil) are characterised by comparatively high Se levels (>0.3 mg/kg; Tan *et al.,* 2002). In contrast, the soils developed under the temperate (warm) steppe and desert conditions (chernozem, chestnut soil, calcic brown soil, desert soil and solonchak) are shown to have moderate Se concentrations (0.14-0.30 mg/kg). Finally, such soils as brown earth, drab soil, dark brown soil, loessial soils, purple soil, red drab soil, developed under the temperate (warm) humid/sub-humid conditions are found to be quite poor in Se (Tan *et al.,* 2002). In particular, low Se soils occur in the northeast to the southwest of China.

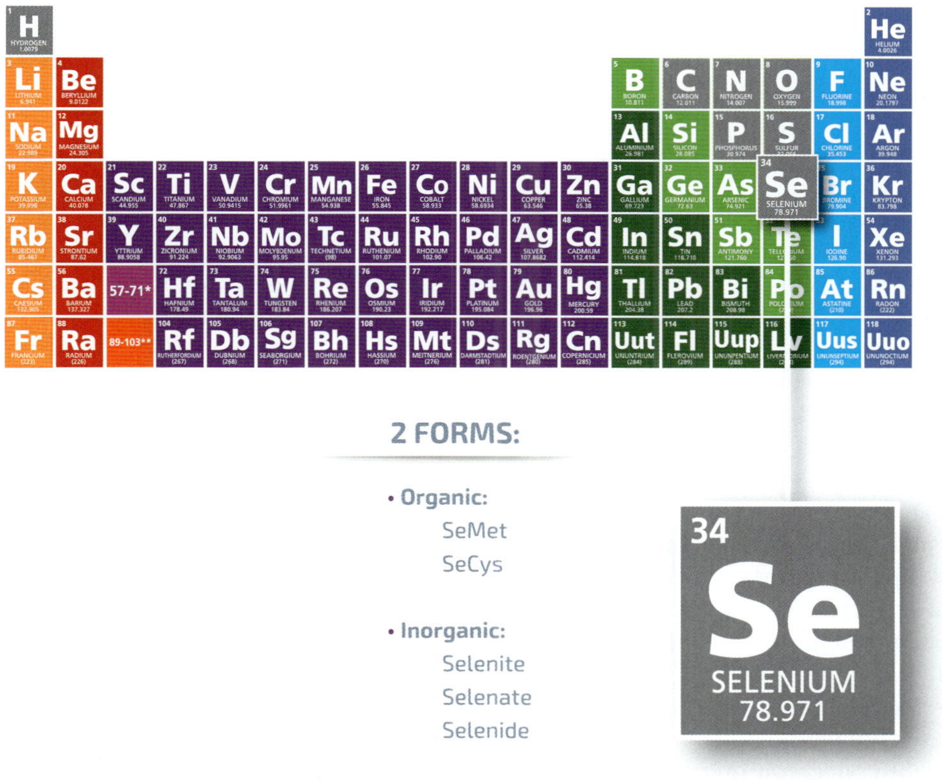

Figure 3.1. Selenium position in the periodic table of elements.

Table 3.1 Selenomethionine (SeMet) in grains.

Grains	SeMet proportion, % total Se	References
Wheat		
Wheat grain	56-83	Whanger, 2002
Wheat	50.4-81.4	Yang *et al.*, 1997
Wheat grain	72-85	Cubadda *et al.*, 2010
Spring wheat grains, Australia	90	Stadlober *et al.*, 2001
Spring wheat grain, India	66	Cubadda *et al.*, 2010
Durum wheat, Austria	62	Stadlober *et al.*, 2001
Winter wheat grain, India	58	Cubadda *et al.*, 2010
Wheat flour, Belgium	52	Moreno *et al.*, 2004
Barley		
Summer barley grains, Austria	77	Stadlober *et al.*, 2001
Corn		
Corn	61-64	Whanger, 2002
Corn	45.5-82.0	Yang *et al.*, 1997
Soybean		
Soybeans	>80	Whanger, 2002
Soybean	62.9-71.8	Yang *et al.*, 1997
Rice		
Rice	68-81	Whanger, 2002
Rice	54.9-86.5	Yang *et al.*, 1997
Basmati rice, India	93	Mar *et al.*, 2009
Jasmine rice, Thailand	96	Mar *et al.*, 2009
White rice, USA	94	Mar *et al.*, 2009

Furthermore, Se availability to plants depends on many different factors including soil pH, the oxidation-reduction potential and mineral composition of the soil, rate of artificial fertilisation, rainfall, etc. Therefore, transfer of the selenium from soil to plant depends on (Munier-Lamy *et al.*, 2007):

- plant species;
- physiological state;
- soil type;
- Se concentration in soil;
- form of Se in soil.

In fact, the bioavailability of Se in soils for plants depends more on its form than on its total concentration (Surai, 2018):

- In the case of acidic soils or poor soil aeration, Se forms insoluble complexes with iron hydroxide and become poorly available. For example, at pH 6, only 47% of labelled Se was transferred from soil to ryegrass leaves, while increasing pH to 7 enhanced Se assimilation to 70% (Haygarth *et al.*, 1995). Indeed, Se in alkaline soils occurs in selenate form being soluble and easily available to plants.

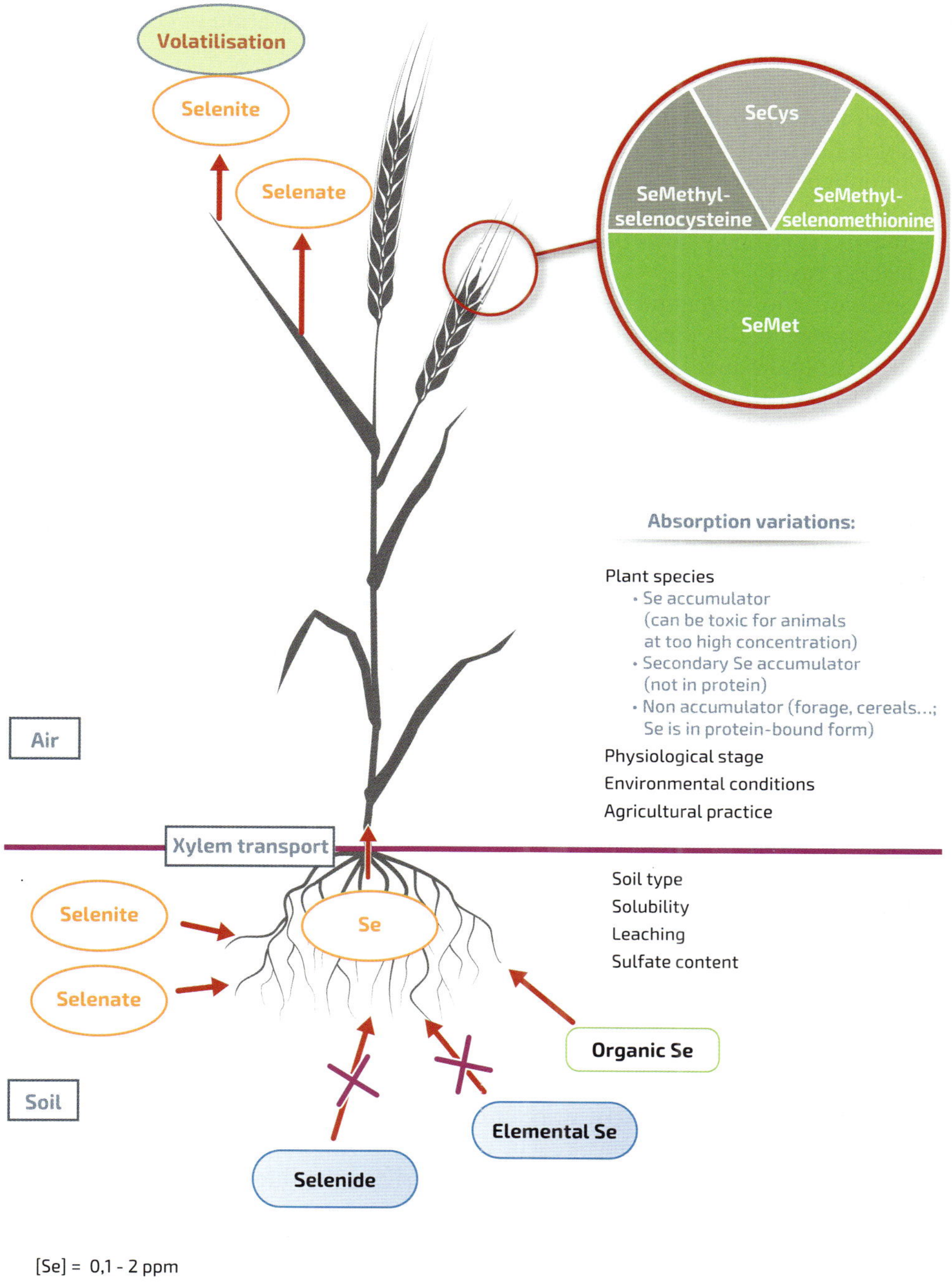

Figure 3.2. Selenium in soils and plants.

- Since sulphate competes with selenate for uptake by the sulphate transporter, high soil sulphate was shown to decrease Se uptake by plants (Terry *et al.*, 2000). This explains low Se availability from soils as a result of application of certain types of fertilisers.
- Selenium can also be leached from the topsoil in areas of high rainfall leading to decreased forage selenium content.
- Solubility is the critical determinant of Se bioavailability to plants, however, the amount of water-soluble Se in soils varies substantially and does not correlate with total soil Se (Combs and Combs, 1986).
- Selenite is strongly absorbed by soils while selenate is only weakly absorbed and leaches easily.
- Selenide and elemental Se are usually found in reducing environments and are unavailable to plants and animals. Indeed, selenite in soils undergoes a lot of microbially mediated transformations leading to the reduction of Se(IV) to insoluble Se(0) or to the production of volatile Se species (Fevrier *et al.*, 2007).
- Selenite is present in mildly oxidising, neutral pH environments and typically humid regions, while selenate is the predominant form under ordinary alkaline and oxidised conditions (Goh and Lim, 2004). The authors also showed that the adsorption of selenite and selenate by soils appeared to be affected by the variable pH-dependent charges on the soil particle surfaces. In particular, phosphate is shown to have more profound effects than sulphate on Se adsorption in the soil.
- Application of gypsum (calcium sulphate) to soils is found to decrease Se availability for plants (NRC, 1983).
- Leaching during the soil development process and irrigation water substantially decrease Se level in plants (NRC, 1983).
- Forage Se is found to be low on sandy soils and lower on mineral upland soils than on organic moorland soils in the British Isles (MacPherson, 2000). In fact, trace metal concentrations in Libyan clay surface soil are higher than the sandy soil (El-Ghawi *et al.*, 2007).
- The main chemical changes under long-term waterlogged conditions are depletion of molecular oxygen, decrease of redox potential, and reduction of Fe (III) to Fe (II) and SeO_3^{-2} to Se^0. This leads to low availability of Se in soils, and subsequently low Se content (29 μg/kg) in brown rice grain produced in this Chinese region (Cao *et al.*, 2001). In addition, selenite binds tightly to iron and aluminium oxides and thus is quite insoluble in soils (Jonnalagadda and Rao, 1993).

Selenium is transported via the xylem to chloroplasts in leaves where it is processed by the sulphur assimilation pathway into organic compounds. It seems likely that, selenate form is transported more easily from root to shoot than is selenite or organic Se (Terry *et al.*, 2000). The uptake of selenium is shown to differ with the plant species. It ranged from 2 to 40% of initial selenium for lettuce, maize and radish and reached 4.8 and 17% for mycorrhizal and non-mycorrhizal ryegrass, respectively (Munier-Lamy *et al.*, 2007). Furthermore, plants differ markedly in their ability to incorporate selenium from soil into tissues; and based on this ability plants can be divided into three major categories:

- Selenium accumulators. Some plant species hyperaccumulate Se in leaves and stems when grown on seleniferous soils. In fact, these plants stimulated the initial Se research, since they caused Se toxicity in animals grazing them. Selenium accumulator genera include such species as *Astragalus, Stanleya, Morinda, Neptunia, Oonopsis*, and *Xylorhiza* (Terry *et al.*, 2000). These plants can accumulate up to several mg Se per g of dry weight and are ultimately very toxic to animals. When these plants were grown hydroponically in the presence of selenate, Se in the older leaves was predominantly inorganic while in young leaves and roots was mainly (90-95%) in organic form (Ellis and Salt, 2003). When other forages are available, a specific odour causes grazing animals to avoid them on pasture.
- Secondary Se accumulators accumulate high Se concentrations even when grown on soils with comparatively low or medium Se content. They include such species of the genera *Aster, Astragalus, Atriplex, Castilleja, Comandra, Grayia, Grindelia, Gutierrezia, Machaeranthera* and *Brassica* (Terry *et al.*, 2000). It is important to underline that in Se-accumulator plants, Se is not incorporated into proteins. This contrasts with non-accumulator plants such as typical forage and cereal grains where Se is found predominantly in protein-bound form (Surai, 2006). This type of Se accumulator plant is more an exception than the rule of Se metabolism in plants.
- Non-accumulator plants. The third category includes most forage, cereal and oilmeal crop plants. Such plants contain less than 25 mg Se/kg dry weight and do not accumulate Se in excess of 100 mg/kg even when grown on seleniferous soils (Terry *et al.*, 2000). These plants typically have Se concentrations in a range of 0.01 to 1.0 mg/kg dry weight. However, there are species-specific differences in Se accumulation from the same soil. For example, lucerne is shown to accumulate more Se than other grasses under conditions of moderately low soil Se concentrations (Van Metre and Callan, 2001) and lucerne contains more selenium than timothy, cocksfoot or brome grass (Ehlig *et al.*, 1968). Furthermore, white clover (Davies and Watkinson, 1966) or tropical legumes (Long and Marshall, 1973) contains less Se than various grasses independently on its soil levels.

After absorption, the distribution of Se in various parts of the plant is shown to depend on species, phase of development and physiological conditions. For example, Se distribution was studied in *Astragalus bisulcatus*, an accumulator species capable of accumulating up to 0.65% of its shoot dry biomass as Se (Pickering *et al.*, 2000). It was shown that plants exposed to 5 μM selenate for 28 days contained predominantly selenate in the mature leaf tissue, whereas the young leaves and the roots contained exclusively organic Se. From this work it is clear that the fate of selenate is dependent on plant tissues and stage of growth. Therefore, chemical reduction of selenate to organic Se in plants is shown to be tissue-specific, inducible and developmentally dependent. It is likely that selenate reduction is a rate-limiting stage in the conversion of Se to organic forms (Pickering *et al.*, 2000).

The plant absorbs Se from the soil in the form of selenite, selenate or organic forms of Se and synthesises selenoamino acids with SeMet representing more than half of the Se in cereal grains (Olson and Palmer, 1976) with Se-methyl-selenomethionine, selenocysteine and Se-methyl-selenocysteine being the other selenocompounds found in plants (Brody, 1994). Therefore, SeMet is shown to be the major selenocompound in cereal grains, grassland legumes and soybeans (Whanger, 2002; Table 3.1). For example, in corn, rice, wheat and soybeans, SeMet comprises 45.5-82, 54.9-86.5, 50.4-81.4 and 62.9-71.8% of total Se, respectively (Yang *et al.*, 1997). SeMet accounted for 72-85% of the sum of the selenium species in wheat grain (Cubadda *et al.*, 2010). Even in wheat grown on seleniferous soils Se (up to 31 mg/kg Se), almost half occurred in the form of SeMet (Olson *et al.*, 1970). It is interesting to note that in three wheat cultivars, SeMet accounted for between 50 and 70% of the total grain Se, however, when N and/or S concentrations in soil were amended, SeMet only accounted for between 30 and 40% of total grain Se (Duncan *et al.*, 2017). Indeed, the majority of the Se is present as SeMet in both rice and corn (Beilstein *et al.*, 1991). Similarly, SeMet was shown to be the main Se-containing amino acid identified in most of the extracts of Indian mustard (*Brassica uncea*), sunflower (*Helianthus annus*), and white lupine (*Lupinus albus*; Ximenez-Embun *et al.*, 2004). SeMet is stored mainly in the grain and the root, while lower concentrations of this amino acid are found in the stems and leaves (Schrauzer, 2003).

The variability in results of Se specification in plant material reflects analytical difficulties. For example, by using SeMet analysis based on its reaction with CNBr, it has been shown that wheat samples, though having a 30-fold range in total Se content, all have about 45% of their total Se values in the form of SeMet (Wolf and Goldschmidt, 2004). However, the authors suggested that additional experiments were needed to verify that all selenomethionine in the wheat samples had been accounted for.

It is generally accepted that environmental conditions and agricultural practises have a major effect on the Se content of various plant feeds. For example, water extractable Se accounted for 60.4-72.6% of the total Se in a plant (*Stanleya pinnata*) extract. Among the soluble Se compounds in the plant extract, Se-amino acids comprised 73-85.5%, Se [VI] ranged from 7.5 to 19.5% and non-amino acid organic Se was less than 7% (Zhang and Frankenberger, 2001). Selenium [IV] in most samples was below the detection limit (1 μg/g). This study showed that considerable amounts of the accumulated Se [VI] in the plant was metabolised to Se-amino acids during growth of the plant. The distribution of selenoamino acids in a Se-tolerant grassland legume species (*Melilotus indica L.*) grown in Se-laden soils was studied using high-resolution gas chromatography and gas chromatography/mass spectrometry (Guo and Wu, 1998). Five selenoamino acids including selenocystine, selenomethionine, selenocysteine, Se-methylselenocysteine, and gamma-glutamyl-Se-methylselenocysteine were identified and measured in plant tissues. SeMet constituted more than 50% of the total selenoamino acid in the plant. It seems likely that rate of Se accumulation in plants and its form depend on the Se form provided. For example, time-dependent kinetic studies in Indian mustard (*B. juncea*) showed that selenate was taken up 2-fold faster than selenite (De Souza *et al.*, 1998). For both selenate- and selenite-supplemented plants, Se accumulation and volatilisation increased linearly with external Se concentration. It is important to note that Se-volatilisation rates were 2- to 3-fold higher in plants supplied with selenite compared with selenate. In fact, the assimilation of selenate by plants appeared to be limited by its reduction, a step that is thought to be mediated by ATP sulfurylase, which is rate limiting for selenate uptake and assimilation (Pilon-Smits *et al.*, 1999).

Furthermore, there was a difference in Se metabolism among plants supplemented with various forms of Se. For example, selenite-supplied plants accumulated organic Se, most likely SeMet, whereas selenate-supplied plants accumulated selenate (De Souza *et al.*, 1998). It seems likely that Se volatilisation from selenate is limited by the rate of selenate reduction, as well as by the availability of Se in roots, as influenced by uptake and translocation. On the other hand, Se volatilisation from selenite was suggested to be limited by selenite uptake and by the conversion of SeMet to dimethylselenide (De Souza *et al.*, 1998).

3.3 Selenium absorption and metabolism

Recent advances in Se biochemistry have provided a background for a deeper understanding of the principal differences in metabolism of the two forms of Se namely inorganic Se (sodium selenite or selenate) and organic Se (mainly SeMet). Organic Se, which can be found in major feed ingredients (grains and forages), is primarily in the form of SeMet and is metabolised in the same way as methionine (Wolfram, 1999). It is actively transported through intestinal membranes during absorption and actively accumulated in such tissues as muscle. It is well known that methionine is not synthesised by farm animals and therefore it is an essential amino acid. The same is true for SeMet, which is not synthesised in animals and must be obtained from feed sources (Schrauzer, 2000, 2003; Schrauzer and Surai, 2009). In contrast, inorganic Se is absorbed as a mineral and little is retained in tissues. Therefore, a large part of inorganic Se is excreted with faeces in ruminants or with urine/urates in non-ruminants with little is stored in the body (Wolfram, 1999).

A number of factors can influence the bioavailability and distribution of selenium in the body (Thomson, 1998) including:

- chemical form of Se;
- other dietary components;
- selenium status;
- physiological status;
- species.

Most of the research devoted to Se biochemistry and metabolism has been using inorganic Se, namely selenite or selenate. For example, the concentrations of ^{82}Se in organs and body fluids and the distributions of their constituents depending on the dose and time after the intravenous administration of ^{82}Se-selenite and -selenate to rats were investigated (Suzuki and Ogra, 2002). Selenite was taken up by red blood cells within several minutes, reduced to selenide by glutathione, and then transported to the plasma, bound selectively to albumin and transferred to the liver. In contrast to selenite, intact selenate was either taken up directly by the liver or excreted into the urine. In fact, the ^{82}Se of selenite origin and that of selenate origin were detected in the forms of the two Se peak materials (A and B) in the liver. The former was methylated to the latter *in vivo* and *in vitro*. The latter was identical with the major urinary metabolite and was identified as Se-methyl-N-acetyl-selenohexosamine (selenosugar). The chemical species-specific metabolic pathway for Se was explained by the metabolic regulation through selenide as the assumed common intermediate for the inorganic and organic Se sources and as the 'checkpoint metabolite' between utilisation for selenoprotein synthesis and methylation for the excretion of Se (Suzuki and Ogra, 2002) (Figure 3.3).

Results of various *in vitro* and *in vivo* experiments with a variety of animal species and model systems have demonstrated that SeMet is readily absorbed through the gut. For example, in dogs this process was two times faster than selenocysteine and four times faster than selenite absorption (Reasbeck *et al.,* 1981). Indeed, SeMet is better absorbed than selenite (Daniels, 1996). However, absorption is not a limiting factor to bioavailability, since organic and inorganic dietary Se are both well absorbed through the intestinal membrane (70-95%) (Finley, 2006). Furthermore, recently it has been shown that in pigs the intestinal tissue may play a crucial role in systemic Se homeostasis and possible intestinal Se retention controlled by intracellular Se saturation awaits further investigation (Dalto and Matte, 2020). The specific role of the chick duodenum in the assimilation of Se (selenate or selenite) was shown by Apsite *et al.* (1993). Selenite is passively absorbed in the intestine with highest concentrations found in duodenum, liver and kidneys (Apsite *et al.,* 1994). Absorption of Se was found to be greatest in the duodenum and anterior ileum of the chicken (Pesti and Combs, 1976). Absorption of ^{75}Se from selenite, selenate, and SeMet was determined in ligated loops from duodena, jejuna, and ilea of Se-deficient rats (0.009 mg/kg Se) or rats fed selenite-supplemented diets (0.20 mg/kg Se) (Vendeland *et al.,* 1992). Selenium deficiency had no effect on absorption of any selenocompound in any intestinal segment. SeMet was absorbed from all segments. In contrast, selenate and selenite were most efficiently absorbed from the ileum. In mice SeMet was absorbed in the entire intestinal tract, most rapidly from the duodenum, with decreasing rate of absorption in the following intestinal segments (Andersen *et al.,* 1994).

Selenomethionine was more rapidly removed from the ligated intestinal segment of chicken and more efficiently retained after oral or parenteral administration (Humaloja and Mykkanen, 1986). The percentage absorption of both Se compounds was greatest from the duodenal segment of the small intestine. The transport of these Se compounds does not appear to depend on the dietary level of Se since the percentage absorption was not altered by feeding the birds diets supplemented with 0.4 or 4.0 mg/kg Se prior to the measurement of absorption (Humaloja and Mykkanen, 1986). In general, the absorption of Se-amino acids is accelerated by the specific amino acid active transport mechanisms in the gut mucosa. Sodium selenite is absorbed more slowly, possibly by simple diffusion through the intestinal mucosa, than the amino acid-bound Se compounds.

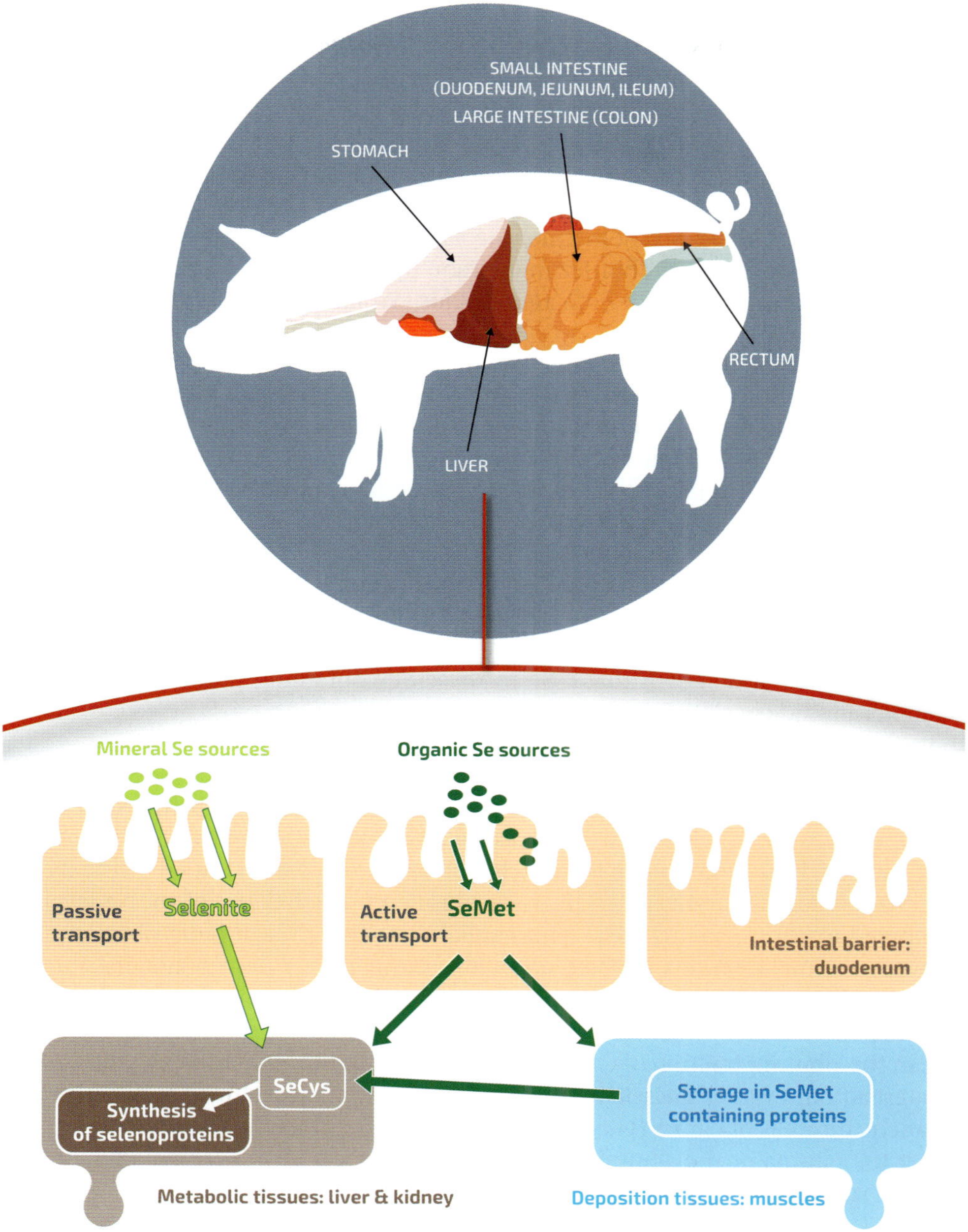

Figure 3.3. Selenium absorption.

Most intensive was the metabolism of SeMet in the kidneys, spleen, testes, and pancreas. Comparatively lower is the metabolism of this agent in the breast, bone and in the brain. Radioactivity of SeMet was lowest in the femoral muscle. As much as 18.2% of the introduced amount of SeMet was excreted for 72 hours through the faeces and urine. The relative efficiency patterns for uptake of different selenocompounds during *in vitro* perfusion and *in vivo* ligated segments was SeMet > selenate > selenite. In contrast, selenite was taken up most rapidly by brush border membrane vesicles, followed by SeMet and selenate in decreasing order. Selenate and SeMet appeared in the vascular effluent largely unchanged, but selenite was metabolised extensively during absorption (Whanger *et al.,* 1996). Selenomethionine as well as methionine were transported across the pig jejunal brush border membrane by a single, Na^+-dependent, carrier-mediated process common for both amino acids (Wolffram *et al.,* 1989).

The metabolism of ^{75}Se-selenite and ^{75}Se-SeMet in chick blood was studied *in vitro*. ^{75}Se from selenite was rapidly taken up by erythrocytes and then subsequently released into the plasma in a protein-bound form. However, ^{75}Se-SeMet showed a more gradual and continuous accumulation in erythrocytes over the 12-hr incubation period, according to a hyperbolic type function. ^{75}Se-selenite was incorporated into GPx, whereas ^{75}Se from SeMet was mostly incorporated into haemoglobin (Ilian and Whanger, 1989). In the same experiment binding of ^{75}Se from either selenite or SeMet to plasma proteins was dependent on the presence of erythrocytes. Addition of reduced glutathione and glutathione reductase to plasma produced the same effects as erythrocytes on binding of ^{75}Se from selenite, but not from SeMet, to plasma proteins. It was also shown that ^{75}Se in these Se-containing proteins is bound in selenotrisulphide bond.

The distribution of Se in plasma fractions was investigated in guinea pigs fed various levels (basal, 0.5, 1.0, 2.0, 4.0, 6.0 and 8.0 mg Se/kg) of dietary SeMet (Gu *et al.,* 1998). There was a corresponding increase of Se concentration in liver, kidney, brain, testis, spleen, heart and muscle with each increase of dietary Se, but GPx activity did not change in liver, brain, testis, heart or muscle in pigs fed any of the Se levels as compared to controls fed a basal commercial diet. There was a redistribution of Se between various fractions in the blood. For example, on a percentage distribution basis, the Se in selenoprotein P decreased, and that in the albumin fraction increased with increased dietary intakes of Se as SeMet. Similarly, the greatest percentage of Se was in the albumin fraction of Chinese people living in the high Se areas, whereas the greatest amount was in the selenoprotein P fraction in subjects living in deficient and Se-adequate areas of China (Gu *et al.,* 1998). Increases in the ratios of Se/albumin in either the plasma or the albumin fraction also occurred with increases of Se intake of these subjects.

It seems likely that ingested Se is firstly bound to albumin which transports the element to the liver, where Se is released and serves for the synthesis of Selenoprotein P which is released into the bloodstream to become itself a Se transporter between the liver and other organs and tissues (Suzuki *et al.,* 2009). In fact, liver and kidney are considered to be the two main places of synthesis for most selenoproteins, including SeP and cellular GPx (liver) and extracellular GPx (kidneys; Suzuki *et al.*, 2009). Okuno *et al.* (2001) indicated that in mouse liver SeMet was directly metabolised to CH_3SeH by an alpha,gamma-elimination enzyme analogous to bacterial L-methionine gamma-lyase, in addition to the generally acceptable pathway via selenocysteine. It has been suggested that L-selenohomocysteine generated from SeMet metabolism can be efficiently recycled to SeMet in mammals (Zhou *et al.*, 2000). When sodium selenite was administered to chickens, the kidney was the most responsive to additional dietary Se, while skeletal (thigh) muscle showed no response to supplemental Se (Guenter and Bragg, 1977). The total Se contents in the kidney and in the liver from rats treated with SeMet, selenocysteine or sodium selenite were markedly higher than those in other organs. The highest accumulation of the total Se was observed in these two organs when rats were given SeMet (Nakamuro *et al.*, 1997).

The number of published studies in animals suggested that the metabolic fate and physiological function of dietary selenite may differ from that of SeMet or of feed/food Se (Figure 3.4). It has been postulated that there are two distinct metabolic pools of Se in the body (Daniels, 1996). The main exchangeable metabolic pool includes all forms of Se derived from inorganic selenite/selenide, including endogenously synthesised selenoproteins (e.g. GPx, selenoprotein P, etc.), excretory Se metabolites (trimethylselenium ion) and various other intermediary products of selenite metabolism. This is an active Se pool providing Se for synthesis of the primary functionally important selenocompounds (Daniels, 1996). The second Se pool consists of SeMet-containing proteins and potentially can contribute to the first pool via participation in selenoprotein synthesis. In fact, Burk *et al.* (2001) demonstrated that Se from SeMet, but not that from selenate or selenocysteine, can be incorporated into albumin, presumably as SeMet in the methionine pool. In another study, albumin was purified from plasma of a human before and after 28 days of supplementation with 400 μg Se/day as SeMet. It was shown that the albumin contained 1 Se atom, presumably as SeMet, per 8,000 methionine residues before supplementation and 1 per 2,800 after supplementation (Hondal *et al.*, 1999).

These findings support the view that SeMet is a non-specific form of Se that is metabolised as a constituent of the methionine pool where it is randomly distributed and is unaffected by specific Se metabolic processes. Therefore SeMet can be considered as a storage form of Se in animals and humans and it is metabolised as a constituent of the methionine pool and is unaffected by specific Se metabolic processes. In contrast, no evidence was obtained for non-specific incorporation of Se into plasma proteins when it was administered as selenate or as selenocysteine. These forms of the element appear to be metabolised by specific Se metabolic processes (Burk *et al.*, 2001).

The skeletal muscles are the major Se-storage organ, accounting for about 46.9% of the total Se in the human body, while kidney contains only 4% of Se reserves (Oster *et al.*, 1988). In humans, whole body Se depends on the regional location and varies from 3-6 mg up to 13-20 mg (Daniels, 1996). GPx activity and deposition of Se were examined in tissues of rats given dietary Se for 7 weeks as either selenite or SeMet with ^{75}Se radiotracer of the same chemical form (Beilstein and Whanger, 1988). The

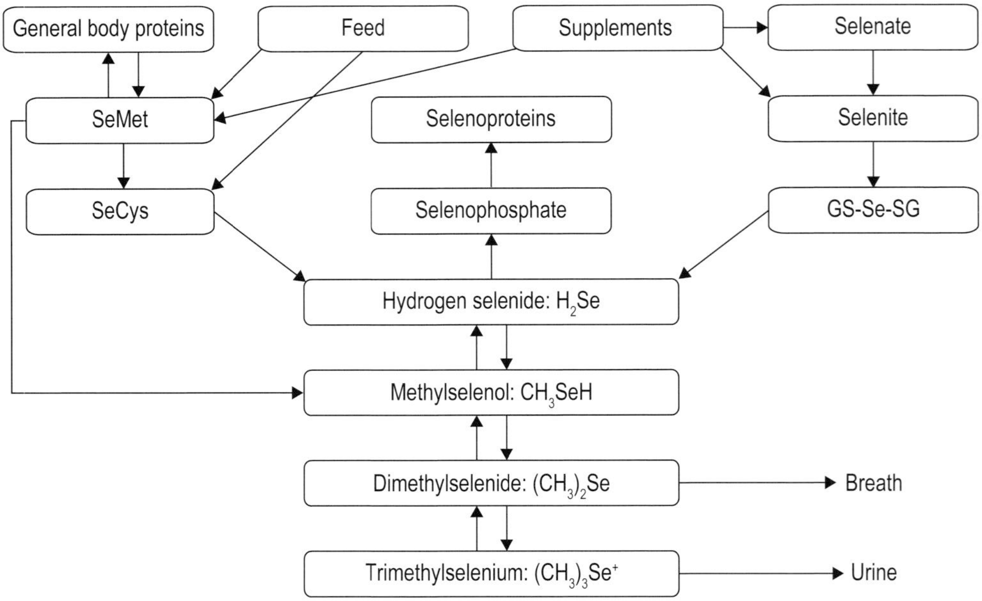

Figure 3.4. Metabolism of selenomethionine, selenite and selenate (adapted from Combs, 2001; Meuillet *et al.*, 2004; Schrauzer, 2000, 2003; Surai, 2018).

authors showed that the proportion of ^{75}Se as SeMet was higher in tissues of rats fed SeMet (highest in muscle and haemoglobin, 70%, and lowest in testes, 16%). In contrast, selenocysteine was the predominant form of Se present in tissues of rats given selenite.

In the chicken breast and leg muscles SeMet comprises 66.7 and 56.1% of total selenium (Bierla *et al.*, 2008), while after enrichment SeMet proportion went up to 99% indicating non-specific incorporation of SeMet into muscle proteins; similar observations should also be done in pigs. The ratio of SeMet to Met is variable and dependent on the amount of SeMet and Met in the diet, with Met always presents in large excess over SeMet. For example, in American adults, the SeMet:Met ratio in skeletal muscle is in the order of 1:7,000, as estimated from the selenium and sulphur contents of human skeletal muscle and is similar to the corresponding ratio of 1:>6,000 found for chicken breast muscle (Schrauzer and Surai, 2009). It is interesting that in the basic starter chicken diet the ratio of SeMet:Met is about 1:60,000; in the growing diet it is 1:50,000 and this ratio is almost the same for breeder birds. After dietary Se supplementation in the form of SeMet at 0.3 mg/kg, the ratio SeMet:Met in the diet changes to 1:15,000 at the start, down to 1: 12-14,000 later in life. As mentioned above, SeMet comprised 53-71% of total Se in the egg albumen and 12-19% in the egg yolk (Lipiec *et al.*, 2010). These findings support the view that SeMet is a non-specific form of Se that is metabolised as a constituent of the methionine pool where it is randomly distributed. Therefore, SeMet can be considered as a storage form of Se in animals and humans and it is unaffected by specific Se metabolic processes.

Similar to chickens, in pigs SeMet supplementation was shown to significantly increase SeMet concentration in longissimus dorsi muscle (Figure 3.5; Zhang *et al.*, 2020) and SeMet proportion in pig muscles comprised 71.2, 75.9, 81.0 and 90.8% in the pigs supplemented with sodium selenite (0.25 mg/kg) or SeMet at 0.25, 0.5 and 2.5 mg/kg respectively.

Since SeMet cannot be synthesised in animals in pigs fed sodium selenite SeMet derived from feed ingredients and its concentration and percentage of total Se significantly increased in pig muscles due to dietary SeMet supplementation (Zhang *et al.*, 2020). Indeed, in pigs, similar to chicken and human, Se in muscles can be considered as a reserve to be used in stress conditions.

In addition, in chickens muscle organic selenium has the longest half-lives (12 days) as well as in brain and lungs (13 days), while the shortest biological half-lives were observed in major metabolic organs, the liver, kidney and pancreas with half-lives close to 4 days (Brandt-Kjelsen *et al.*, 2014). The half-life of ^{75}Se in heart was 9 days and 7 days in blood. Therefore, SeMet accumulated non-specifically in muscle proteins can build Se reserves, which can be used in stress conditions when Se requirement is increased, but feed consumption usually decreased. In stress conditions protein catabolism by proteasomes can release SeMet, which could serve as a source of Se for newly synthesised selenoproteins, such as GPx, thioredoxin reductase, methionine sulfoxide reductase, etc. Those enzymes can deal with overproduction of free radicals and prevent decrease in productive and reproductive performance of farm animals. It is important to mention that selenocysteine provided with the feed had no advantage

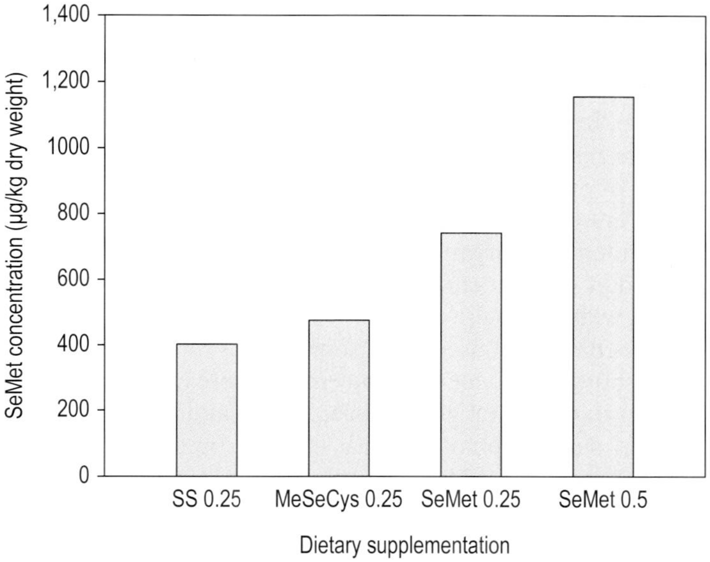

Figure 3.5. Selenomethionine (SeMet) concentration in pig longissimus dorsi muscle depending on dietary Se supplementation, mg.kg (adapted from Zhang *et al.*, 2020). SS = sodium selenite; MeSeCys = methyl selenocysteine.

in comparison to SS in terms of Se concentrations in tissues (Deagen *et al.*, 1987) and it is not able to build any Se reserves in the body. It was proven that Se from both selenite and SeMet is readily available for synthesis of the selenoenzyme GPx in rat tissues (Pierce and Tappel, 1977).

There are several lines of direct and indirect evidence confirming the idea that Se accumulated in tissues in the form of SeMet can be available for selenoprotein synthesis (Alfthan *et al.*, 2000; Ip and Hayes, 1989; Levander, 1983; Pappas *et al.*, 2005; Persson-Moschos *et al.*, 1998; Robinson *et al.*, 1985; Surai, 2000; Surai *et al.*, 2006):

- Studies in our laboratory (Surai, 2000) indicated that chicks hatched from eggs enriched with Se by means of using Se yeast had higher liver GPx activity not only at hatching, but more importantly, even at 5 days posthatch. More recent observations with quail (Surai *et al.*, 2006) and chickens (Pappas *et al.*, 2005) indicate that when organic Se was included in the maternal diet, Se concentration in the liver of the progeny was alleviated up to 2-4 weeks posthatch.
- It was clearly demonstrated that the effects of maternal organic Se supplementation on the expression of GPx in the offspring fed on low Se diet are sustained for several weeks after hatching (Pappas *et al.*, 2005). In fact when the offspring from the two parental groups (high Se and low Se) were both maintained on the low-Se progeny diet, the tissue Se concentrations in chicks originating from the high-Se hens remained significantly higher for 3-4 weeks after hatching, compared with the values in chicks from the low-Se hens. Similarly, tissue GPx activity remained significantly higher in chicks obtained from the high-Se hens for 2 (liver) or 4 (muscles) weeks post-hatch. It seems likely that maternal Se has an epigenetic effect of Se metabolism in progeny chicks, since significantly increased Se concentration in the muscles of 4 week old chickens reflects better assimilation of Se from the diet or lower usage of Se in the body.
- The bioavailability of the Se pool in maintaining liver GPx activity during a period of Se deprivation, following excess selenite or SeMet loading was assessed in rats (Ip and Hayes, 1989). In this study half-life of decay of the enzyme was calculated to be 4.2 and 9.1 days, respectively, in rats that had already been exposed to 3 mg/kg Se as either selenite or SeMet.
- In a human study Persson-Maschos *et al.* (1998) showed that in individuals who had been supplemented with organic Se, the decline in the level of selenoprotein P following a period of supplementation was slower than in individuals who had been supplemented with selenite.
- When wheat and selenate were used as Se sources in a supplementation study in Finnish men it was shown that once the supplements were withdrawn, platelet GPx activity declined less in the group given wheat Se (Levander *et al.*, 1983).
- After several weeks' supplementation with high-Se bread, plasma Se of New Zealand subjects increased from 50-70 ng/ml to 120-175 ng/ml (Robinson *et al.*, 1985). Plasma Se remained elevated when supplementation ceased.
- In SeMet or Se-yeast supplemented mice, liver GPx activities declined more slowly during Se depletion than in mice given selenite (Spallholz and Rafferty, 1987).

- In children the relative bioavailability of Se-yeast versus selenite measured as GPx activity was similar in plasma, red blood cells, and platelets, however, Se-yeast provided a longer lasting body pool of Se (Alfthan *et al.*, 2000).
- Study with broiler chickens fed organic or mineral Se demonstrated that endogenous Se could be released from tissues, and, thus, that organic Se sources were more efficient in maintaining the GPx level (Payne and Southern, 2005). Aforementioned data indicate that protective effect of organic selenium is more pronounced under stressful conditions.
- Se reserves in the body (mainly in the muscles) built in the form of SeMet non-specifically incorporated into the proteins in place of Met could be considered as an important element in increasing adaptive ability of animals/pigs to various stresses. This could increase their productive and reproductive performance.

The mechanisms regulating a SeMet conversion to H_2Se and further to synthesis of SeCys and respective selenoproteins are not clear at present, but is seems likely that changes in redox status of the cells/tissues and activation of proteasomal protein degradation could be involved. Indeed, ATP- and ubiquitin-independent proteolysis by the 20S proteasome is responsible for the selective degradation of oxidised proteins. *In vitro*, the 20S proteasome shows an increased proteolytic activity toward oxidised polypeptides. In fact, a 30% decreased activity of the chymotrypsin-like activity of proteasome in cells overexpressing GPx-1 was shown (Kretz-Remy and Arrigo, 2003). This observation correlated with a 2-fold increase in inhibitor of kappa B alpha half-life, a protein whose basal turnover is 20S proteasome-dependent. Furthermore, following exposure to H_2O_2, human T47D cells overexpressing GPx showed a seleno-dependently decreased accumulation of intracellular ROS and 20S proteasome chymotrypsin-like activity. Moreover, exposure of HeLa cells to antioxidant compounds reduced the proteasome 20S chymotrypsin-like activity. These results suggest that GPx activity or pro-reducing conditions can downregulate basal 20S proteasome activity (Kretz-Remy and Arrigo, 2003). Indeed, the chymotrypsin activity of the proteasome is strongly attenuated by GPx overexpression. This suggests that selenium is a key element that controls proteasome activity. This could be a feedback mechanism of recognition of SeMet as a source of Se for selenoprotein synthesis. In stress conditions some amino acids inside muscle proteins would be oxidised and this will trigger an increase in proteasome activity to degrade such proteins and release SeMet to be available as an additional source of Se for selenoprotein synthesis in stress conditions. When oxidative antioxidant-prooxidant equilibrium is restored increased GPx activity would decrease proteasome activity and protein degradation.

The most probable sequence of events is as follows (Figure 3.6):

- when organic Se is used in the diet, Se reserves in the form of SeMet are non-specifically incorporated in various proteins, for example in muscles;
- under stress conditions, the requirement for selenoproteins to prevent free-radical related damages is increased, but selenium bioavailability is decreased due to decreased feed consumption;
- redox status of muscle cells decreases due to depletion of antioxidants and probably some proteins are oxidised;

- proteosomes are activated and protein degradation is increased;
- SeMet is released from proteins and used for additional synthesis of selenoproteins;
- antioxidant defences are improved and redox status of the cells is changed;
- proteosome activity is decreased and antioxidant-prooxidant equilibrium is established.

Indeed, SeMet is the major selenocompound found initially in animals given this selenoamino acid but is converted with time afterwards to selenocysteine when incorporated in functional selenoproteins (Whanger, 2002). For example, the chemical forms of Se were determined in erythrocyte and liver proteins after injection of ^{75}Se as either sodium selenite or SeMet in male weanling rats. Void volume proteins contained principally selenocysteine (^{75}Se-Cys) in [^{75}Se]selenite-injected animals. This material contained both ^{75}Se-Met and ^{75}Se-Cys 1 d post-injection in ^{75}SeMet-injected animals, but primarily ^{75}Se-Cys at 20 d afterwards (Beilstein and Whanger, 1986). This means that Se-Cys was synthesised from SeMet and with time all SeMet was converted to Se-Cys. In acid hydrolysates of whole liver, ^{75}Se was recovered principally as ^{75}Se-Cys from animals injected with [^{75}Se]selenite. However, for animals injected with ^{75}Se-Met, liver ^{75}Se was present initially as ^{75}SeMet, but after five days the majority of liver ^{75}Se was as Se-Cys. The long-term fate in rats of an oral dose of [^{75}Se]selenocystine was compared with that of an oral dose of [^{75}Se] SeMet. It was shown that intestinal absorption of [^{75}Se]selenocystine was 81% of the administered dose and that of [^{75}Se]SeMet was 86% (Thomson *et al.*, 1975). The initial utilisation of [^{75}Se]selenocystine was different from that of [^{75}Se]SeMet. However, after the first week ^{75}Se from both sources appeared to be metabolised

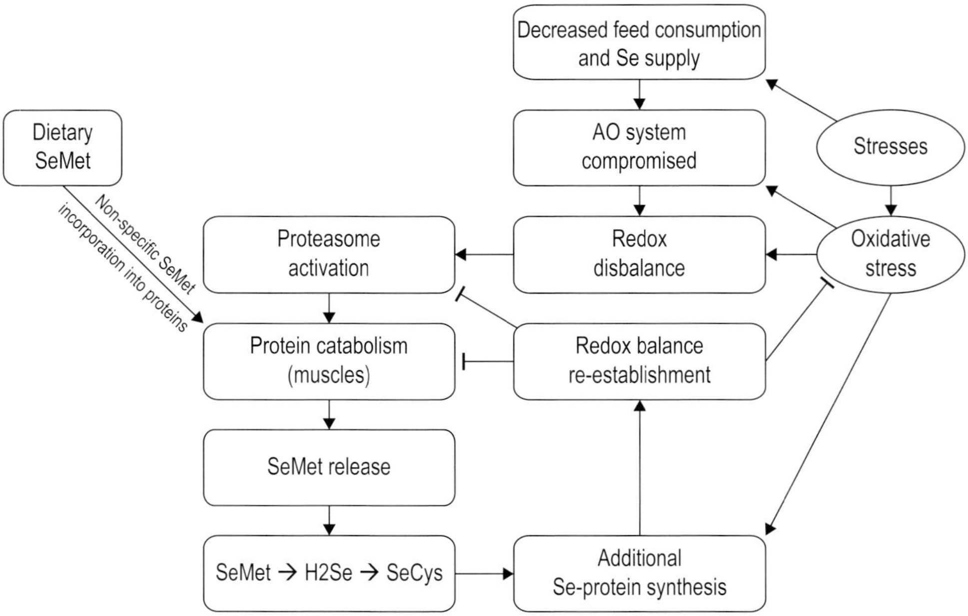

Figure 3.6. Protective effects of selenomethionine (SeMet) in stress conditions (adapted from Surai, 2006).

similarly, suggesting that dietary Se of both forms is ultimately incorporated into the same metabolic pool (Thomson *et al.*, 1975).

Weanling male rats were fed a basal Se-deficient diet or this diet plus 2 mg/kg Se as either selenite, SeCys or SeMet for nine weeks (Deagen *et al.*, 1987). Except for the kidney, the tissue Se concentrations were similar in rats fed selenite or SeCys, but the Se content in testis, muscle, pancreas, heart, spleen, whole blood, erythrocytes and plasma was significantly higher in rats fed SeMet than in those fed either selenite or SeCys. The greatest increase due to SeMet compared with the selenite and SeCys treatments was about 10-fold in the muscle compared with 1.3- to 3.6-fold for the other tissues (Deagen *et al.*, 1987). In general, SeMet has a slower, whole body turnover in comparison to sodium selenite and there is greater efficiency in the re-utilisation of Se from SeMet (Swanson *et al.*, 1991). Indeed, the average whole body half-lives of SeMet and selenite in humans were shown to be 252 and 102 days, respectively, confirming re-utilisation of SeMet in the body (Patterson *et al.*, 1989). It should be noted that only a small proportion of the methionine pool can be replaced by SeMet, since only part of methionine could be replaced by SeMet in the diet. Furthermore, protein turnover prevents accumulation of SeMet to toxic levels in the organism (Schrauzer, 2003).

In fact, rapid turnover of various selenoproteins and dependence of this process on Se status were described. For example, the half-life of GPx is approximately 3 days (Sunde *et al.*, 1989), and 2-iodothyronine deiodinase has a half-life of only 30-45 minutes (Botero *et al.*, 2002; Curcio *et al.*, 2001; Kim *et al.*, 2003), while that of selenoprotein P in plasma is 3-4 hrs (Burk and Hill, 1994). In growth medium there was an increase in TrxR mRNA levels of 2-5-fold at 1 μM Se and an increase in the stability of TrxR mRNA with a half-life for degradation of 21 hrs compared to 10 hrs in the absence of Se (Gallegos *et al.*, 1997). Similarly, the selenoprotein W mRNA half-life in myoblasts is about 57 hrs for cells grown in a low Se medium while Se treatment increased half-life by 2-fold (Gu *et al.*, 2002). Therefore, it is clear that Se reserve development could be an important regulatory mechanism for maintaining effective antioxidant defence during periods of increased demand. Therefore, from a nutritional viewpoint, SeMet is superior to selenite, especially with respect to maintenance of GPx during periods of Se inadequacy (Ip and Hayes, 1989). At physiological levels of Se intake, urine is the most important route of excretion and regulates Se homeostasis (Daniels, 1996). Various Se metabolites were found in urine, including trimethylselenium. There is a great body of evidence indicating that urinary Se is lower when organic Se is used in comparison to selenite.

3.4 Selenium status and bioavailability

To assess Se status of animals and humans various static or functional tests are used. Static tests measure the total quantity of Se in various accessible tissues and body fluids such as hair, nails, blood or its components, and urine. Functional tests measure the activity of Se-dependent enzymes, or a physiological or behavioural function dependent on Se (Gibson, 1989). There is no single test that can describe Se status and a combination of various techniques is preferable, and this explains a great discrepancy in results of comparative evaluation of Se bioavailability from various sources. The problem is that it is difficult to choose a proper end point for such evaluation.

Selenium availability from plant-based sources varies. For example, biological availability of Se for prevention of ED was taken as 100% for sodium selenite. It ranged in other forms of Se from 74% for sodium selenate to 7% for elemental Se; in plant feedstuffs from 210% for lucerne meal to 60% for soybean meal and in animal feedstuffs from 25% for herring meal to 8.5% for fish solubles (Table 3.2; Cantor and Scott, 1974b; Cantor *et al.*, 1975). However, SeMet was four times as effective as either selenite or selenocystine with respect to prevention of pancreatic degeneration and increasing the relative weight and Se concentration of the pancreas (Cantor *et al.*, 1975). Cantor and Scott (1974a) observed that protection against ED was closely related to plasma GPx. It was concluded that relative to selenite the retention of Se in fish meal or solubles was 43%, and relative to SeMet, 31% (Miller *et al.*, 1972).

Newly-hatched White Leghorn chicks with low-Se status were fed a low-Se basal diet for two weeks post-hatch followed by either continued depletion or a repletion period of four weeks with graded levels of Se (0.03, 0.06, 0.09 and 0.12 mg/kg) provided via sodium selenite, wheat or fish meal. The bioavailability of Se in wheat and fish meal in comparison to selenite for increasing the activity of GPx was 78 and 58%, respectively; and for increasing Se whole blood Se concentration was 123 and 107%, respectively (Hassan *et al.*, 1993). Based on the activity of plasma GPx, the biological availability of Se in soybean meal, lucerne, fish meal and SeMet was 33, 85, 82 and 92%, respectively (Ikumo and Yoshida, 1981). On the other hand, Se availability in fish meal, based on the prevention of incidence of ED, was 74%. In contrast to previous data, Se in fish meal was poorly able to prevent deficiency in chickens (Martello and Latshaw, 1982) and the results of Whitacre and Latshaw (1982) clearly showed that the commercial preparation of fish meal significantly decreased Se utilisation. Availability of Se in feeds was estimated in relation to restoring blood serum GPx activity in Se-depleted chickens. The availability of the Se (relative to Se in selenite) in capelin fish meal was 48.0 (38.5-60.0), mackerel fish meal, 34.1 (32.3-35.8), soybean meal, 17.5, maize gluten meal, 25.7, and SeMet, 78.3% (Gabrielsen and Opstvedt, 1980). These data on Se availability from fish sources are in line with others studied, but data on soybean meal are somewhat low. It is possible that thermal treatment of processed soya could affect Se availability.

In Se-deficient chicks, when providing Se at 10 μg/kg diet, selenite and selenocystine were about equal in promoting weight gain and preventing ED, while SeMet was less effective. Tissues from chicks given the Se sources providing 60 μg/kg diet for four weeks were analysed for Se. The content of tissues from chicks given selenite or selenocystine was similar. Chicks given SeMet had higher concentrations of Se in pancreas and breast muscle than the others, but lower concentrations in kidney, liver and heart (Osman, and Latshaw, 1976). Bioavailability of Se in oats, meat meal and SeMet based on efficacy in preventing ED in Se-depleted chicks was 41, 30 and 77%, respectively, while based on Se biopotency in elevating GPx activity in plasma, it was 33, 21 and 77%, respectively. When the increase in Se concentration in cardiac

Table 3.2. Bioavailability of selenium (Se) in feedstuffs (adapted from Cantor, 1997).[1]

Feedstuff	Biological availability, %
Dehydrated alfalfa meal	210
Brewer's yeast	89
Cottonseed meal	86
Corn	86
Brewer's grains	80
Wheat	71
Distiller's dried grains and solubles	65
Soybean meal	60
Herring meal	25
Tuna meal	22
Poultry by-product meal	18
Menhaden fish meal	16
Meat and bone meal	15
Fish solubles	9

[1] Sodium selenite was used as standard and prevention of exudative diathesis in chicks was used as an index of Se bioavailability.

muscle was used as the indicator of bioavailability, the Se bioavailabilities were 60, 42 and 114% for oats, meat meal and SeMet, respectively. Thus, SeMet was superior to sodium selenite in increasing Se concentration in tissues, but inferior to it in protecting chicks against ED (Hassan, 1987). By taking both the Se level and the GPx activity into consideration, organic Se was judged to be more bioavailable than selenite and selenate for humans (Clausen and Nielsen, 1988). Therefore Se from yeast, wheat and lucerne is highly available while Se in most other plant sources is moderately available. Selenium in fish meal is poorly available (Cantor, 1997), despite the fact that Se content of fish meal can be quite high. Indeed Se availability depends not only on feed ingredient but also on the methods used for evaluation.

3.5 Effectors of selenium absorption, metabolism and bioavailability

Animal studies reveal that various factors influence Se bioavailability. Animals cannot synthesise SeMet or distinguish it from methionine and as a result it is non-specifically incorporated into a wide range of proteins (Daniels, 1996). SeMet is retained in tissue proteins to a greater extent than selenocysteine and the inorganic forms. However, a number of other factors besides chemical form, may also influence the bioavailability and distribution of Se, including other dietary components, Se status, physiological status and species (Thomson, 1998). For example, Se is better absorbed from a high protein diet (Daniels, 1996). In particular, it has been suggested that SeMet is preferentially incorporated into body protein when dietary Met is limited

(Tian *et al.*, 2001). Se bioavailability also depends on many other nutrients in the diet, including Cu, Zn, Mg, vitamins B2 and B6, etc. (Surai, 2006). It seems likely that Se can affect absorption of other nutrients. For example, in studies investigating the relationship between vitamin E/Se and glucose absorption, Se plus vitamin E or Se alone modified jejunal glucose absorption, depending on the duration of feeding and age of chickens. In younger chickens, providing the supplements for 11 days depressed glucose absorption, but when they were given the supplements for 18 days glucose absorption was increased. In older chickens Se alone or with vitamin E given for 13 days increased glucose absorption (Giurgea and Roman, 1992).

Experiments were conducted to determine the initial effects of oral Se administration on Se-deficient chicks. Administration of 5 μg Se as seleno-DL-Met increased voluntary feed consumption within 2-3 hrs, whereas selenite did not have a significant effect until 3-4 hrs (Bunk and Combs, 1980). Spontaneous activity, body weight gain and plasma glucose concentration increased 6-8 hrs after Se administration. The earliest response in the specific activity of Se-dependent GPx occurred in plasma at 8 hrs and in liver at 24 hrs after Se administration. The onset of pancreatic atrophy, however, was not affected by the level of feed intake suggesting that the effect of Se upon appetite may be distinct from the involvement of Se in nutritional pancreatic atrophy and fibrosis. Inorganic Se compounds were shown to be anti-feed intake factors whereas the organic Se compounds tested had little such activity.

An interesting study was conducted with an artificial diet to examine the feeding responses of a generalist herbivore, *Spodoptera exigua* (Hubner) (Lepidoptera: Noctuidae), to various forms and concentrations of Se (Vickerman and Trumble, 1999). Tests initiated with neonates showed larvae significantly preferred the control diet over one containing sodium selenate, sodium selenite, or selenocystine, but at most concentrations showed no preference between the SeMet-containing and control diets. Choice tests initiated with third instars demonstrated a preference for the control diet over sodium selenate and sodium selenite treatments. In contrast, no significant responses were found in tests initiated with third instars offered a choice between selenocystine or SeMet and untreated controls. Similarly, when Se-enriched yeast was used as a source of organic Se, young Se-deficient laying hens reduced their Se deficit by preferentially selecting the high-Se diet (Zuberbuehler *et al.*, 2002).

3.6 Selenium sources for pigs

As mentioned above, Se content of feed ingredients greatly varies (Table 3.3) depending on many different factors. For example, Se concentration in corn and rice grown in normal and high Se areas can vary 100-500-fold. Indeed, average data on Se content in feedstuffs presented in various tables are not suitable for diet balancing and Se supplementation is a routing practice in commercial animal production. In fact, FDA approved Se supplements for poultry and pigs in 1974 in the form of selenite or selenate.

Table 3.3. Selenium concentrations in various feed ingredients, mg/kg (adapted from NRC, 1983).

Ingredient	USA	Canada
Alfalfa meal	0.01-2.00	0.02-0.27
Barley	0.05-0.5	0.02-0.99
Brewer's grains	0.15-1.00	0.29-1.10
Corn	0.01-1.00	0.01-0.33
Fish meal	1.0-5.0	1.3-3.4
Linseed meal	0.5-1.2	0.7-1.5
Meat meal	0.08-0.5	0.2-0.81
Oats	0.01-1.00	0.01-1.10
Poultry by-product	0.5-.10	-
Soybean meal	0.06-1.00	0.04-0.78
Wheat	0.1-3.00	0.02-1.5
Wheat bran	0.1-3.0	0.24-1.3
Wheat middling	0.15-1.0	0.41-0.89
Whole soybeans	0.07-0.90	-

While Se form was not rigorously considered in the initial research on Se nutrition, for the last 40 years information has accumulated indicating that the natural form of Se in plant-based feed ingredients consists of various selenoamino acids with SeMet being major form of Se in grains, oil seeds and other important feed ingredients. Therefore, organic Se is the natural form of Se to include in feed formulations. However, sodium selenite remains in use in many animal feeds. The limitations of using inorganic Se are well known and include toxicity, interactions with other minerals and vitamins (Figure 3.7), low efficiency of transfer to milk, meat and eggs and an inability to build and maintain Se reserves in the body. As a result, a high proportion of the element consumed in inorganic form is simply excreted. Further, a pro-oxidant effect of the selenite ion (Spallholz, 1997) is a great disadvantage as well, particularly when shelf life of food animal products is considered.

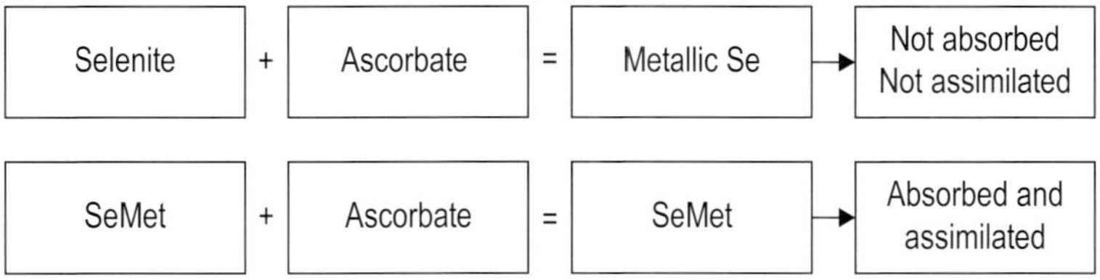

Figure 3.7. Selenium-ascorbate interactions in premixes and digestive tract (adapted from Surai, 2018).

Ingredient interactions should be carefully considered. It has been shown that selenite can be dissolved when dispersed in feeds of relatively high water activity. When dissolved, it may form selenious acid and disperse as a vapor losing its biological activity and nutritional function (Eisenberg, 2007). When the premix contains sodium selenite and ascorbic acid, the chemical reaction between them causes selenite reduction to elemental Se, which is not absorbed in the digestive tract of animals, and ascorbic acid is also oxidised thereby losing biological activity. Therefore, in such a situation, both nutrients are lost. Pink particles in the premix very often represent elemental Se produced in a way mentioned above. This could happen in the premix/feed during storage or in the digestive tract during absorption. In fact, when solutions of selenite and ascorbic acid were mixed (proportion between AA and selenite was similar to those in supplements) – after about 2 h from the preparation is present at percentages of about 50% of that expected and progressively decreases to practically disappear after 24 h (Gosetti *et al.*, 2007). Furthermore, transport of ^{75}Se was inhibited when ascorbic acid and selenite were injected directly into ligated duodenal loops of anaesthetised chickens. The mechanism of the inhibition seemed to be precipitation of Se within the intestinal lumen, since less ^{75}Se was found in the supernatant fraction of the luminal fluid when ascorbic acid was present (Mykkänen and Mutanen, 1983). Therefore, vitamin C in premixes is compatible with organic Se but incompatible with selenate or selenite. This issue is extremely important for anti-stress premixes containing increased levels of ascorbic acid. For example, an interaction of 0.5% vitamin C with either selenite or seleno-DL-Met (SeMet, 3 mg/kg) was studied in rats (Ip, 1986). Results showed that the protective effect of selenite in tumorigenesis was nullified by vitamin C, whereas the chemopreventive action of SeMet was not affected. The authors suggested that selenite is reduced by vitamin C to elemental Se and was not available for uptake by tissues. Similarly, the availability of Se was reduced almost to zero when selenite and 1 g ascorbic acid were taken together well before the meal (Robinson *et al.*, 1985). Similarly, 0.5 or 0.25% of vitamin C in the diet completely negated in blood, liver and mammary gland the accumulation of Se induced by 3 mg/kg of selenite supplementation. Indeed, selenite is a hygroscopic compound and can be dissolved when dispersed in feeds of relatively high water activity. It is also interesting that other compounds in the premix can similarly reduce Se in selenite to produce elemental Se, which is not absorbed from the feed. For example, a reference from Michigan State University (Groce *et al.*, 1973) indicates changes in odour and/or colour when premixes containing 200 mg/kg Se were prepared with glucose monohydrate, corn starch or sucrose and then stored at room temperature. The odour was musty and sweetish in character, while the colour changes involved appearance of pink to dark red particles in the original white matrix. Indeed, Se retention was decreased and Se excretion was increased as a result of old premix inclusion in pig diets. In contrast, ascorbic acid enhances SeMet assimilation from the diet. Furthermore, in various *in vitro* model systems SeMet itself is considered to possess antioxidant properties (Schrauzer, 2000).

Thus, the use of sodium selenite in animal diets has recently been questioned (Fisinin *et al.*, 2008; Mahan and Peters, 2004; Ortman and Pehrson, 1997, 1998; Surai, 2002, 2006, 2018; Surai and Fisinin, 2014, 2015, 2016a,b). Prooxidant properties of selenite and its interactions with other nutrients, including vitamin C, put pressure on feed manufacturers to find new, more effective sources of supplemental Se. Therefore, the simplest idea was to use Se forms produced by plants.

3.7 Selenium-enriched yeast: pluses and minuses

It is well known that chemical and physical properties of Se and sulphur are very similar, reflecting similar outer-valence-shell electronic configurations and atomic sizes (Combs and Combs, 1984). Therefore, plants cannot distinguish between these two elements when synthesising amino acids. As a result, they can synthesise SeMet when Se is available. This biological feature was the basis for development of the commercial technology of organic Se production from yeast. Indeed, various commercial forms of Se-Yeast found their way to the market place and showed to be effective sources of Se for animal production (for review see Fisinin *et al.*, 2008; Surai, 2006; Surai and Fisinin, 2014, 2015, 2016a,b; Surai *et al.*, 2010).

The legal definition of Se-yeast is as follows:

> Selenium yeast is a dried, non-viable yeast (*Saccharomyces cerevisiae*) cultivated in a fed-batch fermentation which provides incremental amounts of cane molasses and selenium salts in a manner which minimises the detrimental effects of selenium salts on the growth rate of the yeast and allows for optimal incorporation of inorganic selenium into cellular organic material. Residual inorganic selenium is eliminated in a rigorous washing process and must not exceed 2% of the total selenium content in the final selenium yeast product (Federal Register, 2007).

Therefore, only total Se level and inorganic Se proportion in Se-Yeast is officially regulated.

However, there are several points to be addressed in relation to commercial usage of Se-Yeast. First of all, it is necessary to mention that yeast is a live organism and its composition will depend on the genetics and conditions of growing, including temperature, pH, oxygen concentration, etc. It seems likely that selenoamino acid composition of the yeast depends on various factors, including yeast species, growth conditions as well as analytical techniques used (Figure 3.8). For example, when three different commercial yeast products were analysed, results showed that the proportion of water-soluble Se varied from 11.5 up to 28.0% and water insoluble polysaccharide bound Se proportion varied from15.5 up to 72% (Encinar *et al.*, 2003). In fact, Se-yeast has been reported to contain over 60 (Arnaudguilhem *et al.*, 2012) or 109 (Gilbert-López *et al.*, 2017) unique selenium species and this number recently have increased to 188 organo-selenium metabolites detected in the Se-yeast by new techniques (Ward *et al.*, 2019) with SeMet being the major selenocompound in Se-enriched yeast. However, its proportion greatly varies. Rayman (2004) presented literature review analysing Se-speciation in different Se-yeast supplements and the percentages of SeMet in 7 supplements were as follows: 84, 69, 75, 81, 83, 61 and 60. In particular, the author presented data on SeMet percentage of individual Se-Yeast products. The values were 58-65% of total Se in LALMINe (Lallemand, Montreal, Canada); 60-70% of total Se in SelenoExcelle (Cypress Systems, Fresno, CA, USA); 54-60% of total Se in Seleno Precisee (Pharma Nord, Vejle, Denmark) and 62-74% in Sel-Plex (Alltech, Kentucky). Indeed, the presented data depend not only on the technology of Se-Yeast production (strain of yeast, source of Se, temperature, oxygen concentration, etc.), but also are dependent on the extraction efficiency of the technique used by the analytical laboratory. The author concluded that commercial Se-Yeast products found on the market contained almost all their Se in organic form and about 55-75% of it (allowing for extraction efficiency) present as SeMet (Rayman, 2004). In general, Bierla *et al.* (2012) mentioned that the typical criteria for Se-yeast of industrial use is >60% SeMet and <2% of selenite and selenate which could indicate a lower quality Se-yeast (Bierla *et al.*, 2012) and a great variability in SeMet content of Se-Yeast products has been reported.

It should be mentioned that SeMet extraction and analysis in the yeast are a quite difficult procedures required specific expertise and could be an additional variable in terms of SeYeast composition. For example, SeMet comprised 79% of the extracted selenium and 64% of total selenium in Se-Yeast; the selenium levels were at the 2-3 mg/g (McSheehy *et al.*, 2005). SeMet in yeast and nuts comprised respectively 65 and 75% of total Se (Wrobel *et al.*, 2003). Similarly, a proteolytic enzyme extract of Se-Yeast was found to contain Se as SeMet (74.8%), selenocystine (9.9%), selenite (5.1%) and as at least three unknown Se compounds (10.2%, Yoshida *et al.*, 2002). SeMet comprised about 85% of total Se compounds found in Se-Yeast used for

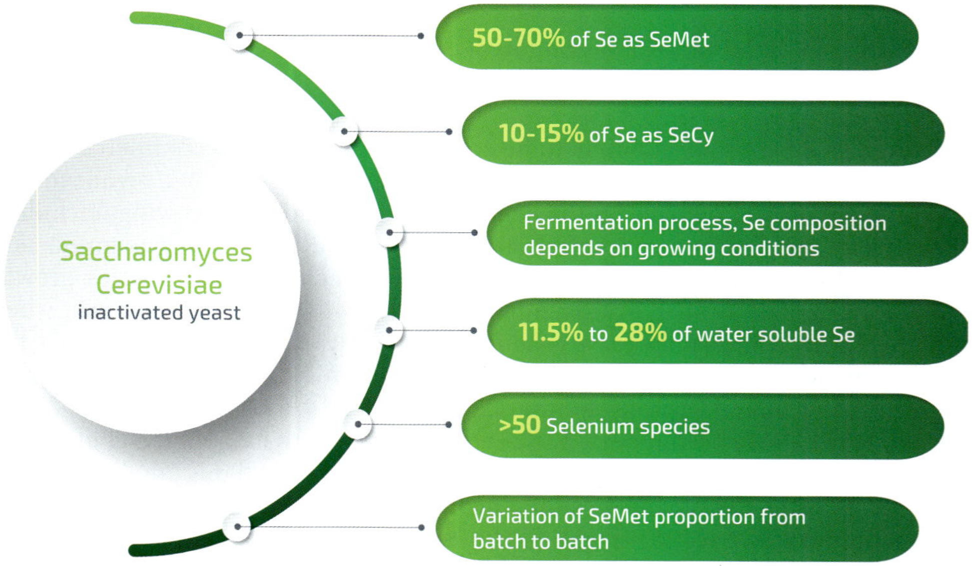

Figure 3.8. Selenium-enriched yeast.

human trials (Ip *et al.*, 2000). Similarly, Se-Yeast, which were used as a source of Se in the PRECISE and other trials, contained SeMet at 54-60% of the total selenium (Larsen *et al.*, 2004). A commercial source of Se-enriched yeast tablets containing 210 g Se/g was found to contain 73% of the total Se as SeMet (Wolf *et al.*, 2001). Samples of five leading manufactures worldwide [Alltech (KY), Angel (China), Biorigin (Brazil), Lallemand (Canada), Lesaffre (France)] were investigated (Casal *et al.*, 2010). The proportion of water-extractable selenium varied from 16 to 35% and proportion of SeMet varied from 44±7 to 93±12%. The authors also showed between batch variations. For example, one sample was analysed as two different production batches and showed 27 and 35% of water-extractable selenium and SeMet comprised 93±12 and 86±25% of total Se respectively. Recently, a considerable incorporation of selenocysteine (SeCys) in proteins of the yeast proteome despite the absence of the UGA codon was demonstrated (Bierla *et al.*, 2013). The authors concluded that 10-15% of selenium present in Se-enriched yeast is in the form of selenocysteine. This means, that if all Se in Se-Yeast is accounted for, the maximum SeMet proportion would not exceed 85%, but in many cases will be lower than that. By using X-ray absorption near-edge structure (XANES) spectroscopy, it was reported that a dietary Se-Yeast supplement contained SeMet (75±10%), SeCys (18±10%; Prange *et al.*, 2019).

It has been shown that the main advantage of organic Se in animal production is related to its ability to build Se reserves in the body in the form of SeMet which can be used in times of stress to help additional selenoprotein production and increased resistance to stress (Fisinin *et al.*, 2008; Surai, 2006; Surai and Fisinin, 2014, 2015, 2016a,b; Surai *et al.*, 2010). Therefore, ideally, it would be necessary to specify the percentage of SeMet (active compound of the Se-Yeast) in the yeast product and price would depend on the SeMet level in the product. This could be achievable if Se assimilation from all yeast products would be the same, but this is not the case judging on Se location in different Se-Yeast compounds. For example, the non-soluble protein fraction accounted for up to 40% of the total selenium in the yeast (Encinar *et al.*, 2003). It should be noted that absorption of dietary Se (organic Se) is generally believed to be good (about 80% absorption) (Reilly, 2006). Absorption and retention of Se from Se-Yeast, measured in twelve volunteers fed ^{77}Se-labelled SelenoPrecise yeast (Pharmanord, Velje, Denmark), was between 75 and 90% (Sloth *et al.*, 2003). Other Se yeasts gave different results (between 50 and 60%). The same Se-Yeast was analysed by Zheng *et al.* (2003) using enzymatic digestion with the recovery of Se of up to 93%, with SeMet to be the predominant species (78% of the total Se), while selenite (2%), selenocystine (1.9%), the major unknown species (10%) and the other unknowns (7.8%) accounted for the remaining selenium in the enzymatic extracts. Recently two commercially-available SeYeast preparations, containing SeMet at 63 and 56.7% have been tested in broilers (Simon *et al.*, 2013). It has been shown that differences in nutritional efficacy of the preparations were proportional to SeMet content.

Therefore, it is important, to know under which form Se is consumed with regard to the bioavailability of the different forms of Se. Furthermore, at present the task of guarantee the SeMet level in the product is almost impossible to achieve. This relates to variability of SeMet proportions in the yeast products due to different conditions of their production as well as analytical difficulties existing until now in Se speciation in the Se-Yeast. In fact, a part of SeMet in the Se-Yeast is bound in resistant membrane hydrophobic proteins and difficult to release during sample preparation. Furthermore, different selenium-enriched yeast products, produced by similar methods but using different yeast strains were shown to have different selenocompound compositions (Casal *et al.,* 2010). Furthermore, when recently the 129 compounds identified in 5 Se-yeast products were grouped, the results showed that there were more differences than similarities between them. In fact, there were only 9 selenium containing compounds in common across the 5 yeast products which included SeMet (Ward *et al.,* 2019). The authors suggested that various Se-Yeast could vary in efficacy due to differences in Se compound composition. Indeed, Se-Yeast can be characterised by the selenium metabolic profile (selenometabolome) reflecting yeast strain and fermentation parameters being a precious fingerprint of the origin of preparations available on the market and of the reproducibility of the production process (Lobinski *et al.,* 2000). Recently, a two-dimensional size-exclusion – strong cation-exchange HPLC with the parallel ICP MS and ESI MS detection was developed as an advanced tool for selenometabolomic studies of yeast (Casal *et al.,* 2010). Furthermore, advances in electrospray mass spectrometry for the selenium speciation in Se-Yeast have been described (Bierla *et al.,* 2018).

Apart from the technology of Se-Yeast production based on Saccharomyces, yeasts belonging to the genus Candida (Torula yeast) have also been shown to have the ability to utilise sodium selenite and convert it into organic Se compounds (Wang *et al.,* 2012). The authors showed that 81% of sodium selenite was assimilated and transformed into organic selenium by *Candida utilis* under optimal conditions and Se concentration in the yeast was >900 mg/kg (Yang *et al.,* 2013). However, Torula yeast (*C. utilis)* was shown to metabolise selenium in a totally different way in comparison to Brewer's yeast (*S. cerevisiae*) leading to the biosynthesis of selenohomolanthionine (SeHLan), a major selenium compound accounting for 60-80% of the total selenium (Bierla *et al.,* 2017). In general, based on recovered urinary selenometabolites and selenoproteins from urine and serum of Se-deficient rats after SeHLan supplementation, it was concluded that this form of Se is bioavailable for animals (Takahashi *et al.,* 2018; Tsuji *et al.,* 2010). However, molecular mechanisms of SeHLan metabolism and its action to affect Se protein synthesis are not clear at present and more research is required in this area. Furthermore, SeHLan is not converted to SeMet, the only form of Se able to build Se reserves in muscles, and therefore all the aforementioned advantages of SeMet described above are not valid for SeHLan.

3.8 SeMet and OH-SeMet

Another option to improve Se status of farm animals, including pigs, would be to use pure SeMet as a dietary supplement (Schrauzer, 2000, 2001, 2003; Schrauzer and Surai, 2009) (Figure 3.9). There are some respectable publications showing beneficial effect of organic Se in the form of SeMet in the pig diets (Falk *et al.,* 2019; 2020;

Hu *et al.*, 2011; Silva *et al.*, 2019; Wang and Xu, 2008; Wang *et al.*, 2011; Yuan *et al.*, 2011, 2013; Zhang *et al.*, 2020). It has been determined whether SeMet or Se-Yeast acts with different potency on six bio-chemical markers including intraprostatic dihydrotestosterone (DHT), testosterone (T), DHT:T, and epithelial cell DNA damage, proliferation, and apoptosis (Waters *et al.*, 2012). By analysing dogs supplemented with SeMet or Se-Yeast that achieved equivalent intraprostatic selenium concentration after supplementation, there was no significant differences in potency of either selenium form on any of the six parameters over three different ranges of target tissue selenium concentration. However, SeMet in purified form is unstable and easily oxidised. For example, it has been shown that in the freeze-dried samples of oyster total Se and the Se species evaluated are stable for at least 12 months, under all the conditions tested. However, after purification of Se species, including SeMet, in the enzymatic extracts they are only stable for 10 days if stored at 4 °C in Pyrex containers (Moreno *et al.*, 2002). After storage of SeMet water solution for 30 days at 20 °C, less than 80% SeMet was recovered (Lindemann *et al.*, 2000). Potentially bioavailable selenium-containing compounds in the Se-Yeast were investigated using candidate reference material SEAS (Reyes *et al.*, 2006). SeMet was the major compound identified in the gastrointestinal extract while SeMet selenoxide was its main degradation product formed after medium and long-term sample storage, respectively. Indeed, pure SeMet is chemically oxidised to SeMetO under the oxic conditions in the small intestine (Lavu *et al.*, 2016). SeMet was shown to react with myeloperoxidase-produced strong oxidants to form SeMetO (Carroll *et al.*, 2017).

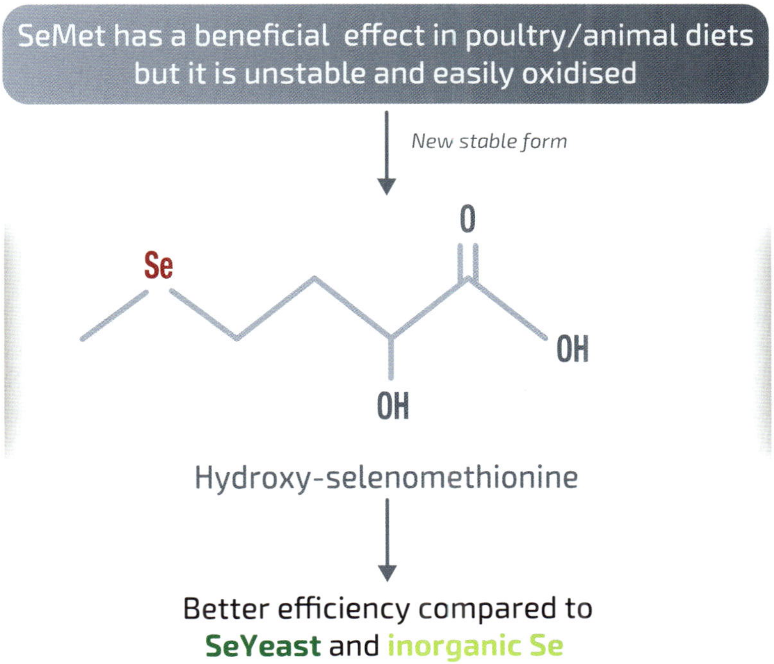

Figure 3.9. SeMet and OH-SeMet.

The oxidability of SeMet during storage could explain different results in terms of gene expression between SeMet-supplemented and Se-Yeast-supplemented group of mice (Barger *et al.*, 2012). An alternate form of Se dietary supplementation could be algae biomass specially enriched with selenium. Indeed, when the algae are exposed to Se in the form of selenite, they are able to synthesise organic selenium species. Technology of heterotrophic fed-batch cultivation of the microalga *Chlorella* enriched by organically bound Se was developed. In fact, spray-dried Se-*Chlorella* biomass is shown to have similar potency as a dietary Se source for poultry and farm animals as Se-Yeast (Doucha *et al.*, 2009).

Recently a new stable organic Se source called Selisseo® (SO) has been developed which is a selenomethionine hydroxyanalogue, 2-hydroxy-4-methylselenobutanoic acid or HMSeBA (Briens *et al.*, 2013, 2014). A range of experiments were conducted with chickens, turkey, piglets, sows and ruminants showing beneficial effect of SO in comparison to sodium selenite. For example, two experiments were conducted on broiler chickens to compare the effect of HMSeBA (SO), with two practical Se additives, SS and Se-Yeast. The different Se sources and levels improved muscle Se concentration compared with the NC, with a significant source effect in the following order: SS, Se-Yeast and SO ($P<0.05$). In fact, the relative muscle Se enrichment comparison, using linear regression slope ratio, indicated an average of 1.48 (95% CI 1.38, 1.58) fold higher selenium deposition in muscle for SO compared to SY (Briens *et al.*, 2014).

Seleno-amino acid speciation results for Se-Yeast and SO at 0.3 mg Se/kg feed indicated that muscle Se was only present as SeMet or SeCys, showing a full conversion of Se by the bird. The results confirmed the higher bioavailability of organic Se sources compared with the mineral source and demonstrated a significantly better efficiency of HMSeBA compared with Se-Yeast for muscle Se enrichment. In particular, the authors showed that Se muscle concentrations significantly improve with SO, increasing the relative bioavailability for total Se by 39% compared with SY. From one hand this could be a reflection of higher SeMet level in the diet (almost 100% SeMet in SO vs 60-70% SeMet in Se-Yeast). On the other hand, there could be other biochemical differences in the Se metabolism, since SO increased SeCys level in the muscle. It would be interesting to note that hens fed the diet with HMSeBA-0.2 accumulated more Se in their eggs (+28.8%) and muscles (+28%) than those fed the diet supplemented with SY-0.2 (Jlali *et al.*, 2013). After 21 days, organic Se sources maintained (SeYeast) or increased (Hydroxy-SeMet) breast muscle Se concentration compared to hatch value whereas inorganic source (Sodium selenite) or non-supplemented group (NC) showed a significant decrease in tissue Se concentration (Couloigner *et al.*, 2015). Furthermore, HMSeBA in turkey diet improved GPx activity in thigh muscles and decreased lipid peroxidation (Briens *et al.*, 2016). These results showed the greater ability of HMSeBA to increase Se deposition in eggs and breast muscle of laying hens, which could be of great importance for breeding birds and newly developing chicks.

Recently, EU decided to limit the maximum supplementation with selenised yeast to 0.2 mg Se/kg complete feed for reasons of consumer safety (European Commission Implementing Regulation No. 427/2013 of 8 May 2013; EC, 2013). At this comparatively low level of supplementation advantages of organic Se in the form of Se-Yeast will be less pronounced, and alternative effective sources of organic selenium with higher efficiency of transfer to the egg and animal tissues would play bigger role in poultry reproduction. Therefore, aforementioned results indicated that a new source of organic selenium in the form of 2-hydroxy-4-methylselenobutanoic acid supplied in the same dose as Se-Yeast in the chicken diet could provide additional benefit in terms of Se reserves in the muscles as well as Se transfer to the egg and probably to the developing embryo. This potentially can be translated into better antioxidant protection in stress conditions of commercial pig production.

Similar beneficial effect of OH-SeMet was confirmed in a study with pigs. In fact, pigs were fed basal diets unsupplemented or supplemented either with sodium selenite (SS), Se-Yeast (SY), or OH-SeMet each at 0.1 or 0.3 mg Se/kg of diet for 32 d (Jlali *et al.,* 2014). There were no differences in final BW, ADG, ADFI, and G:F among dietary treatments. As expected, all Se-supplemented groups demonstrated greater total Se contents in plasma and liver compared with unsupplemented control group. However, Se concentration in psoas major muscle reflecting Se reserves in the body, was improved only when organic Se (SY or OH-SeMet) was added to diets. Furthermore, independently on the Se level in the diet, the Se concentration in muscle was greater in pigs supplemented with OH-SeMet than those supplemented with SY. It was calculated that the relative bioavailability of Se from OH-SeMet for plasma, liver, and muscle Se response was 170, 141, and 162%, respectively, for SY (Jlali *et al.,* 2014). Recently, it was shown that maternal supplementation of OH-SeMet during pregnancy improved antioxidant capacities and reduced inflammation level in placenta, and fetus. Furthermore, there was significantly decreased percentage of stillborn piglets with a trend in increasing the number of piglets born alive (Mou *et al.,* 2020). For more detailed information see Chapter 5 (Surai, 2021).

It seems likely that stable form of SeMet (OH-SeMet) could have beneficial effects on other farm animals. For example, pregnant heifers supplemented with OH-SeMet were characterised by increased SeMet concentration in their plasma (Juniper *et al.,* 2019). Furthermore, OH-SeMet-supplemented cows had improved serum GPx activity, total antioxidant capacity, and SOD activity when compared to SS-supplemented cows (Sun *et al.,* 2017), and cows supplemented with OH-SeMet during a period of heat stress had increased Se concentrations in serum and milk, increased total antioxidant capacity and decreased serum MDA, hydrogen peroxide, and nitric oxide concentrations when compared with SS-fed controls (Sun *et al.,* 2019). It could well be, that organic Se can beneficially affect pig microbiota. For example, organic Se has been reported to alter rumen fermentation characteristics, feed digestion, milk yields, and milk Se (Wang *et al.,* 2009), and the authors suggested that organic Se stimulated digestive microorganisms in a dose-dependent manner. More recently, a comparison between SS and OH-SeMet demonstrated that OH-SeMet altered rumen fermentation characteristics, improved apparent nutrient digestibility with respect to crude protein, neutral detergent and acid detergent fibre, and improved selenium absorption (Wei *et al.*, 2019). Major differences between organic selenium and sodium selenite are shown in Table 3.4.

Table 3.4. Major differences between organic selenium (Se) and selenite.[1]

	Organic selenium	Selenite
Absorption	Similar to methionine with active transport in the gut	Similar to other mineral with passive transport in the gut
Accumulation	Building Se reserves by non-specific incorporation of SeMet into the proteins	Not accumulated in the body
Toxicity	At least 3 times less toxic than selenite	Highly toxic, can penetrate via skin causing problems
Bioavailability	Higher bioavailability in comparison to selenite to animals	Lower bioavailability in comparison to SeMet
Indirect AO activity via selenoproteins	A source of Se for selenoprotein synthesis	A source of Se for selenoprotein synthesis
Direct antioxidant activity independent on selenoproteins	Possesses antioxidant properties per se and could scavenge NO and other radicals	Possesses pro-oxidant properties and could stimulate free radical production when reacting with GSH
Effect on DNA	Stimulate DNA-repair enzymes	Causes DNA damage
Transfer to eggs and muscles	Transferred to egg and muscles giving an opportunity to produce Se-eggs and Se-meat	Poorly transferred to eggs and muscles
Reactions with other elements	Neutral, ascorbic acid promotes SeMet assimilation from the diet	Highly reactive, reduced to metallic, unavailable selenium by ascorbic acid
Protective effect in stress conditions	Provide additional protection due to Se reserves in the body	Cannot provide additional protection due to absence Se reserves in the body
Effect on drip loss	Decrease drip loss	Does not affect drip loss
Environmental issues	Better retention in tissues, less released with faeces and urine	Low retention in tissues and high release with faeces and urine
Stability	Stable	Stable
Classification based on the mode of action	Feed additive	Drug

[1] AO = antioxidant; GSH = glutathione; NO = nitric oxide; SeMet = selenomethionine.

3.9 Chelated selenium products

There is a range of products on the market claiming to contain chelated Se (Se-glycinates, Se-proteinates, Se-amino acids complexes, etc.), however, chemical position of Se in the Periodic table of elements indicates that Se is not a true metal and therefore its chelating ability is in question. Indeed, attempts to determine chelated Se in such products ended up with a detection of only inorganic Se (selenite and selenate, Amoako *et al.*, 2009; Kubachka *et al.*, 2017). This could explain results of Givens *et al.* (2004) showing that a chelated selenium-amino acid complex in the cow diet was not different from sodium selenite when efficacy of Se transfer to the milk was assessed. Indeed, chelated Se products are not related to SeMet and, probably, should not be included into organic Se category. Recently, it has been also shown that neither sodium selenite nor soy-chelated Se showed consistent antioxidant effects that were comparable to SY in protection against diquat-induced oxidative stress in weaned pigs (Doan *et al.*, 2020). Clearly, there is a need to clarify what kind of Se compounds are formed as a result of reaction (if any) between sodium selenite and soya protein/ hydrolysate including chemical characteristics, stability, mechanisms of absorption, metabolism, etc. Furthermore, this product should not be called 'organic Se', since this term in animal industry is related to SeMet and SeMet-related products (Surai, 2006, 2018; Surai *et al.*, 2018).

3.10 Nano-selenium products

Recently selenium nanoparticles (SeNPs, nano-Se) have received substantial attention as possible novel nutritional supplements because of their lower toxicity and ability to gradually release selenium after ingestion (Skalickova *et al.*, 2017). It has been suggested that nano-Se can serve as an antioxidant with reduced risk of selenium toxicity and as a potential chemopreventive agent if the induction of GST by selenium is a crucial mechanism for its chemopreventive effect (Wang *et al.*, 2007). The authors also showed that nano-Se has comparable to SeMet ability to increase selenoenzymes activity. However, the question still is how elemental Se can be involved in SeCys synthesis and selenoprotein expression. There are several reports of successful testing of nano-Se in broiler nutrition (Cai *et al.*, 2012; Gulyas *et al.*, 2017; Wang, 2009; Zhou and Wang, 2011).

In many cases low nano-Se toxicity is considered as its main advantage. However, one should also realise that Se toxicity is not a major problem in pig industry and Se in the form of sodium selenite or organic Se (SeMet, Se-Yeast or other preparations) is an essential part of premixes produced worldwide. It seems likely that nano-Se could be a new chemopreventive agent for treatment of various diseases (Zhang *et al.*, 2008), including cancer, but its nutritional value as a feed supplement for pig industry is questionable. Indeed, increased availability in comparison to sodium selenite could be a great advantage in some medical applications of Se where direct (non-selenoprotein) action is key for its efficacy (Menon *et al.*, 2018). However, the same features of nano-Se metabolism and assimilation could be disadvantageous in animal nutrition where the main mechanism of biological activity of Se is mediated via selenoprotein synthesis (Surai, 2018). First of all, uncontrolled and enhanced nano-Se penetration via cellular barriers could have negative consequences for animal health. Indeed, when considering nano-Se absorption and its effects on gut health, it is necessary to take into account the so-called Trojan horse effect, when nanoparticles may have permeation-enhancing properties for other substances in the gut (Fröhlich and Roblegg, 2012). This could be damaging for the gut and organism in general, since there is a range of 'unwanted' compounds/chemicals in the feed and the gut. Therefore, nano-Se behaviour in the gut warrants further investigation. Secondly, it is not known in which form Se is accumulated in tissues due to nano-Se supplementation and this raises a question about the long-term consequences of such supplementation. Thirdly, it seems unlikely that nano-Se can build Se reserves in the body in the form of SeMet and therefore the whole concept of organic Se advantages in farm animal/pigs (Surai *et al.*, 2018) cannot be applied to nano-Se. Fourthly, as it was mentioned above Se toxicity, in general, is not a problem in modern animal nutrition. Fifthly, nano size of Se particles in the supplement could be a great health hazard for feed mill personnel (Surai, 2018). In addition, based on mode of action nano-Se should be considered as a drug, not a feed/food additive. Finally, Se supplementation in any form to the balanced diet of animals housed in optimal conditions usually did not affect productive and reproductive performance (Surai, 2006).

Furthermore, the promising results and the positive responses of nano-Se in animal nutrition should not overshadow the possible detrimental consequences of its usage. In particular, nanoparticle behaviour in various conditions could differ substantially, and before we understand how to control that behaviour, nano-Se usage on a wide industrial scale is in question. In fact, supplementation of nano-Se was shown to have variable responses compared with the other Se sources on production performance of chickens and layers (Patra and Lalhriatpuii, 2020). Suggested pathways of nano-Se participation in animal health maintenance are shown in Figure 3.10.

For the last 5 years nano-Se has received tremendous attention in terms of its production, characteristic and possible application for medical sciences. Furthermore, nano-Se is shown to be a potential source of Se for pig nutrition. However, there is an urgent need to address the questions related to nano-Se absorption, assimilation and metabolism in animals before it finds its way to animal production as a feed supplement. This subject warrants further investigation.

3.11 Conclusions

Selenium is an essential element in animal nutrition and its dietary supplementation in optimal form and amount is key for maintaining their health, productive and reproductive performance. It has been proven that organic form of Se (mainly SeMet) in the animal/pig diet has a range of advantages in comparison to traditional sodium selenite. In fact organic Se concept considers SeMet as a storage form of Se in pig body. Since pigs are not able to synthesise SeMet its provision with a diet is a

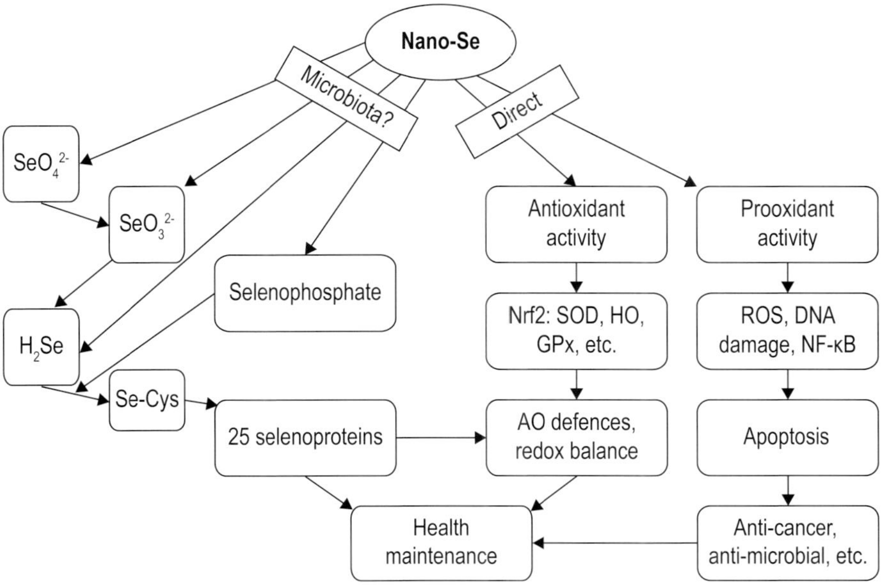

Figure 3.10. Possible pathways of selenium nanoparticles (Nano-Se) participation in in health maintenance (adapted from Surai and Kochish, in press; Surai *et al.*, 2017).

key element of a strategy to fight commercially-relevant stresses. Indeed, in stress conditions, when increased selenoprotein expression requires additional Se, but its provision with feed usually decreased due to reduction in feed consumption, Se reserves in the body (mainly in muscles) could help to maintain an effective antioxidant defence. Pig industry is looking for most effective sources of organic Se to be commercially used. Se-yeast received substantial attention and commercial applications as a reliable source of organic Se. However, active Se component of the yeast, SeMet, comprising only 50-70% of the total Se and there is no proof that the rest (30-50%) of Se in the form of SeCys, MeSeCys, etc. has any additional benefits in comparison to sodium selenite. From the production point of view it is difficult to guarantee certain percentage of SeMet in the Se-Yeast, since there is a range of factors (yeast strain, medium composition and selenite concentration in it, temperature, etc.) affecting the aforementioned parameter. Analytical difficulties in precise determination of SeMet amount further complicate the issue. Stability issues with pure SeMet in the pig diets restrict its wide usage. Major sources of organic Se in the pig market are shown in Table 3.5.

It seems likely that a recently developed product (hydroxy-SeMet) combines advantages of both – Se-Yeast (SeMet stability) and pure SeMet (high proportion of SeMet) and could be considered as a next step in Se application in pig nutrition. Indeed, the first generation of Se supplements for pigs includes selenite and selenate. The second generation of organic Se for pigs includes Se-Yeast, Se-Met and Zn-Se-Met products. Their advantages and disadvantages are mentioned above. Indeed,

Table 3.5. Sources of organic selenium (Se) on the market (Surai *et al.*, 2018).

Source	Comments	SeMet[1]
Se-Yeast	Well researched, SeMet 50-70%	Yes
Selenomethionine (SeMet)	Not stable, >95% SeMet	Yes
Zn-SeMet	Not stable, >95% SeMet	Yes
OH-SeMet	Stable, >95% OH-SeMet	Yes
Se-proteinates	Chemistry is not proven	No
Se-glycinate	Chemistry is not proven	No
Se-chelates	Chemistry is not proven	No
Selenohomolathionine	Biochemistry of metabolism is not clear	No
Other (nano-Se, etc.)	In development	No

[1] Since only SeMet can be non-specifically incorporated into proteins in measurable amounts, it is an active component of organic selenium sources.

OH-SeMet can be considered as the next, third generation of Se supplement for pigs combining advantages of previous Se supplements on the market. It is well appreciated that via selenoprotein expression Se is involved in regulation of cell growth, apoptosis and modifying the action of cell signalling systems and transcription factors and therefore its adequate dietary supply is a crucial factor for many physiological processes in animal body. Major advantages of organic Se in pig diets are related to building Se reserves in the body which can be used in stress conditions as additional sources of Se to upregulate selenoproteins and improve antioxidant defences (Figure 3.11).

There is a range of Se-containing compounds in the feed, including SeMet, SeCys, Se-GSH, Se-peptides, etc. and various Se supplements, including selenite, selenate, SeMet, Zn-SeMet, OH-SeMet, Se-yeast, etc. can be included into premix. All those Se forms come to the intestine where initial hydrolysis (Se-Met will be released from SeYeast or ZnSeMet; OH-SeMet will be converted into SeMet) and some metabolic changes will take place. This includes excretion of Se metabolites via bile, faeces and urine. Further, selenite, selenate, SeMet and some other Se forms will be delivered to the liver for metabolic changes and distribution. In parallel, some SeMet will go to free amino acid pool and build Se reserves mainly in muscles. Next step of Se assimilation and metabolism includes conversion of all major forms of Se into H_2Se from which SeCys will be synthesised and incorporated into 25 newly synthesised selenoproteins which are integral part of the antioxidant system of the body.

In stress conditions, protein catabolism will take place which will release some SeMet incorporated into those proteins and that SeMet will be converted into H_2Se and further will be converted into newly synthesised SeCys and 25 selenoproteins. In fact, additional source of Se will be responsible for upregulation of Selenoprotein genes and additional synthesis of selenoproteins which will upregulate antioxidant

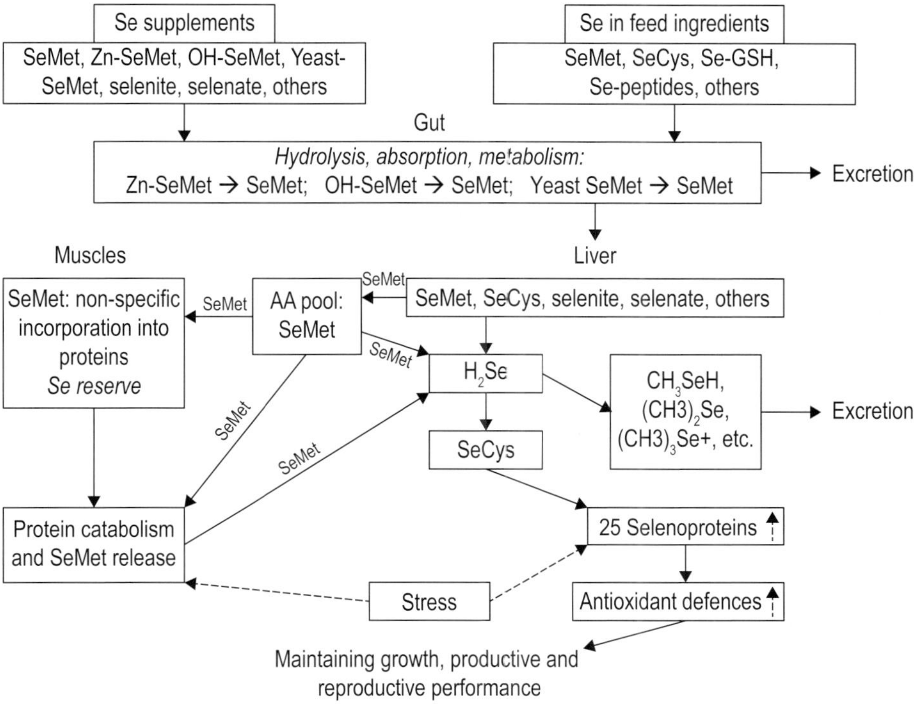

Figure 3.11. Organic selenium in action (adapted from Surai, 2018). Arrows mean conversion of one compound into another; broken arrows mean effect (of stress) on those conversions.

defences and will be able to help the body to adapt to stress and overcome stress with minimal negative consequences. In the case of usage selenite in the diet, Se reserves in the muscles will not be built and therefore the ability of the body to adapt to stress will be restricted.

Acknowledgements

Peter F. Surai and Ivan I. Kochish are supported by a grant of the Government of Russian Federation (Contract No. 14.W03.31.0013).

References

Alfthan, G., Xu, G.L., Tan, W.H., Aro, A., Wu, J., Yang, Y.X., Liang, W.S., Xue, W.L. and Kong, L.H., 2000. Selenium supplementation of children in a selenium-deficient area in China: blood selenium levels and glutathione peroxidase activities. Biological Trace Element Research 73: 113-125.

Amoako, P.O., Uden, P.C. and Tyson, J.F., 2009. Speciation of selenium dietary supplements; formation of S-(methylseleno)cysteine and other selenium compounds. Analytica Chimica Acta 652: 315-323.

Andersen, O., Nielsen, J.B., Sorensen, J.A. and Scherrebeck, L., 1994. Experimental localization of intestinal uptake sites for metals (Cd, Hg, Zn, Se) *in vivo* in mice. Environmental Health Perspectives 102, Suppl. 3: 199-206.

Apsite, M., Pitrans, B. and Atlavin, A., 1994. Absorption of ^{75}Se-selenate and ^{75}Se-selenite in chicks. In: Anke, M. and Meissner, D. (eds) Mengen- und spurenelemente. Arbeitstagung, Jena, Germany, 188 pp.

Apsite, M., Pitrans, B.V. and Atlavin, A.B., 1993. The role of duodenum in selenium assimilation in chick organism. In: Trace elements in man and animals. Proceedings of the 8th International Symposium on Trace Elements in Man and Animals, pp. 392-393.

Arnaudguilhem, C., Bierla, K., Ouerdane, L., Preud'homme, H., Yiannikouris, A. and Lobinski, R., 2012. Selenium metabolomics in yeast using complementary reversed-phase/hydrophilic ion interaction (HILIC) liquid chromatography-electrospray hybrid quadrupole trap/Orbitrap mass spectrometry. Analytica Chimica Acta 757: 26-38.

Barger, J.L., Kayo, T., Pugh, T.D., Vann, J.A., Power, R., Dawson, K., Weindruch, R. and Prolla, T.A., 2012. Gene expression profiling reveals differential effects of sodium selenite, selenomethionine, and yeast derived selenium in the mouse. Genes and Nutrition 7: 155-165.

Beilstein, M.A. and Whanger, P.D., 1986. Deposition of dietary organic and inorganic selenium in rat erythrocyte proteins. Journal of Nutrition 116: 1701-1710.

Beilstein, M.A. and Whanger, P.D., 1988. Glutathione peroxidase activity and chemical forms of selenium in tissues of rats given selenite or selenomethionine. Journal of Inorganic Biochemistry 33: 31-46.

Beilstein, M.A., Whanger, P.D. and Yang, G.Q., 1991. Chemical forms of selenium in corn and rice grown in a high selenium area of China. Biomedical and Environmental Sciences 4: 392-398.

Bierla, K., Bianga, J., Ouerdane, L., Szpunar, J., Yiannikouris, A. and Lobinski, R., 2013. A comparative study of the Se/S substitution in methionine and cysteine in Se-enriched yeast using an inductively coupled plasma mass spectrometry (ICP MS)-assisted proteomics approach. Journal of Proteomics 87: 26-39.

Bierla, K., Dernovics, M., Vacchina, V., Szpunar, J., Bertin, G. and Lobinski, R., 2008. Determination of selenocysteine and selenomethionine in edible animal tissues by 2D size-exclusion reversedphase HPLC-ICP MS following carbamidomethylation and proteolytic extraction. Analytical and Bioanalytical Chemistry 390: 1789-1798.

Bierla, K., Godin, S., Lobinski, R. and Szpunar, J., 2018. Advances in electrospray mass spectrometry for the selenium speciation: focus on Se-rich yeast. Trends in Analytical Chemistry 104: 87-94.

Bierła, K., Suzuki, N., Ogra, Y., Szpunar, J. and Łobiński, R., 2017. Identification and determination of selenohomolanthionine – the major selenium compound in Torula yeast. Food Chemistry 237: 1196-1201.

Bierla, K., Szpunar, J., Yiannikouris, A. and Lobinski, R., 2012. Comprehensive speciation of selenium in selenium-rich yeast. TrAC Trends in Analytical Chemistry 41: 122-132.

Botero, D., Gereben, B., Goncalves, C., De Jesus, L.A., Harney, J.W. and Bianco, A.C., 2002. Ubc6p and ubc7p are required for normal and substrate-induced endoplasmic reticulum-associated degradation of the human selenoprotein type 2 iodothyronine monodeiodinase. Molecular Endocrinology 16: 1999-2007.

Brandt-Kjelsen, A., Govasmark, E., Haug, A. and Salbu, B., 2014. Turnover of Se in adequately fed chickens using Se-75 as a tracer. Journal of Animal Physiology and Animal Nutrition 98: 547-558.

Briens, M., Faure, M., Coiloigner, F., Garet, J., Maucotel, T., Tommasino, N., Gatelier, P., Durand, D., Garaert, P.A. and Mercier, Y., 2016. Hydroxy-selenomethionine contributes to maintain color stability of turkey meat. In: Proceedings of the Australian Poultry Science Symposium. Sydney, Australia, pp. 149-152.
Briens, M., Mercier, Y., Rouffineau, F. and Geraert, P.A., 2014. 2-hydroxy-4-methylselenobutanoic acid induces additional tissue selenium enrichment in broiler chicken compared to other selenium sources. Poultry Science 93: 85-93.
Briens, M., Mercier, Y., Rouffineau, F., Vacchina, V. and Geraert, P.A., 2013. Comparative study of a new organic selenium source v. seleno-yeast and mineral selenium sources on muscle selenium enrichment and selenium digestibility in broiler chickens. British Journal of Nutrition 110: 617-624.
Brody, T., 1994. Nutritional biochemistry. Academic Press Inc., New York, NY, USA.
Bunk, M.J. and Combs Jr., G.F., 1980. Effect of selenium on appetite in the selenium-deficient chick. Journal of Nutrition 110: 743-749.
Burk, R.F. and Hill, K.E., 1994. Selenoprotein P. A selenium-rich extracellular glycoprotein. Journal of Nutrition 124: 1891-1897.
Burk, R.F., Hill, K.E. and Motley, A.K., 2001. Plasma selenium in specific and non-specific forms. Biofactors 14: 107-114.
Cai, S.J., Wu, C.X., Gong, L.M., Song, T., Wu, H. and Zhang, L.Y., 2012. Effects of nano-selenium on performance, meat quality, immune function, oxidation resistance, and tissue selenium content in broilers. Poultry Science 91: 2532-2539.
Cantor, A.H., 1997. The role of selenium in poultry nutrition. In: Lyons, T.P. and. Jacques, K.A. (eds) Biotechnology in the feed industry. Proceedings of 13th Alltech's Annual Symposium. Nottingham University Press, Nottingham, UK, pp. 155-164.
Cantor, A.H. and Scott, M.L., 1974a. Biological activity of selenium compounds in chickens. In: Proceedings of Cornell Nutrition Conference for Feed Manufacturers. October 29-31, 1974. Statler-Hilton Hotel, Buffalo, NY, USA, pp. 63-68.
Cantor, A.H. and Scott, M.L., 1974b. The effect of selenium in the hen's diet on egg production, hatchability, performance of progeny and selenium concentration in eggs. Poultry Science 53: 1870-1880.
Cantor, A.H., Scott, M.L. and Noguchi, T., 1975. Biological availability of selenium in feedstuffs and selenium compounds for prevention of exudative diathesis in chicks. Journal of Nutrition 105: 96-105.
Cao, Z.H., Wang, X.C., Yao, D.H., Zhang, X.L. and Wong, M.H., 2001. Selenium geochemistry of paddy soils in Yangtze River Delta. Environment International 26: 335-339.
Carroll, L., Pattison, D.I., Fu, S., Schiesser, C.H., Davies, M.J. and Hawkins, C.L., 2017. Catalytic oxidant scavenging by selenium-containing compounds: reduction of selenoxides and N-chloramines by thiols and redox enzymes. Redox Biology 12: 872-882.
Casal, S.G., Far, J., Bierla, K., Ouerdane, L. and Szpunar, J., 2010. Study of the Se-containing metabolomes in Se-rich yeast by size-exclusion-cation-exchange HPLC with the parallel ICP MS and electrospray orbital ion trap detection. Metallomics 2: 535-548.
Clausen, J. and Nielsen, S.A., 1988. Comparison of whole blood selenium values and erythrocyte glutathione peroxidase activities of normal individuals on supplementation with selenate, selenite, L-selenomethionine, and high selenium yeast. Biological Trace Element Research 15: 125-138.
Combs Jr., G.F., 2001. Selenium in global food systems. British Journal of Nutrition 85: 517-547.
Combs Jr., G.F. and Combs, S.B., 1984. The nutritional biochemistry of selenium. Annual Review of Nutrition 4: 257-280.

Combs Jr., G.F. and Combs, S.B., 1986. The role of selenium in nutrition. Academic Press Inc., New York, NY, USA.

Couloigner, F., Jlali, M., Briens, M., Rouffineau, F., Geraert, P.A. and Mercier, Y., 2015. Selenium deposition kinetics of different selenium sources in muscle and feathers of broilers. Poultry Science 94: 2708-2714.

Cubadda, F., Aureli, F., Ciardullo, S., D'Amato, M., Raggi, A., Acharya, R., Ramana, A.V., Tejo Prakash, R. and Tejo Prakash, N., 2010. Changes in selenium speciation associated with increasing tissue concentrations of selenium in wheat grain. Journal of Agricultural and Food Chemistry 58: 2295-2301.

Curcio, C., Baqui, M.M., Salvatore, D., Rihn, B.H, Mohr, S., Harney, J.W., Larsen, P.R. and Bianco, A.C., 2001. The human type 2 iodothyronine deiodinase is a selenoprotein highly expressed in a mesothelioma cell line. Journal of Biological Chemistry 276: 30183-30187.

Dalto, D.B. and Matte, J.J., 2020. Effects of different sources and levels of dietary iron and selenium on the postprandial net portal appearance of these minerals in growing pigs. Journal of Animal Science 98(3): skaa063. https://doi.org/10.1093/jas/skaa063

Daniels, L.A., 1996. Selenium metabolism and bioavailability. Biological Trace Element Research 54: 185-199.

Davies, E.B. and Watkinson, J.H., 1966. Uptake of native and applied selenium by pasture species. I. Uptake of selenium by brown top, ryegrass, cocksfoot, and white clover from Atiamury sand. New Zealand Journal of Agricultural Research 9: 317-327.

De Souza, M.P., Pilon-Smits, E.A., Lytle, C.M., Hwang, S., Tai, J., Honma, T.S., Yeh, L. and Terry, N., 1998. Rate-limiting steps in selenium assimilation and volatilization by Indian mustard. Plant Physiology 117: 1487-1494.

Deagen, J.T., Butler, J.A., Beilstein, M.A. and Whanger, P.D., 1987. Effects of dietary selenite, selenocystine and selenomethionine on selenocysteine lyase and glutathione peroxidase activities and on selenium levels in rat tissues. Journal of Nutrition 117: 91-98.

Doan, N., Liu, Y., Xiong, X., Kim, K., Wu, Z., Bravo, D.M., Blanchard, A. and Ji, P., 2020. Organic selenium supplement partially alleviated diquat-induced oxidative insults and hepatic metabolic stress in nursery pigs. British Journal of Nutrition 124(1): 23-33. https://doi.org/10.1017/S0007114520000689

Doucha, J., Lívansky, K., Kotrbácek, V. and Zachleder, V., 2009. Production of *Chlorella* biomass enriched by selenium and its use in animal nutrition: a review. Applied Microbiology and Biotechnology 83: 1001-1008.

Duncan, E.G., Maher, W.A., Jagtap, R., Krikowa, F., Roper, M.M. and O'Sullivan, C.A., 2017. Selenium speciation in wheat grain varies in the presence of nitrogen and sulphur fertilisers. Environmental Geochemistry and Health 39: 955-966.

Ehlig, C.F., Allaway, W.H., Carey, E.E. and Kubota, J., 1968. Differences among plant species in selenium accumulation from soils low in available selenium. Agronomy Journal 60: 43-47.

Eisenberg, S., 2007. Relative stability of selenites and selenates in feed premixes as a function of water activity. Journal of AOAC International 90: 349-353.

El-Ghawi, U.M., Al-Fakhri, S.M., Al-Sadeq, A.A., Bejey, M.M. and Doubali, K.K., 2007. The level of selenium and some other trace elements in different Libyan arable soils using instrumental neutron activation analysis. Biological Trace Element Research 119: 89-96.

Ellis, D.R. and Salt, D.E., 2003. Plants, selenium and human health. Current Opinion in Plant Biology 6: 273-279.

Encinar, J.R., Ruzik, R., Buchmann, W., Tortajada, J., Lobinski, R. and Szpunar, J., 2003. Detection of selenocompounds in a tryptic digest of yeast selenoprotein by MALDI time-of-flight MS prior to their structural analysis by electrospray ionization triple quadrupole MS. Analyst 128: 220-224.

European Commission (EC), 2013. Commission implementing regulation (EU) No 427/2013 of 8 May 2013 concerning the authorisation of selenomethionine produced by *Saccharomyces cerevisiae* NCYC R646 as a feed additive for all animal species and amending Regulations (EC) No 1750/2006, (EC) No 634/2007 and (EC) No 900/2009 as regards the maximum supplementation with selenised yeast. Official Journal of the European Union L 127: 20-22.

Falk, M., Bernhoft, A., Reinoso-Maset, E., Salbu, B., Lebed, P., Framstad, T., Fuhrmann, H. and Oropeza-Moe, M., 2020. Beneficial antioxidant status of piglets from sows fed selenomethionine compared with piglets from sows fed sodium selenite. Journal of Trace Elements in Medicine and Biology 58: 126439.

Falk, M., Lebed, P., Bernhoft, A., Framstad, T., Kristoffersen, A.B., Salbu, B. and Oropeza-Moe, M., 2019. Effects of sodium selenite and L-selenomethionine on feed intake, clinically relevant blood parameters and selenium species in plasma, colostrum and milk from high-yielding sows. Journal of Trace Elements in Medicine and Biology 52: 176-185.

Federal Register, 2007. Food additives permitted in feed and drinking water of animals; selenium yeast. Document 72 FR 39560. Available at: https://tinyurl.com/y5g42rnn.

Février, L., Martin-Garin, A. and Leclerc, E., 2007. Variation of the distribution coefficient (Kd) of selenium in soils under various microbial states. Journal of Environmental Radioactivity 97: 189-205.

Finley, J.W., 2006. Bioavailability of selenium from foods. Nutrition Reviews 64: 146-151.

Fisinin, V.I., Papazyan, T.T. and Surai, P.F., 2008. Selenium in poultry nutrition. In: Surai, P.F. and Taylor-Pickard, J. (eds) Current advances in Se research and applications. Vol. 1. Wageningen Academic Publishers, Wageningen, the Netherlands, pp. 221-261.

Fröhlich, E. and Roblegg, E., 2012. Models for oral uptake of nanoparticles in consumer products. Toxicology 291: 10-17.

Gabrielsen, B.O. and Opstvedt, J., 1980. Availability of selenium in fish meal in comparison with soybean meal, corn gluten meal and selenomethionine relative to selenium in sodium selenite for restoring glutathione peroxidase activity in selenium-depleted chicks. Journal of Nutrition 110: 1096-1100.

Gallegos, A., Berggren, M., Gasdaska, J.R. and Powis, G., 1997. Mechanisms of the regulation of thioredoxin reductase activity in cancer cells by the chemopreventive agent selenium. Cancer Research 57: 4965-4970.

Gibson, R.S., 1989. Assessment of trace element status in humans. Progress in Food and Nutrition Science 13: 67-111.

Gilbert-López, B., Dernovics, M., Moreno-González, D., Molina-Díaz, A. and García-Reyes, J.F., 2017. Detection of over 100 selenium metabolites in selenized yeast by liquid chromatography electrospray time-of-flight mass spectrometry. Journal of Chromatography B 1060: 84-90.

Giurgea, R. and Roman, I., 1992. Selenium and vitamin E effect upon glucose absorption in chicken jejunum. Revue Roumaine de Biologie. Serie de Biologie Animale 37: 103-105.

Givens, D.I., Allison, R., Cottrill, B. and Blake, J.S., 2004. Enhancing the selenium content of bovine milk through alteration of the form and concentration of selenium in the diet of the dairy cow. Journal of the Science of Food and Agriculture 84: 811-817.

Goh, K.H. and Lim, T.T., 2004. Geochemistry of inorganic arsenic and selenium in a tropical soil: effect of reaction time, pH, and competitive anions on arsenic and selenium adsorption. Chemosphere 55: 849-859.

Gosetti, F., Frascarolo, P., Polati, S., Medana, C., Gianotti, V., Palma, P., Aigotti, R., Baiocchi, C. and Gennaro, M.C., 2007. Speciation of selenium in diet supplements by HPLC-MS/MS methods. Food Chemistry 105: 1738-1747.

Groce, A.W., Miller, E.R., Hitchcock, J.P., Ullrey, D.E. and Magee, W.T., 1973. Selenium balance in the pig as affected by selenium source and vitamin E. Journal of Animal Science 37: 942-947.

Gu, Q.P., Ream, W. and Whanger, P.D., 2002. Selenoprotein W gene regulation by selenium in L8 cells. Biometals 15: 411-420.

Gu, Q.P., Xia, Y.M., Ha, P.C., Butler, J.A. and Whanger, P.D., 1998. Distribution of selenium between plasma fractions in guinea pigs and humans with various intakes of dietary selenium. Journal of Trace Elements in Medicine and Biology 12: 8-15.

Guenter, W. and Bragg, D.B., 1977. Response of broiler chick to dietary selenium. Poultry Science 56: 2031-2038.

Gulyas, G., Csosz, E., Prokisch, J., Javor, A., Mezes, M., Erdelyi, M., Balogh, K., Janaky, T., Szabo, Z., Simon, A. and Czegledi, L., 2017. Effect of nano-sized, elemental selenium supplement on the proteome of chicken liver. Journal of Animal Physiology and Animal Nutrition 101: 502-510.

Guo, X. and Wu, L., 1998. Distribution of free seleno-amino acids in plant tissue of *Melilotus indica* L. grown in selenium-laden soils. Ecotoxicology and Environmental Safety 39: 207-214.

Hassan, S., 1987. Comparative effects of selenium in oats, meat meal, selenomethionine and sodium selenite for prevention of exudative diathesis in chicks. Zentralblatt fur Veterinarmedizin. Reihe A 34: 204-215.

Hassan, S., Hakkarainen, R.V., Lindberg, P.O. and Sankari, S., 1993. Response of whole blood glutathione peroxidase and selenium in chicks fed with sodium selenite, wheat and fish meal. Animal Feed Science and Technology 41: 103-111.

Haygarth, P.M., Harrison, A.F. and Jones, K.C., 1995. Plant selenium from soil and the atmosphere. Journal of Environmental Quality 24: 768-771

Hondal, R.J., Motley, A.K., Hill, K.E. and Burk, R.F., 1999. Failure of selenomethionine residues in albumin and immunoglobulin G to protect against peroxynitrite. Archives of Biochemistry and Biophysics 371: 29-34.

Hu, H., Wang, M., Zhan, X., Li, X. and Zhao, R., 2011. Effect of different selenium sources on productive performance, serum and milk Se concentrations, and antioxidant status of sows. Biological Trace Element Research 142: 471-480.

Humaloja, T. and Mykkänen, H.M., 1986. Intestinal absorption of ^{75}Se-labeled sodium selenite and selenomethionine in chicks: effects of time, segment, selenium concentration and method of measurement. Journal of Nutrition 116: 142-148.

Ikumo, H. and Yoshida, M., 1981. Selenium content of feedstuffs and biological availability of selenium in chicks. Japanese Poultry Science 18: 307-311.

Ilian, M.A. and Whanger, P.D., 1989. *In vitro* metabolism of ^{75}Se-selenite and ^{75}Se-selenomethionine in chick blood. Journal of Trace Elements and Electrolytes in Health and Disease 3: 9-16.

Ip, C., 1986. Interaction of vitamin C and selenium supplementation in the modification of mammary carcinogenesis in rats. Journal of the National Cancer Institute 77: 299-303.

Ip, C. and Hayes, C., 1989. Tissue selenium levels in selenium-supplemented rats and their relevance in mammary cancer protection. Carcinogenesis 10: 921-925.

Ip, C., Birringer, M., Block, E., Kotrebai, M., Tyson, J.F., Uden, P.C. and Lisk, D.J., 2000. Chemical speciation influences comparative activity of selenium-enriched garlic and yeast in mammary cancer prevention. Journal of Agricultural and Food Chemistry 48: 2062-2070.

Jlali, M., Briens, M., Rouffineau, F., Geraert, P.A., Mercier, Y., 2014. Evaluation of the efficacy of 2-hydroxy-4-methylselenobutanoic acid on growth performance and tissue selenium retention in growing pigs. Journal of Animal Science 92: 182-188.

Jlali, M., Briens, M., Rouffineau, F., Mercerand, F., Geraert, P.A. and Mercier, Y., 2013. Effect of 2-hydroxy-4-methylselenobutanoic acid as a dietary selenium supplement to improve the selenium concentration of table eggs. Journal of Animal Science 91: 1745-1752.

Jonnalagadda, S.B. and Rao, P.V., 1993. Toxicity, bioavailability and metal speciation. Comparative Biochemistry and Physiology 106C: 585-595.

Juniper, D.T., Rymer, C. and Briens, M., 2019. Bioefficacy of hydroxy-selenomethionine as a selenium supplement in pregnant dairy heifers and on the selenium status of their calves. Journal of Dairy Science 102: 7000-7010.

Kim, B.W., Zavacki, A.M., Curcio-Morelli, C., Dentice, M., Harney, J.W., Larsen, P.R. and Bianco, A.C., 2003. ER-associated degradation of the human type 2 iodothyronine deiodinase (D2) is mediated via an association between mammalian UBC7 and the carboxyl region of D2. Molecular Endocrinology 17: 2603-2612.

Kretz-Remy, C. and Arrigo, A.P., 2003. Modulation of the chymotrypsin-like activity of the 20S proteasome by intracellular redox status: effects of glutathione peroxidase-1 overexpression and antioxidant drugs. Journal of Biological Chemistry 384(4): 589-595.

Kubachka, K.M., Hanley, T., Mantha, M., Wilson, R.A., Falconer, T.M., Kassa, Z., Oliveira, A, Landero, J. and Caruso, J., 2017. Evaluation of selenium in dietary supplements using elemental speciation. Food Chemistry 218: 313-320.

Larsen, E.H., Hansen, M., Paulin, H., Moesgaard, S., Reid, M. and Rayman, M., 2004. Speciation and bioavailability of selenium in yeast-based intervention agents used in cancer chemoprevention studies. Journal of AOAC International 87: 225-232.

Lavu, R.V., Van de Wiele, T., Pratti, V.L., Tack, F. and Du Laing, G., 2016. Selenium bioaccessibility in stomach, small intestine and colon: comparison between pure Se compounds, Se-enriched food crops and food supplements. Food Chemistry 197A: 382-387.

Levander, O.A., 1983. Bio-availability of selenium to Finnish men as assessed by platelet glutathione peroxidase activity and other blood parameters. American Journal of Clinical Nutrition 37: 887-897.

Levander, O.A., Alfthan, G., Arvilommi, H., Gref, C.G., Huttunen, J.K., Kataja, M., Koivistoinen, P. and Pikkarainen, J., 1983. Bioavailability of selenium to Finnish men as assessed by platelet glutathione peroxidase activity and other blood parameters. American Journal of Clinical Nutrition 37: 887-897.

Lindemann, T., Prange, A., Dannecker, W. and Neidhart, B., 2000. Stability studies of arsenic, selenium, antimony and tellurium species in water, urine, fish and soil extracts using HPLC/ICP-MS. Fresenius Journal of Analytical Chemistry 368: 214-220.

Lipiec, E., Siara, G., Bierla, K., Ouerdane, L. and Szpunar, J., 2010. Determination of selenomethionine, selenocysteine, and inorganic selenium in eggs by HPLC-inductively coupled plasma mass spectrometry. Analytical and Bioanalytical Chemistry 397: 731-741.

Lobinski, R., Edmonds, J.S., Suzuki, K.T. and Uden, P.C., 2000. Species-selective determination of selenium compounds in biological materials. Pure and Applied Chemistry 72: 447-461.

Long, M.I.E. and Marshall, B., 1973. The selenium status of pastures in Uganda. Tropical Agriculture 50: 121-128.

MacPherson, A., 2000. Trace mineral status of forages. In: Givens, D.I., Owen, E., Omed, H.M. and Axford, R.F.E. (eds) Forage evaluation in ruminant nutrition. CAB International, Wallingford, UK, pp. 345-371.

Mahan, D.C. and Peters, J.C., 2004. Long-term effects of dietary organic and inorganic selenium sources and levels on reproducing sows and their progeny. Journal of Animal Science 82: 1343-1358.

Mar, J.L., Reyes, L.H., Rahman, G.M. and Kingston, H.M., 2009. Simultaneous extraction of arsenic and selenium species from rice products by microwave-assisted enzymatic extraction and analysis by ion chromatography-inductively coupled plasma-mass spectrometry. Journal of Agricultural and Food Chemistry 57: 3005-3013.

Martello, M.A. and Latshaw, J.D., 1982. Utilization of dietary selenium as indicated by prevention of selenium deficiency and by retention in eggs. Nutrition Reports International 26: 43-50.

McSheehy, S., Yang, L., Sturgeon, R. and Mester, Z., 2005. Determination of methionine and selenomethionine in selenium-enriched yeast by species-specific isotope dilution with liquid chromatography-mass spectrometry and inductively coupled plasma mass spectrometry detection. Analytical Chemistry 77: 344-349.

Menon, S., Shrudhi Devi, K.S., Santhiya, R., Rajeshkumar, S. and Kumar, V., 2018 Selenium nanoparticles: a potent chemotherapeutic agent and an elucidation of its mechanism. Colloids and Surfaces B: Biointerfaces 170: 280-292.

Meuillet, E., Stratton, S., Prasad Cherukuri, D., Goulet, A.C., Kagey, J., Porterfield, B. and Nelson, M.A., 2004. Chemoprevention of prostate cancer with selenium: an update on current clinical trials and preclinical findings. Journal of Cell Biochemistry 91: 443-458.

Miller, D., Soares Jr., J.H., Bauersfeld Jr., P. and Cuppett, S.L., 1972. Comparative selenium retention by chicks fed sodium selenite, selenomethionine, fish meal and fish solubles. Poultry Science 51: 1669-1673.

Moreno, P., Quijano, M.A., Gutierrez, A.M., Perez-Conde, M.C. and Camara, C., 2004. Study of selenium species distribution in biological tissues by size exclusion and ion exchange chromatography inductively coupled plasma-mass spectrometry. Analytica Chimica Acta 524: 315-327.

Moreno, P., Quijano, M.A., Gutierrez, A.M., Perez-Conde, M.C. and Camara, C., 2002. Stability of total selenium and selenium species in lyophilised oysters and in their enzymatic extracts. Analytical and Bioanalytical Chemistry 374: 466-476.

Mou, D., Ding, D., Yan, H., Qin, B., Dong, Y., Li, Z., Che, L., Fang, Z., Xu, S., Lin, Y., Zhuo, Y., Li, C., Huang C, Zou Y., Li, L., Briens, M., Wu, D. and Feng, B., 2020. Maternal supplementation of organic selenium during gestation improves sows and offspring antioxidant capacity, inflammatory status, and promotes embryo survival. Food and Function 11: 7748-7761. https://doi.org/10.1039/D0FO00832J

Munier-Lamy, C., Deneux-Mustin, S., Mustin, C., Merlet, D., Berthelin, J. and Leyval, C., 2007. Selenium bioavailability and uptake as affected by four different plants in a loamy clay soil with particular attention to mycorrhizae inoculated ryegrass. Journal of Environmental Radioactivity 97: 148-158.

Mykkänen, H.M. and Mutanen, M.L., 1983. Effect of ascorbic acid on the intestinal absorption of ^{75}Se selenite in chicks. Nutrition Reports International 28: 67-73.

Nakamuro, K., Nakanishi, K., Okuno, T., Hasegawa, T. and Sayato, Y., 1997. Composition of methylated selenium metabolites in rats after oral administration of various selenium compounds. Japanese Journal of Toxicology and Environmental Health 43: 182-189.

National Research Council (NRC), 1983. Selenium in nutrition. The National Academies Press, Washington, DC, USA, 174 pp.

Okuno, T., Kubota, T., Kuroda, T., Ueno, H. and Nakamuro, K., 2001. Contribution of enzymic alpha, gamma-elimination reaction in detoxification pathway of selenomethionine in mouse liver. Toxicology and Applied Pharmacology 176: 18-23.

Olson, O.E. and Palmer, I.S., 1976. Selenoamino acids in tissues of rats administered inorganic selenium. Metabolism 25: 299-306.

Olson, O.E., Novacek, E.J., Whitehead, E.I. and Palmer, I.C., 1970. Investigations on selenium in wheat. Phytochemistry 9: 1181-1188.

Ortman, K. and Pehrson, B., 1997. Selenite and selenium yeast as feed supplements for dairy cows. Zentralblatt fur Veterinarmedizin A 44: 373-380.

Ortman, K. and Pehrson, B., 1998. Selenite and selenium yeast as feed supplements to growing fattening pigs. Zentralblatt fur Veterinarmedizin, Reihe A 45: 551-557.

Osman, M. and Latshaw, J.D., 1976. Biological potency of selenium from sodium selenite, selenomethionine, and selenocystine in the chick. Poultry Science 55: 987-994.

Oster, O., Schmiedel, G. and Prellwitz, W., 1988. The organ distribution of selenium in German adults. Biological Trace Element Research 15: 23-45.

Pappas, A.C., Karadas, F., Surai, P.F. and Speake, B.K., 2005. The selenium intake of the female chicken influences the selenium status of her progeny. Comparative Biochemistry and Physiology B 142: 465-474.

Patra, A. and Lalhriatpuii, M., 2020. Progress and prospect of essential mineral nanoparticles in poultry nutrition and feeding – a review. Biological Trace Element Research 197: 233-253. https://doi.org/10.1007/s12011-019-01959-1

Patterson, B.H., Levander, O.A., Helzlsouer, K., McAdam, P.A., Lewis, S.A., Taylor, P.R., Veillon, C. and Zech, L.A., 1989. Human selenite metabolism: a kinetic model. American Journal of Physiology 257: R556-567.

Payne, R.L. and Southern, L.L., 2005. Changes in glutathione peroxidase and tissue selenium concentrations of broilers after consuming a diet adequate in selenium. Poultry Science 84: 1268-1276.

Persson-Moschos, M., Alfthan, G. and Akesson, B., 1998. Plasma selenoprotein P levels of healthy males in different selenium status after oral supplementation with different forms of selenium. European Journal of Clinical Nutrition 52: 363-367.

Pesti, G.M. and Combs Jr., G.F., 1976. Studies on the enteric absorption of selenium in the chick using localized coccidial infections. Poultry Science 55: 2265-2274.

Pickering, I.J., Prince, R.C., Salt, D.E. and George, G.N., 2000. Quantitative, chemically specific imaging of selenium transformation in plants. Proceedings of the National Academy of Sciences of the USA 97: 10717-10722.

Pierce, S. and Tappel, A.L., 1977. Effects of selenite and selenomethionine on glutathione peroxidase in the rat. Journal of Nutrition 107: 475-479.

Pilon-Smits, E.A.H., De Souza, M.P., Hong, G., Amini, A., Bravo, R.C., Payabyab, S.T. and Terry, N., 1999. Selenium volatilization and accumulation by twenty aquatic plant species. Journal of Environmental Quality 28: 1011-1018.

Prange, A., Sari, M., Von Ameln, S., Hajdu, C., Hambitzer, R., Ellinger, S. and Hormes, J., 2019. Characterization of selenium speciation in selenium-enriched button mushrooms (*Agaricus bisporus*) and selenized yeasts (dietary supplement) using X-ray absorption near-edge structure (XANES) spectroscopy. Journal of Trace Elements in Medicine and Biology 51: 164-168.

Rayman, M.P., 2004. The use of high-selenium yeast to raise selenium status: how does it measure up? British Journal of Nutrition 92: 557-573.

Reasbeck, P.G., Barbezat, G.O., Robinson, M.F. and Thompson, C.D., 1981. Direct measurement of selenium absorption *in vivo*: triple-lumen gut perfusion in the conscious dog. In: Proceedings New Zealand Workshop on Trace Elements, University Otago, Dunedin, NZ, p.107.

Reilly, C., 2006. Selenium in food and health. Springer, New York, NY, USA.
Reyes, L.H., Marchante-Gayón, J.M., Alonso, J.I.J. and Sanz-Medel, A., 2006. Application of isotope dilution analysis for the evaluation of extraction conditions in the determination of total selenium and selenomethionine in yeast-based nutritional supplements. Journal of Agricultural and Food Chemistry 54: 1557-1563.
Robinson, M.F., Thomson, C.D. and Huemmer, P.K., 1985. Effect of a megadose of ascorbic acid, a meal and orange juice on the absorption of selenium as sodium selenite. New Zealand Medical Journal 98: 627-629.
Schrauzer, G.N., 2000. Selenomethionine: a review of its nutritional significance, metabolism and toxicity. Journal of Nutrition 130: 1653-1656.
Schrauzer, G.N., 2001. Nutritional selenium supplements: product types, quality, and safety. Journal of the American College of Nutrition 20: 1-4.
Schrauzer, G.N., 2003. The nutritional significance, metabolism and toxicology of selenomethionine. Advances in Food and Nutrition Research 47: 73-112.
Schrauzer, G.N. and Surai, P.F., 2009. Selenium in human and animal nutrition: resolved and unresolved issues. A partly historical treatise in commemoration of the fiftieth anniversary of the discovery of the biological essentiality of selenium, dedicated to the memory of Klaus Schwarz (1914-1978) on the occasion of the thirtieth anniversary of his death. Critical Reviews in Biotechnology 29(1): 2-9.
Silva, V.A., Bertechini, A.G., Clemente, A.H.S., De Freitas, L.F.V.B., Nogueira, B.R.F., De Oliveira, B.L. and Ramos, A.L.S., 2019. Different levels of selenomethionine on the meat quality and selenium deposition in tissue of finishing pigs. Journal of Animal Physiology and Animal Nutrition 103: 1866-1874.
Simon, E., Ballet, N., Francesch, M. and Brufau, J., 2013. Comparison between the dietary supplementation of sodium selenite and selenium-yeast on meat Se accumulation in broiler. Worlds Poultry Science Journal 69: 1-5.
Skalickova, S., Milosavljevic, V., Cihalova, K., Horky, P., Richtera, L. and Adam, V., 2017. Selenium nanoparticles as a nutritional supplement. Nutrition 33: 83-90.
Sloth, J.J., Larsen, E.H., Bügel, S.H. and Moesgaard, S., 2003. Determination of total selenium and ^{77}Se in isotopically enriched human samples by ICP-dynamic reaction cell-MS. Journal of Analytical Atomic Spectrometry 18: 317-322.
Spallholz, J.E., 1997. Free radical generation by selenium compounds and their prooxidant toxicity. Biomedical and Environmental Sciences 10: 260-270.
Spallholz, J.E. and Rafferty, A., 1987. Nutritional, chemical and toxicological evaluation of high-selenium yeast. In: Combs, G.F., Spallholz, J.E., Levander, O.A. and Oldfield, J.E. (eds) Selenium in biology and medicine, Part A. Avi-Van Nostrand Reinhold, New York, NY, USA, pp. 516-529.
Stadlober, M., Sager, M. and Irgolic, K.J., 2001. Effects of selenate supplemented fertilisation on the selenium level of cereals: identification and quantification of selenium compounds by HPLC-ICPMS. Food Chemistry 73: 357-366.
Sun, L.L., Gao, S.T., Wang, K., Xu, J.C., Sanz-Fernandez, M.V., Baumgard, L.H. and Bu, D.P., 2019. Effects of source on bioavailability of selenium, antioxidant status, and performance in lactating dairy cows during oxidative stress-inducing conditions. Journal of Dairy Science 102: 311-319.
Sun, P., Wang, J., Liu, W., Bu, D.P., Liu, S.J. and Zhang, K.Z., 2017. Hydroxy-selenomethionine: a novel organic selenium source that improves antioxidant status and selenium concentrations in milk and plasma of mid-lactation dairy cows. Journal of Dairy Science 100: 9602-9610.

Sunde, R.A., Saedi, M.S., Knight, S.A.B., Smith, C.G. and Evenson, J.K., 1989. Regulation of expression of glutathione peroxidase by selenium. In: Wendel, A. (ed.) Selenium in biology and medicine. Springer-Verlag, Heidelberg, Germany, pp. 8-13.

Surai, P.F., 2000. Effect of the selenium and vitamin E content of the maternal diet on the antioxidant system of the yolk and the developing chick. British Poultry Science 41: 235-243.

Surai, P.F., 2002. Natural antioxidants in avian nutrition and reproduction. Nottingham University Press, Nottingham, UK.

Surai, P.F., 2006. Selenium in nutrition and health. Nottingham University Press, Nottingham, UK.

Surai, P.F., 2018. Selenium in poultry nutrition and health. Wageningen Academic Publishers, Wageningen, the Netherland, 430 pp.

Surai, P.F., 2021. Selenium in sow nutrition. Chapter 5. In: Surai, P.F. (ed.) Selenium in pig nutrition and health. Wageningen Academic Publishers, Wageningen, the Netherlands, pp. 133-172. https://doi.org/10.3920/978-90-8686-912-1_5

Surai, P.F. and Fisinin, V.I., 2014. Selenium in poultry breeder nutrition: an update. Animal Feed Science and Technology 191: 1-15.

Surai, P.F. and Fisinin, V.I., 2015. Selenium in pig nutrition and reproduction: boars and semen quality – a review. Asian-Australasian Journal of Animal Science 28: 730-746.

Surai, P.F. and Fisinin, V.I., 2016a. Selenium in livestock and other domestic animals. In: Hatfield, D.L., Schweizer, U., Tsuji, P.A. and Gladyshev, V.N. (eds) Selenium – its molecular biology and role in human health. Springer International Publishing, New York, NY, USA, pp. 595-606.

Surai, P.F. and Fisinin, V.I., 2016b. Selenium in sow nutrition. Animal Feed Science and Technology 211: 18-30.

Surai, P.F. and Kochish, I.I., in press. Food for thought: nano-selenium in poultry nutrition and health. Animal Health Research Reviews. https://doi.org/10.1017/S1466252320000183

Surai, P.F., Karadas, F., Pappas, A.C. and Sparks, N.H., 2006. Effect of organic selenium in quail diet on its accumulation in tissues and transfer to the progeny. British Poultry Science 47: 65-72.

Surai, P.F., Kochish, I.I. and Velichko, O.A., 2017. Nano-Se assimilation and action in poultry and other monogastric animals: is gut microbiota an answer? Nanoscale Research Letters 12: 612.

Surai, P.F., Kochish, I.I., Fisinin, V.I. and Velichko, O.A., 2018. Selenium in poultry nutrition: from sodium selenite to organic Se sources. Journal of Poultry Science 55: 79-93.

Surai, P.F., Pappas, A.C., Karadas, F., Papazyan, T.T. and Fisinin, V.I., 2010. Selenium enigma: health implications of an inadequate supply. In: De Meester, F., Zibadi, S. and Ross Watson, R. (eds) Modern dietary fat intakes in disease promotion. Humana Press, New York, NY, USA, pp. 379-403.

Suzuki, K.T. and Ogra, Y., 2002. Metabolic pathway for selenium in the body: speciation by HPLC-ICP MS with enriched Se. Food Additives and Contaminants 19: 974-983.

Suzuki, Y., Hashiura, Y., Matsumura, K., Matsukawa, T., Shinohara, A. and Furuta, N., 2009. Dynamic pathways of selenium metabolism and excretion in mice under different selenium nutritional stresses. Metallomics 2: 126-132.

Swanson, C.A., Patterson, B.H., Levander, O.A., Veillon, C., Taylor, P.R., Helzlsouer, K., McAdam, P.A. and Zech, L.A., 1991. Human [^{74}Se] selenomethionine metabolism: a kinetic model. American Journal of Clinical Nutrition 54: 917-926.

Takahashi, K., Suzuki, N. and Ogra, Y., 2018. Effect of administration route and dose on metabolism of nine bioselenocompounds. Journal of Trace Elements in Medicine and Biology 49: 113-118.

Tan, J., Zhu, W., Wang, W., Li, R., Hou, S., Wang, D. and Yang, L., 2002. Selenium in soil and endemic diseases in China. Science of the Total Environment 284: 227-235.

Terry, N., Zayed, A.M., De Souza, M.P. and Tarun, A.S., 2000. Selenium in higher plants. Annual Review of Plant Physiology and Molecular Biology 51: 401-432.

Thomson, C.D., 1998. Selenium speciation in human body fluids. Analyst 123: 827-831.

Thomson, C.D., Robinson, B.A., Stewart, R.D. and Robinson, M.F., 1975. Metabolic studies of [^{75}Se] selenocystine and [^{75}Se]selenomethionine in the rat. British Journal of Nutrition 34: 501-509.

Tian, Y., Qu, N., Zhou, Y. and Xu, J., 2001. Effect of methionine supplementation on the selenium bioavailability in rats fed on grains from Keshan disease endemic area. Wei Sheng Yan Jiu 30: 55-57.

Tsuji, Y., Mikami, T., Anan, Y. and Ogra, Y., 2010. Comparison of selenohomolanthionine and selenomethionine in terms of selenium distribution and toxicity in rats by bolus administration. Metallomics 2: 412-418.

Van Metre, D.C. and Callan, R.J., 2001. Selenium and vitamin E. Veterinary Clinics of North America, Food Animal Practice 17: 373-402.

Vendeland, S.C., Deagen, J.T. and Whanger, P.D., 1992. Uptake of selenotrisulfides of glutathione and cysteine by brush border membranes from rat intestines. Journal of Inorganic Biochemistry 47: 131-140.

Vickerman, D.B. and Trumble, J.T., 1999. Feeding preferences of *Spodoptera exigua* in response to form and concentration of selenium. Archives of Insect Biochemistry and Physiology 42: 64-73.

Wang, C., Liu, Q., Yang, W.Z., Dong, Q., Yang, X.M., He, D.C., Zhang, P., Dong, K.H. and Huang, Y.X., 2009. Effects of selenium yeast on rumen fermentation, lactation performance and feed digestibilities in lactating dairy cows. Livestock Science 126: 239-244.

Wang, D., Yang, B., Wei, G., Liu, Z. and Wang, C., 2012. Efficient preparation of selenium/glutathione-enriched *Candida utilis* and its biological effects on rats. Biological Trace Element Research 150(1-3): 249-57.

Wang, H., Zhang, J. and Yu, H., 2007. Elemental selenium at nano size possesses lower toxicity without compromising the fundamental effect on selenoenzymes: comparison with selenomethionine in mice. Free Radical Biology and Medicine 42: 1524-1533.

Wang, Y., 2009. Differential effects of sodium selenite and nano-Se on growth performance, tissue se distribution, and glutathione peroxidase activity of avian broiler. Biological Trace Element Research 128: 184-190.

Wang, Y.-B., Xu, B.-H., 2008. Effect of different selenium source (sodium selenite and selenium yeast) on broiler chickens. Animal Feed Science and Technology 144: 306-314.

Wang, Y.X., Zhan, X.A., Yuan, D., Zhang, X.W. and Wu, R.J., 2011. Influence of dietary selenomethionine supplementation on performance and selenium status of broiler breeders and their subsequent progeny. Biological Trace Element Research 143: 1497-1507.

Ward, P., Chadha, M., Connolly, C., Stalcup, A. and Murphy, R., 2019. A comparative assessment of water-soluble selenium metabolites in commercial selenised yeast supplements by liquid chromatography-electrospray ionisation QTOF-MS. International Journal of Mass Spectrometry 439: 42-52.

Waters, D.J., Shen, S., Kengeri, S.S., Chiang, E.C., Combs Jr., G.F., Morris, J.S. and Bostwick, D.G., 2012. Prostatic response to supranutritional selenium supplementation: comparison of the target tissue potency of selenomethionine vs. selenium-yeast on markers of prostatic homeostasis. Nutrients 4: 1650-1663.

Wei, J.Y., Wang, J., Liu, W., Zhang, K.Z. and Sun, P., 2019. Short communication: Effects of different selenium supplements on rumen fermentation and apparent nutrient and selenium digestibility of mid-lactation dairy cows. Journal of Dairy Science 102: 3131-3135.

Whanger, P.D., 2002. Selenocompounds in plants and animals and their biological significance. Journal of the American College of Nutrition 21: 223-232.

Whanger, P., Vendeland, S., Park, Y.C. and Xia, Y., 1996. Metabolism of subtoxic levels of selenium in animals and humans. Annals of Clinical Laboratory Science 26: 99-113.

Whitacre, M. and Latshaw, J.D., 1982. Selenium utilization from menhaden fish meal as affected by processing. Poultry Science 61: 2520-2522.

Wolf, W.R. and Goldschmidt, R.J., 2004. Selenomethionine contents of NIST wheat reference materials. Analytical and Bioanalytical Chemistry 378: 1175-1181.

Wolf, W.R., Zainal, H. and Yager, B., 2001. Selenomethionine content of candidate reference materials. Fresenius' Journal of Analytical Chemistry 370: 286-290.

Wolffram, S., 1999. Absorption and metabolism of selenium: difference between inorganic and organic sources. In: Lyons, T.P. and Jacques, K.A. (eds) Biotechnology in the feed industry. Proceedings of 15th Alltech's Annual Symposium. Nottingham University Press, Nottingham, UK, pp. 547-566.

Wolffram, S., Berger, B., Grenacher, B. and Scharrer, E., 1989. Transport of selenoamino acids and their sulfur analogues across the intestinal brush border membrane of pigs. Journal of Nutrition 119: 706-712.

Wrobel, K., Kannamkumarath, S.S., Wrobel, K. and Caruso, J.A., 2003. Hydrolysis of proteins with methanesulfonic acid for improved HPLC-ICP-MS determination of seleno-methionine in yeast and nuts. Analytical and Bioanalytical Chemistry 375: 133-138.

Ximenez-Embun, P., Alonso, I., Madrid-Albarran, Y. and Camara, C., 2004. Establishment of selenium uptake and species distribution in lupine, Indian mustard, and sunflower plants. Journal of Agricultural and Food Chemistry 52: 832-838.

Yang, B., Wang, D., Wei, G., Liu, Z. and Ge, X., 2013. Selenium-enriched *Candida utilis*: efficient preparation with l-methionine and antioxidant capacity in rats. Journal of Trace Elements in Medicine and Biology 27: 7-11.

Yang, X., Tian, Y., Ha, P. and Gu, L., 1997. Determination of the selenomethionine content in grain and human blood. Wei Sheng Yan Jiu 26: 113-116.

Yoshida, M., Abe, M., Fukunaga, K. and Kikuchi, K., 2002. Bioavailability of selenium in the defatted dark muscle of tuna. Food Additives and Contaminants 19: 990-995.

Yuan, D., Zhan, X. and Wang, Y., 2011. Effects of selenium sources and levels on reproductive performance and selenium retention in broiler breeder, egg, developing embryo, and 1-day-old chick. Biological Trace Element Research 144: 705-714.

Yuan, D., Zheng, L., Guo, X.Y., Wang, Y.X. and Zhan, X.A., 2013. Regulation of selenoprotein P concentration and expression by different sources of selenium in broiler breeders and their offspring. Poultry Science 92: 2375-2380.

Zhang, J., Wang, X. and Xu, T., 2008. Elemental selenium at nano size (nano-Se) as a potential chemopreventive agent with reduced risk of selenium toxicity: comparison with semethylselenocysteine in mice. Toxicology Science 101: 22-31.

Zhang, K., Zhao, Q., Zhan, T., Han, Y., Tang, C. and Zhang, J., 2020. Effect of different selenium sources on growth performance, tissue selenium content, meat quality, and selenoprotein gene expression in finishing pigs. Biological Trace Element Research 196: 463-471. https://doi.org/10.1007/s12011-019-01949-3

Zhang, Y. and Frankenberger Jr., W.T., 2001. Speciation of selenium in plant water extracts by ion exchange chromatography-hydride generation atomic absorption spectrometry. Science of the Total Environment 269: 39-47.

Zheng, J., Shibata, Y. and Furuta, N., 2003. Determination of selenoamino acids using two-dimensional ion-pair reversed phase chromatography with on-line detection by inductively coupled plasma mass spectrometry. Talanta 59: 27-36.

Zhou, X. and Wang, Y., 2011. Influence of dietary nano elemental selenium on growth performance, tissue selenium distribution, meat quality, and glutathione peroxidase activity in Guangxi Yellow chicken. Poultry Science 90: 680-686.

Zhou, Z.S., Smith, A.E. and Matthews, R.G., 2000. L-selenohomocysteine: one-step synthesis from L-selenomethionine and kinetic analysis as substrate for methionine synthases. Bioorganic and Medicinal Chemistry Letters 10: 2471-2475.

Zuberbuehler, C.A., Messikommer, R.E. and Wenk, C., 2002. Choice feeding of selenium-deficient laying hens affects diet selection, selenium intake and body weight Journal of Nutrition 132: 3411-3417.

Chapter 4
Selenium deficiency and requirement

P.F. Surai[1-7], I.I. Kochish[2] and V.I. Fisinin[8]
[1]Vitagene and Health Research Centre, Bristol, BS4 2RS, United Kingdom; [2]Department of Hygiene and Poultry Sciences, Moscow State Academy of Veterinary Medicine and Biotechnology Named After K.I. Skryabin, 109472 Moscow, Russia; [3]Department of Microbiology and Biochemistry, Faculty of Veterinary Medicine, Trakia University, 6000 Stara Zagora, Bulgaria; [4]Department of Animal Nutrition, Faculty of Agricultural and Environmental Sciences, Szent Istvan University, 2103 Gödöllo, Hungary; [5]Saint-Petersburg State Academy of Veterinary Medicine, 196084 St. Petersburg, Russia; [6]Sumy National Agrarian University, 40000 Sumy, Ukraine; [7]Odessa National Academy of Food Technologies, 65039 Odessa, Ukraine; [8]All-Russian Institute of Poultry Husbandry, 141311 Sergiev Posad, Russia; psurai@feedfood.co.uk

Abstract

In accordance with Nutrient Requirements of Swine (NRC, 2012) pig requirement in Se is defined at the level of 0.15-0.3 ppm. However, it is necessary to take into account that they are minimal Se requirements and in commercial conditions Se requirements could be substantially higher. Se deficiency in pigs is characterised by hepatic necrosis, ulcers, mulberry heart disease, muscular dystrophy, impaired farrowing and immunity, problems with reproduction, increased mastitis, metritis and agalactia, decreased performance, tissue damages and decreased tolerance to iron. Clinical Se deficiency is a rare event in commercial pig production; however, sub-clinical Se deficiency could be responsible for decreased productive and reproductive performance of pigs. From the summary of negative consequences of Se deficiency/inadequacy in pigs/sows it is clear that meeting Se requirement of growing pigs and producing/reproducing animals is an important task for pig nutritionist. Indeed, only optimal Se status of farm animals, including pigs can give them an opportunity to grow and reproduce to the genetically programmed levels.

Keywords: selenium, Se deficiency, Se requirement, pigs

4.1 Introduction

Selenium is an essential trace element in pig nutrition. It participates in a regulation of a range of various biochemical pathways as an integral part of 25 selenoproteins. Since Se content in feed ingredients can substantially vary and in most cases is low, animal nutrition is based on a usage of balanced diets containing mineral or vitamin-mineral premixes delivering main minerals in excess of requirement, including selenium. Therefore, it is difficult to find commercial pig producing units were clinical

Peter F. Surai (ed.) **Selenium in pig nutrition and health**
DOI 10.3920/978-90-8686-912-1_4, © Wageningen Academic Publishers 2021

signs of Se deficiency in growing pigs and/or producing/reproducing animals can be observed. However, Se inadequacy due to high stress conditions or ineffective form of supplemental Se still can occur in commercial pig production leading to decreased productive and reproductive performance of pigs and compromised their health. Therefore, an update on Se deficiency and requirement in pigs will be presented in this chapter.

4.2 Selenium-deficiency

Clinical Se deficiency is a rare event in commercial pig production; however, subclinical Se deficiency could be responsible for decreased productive and reproductive performance of pigs. Indeed, the incidence of selenium deficiency is seen most frequently in the neonatal, postnatal and post-weaning periods (Close, 2003). It is known that most production of maize and soya in the USA is concentrated in the Midwest region with poor Se level in the soils. Therefore, very often the corn and soybean meal used in American pig diets is Se-deficient (Surai, 2006).

There have been several important changes in pig industry for the last 30-50 years that have contributed to Se-related problems (Mahan, 1991b; Surai, 2006):
- pig confinement has eliminated soil as an additional source of selenium for pigs;
- pig confinement and progress in pig selection have resulted in increased growth rate and increased Se requirement;
- increased crop yields depleted the soil of Se for incorporation into plants and usage of synthetic fertilisers containing sulphur and phosphorus decreased Se availability for plants;
- modern agricultural practises decreased soil pH and further decreased Se availability from soil to plants;
- increased usage of simplified grain-soybean-based diets excluded some other sources of Se from the pig diet;
- earlier weaning decreased Se reserved accumulated in the body of young piglets by colostrum and milk consumption;
- increased sow productivity, early rebreeding, longer retention of sows in the herd depleted Se maternal reserves.

Furthermore, it is necessary to take into account a complexity of selenium-vitamin E relationship in the body (Surai, 2006). Indeed, in many cases clinical signs of Se deficiency are complicated by simultaneous vitamin E deficiency. It seems likely, that in many cases the compromised antioxidant system of the body (see Chapter 1; Surai, 2021a) and oxidative stress with damages to PUFAs, proteins and DNA are responsible for those clinical signs of Se deficiency.

There is great body of information showing detrimental consequences of Se deficiency in pigs (Figure 4.1). Indeed, the main consequences of selenium deficiency in pigs are similar to those of vitamin E deficiency and include (Jenkins and Hidiroglou, 1972; Jonsson, 1993; NRC, 1998, 2012; Reid, 2001; Surai, 2006):

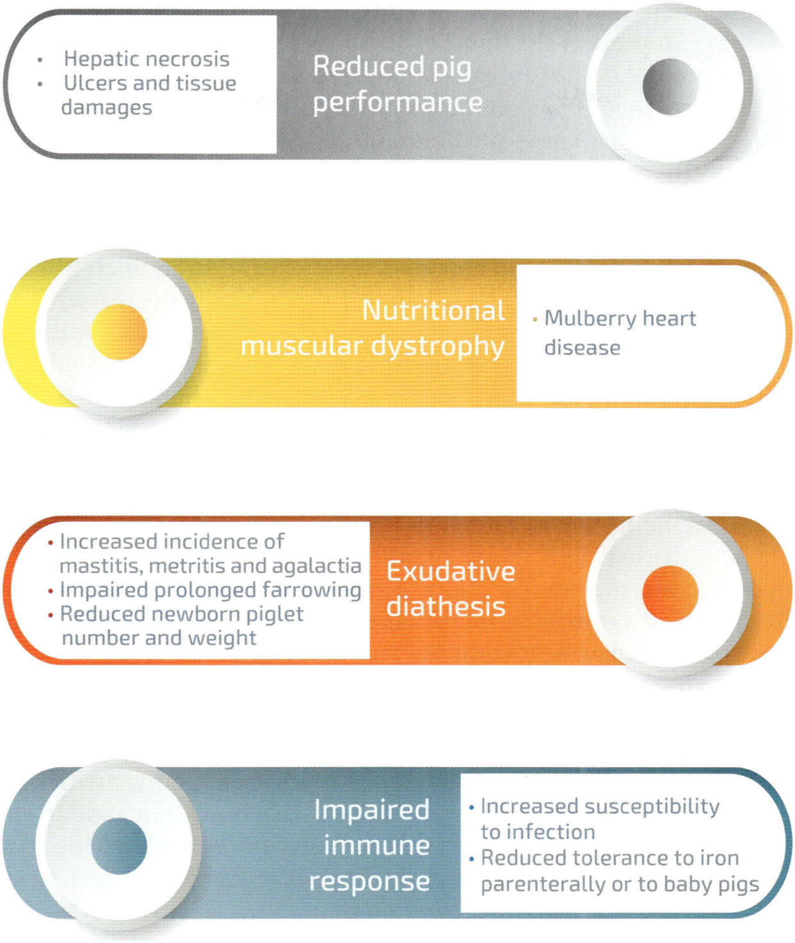

Figure 4.1. Selenium deficiency in pigs – development of diseases.

- Reduced pig performance (growth rate, feed intake, feed utilisation and the apparent digestion for dry matter, ether extract and nitrogen (Glinke and Ewan, 1977). For example, Se-deficient pigs (Se in the feed was 0.019 mg/kg) had a significantly lower daily food intake and daily weight gain than control pigs (0.25 mg/kg Se), whereas, food conversion was identical within both groups. Selenium concentrations in liver, muscle and serum, as well as activity of GPx in serum, was significantly lower in selenium-deficient pigs (Kirchgessner *et al.*, 1995).
- Hepatic necrosis (hepatosis dietetica); an acute type with liver failure and sudden death or a subacute type with ascites and jaundice with edema and cardiomyopathy. The disease occurs most frequently in young pigs up to 3 months of age. Pigs often dying without clinical signs of deficiency (Jenkins and Hidiroglou, 1972). Massive necrosis can be seen in the liver of affected pigs and selenium concentration in

the liver substantially decreased (Figure 4.2). Selenium inclusion into the feed or injection can prevent the disease. The disease can occur in combination with nutritional muscular dystrophy and mulberry heart disease.

- Ulcers in the cutaneous mucous membrane of the stomach and oesophagus, caecal and colonic haemorrhages, possibly gut oedema and post-weaning diarrhoea (Mahan, 1991b) and bloody faeces.
- Nutritional muscular dystrophy (NMD, White muscle disease) is characterised by bilateral paleness and dystrophy of the skeletal muscles and it is frequently observed in combination with hepatosis dietetica and/or mulberry heart disease. Selenium concentration in major tissues is dramatically decreased (Figure 4.2). Affected pigs display generalised muscular weakness and walk with an unsteady gait. There is hyaline degeneration with loss of striations, vacuolisation, and disruption of muscle fibre groups. The longissimus, diaphragm and adductor muscles are most commonly affected by NMD (Jenkins and Hidiroglou, 1972).
- Mulberry heart disease (MHD) is dietetic microangiopathy and often occurs in rapidly growing pigs and may be combined with other Se-responsive conditions such as subacute hepatosis dietetica. It is well known that the MHD of the pig represents a disease with major manifestation at the heart, with vascular and myocytic lesions and transudation to the serous cavities. The vascular lesions include fibrinoid necrosis in intramyocardial small arteries and arterioles with fibrin microthrombi in the myocardial capillaries (Jonsson, 1993). Affected pigs are usually fast growing young animals in good body condition and they die suddenly from acute cardiac failure, without premonitory signs (Jenkins and Hidiroglou, 1972). Clinical signs of the disease include dyspnoea and severe muscle weakness and the heart lesions vary considerably affecting atria and ventricles bilaterally. Hemorrhages can be seen in the epi-, endo- and myocardium, and in some cases pale areas of necrosis are visible in the myocardium (Oropeza-Moe *et al.*, 2015). Selenium concentration in the liver and heart is reduced (Figure 4.3).
  In a later study, hepatic vitamin E concentrations were shown to be deficient (below 2 mg/kg) in 25% of the pigs with gross and microscopic lesions of MHD, while liver selenium concentrations were adequate in all pigs (Pallarés *et al.*, 2002). The data generated from a recent study provided evidence for the association of MHD with decreased, but still within the currently defined normal range, selenium levels in tissues (Shen *et al.*, 2011), raising the question if biological Se requirements in high-yielding pigs are met with current feeding strategies (Oropeza-Moe *et al.*, 2015). Furthermore, significantly decreased Se concentrations were found in cardiac samples as well as almost all skeletal muscles of pigs with MHD ($P<0.05$; Oropeza-Moe *et al.*, 2019): Se concentrations in muscles of pigs with MHD were shown to be 0.34 mg/kg DM compared with 0.65 mg/kg DM in control pigs ($P<0.0001$). Additionally, significantly lower Se concentrations compared with controls were found in the cardiac samples, liver as well as the caecum, duodenum, gastric ventricle, jejunum, kidney, skin and thymus samples (Oropeza-Moe *et al.*, 2019).
- Impaired reproduction. Total number and number of liveborn piglets and number of piglets alive at weaning were significantly lower for sows in the Se-deficient group in comparison to the sow group supplemented with 0.1 mg/kg Se. Average litter weights at birth and weaning were also lower in the Se-deficient group (Mihailovic

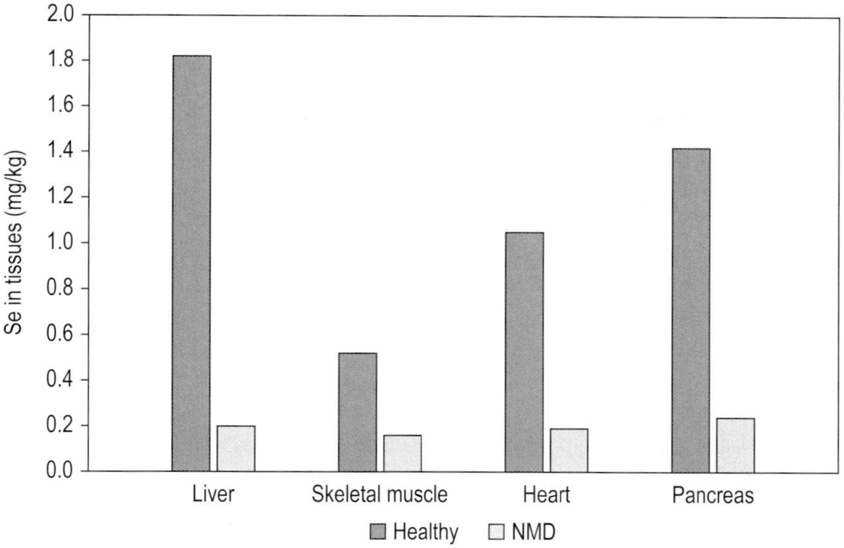

Figure 4.2. Selenium (Se) concentrations in tissues of normal pigs and pigs with nutritional muscular dystrophy (NMD), mg/kg DM (adapted from Lindberg, 1968).

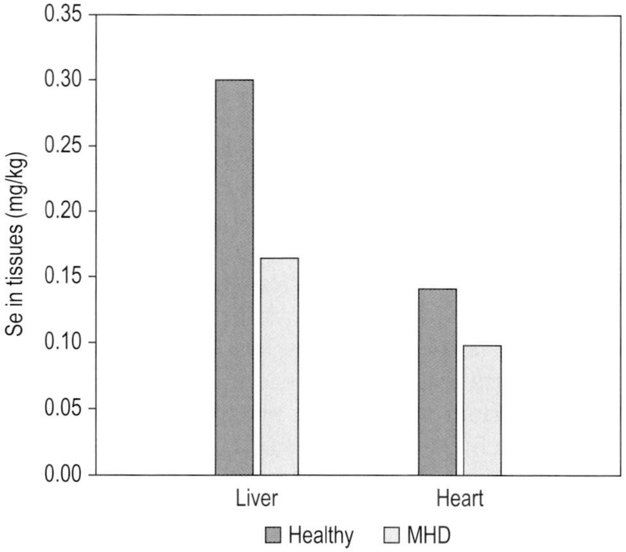

Figure 4.3. Selenium (Se) concentrations in tissues of normal pigs and pigs with Mulberry heart disease (MHD), mg/kg wet weight (adapted from Pedersen and Simensen, 1977).

et al., 1989). Adding sodium selenite to Se-deficient diet during gestation (from 60 days before farrowing) and lactation (28 days) improved weight gain by 7.5-12% and feed utilisation by 14-20% (Bonomi, 2001). Decreased semen quality (lower motility and increased percentage of abnormal sperm, lower fertilising rate of oocytes, Marin-Guzman *et al.*, 1997; decreased number of spermatozoal reserves and Sertoli cells, Marin-Guzman *et al.,* 2000) and reduced the conception rate of gilts to first service (Edwards *et al.,* 1977) were characteristic of Se deficiency. Furthermore, decreased litter size (Mahan *et al.,* 1974) and increased piglet mortality (Nielsen *et al.,* 1979) were observed as a result of Se deficiency. Progeny from Se and vitamin E-deficient sows are weak at birth with a reduced desire to nurse the dam (Mahan, 1991b) and smaller litter size was related to low Se-diet (Mahan and Peters, 2004).
- Increased incidences of mastitis, metritis and agalactia and reduced milk production. These are post-parturient diseases in sows with lactation failure (Mahan, 1991a, 1994, 1999).
- Impaired prolonged farrowing due to poor muscle tone with increased piglet mortality, lethargic and weak piglets (Acda and Chae, 2002; Close, 2003) and reduced muscle tone and strength with an increased incidence of stillborn piglets (Close, 1998).
- Impaired immune response and increased susceptibility to infection (Chapter 9; Surai, 2021b).
- Reduced tolerance to parenterally administered iron and iron toxicity in baby pigs (Surai, 2006).
- Tissue damage (elevated activities of lactate dehydrogenase, glutamic- oxaloacetic transaminase and glutamic-pyruvic transaminase in serum; Surai, 2006).

It seems likely that Se deficiency can affect fatty acid profile of some tissues. For example, Se-deficient pigs were characterised by increased levels of total and n-6 PUFAs in liver neutral lipids and lower levels of total saturated fatty acids in muscle total lipids than control pigs, whereas the fatty acid composition of serum, erythrocytes and backfat was not affected by selenium deficiency. Increased lipid unsaturation, simultaneously with compromised antioxidant defences in Se deficient pigs could trigger a chain of events related to oxidative stress associated with lipid peroxidation, protein oxidation and damages to DNA and leading to various lesions described in literature. Furthermore, various stress in pig production have been suggested to increase the susceptibility to clinical manifestations of Se deficiency (NRC, 2012). Impacts of Se deficiency on pig production are summarised in Figure 4.4.

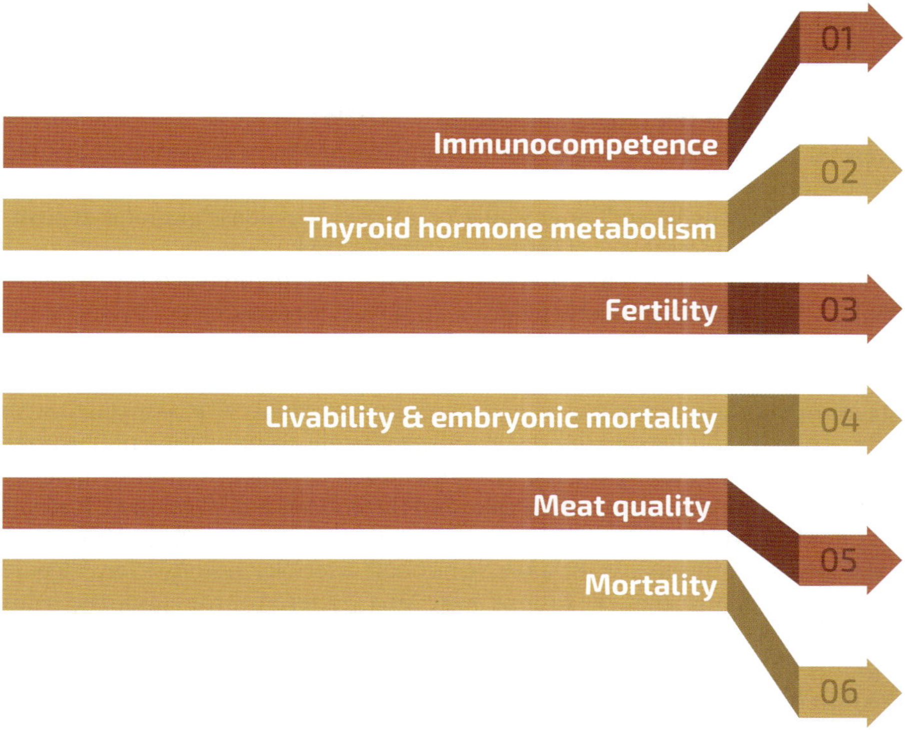

Figure 4.4. Selenium deficiency in pigs – impact on production.

4.3 Evaluation of selenium deficiency

There are several studies conducted to evaluate details of clinical, morphological, microscopic and biochemical changes in Se or Se/Vitamin E deficiency in pigs. For example, twenty two newborn pigs from 2 weeks of age to weaning were fed a corn-torula yeast creep feed containing Se at the concentration of 0.03 mg/kg, and from weaning to slaughter, a corn-soybean meal ration containing Se at the concentration of 0.07 mg/kg and vitamin E at the concentration of 15.7 mg/kg (Van Vleet *et al.,* 1975). Subclinical Se-E deficiency was developed and characterised by subtle muscular stiffness, significant increases in plasma activities of glutamic oxaloacetic transaminase (GOT) and creatine phosphokinase, and typical residual lesions in heart and skeletal muscle, but not in liver, at slaughter at 165 days of age. The major alterations in injured fibres progressed from hyaline degeneration, with subsequent macrophagic invasion and phagocytosis of disrupted sarcoplasm, to muscle fibre regeneration by myoblastic proliferation, fusion, and differentiation into fibres with mature myofibrils. When electron microscopic techniques were used, myofibrillar lysis and disruption was observed. In particular, disruption of mitochondria, sarcoplasmic reticulum, and plasma membranes was evident in fibres with myofibrillar alterations (Van Vleet *et al.,* 1976).

Selenium-vitamin E deficiency was produced in weanling pigs by feeding a semi-synthetic basal diet for 13 to 59 days. In hearts with vascular damage, gross haemorrhages were observed in the myocardium and serosal surfaces (Van Vleet *et al.*, 1977b). The authors indicated that:

- myocardial arteriolar damage was characterised by segmental fibrinoid accumulation in vessel walls and by scattered fibrin thrombi;
- extensive subendothelial and inner wall accumulations of dense granular deposits of serum proteins and masses of fibrin in arterioles took place;
- endothelial cells of affected arterioles were loosely attached to each other and in arterioles with fibrin thrombi, the endothelium was disrupted.

The changes in arterioles depended on the severity of damage. From the one hand, in mildly injured arterioles, increased endothelial permeability was associated with insudation of blood proteins into the vessel wall producing accumulation of fibrinoid. On the other hand, in severely injured vessels, endothelial integrity was completely destroyed, smooth muscle cells were necrotic and thrombosis had developed. It seems likely, that lipid peroxidation and alteration in regulation of antioxidant defences in endothelial cells are responsible for aforementioned arteriolar lesions (Van Vleet *et al.*, 1977b). It was shown that the lesions were also scattered throughout the heart but were most severe in the atria. It is interesting that the damaged fibres had many features of myofibrillar degeneration with hypercontraction bands, myofibrillar lysis, and mitochondrial swelling, disruption, and mineralisation (Van Vleet and Ferrans, 1977; Van Vleet *et al.*, 1977a). Immune system was also involved in the development of lesions. For example, numerous macrophages passed through focal disruptions in the external lamina of the muscle cells and engulfed sarcoplasmic and nuclear debris. Although vascular lesions were present in the hearts of selenium and vitamin E deficient pigs, it was concluded that the fibre alterations developed independently of the vascular changes. The pathogenesis of this cardiomyopathy induced by selenium and vitamin E deficiency is thought to be a result of the antioxidant system alteration (Van Vleet *et al.*, 1977a). In fact, in pigs fed on diets depleted of vitamin E and Se, increases in concentrations of markers of lipid peroxidation (4-hydroxynonenal and hexanal) were observed (Nolan *et al.*, 1995). However, skeletal myopathy and increased lipid peroxidation in heart and supraspinatus muscle were only observed in extra-stressed pigs, fed on diets depleted of vitamin E and Se and supplemented with extra n-3 PUFAs in the form of vitamin E-stripped linseed oil.

An evaluation of the coagulation system has been conducted in vitamin E and/or selenium deficient pigs and confirmed that in selenium deficiency, the prothrombin time was shortened (Fontaine *et al.*, 1977b). The platelet count and platelet turnover were also greatly decreased by selenium deficiency and this was associated with decreased half-life of fibrinogen labelled with ^{75}Se-selenomethionine and increased turnover of fibrinogen. From these fibrinogen kinetic findings, it was concluded that chronic low grade disseminated intravascular coagulation could occur in selenium deficient animals and it is related to the development of hepatosis dietetica or widespread microvascular damage. Furthermore, some other factors, such as increased fibrinogenolysis also were considered to be involved in the development of lesions of Se deficiency. Furthermore, erythropoiesis appeared to be only slightly decreased in selenium deficient pigs (Fontaine *et al.*, 1977a). However, these alterations of erythropoiesis depend on the severity of Se/vitamin E deficiency. For example, sows of Swedish Landrace × Yorkshire breed were fed a diet extremely deficient in vitamin E and selenium during the last six weeks of pregnancy and compared to sows of the same breed and age fed a normal commercial diet (Thoren-Tolling, 1984). Selenium-vitamin E deficient piglets showed significantly lower haemoglobin values at birth than the normal piglets. Serum creatine phosphokinase activities were found to be increased in association with selenium deficiency (Van Vleet and Ferrans, 1977) and it reflects the occurrence of sub-clinical muscular dystrophy.

It is possible that disturbances in antioxidant defences are responsible for the development of microangiopathy (MAP, mulberry heart disease) in pigs. However, there is some controversy among data related to this disease and it is not clear at present what role Se deficiency plays in the development of this disease. As mentioned above, the mulberry heart disease of the pig represents a disease with major manifestation at the heart, primarily affecting young animals with vascular and myocytic lesions. The vascular lesions include fibrinoid necrosis in intramyocardial small arteries and arterioles with fibrin microthrombi in the myocardial capillaries (Jonsson, 1993). However, it seems likely that apart from the characteristic haemorrhagic appearance of the heart, the brain can also be affected. In particular, histopathological changes include marked alteration of the small cerebral blood vessels with fibrinoid degeneration, necrosis or infiltration with pleomorphic cells (Stierstorfe *et al.*, 2001). Furthermore, the vessel lumina are frequently obliterated by thrombi, and as a consequence, large areas of malacia are found in brain parenchyma. Indeed, diet with Se deficiency was shown to activate the iNOS/NF-κB pathway to upregulate the expression of inflammatory cytokines, thereby leading to inflammatory lesions in the pig's brain (Zhang *et al.*, 2019). It was reported that three 5-week-old pigs died of the cardiac form MAP of selenium-vitamin E deficiency. It was observed that the hearts had prominent hydropericardium and extensive serosal and myocardial haemorrhage (Van Vleet and Ferrans, 1977).

Histopathologic observations demonstrated marked myocardial oedema, congestion, and haemorrhage, with numerous capillary hyaline microthrombi that stained red with the periodic-acid Schiff procedure. Ultrastructurally, these thrombi were composed of elongated, dense masses of intertwined fibrin strands with occasional entrapped erythrocytes. It was also shown that MAP persisted among young pigs in Denmark although the diets of sows and pigs were supplemented with selenium and vitamin E (Nielsen *et al.*, 1989). Furthermore, the concentrations of selenium and vitamin E in the liver and heart tissues of young pigs, which had died suddenly, and had the characteristic lesions of mulberry heart disease post mortem, were not significantly different from the concentrations found in pigs of the same age, which had died suddenly for other reasons. Similarly, pigs with dietetic MAP had lower heart and liver α-tocopherol concentrations than did control pigs. However, heart and kidney Se concentrations and heart and liver PUFA concentrations were similar in pigs of either group (Rice and Kennedy, 1989). Indeed, in spite of apparently adequate amount of dietary α-tocopherol, pigs with MAP had lower tissue α-tocopherol concentration than did control pigs. The authors concluded that spontaneous MAP is associated with altered α-tocopherol metabolism but is unrelated to alterations in tissue Se and PUFA concentrations and GPx activity. In another study, the significance of selenium deficiency was also investigated in pigs that died suddenly of MAP. It was shown that hepatic selenium concentration in pigs with MAP (1.04 μg/g dry weight) was only slightly lower than in healthy pigs (1.23 μg/g dry weight; Korpela *et al.*, 1990). However, in 22.2% of MAP pigs hepatic selenium concentration was below 0.5 μg/g which reflects selenium deficiency. Thus, there is a possibility that the low selenium status together with vitamin E deficiency increases oxidative stress and thus contributes to the development of oxidative damage.

During a 2-year period from January 1, 1999, to December 31, 2000, 77 diagnoses of mulberry heart disease were documented at the Iowa State University Veterinary Diagnostic Laboratory (Pallares *et al.*, 2002). The level of vitamin E in the liver of pigs with the disease was significantly decreased, however, liver selenium concentrations were adequate in all pigs. It has been suggested that heredity could be involved in the development of such a disease (Poulsen, 1993). Clearly, there is a need for further research to clarify causes and molecular mechanisms of the development of mulberry heart disease in pigs.

Se dietary supplementation is an effective way to deal with Se deficiency. In some instances of severe Se deficiency an intramuscular administration of this element is used (Van Vleet *et al.*, 1973; Mahan *et al.*, 1973). It is interesting that even after 24 days post-injection, Se concentration in serum and tissues are still elevated (Table 4.1).

Table 4.1. Effect of selenium (Se) injection on its concentration in pig tissues, mg/kg dry matter (adapted from Diehl *et al.*, 1975).

Collection period	Se injection, mg Se/kg BW	Kidney	Liver	Longissimus muscle	Serum
Initial	Control	1.24	0.43	0.24	-
24 days post-injection	Control	2.62	0.43	0.17	0.025
	0.275	4.06	0.72	0.32	0.053
	0.55	5.17	0.91	0.37	0.078
	1.10	5.30	1.05	0.31	0.093
60 days post-injection	Control	3.97	0.38	0.17	0.030
	0.275	5.99	0.54	0.20	0.047
	0.55	6.99	0.60	0.28	0.062
	1.10	6.54	0.71	0.32	0.060

However, forms of selenium accumulated and its availability for synthesis of functional selenoproteins have not been elucidated. It seems likely, that usage of organic form of selenium in the sow's and pig's diets could help meeting Se requirements and there would not be a need for Se injections.

From information provided above it is obvious that manifestations of Se deficiency depend on a range of nutritional and environmental factors such as degree of deficiency, presence of other antioxidants and pro-oxidants, activities of tissues at time of deficiency and some others factors (Jonsson, 1993):

- myopathy is more prevalent in young growing animals with active myogenesis;
- liver necrosis is observed in pigs reared in confinement;
- foetal placenta, uterus and mammary glands are affected in reproduction-lactating animals.

Indeed, alterations in antioxidant defences and oxidative stress are responsible for different metabolic and morphological changes on cellular and sub-cellular levels leading to clinical manifestation of Se deficiency in pigs. Indeed, sudden heart failure caused by single Se deficiency or general antioxidant deficiency can occur at various stages of pig development and related to negative balance between antioxidants and pro-oxidants in stress conditions of modern pig production. Indeed, movement of pigs from one pen to another, the stress of weaning, stress of transportation or stress of farrowing (Close and Cole, 2000) could be important triggers for various detrimental consequences in pigs including sudden death (Surai, 2006).

Recent experimental results confirmed that selenium deficiency upregulated the expression of necrosis-related genes in pigs liver tissues, and necrosis-related phenomena were observed (Zhang *et al.*, 2020). Indeed, Se deficiency was associated with almost 20-fold decrease in serum Se (Figure 4.5) and was found to lead to necroptosis and apoptosis in liver tissues and oxidative stress damage was shown to occurs in the liver tissues of Se-deficient pigs with simultaneous upregulation of the expression levels of necroptosis- and apoptosis-related genes.

A relationship between selenium and antioxidant defences in piglets is summarised in Figure 4.6.

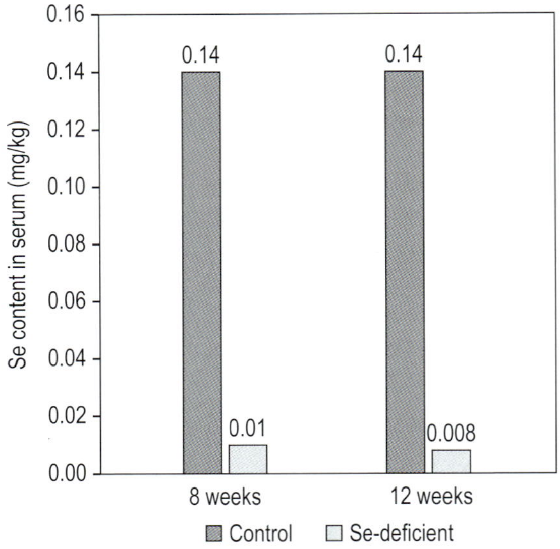

Figure 4.5. Effect of selenium deficient diet fed for 8 and 12 weeks on selenium concentration in pig serum (adapted from Zhang *et al.*, 2020).

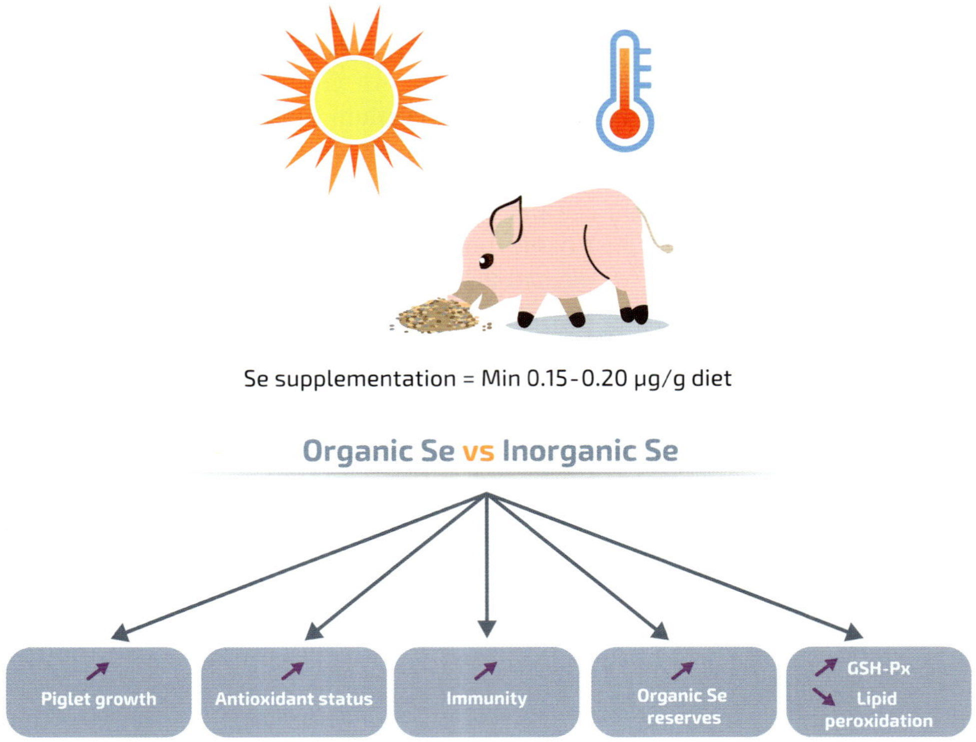

Figure 4.6. Selenium for pigs – selenium and antioxidant defences.

4.4 Establishment of selenium requirement of pigs

It has been suggested that most, if not all, aspects of pig production are affected by feed levels of Se and vitamin E (Jenkins and Hidiroglou, 1972). After research work showing that many regions in the USA are deficient in Se, great efforts have been made to gain FDA approval for Se supplementation of animal diets (Ullrey, 1992). For example, results of a cooperative survey of the dietary selenium fed growing pigs in 13 states (Ku *et al.*, 1972) showed that the highest Se concentration was in South Dakota (0.493 mg/kg) and lowest in Virginia (0.027 mg/kg) being more than 15-fold lower (Table 4.2).

In 1974 U.S. FDA approved the addition of 0.1 mg/kg of selenium to all pig diets. The approval of selenite for use in animal diets followed in other countries as well. For example, addition of selenium to pig feed (max. 0.1 mg/kg) was legalised in Denmark in 1975 (Pedersen and Simesen, 1977). In 1991 the Denmark recommendation was changed to 0.4 mg/kg for pre-starter diets and 0.22 mg/kg for both growing and breeding pigs (Poulsen, 1993). The current maximum accepted total dietary Se level in Denmark is 0.5 mg/kg. Indeed, the inability to supplement selenium-deficient diets prior to FDA approval has cost hundreds of millions of dollars. For example, annual losses to the beef cattle, dairy cattle and sheep industry were estimated in 1975 at nearly $545 million (Ullrey, 1980). Later, in 1982, FDA reconsidered a level of Se supplementation and allowed the addition of 0.3 mg/kg of selenium to diets for pigs up to 20 kg (NRC, 1998). Ultimately, in 1987, levels of supplemental Se in diets for chickens, turkeys, ducks, pigs, sheep, and cattle were approved at 0.3 mg/kg (Ullrey, 1992). Therefore, currently 0.3 mg/kg of selenium in the diet is allowed for all pigs. This dietary Se allowance for pigs was based on tissue Se concentrations, Se balance

Table 4.2 Selenium content in diets and in Longissimus muscle of pigs in the USA (adapted from Ku *et al.*, 1972).

State	Dietary selenium, mg/kg	Selenium in longissimus muscle, mg/kg (wet weight)
South Dakota	0.493	0.521
North Dakota	0.412	0.386
Nebraska	0.330	0.313
Iowa	0.235	0.278
Wisconsin	0.178	0.125
Wyoming	0.158	0.311
Arkansas	0.152	0.212
Idaho	0.086	0.111
Indiana	0.052	0.063
Michigan	0.040	0.052
New York	0.036	0.046
Illinois	0.036	0.059
Virginia	0.027	0.034

Table 4.3 Selenium concentrations (mg/kg wet weight) in tissues of pigs and cattle in Finland (adapted from Venalainen *et al.*, 1997).[1]

Year	Muscle		Liver	
	pig	cattle	pig	cattle
1974	0.11	-	0.45	0.10
1981	0.08	0.04	0.56	0.16
1985	0.08	0.07	0.49	0.28
1986	0.12	0.10	0.56	0.34
1987	0.23	0.16	0.64	0.42
1988	0.26	0.16	0.7	0.45
1989	0.30	0.21	0.73	0.51
1990	0.29	0.19	0.68	0.46
1991	0.26	0.16	0.62	0.40
1992	0.19	0.16	0.57	0.37
1993	0.19	0.14	0.56	0.35
1994	0.13	0.09	0.50	0.32
1995	0.15	0.10	0.52	0.28

[1] Selenium was introduced to all multi-nutrient fertilisers used in Finland after July 1984 at 16 mg/kg fertilisers for cereal crops and 6 mg/kg for grassland crops. In 1990 the level of selenium inclusion to the all fertilisers was decreased to 6 mg/kg.

and the activity response of cellular GPx. In addition to dietary Se supplementation, other solutions of Se deficiency are also available. For example, inclusion of selenite into fertilisers are shown to be effective in increasing Se level in grains, animal diets and animal tissues (Table 4.3).

However, this application found its commercial use only in Finland and New Zealand. Indeed, environmental concerns of potential Se pollution from animal production re-opened a question whether the Se supplementation was in excess of dietary needs (Ullrey, 1992), but there was no sufficient evidence to change Se supplementation and the level of 0.3 mg/kg has been maintained.

There is a great deal of difficulties to establish an optimal Se requirement of pigs:
- Data obtained in the laboratory low-stress conditions cannot be the same in commercial higher-stress conditions. For example, the size of groups in which pigs are tested in experimental conditions is usually much smaller than in commercial conditions. Health status of pigs and microbial population in pig houses could also substantially differ between laboratory and commercial conditions.
- The level of other antioxidants and pro-oxidants in the diets are variable and since Se is an essential part of antioxidant defences its requirement could substantially vary depending on the diet used.

- The data obtained with average-productive animals cannot be transferred directly to high productive stress-sensitive animals. Indeed, pig genotype determines their physiological features including sensitivity to stresses.
- The choice of the end points to assess requirement. Indeed, most of the data are related to Se level in plasma, whole blood, tissues or GPx activity. However, GPx is only one from 25 various selenoproteins identified in animal and human bodies. The usage of GPx activity in plasma is not the best option to assess Se requirement and to assess physiological consequences of Se deprivation or supplementation. Technically, it is the easiest option, but its value is questionable. GPx activity in the target tissues and a combination of various functional tests would be more physiologically relevant tests.
- Form of selenium used. In most of studies sodium selenite was used and it is known that organic selenium is more efficiently assimilated from the diet building Se reserves in tissues.
- Se requirement to maintain various physiological functions could vary substantially. For example, to maintain high immunocompetence, Se requirement could be higher than for growth and development.
- In accordance with NRC (2012) pig requirement in Se is defined at the level of 0.15-0.3 mg/kg (Table 4.4). However, it is necessary to take into account that they are minimal Se requirements and in commercial conditions Se requirements could be substantially higher.

Selenium supplementation of the deficient diet could affect various parameters of pig growth and development; however, the final results depend on many factors. For example, pigs from sows fed a diet deficient in Se and low in vitamin E were fed a Torula yeast diet supplemented with 100 IU dl-alpha-tocopheryl acetate/kg of diet and with Se at 0, 0.025, 0.050, 0.075 or 0.100 mg/kg (Adkins and Ewan, 1984). Selenium supplement had no significant effect on average daily gain, feed intake or

Table 4.4. Selenium requirements of pigs (adapted from NRS, 2012).

Pigs, sows and boars	Amount per kg of diet, mg	Amount per day, mg
Pigs with bodyweight 5-7 kg	0.3	0.08
Pigs with bodyweight 7-11 kg	0.3	0.14
Pigs with bodyweight 11-25 kg	0.25	0.23
Pigs with bodyweight 25-50 kg	0.20	0.30
Pigs with bodyweight 50-75 kg	0.15	0.32
Pigs with bodyweight 75-100 kg	0.15	0.38
Pigs with bodyweight 100-135 kg	0.15	0.42
Gestating sows[1]	0.15	0.31
Lactating sows[1]	0.15	0.89
Sexually active boars	0.30	0.71

[1] Recommendations by Close and Cole (2000) are 0.3 mg Se/kg diet.

gain to feed ratio for the 4-wk experiment. Serum Se increased linearly with increasing supplemental Se and serum GPx activity also increased linearly with increasing supplemental Se. The correlation between serum Se concentration and GPx activity was 0.81 (Adkins and Ewan, 1984). Meyer *et al.* (1981) showed that 0.3 mg Se/kg was required by weanling pigs based on breakpoint regression analysis of liver GPx1 and plasma GPx3 activities.

It is necessary to take into account that relative activity distribution and age-related changes are different between GPx1 and GPx4 in pigs (Lei *et al.*, 1997). In particular, it was found that 0.1 mg Se/kg was insufficient, and 0.3 mg Se/kg was required for young pigs to express the full activity of GPx1 in liver. However, the authors did not test Se level at 0.2 mg Se/kg in their study. In an experiment conducted by Lei *et al.* (1998) eight different tissues were collected from Se-adequate male pigs aged 1, 28 and 180 days, and PH-GPx, and GPx activities were assayed. Both mRNA and activity expression of PH-GPx in most assayed tissues was increased with age but increases in PH-GPx mRNA levels between ages did not fully account for all changes in activity (Lei *et al.,* 1998). Expression of GPx activity was increased more than that of PH-GPx between day 1 and day 28. There was no difference in GPx activities between pigs fed 0.2 and 0.3 or 0.5 mg Se/kg in the experiment of Lei *et al.* (1998). Therefore, it was concluded that full activity expression of GPx1 and GPx4 in various tissues of weanling pigs requires 0.20 mg Se/kg in the corn-soy-based diets (Lei *et al.,* 1998).

As mentioned above, pig Se requirement depends on the experimental conditions, especially on amounts of stresses caused by environmental and technological factors and in experimental laboratory conditions it seems to be lower than in practical conditions of pig producing units. Indeed, in physiological conditions Se requirements of the growing pigs are quite low. For example, using serum GPx activity as the measurement criterion, Mahan *et al.* (1999) showed that the supplemental dietary Se requirement did not exceed 0.10 and 0.05 mg Se/kg diet for the growing and finishing phases, respectively, when added to a basal diet containing 0.06 mg Se/kg. Grower pigs fed sodium selenite had serum GPx activity that reached a plateau at 0.1 mg/kg Se and 0.3 mg/kg when the Se-enriched yeast source was fed, but the interaction response was not significant. On the other hand, during the finisher period, serum GPx activity reached a plateau at 0.1 mg/kg Se for both Se sources (Mahan and Parrett, 1996). In earlier work Se requirement to maintain an optimal GPx activity was shown to be higher. In a trial with sows, different dietary selenium levels (0.1; 0.3 and 0.5 mg/kg) were fed and GPx activity in plasma was determined at day 60, 90 and 110 of gestation, day 5, 15 and 25 of lactation and day 7 post weaning. It was shown that GPx activities dropped within the treatments, independent of dietary selenium, from day 60 of gestation to a minimum at day 5 and 15, respectively, of lactation and increased again after weaning (Neumann *et al.*, 1989).

Highest GPx activities were observed at the 0.5 mg/kg Se level in the diet. However, reproductive performance ameliorated with increasing dietary selenium. For example, piglet's gain from birth to weaning and litter weight at weaning increased significantly. The relation between selenium intake and plasma GPx activity was investigated in 178

German Landrace sows fed on barley-oat concentrate with Se 0.1, 0.5 or 0.9 mg/kg diet. It is concluded that the optimal Se supply for pregnant and non-pregnant sows in concentrate is 0.5 mg/kg. In particular, lowest percent of stillborn piglets was seen in the group fed on the diet containing 0.5 mg/kg selenium (Neumann and Bronsch, 1988; Neumann *et al.*, 1989). The authors concluded that the requirement of the sow during the reproductive cycle is 0.5 mg selenium/kg diet.

4.5 Conclusions

Clinical Se deficiency is a rear event in commercial pig production; however, sub-clinical Se deficiency could be responsible for decreased productive and reproductive performance of pigs.

From the summary of negative consequences of Se deficiency/inadequacy in pigs/sows shown in Figure 4.7 it is clear that meeting Se requirement of growing pigs and producing/reproducing animals is an important task for pig nutritionist. Indeed, only optimal Se status of farm animals, including pigs can give them an opportunity to grow and reproduce to the genetically programmed levels.

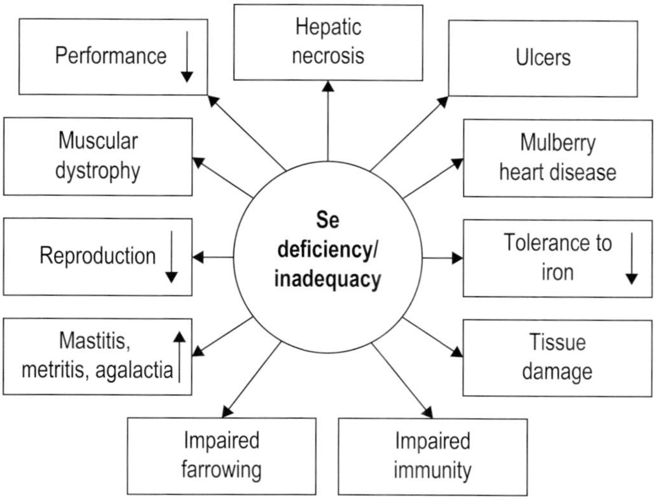

Figure 4.7. Summary of effects of selenium deficiency/inadequacy on pigs.

Acknowledgements

Peter F. Surai and Ivan I. Kochish are supported by a grant of the Government of Russian Federation (Contract No. 14.W03.31.0013).

References

Acda, S.P. and Chae, B.J., 2002. A review of the applications of organic trace minerals in pig nutrition. Pakistan Journal of Nutrition 1: 25-30.

Adkins, R.S. and Ewan, R.C., 1984. Effect of supplemental selenium on pancreatic function and nutrient digestibility in the pig. Journal of Animal Science 58: 351-355.

Bonomi, A., 2001. Effects of selenium used as feed integrator on productive and reproductive efficiency in swine. Rivista di Suinicoltura 42: 115-120.

Close, W.H., 1998. The role of trace mineral proteinates in pig nutrition. In: Lyons T.P. and Jacques, K.A. (eds) Biotechnology in the feed industry. Nottingham University Press, Nottingham, UK, pp. 469-483.

Close, W.H., 2003. Organic selenium may improve sow, piglet performance. Feedstuffs 75: 8.

Close, W.H. and Cole, D.J.A., 2000. Nutrition of sows and boars. Nottingham University Press, Nottingham, UK.

Diehl, J.S., Mahan, D.C. and Moxon, A.L., 1975. Effects of single intramuscular injections of selenium at various levels to young swine. Journal of Animal Science 40: 844-850.

Edwards, M.J., Hartley, W.J. and Hansen, E.A., 1977. Selenium and lowered reproductive efficiency in pigs. Australian Veterinary Journal 53: 553-554.

Fontaine, M., Valli, V.E. and Young, L.G., 1977a. Studies on vitamin E and selenium deficiency in young pigs. III. Effect on kinetics of erythrocyte production and destruction. Canadian Journal of Comparative Medicine 41: 57-63.

Fontaine, M., Valli, V.E. and Young, L.G., 1977b. Studies on vitamin E and selenium deficiency in young pigs. IV. Effect on coagulation system. Canadian Journal of Comparative Medicine 41: 64-76.

Glinke, L.R. and Ewan, R.C., 1977. Selenium deficiency in the young pig. Journal of Animal Science 45: 1334-1340.

Jenkins, K.J. and Hidiroglou, M., 1972. A review of selenium/vitamin E responsive problems in livestock: a case for selenium as a feed additive in Canada. Canadian Journal of Animal Science 52: 591-620.

Jonsson, L., 1993. The pathology of diseases and diffuse disorders due to selenium deficiency in non-ruminants. Norwegian Journal of Agricultural Sciences. Suppl. 11: 95-103.

Kirchgessner, M., Hartmann, S. and Eder K., 1995. The effect of selenium deficiency on the fatty acid composition of various tissues and the osmotic fragility of erythrocytes in the growing pig. Journal of Animal Physiology and Animal Nutrition 73: 77-85.

Korpela, H., 1990. Hepatic selenium concentration in pigs with microangiopathy (mulberry heart disease)-an animal model for the study of oxidative damage. International Journal for Vitamin and Nutrition Research 60: 156-158.

Ku, P.K., Ely, W.T., Groce, A.W. and Ullrey, D.E., 1972. Natural dietary selenium, α-tocopherol and effect on tissue selenium. Journal of Animal Science 34: 208-211.

Lei, X.G., Dann, H.M., Ross, D.A., Cheng, W.H., Combs, G.F. and Roneker, K.R., 1998. Dietary selenium supplementation is required to support full expression of three selenium-dependent glutathione peroxidases in various tissues of weanling pigs. The Journal of Nutrition 128: 130-135.

Lei, X.G., Ross, D.A. and Roneker, K.R., 1997. Comparison of age-related differences in expression of phospholipid hydroperoxide glutathione peroxidase mRNA and activity in various tissues of pigs. Comparative Biochemistry and Physiology 117B: 109-114.

Lindberg, P., 1968. Selenium determination in plant and animal material, and in water. A methodological study. Acta Veterinaria Scandinavica. Suppl. l 23: 1-48.

Mahan, D.C., 1991a. Assessment of the influence of dietary vitamin E on sows and offspring in three parities: reproductive performance, tissue tocopherol, and effects on progeny. Journal of Animal Science 69: 2904-2917.

Mahan, D.C., 1991b. Vitamin E and selenium in swine nutrition. In: Miller E.R., Ullrey D.E. and Lewis A.J. (eds) Swine nutrition. Butterworth-Heineman, Boston, MA, USA, pp. 193-214.

Mahan, D.C., 1994. Effects of dietary vitamin E on sow reproductive performance over a five-parity period. Journal of Animal Science 72: 2870-2879.

Mahan, D.C., 1999. Selenium nutrition in swine – the emerging value of organic selenium. In: Hogan, J. (ed.) Proceedings The Alvin Lloyd Moxon Honorary Lectures on Selenium and Vitamin E. The Ohio State University, USA, pp.37-47.

Mahan, D.C. and Parrett, N.A., 1996. Evaluating the efficacy of selenium-enriched yeast and sodium selenite on tissue selenium retention and serum glutathione peroxidase activity in grower and finisher swine. Journal of Animal Science 74: 2967-2974.

Mahan, D.C. and Peters, J.C., 2004. Long-term effects of dietary organic and inorganic selenium sources and levels on reproducing sows and their progeny. Journal of Animal Science 82: 1343-1358.

Mahan, D.C., Cline, T.R. and Richert, B., 1999. Effects of dietary levels of selenium enriched yeast and sodium selenite as selenium sources fed to growing-finishing pigs on performance, tissue selenium, serum glutathione peroxidase activity responses, semen quality, and subsequent fertilization rates in mature gilts. Journal of Animal Science 75: 2994-3003.

Mahan, D.C., Jones, J.E., Cline, J.H., Cross, R.F., Teague, H.S. and Grifo, A.P. Jr., 1973. Efficacy of selenium and vitamin E injections in the prevention of white muscle disease in young swine. Animal Science 36: 1104-1108.

Mahan, D.C., Penhale, L.H., Cline, J.H., Moxon, A.L., Fetter, A.W. and Yarrington, J.T., 1974. Efficacy of supplemental selenium in reproductive diets on sow and progeny performance. Journal of Animal Science 39: 536-543.

Marin-Guzman, J., Mahan, D.C. and Pate, J.L., 2000. Effect of dietary selenium and vitamin E on spermatogenic development in boars. Journal of Animal Science 78: 1537-1543.

Marin-Guzman, J., Mahan, D.C., Chung, Y.K., Pate, J.L. and Pope, W.F., 1997. Effects of dietary selenium and vitamin E on boar performance and tissue carcass characteristics, and loin quality. Journal of Animal Science 77: 2172-2179.

Meyer, W.R., Mahan, D.C. and Moxon, A.L., 1981. Value of dietary selenium and vitamin E for weanling swine as measured by performance and tissue selenium and glutathione peroxidase activities. Journal of Animal Science 52: 302-311.

Mihailovic, M., Velickovski, S. and Ilic, V., 1989. Reproductive performance in sows fed a diet low in selenium. Acta Veterinaria (Beograd) 39: 19-24.

National Research Council (NRC), 1998. Nutrient requirements of swine. 10th revised edition. National Academies Press, Washington, DC, USA.

National Research Council (NRC), 2012. Selenium. In: Nutrient requirements of swine. 11th ed. National Academies Press, Washington, DC, USA.

Neumann, U.F. and Bronsch, K., 1988. Studies on the optimum selenium supplementation of non gravid and gravid sows. Journal of Veterinary Medicine Series A 35: 673-682.

Neumann, U.F, Bronsch, K. and Schneider, D., 1989. The optimal selenium supplementation in reproducing sows. Zentralblatt fur Veterinarmedizin Reihe A 36: 175-187.

Nielsen, H.E., Danielsen, V., Simesen, M.G., Gissel-Nielsen, G., Hjarde, W., Leth, T. and Basse, A., 1979. Selenium and vitamin E deficiency in pigs. I. Influence on growth and reproduction. Acta Veterinaria Scandinavica 20: 276-288.

Nielsen, T.K., Wolstrup, C., Schirmer, A.L. and Jensen, P.T., 1989. Mulberry heart disease in young pigs without vitamin E and selenium deficiency. Veterinary Record 124: 535-537.

Nolan, M.R., Kennedy, S., Blanchflower, W.J. and Kennedy, D.G., 1995. Lipid peroxidation, prostacyclin and thromboxane A2 in pigs depleted of vitamin E and selenium and supplemented with linseed oil. British Journal of Nutrition 74: 369-380.

Oropeza-Moe, M., Falk, M., Vollset, M., Wisløff, H., Bernhoft, A., Framstad, T. and Salbu, B., 2019. A descriptive report of the selenium distribution in tissues from pigs with mulberry heart disease (MHD). Porcine Health Management 5: 17.

Oropeza-Moe, M., Wisløff, H. and Bernhoft, A., 2015. Selenium deficiency associated porcine and human cardiomyopathies. Journal of Trace Elements in Medicine and Biology 31: 148-156.

Pallares, F.J., Yaeger, M.J., Janke, B.H., Fernandez, G. and Halbur, P.G., 2002. Vitamin E and selenium concentrations in livers of pigs diagnosed with mulberry heart disease. Journal of Veterinary Diagnostic Investigation 14: 412-414.

Pedersen, K.B. and Simesen, M.G., 1977. Feed supplementation with selenium in relation to the vitamin E-selenium deficiency syndrome in pigs. Nordisk Veterinaermedicin 29: 161-165.

Poulsen, H.D., 1993. Selenium supplementation in pigs. Norwegian Journal of Agricultural Sciences. Suppl. 11: 189-197.

Reid, G.M., 2001. Sudden infant death syndrome and enteric infection. Medical Hypotheses 57: 580-582.

Rice, D.A. and Kennedy, S., 1989. Vitamin E, selenium, and polyunsaturated fatty acid concentrations and glutathione peroxidase activity in tissues from pigs with dietetic microangiopathy (mulberry heart disease). American Journal of Veterinary Research 50: 2101-2104.

Shen, H., Thomas, P.R., Ensley, S.M., Kim, W.I., Loynachan, A.T., Halbur, P.G. and Opriessnig T., 2011. Vitamin E and selenium levels are within normal range in pigs diagnosed with mulberry heart disease and evidence for viral involvement in the syndrome is lacking. Transboundary and Emerging Diseases 58: 483-491.

Stierstorfer, B., Muller, M., Hermanns, W. and Schmahl, W., 2001. Central nervous manifestation of mulberry heart disease in pigs. Wiener Tierarztliche Monatsschrift 88: 187-191.

Surai, P.F., 2006. Selenium in nutrition and health. Nottingham University Press, Nottingham, UK.

Surai, P.F., 2021a. Antioxidant defence systems in animal health. Chapter 1. In: Surai, P.F. (ed.) Selenium in pig nutrition and health. Wageningen Academic Publishers, Wageningen, the Netherlands, pp. 17-38. https://doi.org/10.3920/978-90-8686-912-1_

Surai, P.F., 2021b. Selenium and immunity. Chapter 9. In: Surai, P.F. (ed.) Selenium in pig nutrition and health. Wageningen Academic Publishers, Wageningen, the Netherlands, pp. 261-347. https://doi.org/10.3920/978-90-8686-912-1_9

Thoren-Tolling, K., 1984. Studies of serum-transferrin and some hematologic parameters in vitamin E and selenium deficient pigs. Annals of Veterinary Research 15: 387-394.

Ullrey, D.E., 1980. Regulation of essential nutrient additions to animal diets (selenium – a model case). Journal of Animal Science 51: 645-651.

Ullrey, D.E., 1992. Basis for regulation of selenium supplements in animal diets. Journal of Animal Science 70: 3922-3927.

Van Vleet, J.F. and Ferrans, V.J., 1977. Ultrastructure of hyaline microthrombi in myocardial capillaries of pigs with spontaneous 'mulberry heart disease'. American Journal of Veterinary Research 38: 2077-2080.

Van Vleet, J.F., Ferrans, V.J. and Ruth, G.R., 1977a. Ultrastructural alterations in nutritional cardiomyopathy of selenium-vitamin E deficient swine. I. Fiber lesions. Laboratory Investigation 37: 188-200.

Van Vleet, J.F., Ferrans, V.J. and Ruth, G.R., 1977b. Ultrastructural alterations in nutritional cardiomyopathy of selenium-vitamin E deficient swine. II. Vascular lesions. Laboratory Investigation 37: 201-211.

Van Vleet, J.F., Meyer, K.B. and Olander, H.J., 1973. Control of selenium-vitamin E deficiency in growing swine by parenteral administration of selenium-vitamin E preparations to baby pigs or to pregnant sows and their baby pigs. Journal of the American Veterinary Medical Association 163: 452-456.

Van Vleet, J.F., Meyer, K.B., Olander, H.J. and Ruth, G.R., 1975. Efficacy and safety of selenium-vitamin E injections in newborn pigs to prevent subclinical deficiency in growing swine. American Journal of Veterinary Research 36: 387-393.

Van Vleet, J.F., Ruth, G. and Ferrans, V.J., 1976. Ultrastructural alterations in skeletal muscle of pigs with selenium-vitamin E deficiency. American Journal of Veterinary Research 37: 911-922.

Venalainen, E.R., Hirvi, T. and Hirn, J., 1997. Effect of selenium supplementation on the selenium content in muscle and liver of Finish pigs and cattle. Journal of Agricultural and Food Chemistry 45: 810-813.

Zhang, Y., Cui, J., Lu, Y., Huang, C., Liu, H. and Xu, S., 2019. Selenium deficiency induces inflammation via the iNOS/NF-κB pathway in the brain of pigs. Biological Trace Element Research 196: 103-109. https://doi.org/10.1007/s12011-019-01908-y

Zhang, Y., Yu, D., Zhang, J., Bao, J., Tang, C. and Zhang, Z., 2020. The role of necroptosis and apoptosis through the oxidative stress pathway in the liver of selenium-deficient swine. Metallomics 12: 607-616. https://doi.org/10.1039/C9MT00295B

Chapter 5
Selenium in sow nutrition

P.F. Surai[1-7]
[1]*Vitagene and Health Research Centre, Bristol, BS4 2RS, United Kingdom;* [2]*Department of Hygiene and Poultry Sciences, Moscow State Academy of Veterinary Medicine and Biotechnology Named After K.I. Skryabin, 109472 Moscow, Russia;* [3]*Department of Microbiology and Biochemistry, Faculty of Veterinary Medicine, Trakia University, 6000 Stara Zagora, Bulgaria;* [4]*Department of Animal Nutrition, Faculty of Agricultural and Environmental Sciences, Szent Istvan University, 2103 Gödöllo, Hungary;* [5]*Saint-Petersburg State Academy of Veterinary Medicine, 196084 St. Petersburg, Russia;* [6]*Sumy National Agrarian University, 40000 Sumy, Ukraine;* [7]*Odessa National Academy of Food Technologies, 65039 Odessa, Ukraine; psurai@feedfood.co.uk*

Abstract

Oxidative stress in commercial sows is an important risk factor associated with decreased productive and reproductive performance. Selenium is shown to be an important dietary element in sow nutrition. Organic selenium has many important advantages in sow nutrition in comparison to traditional sodium selenite including better Se status of sows, especially in the case of advanced parities; improved antioxidant defences of sows; increased Se concentration in colostrum and milk and improved antioxidant status of colostrum and milk; improved Se transfer via placenta; improved Se status of foetus and development of pig embryos. Replacement of sodium selenite by organic Se in the sow's diet improved Se status of newborn piglets. Increased Se concentration in piglet muscles could be considered as an important storage form of Se to be used in stress conditions. Organic Se in sow's diet (vs sodium selenite) provides better Se status and antioxidant status of weaning piglets, as well as improves thyroid metabolism and increases activities of major digestive enzymes in piglet pancreas at time of weaning. Recent results indicated that a new source of organic selenium in the form of 2-hydroxy-4-methylselenobutanoic acid supplied in the same dose as Se-Yeast in the pig diets could provide additional benefit in terms of Se reserves in the muscles as SeMet but also higher levels of SeCys supporting enhanced levels of antioxidant selenoproteins. One can thus expect higher Se levels in sow tissues as well as transfer to the progeny via placenta. An optimal Se form and level in the sow's diet potentially can be translated into better antioxidant protection in stress conditions of commercial pig production and maintain sow reproductive performance and newborn piglet viability at high level. Indeed, additional well-designed organic Se trials with big numbers of sows in commercial conditions are needed to explore a full potential of organic Se in sow nutrition.

Keywords: selenium, sow, SeMet, OH-SeMet, piglets

Peter F. Surai (ed.) **Selenium in pig nutrition and health**
DOI 10.3920/978-90-8686-912-1_5, © Wageningen Academic Publishers 2021

5.1 Introduction

Effect of oxidative stress in animal reproduction has been extensively studied recently. A hypothesis of programming by oxidative stress has been developed (Luo *et al.*, 2006) suggesting that oxidative stress in dams can have long-lasting consequences for the progeny. The authors suggested that mechanisms of oxidative stress programming might be through directly modulating gene expression or indirectly through the effects of certain oxidised molecules. In fact, oxidative stress can be modified during gestation and early postnatal periods by dietary antioxidants and selenium is considered to be a major player in the antioxidant systems of the body (Surai, 2006). The aim of this chapter is to update existing information about Se roles in sow nutrition with a specific emphasis to the oxidative stress and possibilities of decreasing negative consequences of the stresses by using dietary Se supplementation in an optimal form.

5.2 Oxidative stress and antioxidant defences in sows

There is a great body of evidence to indicate that the antioxidant-prooxidant balance is an important regulator of the mammalian reproductive functions, such as ovarian follicular development, ovulation, fertilisation, luteal steroidogenesis, endometrium receptivity and shedding, embryonic development, implantation and early placental growth and development (Al-Gubory *et al.*, 2010). It is well known that reactive oxygen species (ROS) have a dual role in the female reproductive tract: they serve as key signal molecules in physiological processes but can also be considered important elements in pathological processes (Rizzo *et al.*, 2012).

Indeed, oxidative stress plays a central role in the pathophysiology of many different disorders, including complications of gestation (Burton and Jauniaux, 2011). In fact, gestation is considered to be a state of oxidative stress arising from increased placental mitochondrial activity and production of ROS. Furthermore, the placenta also produces ROS which have pronounced effects on placental function including trophoblast proliferation and differentiation and vascular reactivity (Myatt and Cui, 2004). Furthermore, labour is shown to be a powerful inducer of placental oxidative stress and inflammatory reactions (Cindrova-Davies *et al.*, 2007). Interestingly, the placental antioxidant capacity in sows was shown to be improved by isomaltooligosaccharide and Bacillus dietary supplementation during late gestation (Gu *et al.*, 2019). Therefore, the antioxidant defence system is shown to be extremely important in animal reproduction (Surai, 2006; Surai and Fisinin, 2016; Surai *et al.*, 2019a,b).

A sow reproduction stress hypothesis has been proposed (Wen *et al.*, 2009). The authors suggested that the special reproductive tasks such as oestrus, gestation, birth and lactation cause stress, which can be called the reproductive stress. It also includes female passive stress due to hormones released by foetus as well as female intrauterine stress caused by uterine tension and intrauterine environment. In fact, gestation is considered to be a period of constant oxidative stress for the dam (Wisdom *et al.*,

1991). It seems likely that at time of gestation sows are characterised by decreased antioxidant protection and undergo increased DNA damage. For example, there was a decline in serum vitamin E from about 80 to 90 days after insemination through farrowing (Mahan *et al.*, 2007). Similarly, plasma vitamin E concentration was lower by 56% (3.07 vs 7.14 mmol/l) at the end of gestation (d 110 of gestation) as compared with d 30 of gestation (Berchieri-Ronchi *et al.*, 2011). Furthermore, the authors showed that lymphocyte DNA damage significantly increased in the second quarter of gestation and continued throughout the gestational and weaning periods reflecting systemic oxidative stress during the late gestation period. In addition, the reproductive performance of sows was related to their oxidative stress status during gestation and lactation (Zhao *et al.*, 2013) and there is a risk that increased oxidative damages in sows during late gestation may negatively affect the growth and health of foetuses as well as postpartum growth of piglets (Zhao *et al.*, 2013). Mahan *et al.* (2007) demonstrated that the sow serum level of ascorbic acid was relatively constant during the initial 60 to 80 days of gestation, whereupon there was a decline in its concentration to farrowing and throughout lactation. Therefore, the sow ascorbic acid status may be low during the period of time of extensive stress and serum Se begins declining around 60 day post-coitum and decreases even more rapidly as approaching parturition. At the same time GPx activity also declines (Mahan *et al.*, 2007). A marked decrease of GST activity and increase of MDA concentration on day 2 and 7 were observed in sows as compared to the value of day -15 ($P<0.05$). Furthermore, the NOx concentration was significantly elevated on day -7, 2 and 7 of farrowing (Gaykwad *et al.*, 2019) confirming oxidative stress in sows at the end of gestation and after farrowing. It is generally accepted that environmental stress can compromise the antioxidant system of the sow. For example, increased oxidative damage to lipid, protein, and DNA was one of the major contributing factors for reduced reproductive performance of sows due to heat stress (Zhao *et al.*, 2011). Furthermore, inclusion of fish oil (2%) into the diet can also cause an oxidative stress and increases lipid peroxidation in sows (Tanghe *et al.*, 2015).

It should be mentioned that oxidative stability of erythrocytes decreased linearly with increasing amounts of fish oil in the diet and there was a decrease of both piglet survival and weaning weight with increasing quantities of fish oil supplemented (Cools *et al.*, 2011). Similarly, maternal fish oil supplementation was found to impose oxidative stress as evidenced by increased MDA concentration in sow plasma, colostrum, milk and piglet plasma. Interestingly, adaptive increase in total antioxidant capacity (T-AOC) and activities of GPx and SOD in sow plasma were not able to prevent lipid peroxidation (Shen *et al.*, 2015). Furthermore, in sows fed oxidised corn oil the placental antioxidant system was also shown to have an adaptive response to oxidative stress through increasing gene expression of antioxidant enzymes (Su *et al.*, 2017). Indeed, the fat-rich diets fed to sows during gestation was shown to have beneficial effects on reproductive performance, but aggravated the oxidative stress in sow and piglets as evidenced by decreased serum activities of SOD and GPx and increased serum MDA levels (Wang *et al.*, 2016). Recent results provide evidence that acutely elevated cortisol concentrations can have long term effects on the growth of offspring and support the possibility that maternal nutrition may alter reproductive function of male offspring (Mack *et al.*, 2014). Interestingly, sows in both high social rank and low social rank groups were shown to have increased oxidative damage during late gestation and lactation (Zhao *et al.*, 2013). Therefore, oxidative stress associated with increased metabolic activity, negative energy balance and low antioxidant defences in gestating sows is a risk factor of decreased sow productive and reproductive performance and could detrimentally affect progeny.

5.3 Oxidative stress and antioxidant defence in newly born piglets

During the foetal-neonatal transition and at birth important circulatory and respiratory changes associated with an increase in the O_2 partial pressure in tissues occur. This causes a physiological oxidative stress (Pallardo *et al.*, 1991; Sastre *et al.*, 1994; Saugstad, 1990). Therefore, newborns are at high risk of oxidative stress and they are very susceptible to free radical oxidative damage. Several lines of evidence support this assumption including their exposure to high oxygen concentrations, reduced antioxidant defence and free iron which enhances the Fenton reaction leading to production of highly toxic hydroxyl radicals (Saugstad, 2003, 2005). Indeed, there is evidence of an imbalance between antioxidant- and oxidant-generating systems which causes oxidative damage (Buonocore *et al.*, 2001). Furthermore, Yin *et al.* (2013) showed that newborn piglets suffered seriously from birth oxidative stress because of the naive antioxidant system. For example, transferring piglets to a nursery at d3 was shown to impose oxidative stress in the gut as evidenced by elevated GSSG/GSH ratio in the distal small intestine at d5 associated with villus atrophy. At the same time GSH concentration in the liver also decreased (Vergauwen *et al.*, 2017). Furthermore, maternal diet can affect antioxidant defences of progeny piglets. For example, pregnant sows fed a low-energy (LE) diet predisposed foetal offspring muscles to oxidative stress manifested by decreased expression and activity of Cu, Zn-SOD and catalase, two major antioxidant enzymes (Zou *et al.*, 2016).

During mammalian prenatal development, the antioxidant system is considered as immature (Allen and Venkatraj, 1992; Fantel, 1996) with respective maturation in the postnatal life. For example, in the liver of newborn rats GPx and glutathione transferase (GST) activities represent only 20 and 15% of the values in the adult animals (Pallardo *et al.*, 1991). Indeed, the first week of the postnatal development of the piglet is the most important time for the antioxidant system development. In fact, SOD and GPx activity in the piglet plasma significantly increased for the first week after birth. At the same time lipid and protein oxidation in piglet plasma significantly decreased. The authors also indicated significant increase in expression of ileal Nrf2 signalling pathway during first week of the postnatal development (Yin *et al.*, 2013).

It has been shown that due to limited placental transfer, piglets are born with low vitamin E content, including its low levels in liver, brain, heart, kidney at birth (Hidiroglou *et al.*, 1993). It is interesting to note that in the case of high dietary intake of vitamin E by the gestating sow, there was a significant increase of vitamin E concentration in sow's plasma, but its transfer via placenta to the piglet was still low (Lauridsen *et al.*, 2002; Mahan and Vallet, 1997; Pinelli-Saavedra and Scaife, 2005) resulting in low vitamin E content at birth in all tissues (Pinelli-Saavedra *et al.*, 2008). In addition to low vitamin E status of the newly born piglet, there are some features of vitamin C metabolism which complicate antioxidant defences. For example, during early developmental stages, foetal piglet has the capacity to synthesise vitamin C, however, it loses the ability as gestation advances (Ching, 2000). In fact, the pig foetus has a high rate of ascorbic acid synthesis by mid gestation (i.e. day 60), but its activity declined as gestation progresses (Ching *et al.*, 2001a). It has been shown that the neonatal pig is born with a moderate body reserve and blood concentration of ascorbic acid (Ching *et al.*, 2001b). However, the neonatal pig is not synthesising vitamin C during the first week of life, probably due to inhibition of the process by ascorbic acid from colostrum and milk (Mahan *et al.*, 2004). Therefore, vitamin C status of the piglet is dependent on its supply via placental transfer, and its content in colostrum and milk (Chavez, 1983; Hidiroglou and Batra, 1995). However, AA concentration in colostrum was shown to decrease after two days of farrowing of sows (Lipko-Przybylska and Kankofer, 2012).

There is a significant decrease in vitamin C concentration in pig plasma during first week after birth and further decrease in the next 3 weeks of the postnatal development (Yen and Pond, 1983) and sow's diet supplementation with ascorbic acid did not change this trend. Later it was shown that dietary vitamin C supplementation (1 or 10 g/day) only slightly increased vitamin C in the colostrum (independently on the dose used) and did not affect its level in the milk (Pinelli-Saavedra and Scaife, 2005). There is a range of stress factors elevated detrimental consequences of oxidative stress related to physiological changes in the body. For example, excessive back-fat was shown to induce increased oxidative stress and greater expression of proinflammatory cytokines in sow's placenta (Zhou *et al.*, 2019), aggravate mitochondrial injury in pig term placenta, with potential detrimental consequences on placental function (Tian *et al.*, 2020). Furthermore, sows under a high thermal environment were shown to be characterised by increased oxidative stress during late gestation leading to reduced reproductive performance of animals (Zhao and Kim, 2019). The composition of dietary fibre in sow's pregnancy diet was found to have an important role in improving antioxidant capacity and decreasing inflammatory response of mothers and their offspring through modulating the composition of gut microbiota. In particular, a high insoluble/soluble fibre ratio (9.12 and 12.81) was associated with decreased antioxidant defences in the sow's liver (Nrf2 and HO-1) and increased NF-κB expression leading to inflammation (Li *et al.*, 2019).

Sows suffering excess weight loss during lactation were shown to be characterised by a severe oxidative stress and metabolic damage with excessive protein breakdown and lipids mobilisation at weaning (Hu *et al.*, 2019). It was shown that the Met:Lys ratio in sow's diet during first week of lactation was shown to control their antioxidant defence mechanisms, including GPx activity and total AOA in their plasma (Wei *et al.*, 2019), which could affect resistance to oxidative stress. From the data presented above it is clear that oxidative stress is the main concern for pregnant sows and newly born piglets. Therefore, improvement of antioxidant defences of sows and piglets at those critical periods of time would be of great importance for their health, productive and reproductive performance. Since vitamin E and ascorbic acid status of newly born piglets is comparatively low, the crucial role of selenoproteins participating in antioxidant defences and redox status maintenance is quite obvious.

5.4 Maternal effect on the progeny: organic selenium vs selenite

The importance of Se in sow reproduction has been studied since 1970th. It was shown that sows which were fed the corn-soybean diet from feedstuffs low in selenium gradually depleted their body selenium reserves and produced smaller litters during the second parity (Mahan *et al.*, 1974). The addition of supplemental selenium at the rate of 0.1 mg/kg in the form of SS maintained constant serum levels of selenium, higher tissue selenium levels and larger litters during the second parity (Mahan *et al.*, 1975). Sow serum, milk and tissue Se levels increased as the dietary Se level increased. Tissue and serum Se concentrations also increased in the neonatal, 2- and 4-week-old nursing pig as the supplemental Se level was raised in the sow diet. During the 4-week lactation period the hepatic Se content declined linearly in the progeny of all sow groups while kidney Se concentrations increased. Upon weaning, and particularly during the first week, there was a marked decline in the serum, hepatic and kidney Se concentrations. These results suggested a relatively short carry-over of Se from the dam to the progeny (Mahan *et al.*, 1977). Indeed, the concentration of Se in maternal whole blood and liver decreased during gestation in sows fed a low-Se diet compared to those fed a Se-supplemented diet (Hostetler and Kincaid, 2004a). It is interesting to observe that sodium selenite in the maternal diet was not effective in changing Se concentration in the foetus. Indeed, foetal liver Se content decreased in both treatments as gestation progressed; however, the decrease was greatest in the liver of foetuses from sows fed the low-Se diet. There was a marked decrease in GPx activity in foetal pig liver at parturition in comparison to earlier stages of gestation, but the changes were not associated with supplementation of the maternal diet with sodium selenite (Hostetler and Kincaid, 2004b). It is important to note that, maternal liver GPx activity was approximately 12 times greater than foetal liver GPx activity, regardless of dietary treatment. Furthermore, lipid peroxidation indexes were greater in liver homogenates from gilts fed the low-Se diet, even though there was no difference in GPx activity. This may indicate that selenoproteins other than GPx were affected, decreasing the resistance of the foetus to oxidative stress.

Pehrson *et al.* (2001) reported that the activity of the selenium-dependent enzyme GPx in the serum of piglets was very low during their first week of life even though the sows' feed had been supplemented with 0.35 mg inorganic selenium/kg. The authors concluded that the selenium status of newborn piglets may be more critical for their health and immunity than their vitamin E status. Similar conclusions were reported by Loudenslager *et al.* (1986) who showed that the biological plasma antioxidant status (vitamin E content and GPx activity) of pigs at birth was very low, predisposing them to oxidative stress. The mean concentration of vitamin E (1.77 mg/litre) and GPx (682 U/litre) in the blood plasma of the piglets at birth were indeed low, but had increased significantly by day 3 (vitamin E: 5.31 mg/litre; GPx: 996 U/litre), reflecting the provision of these antioxidant compounds with colostrum and milk (Hakansson *et al.*, 2001). The highest concentration of vitamin E (17.9 mg/kg) and Se (200 μg/litre) was found in colostrum, but there was large variation between sows. One week post-partum, the concentration of vitamin E and Se in sow milk had decreased, on average by 86 and 68%, respectively. In general, it was realised that there are some limitations in usage of SS as a source of Se for sows. In one study (Blodgett *et al.*, 1989) a 9-fold increase in sodium selenite level in sow feed produced only a moderate increase in Se concentration in blood and colostrum. Average Se concentration in sows near parturition was 144.1 and 174.3 μg/litre of blood, 86.0 and 148.0 μg/litre of colostrum and 59.2 and 79.9 μg /litre of blood in piglets, for 0.1 and 0.9 mg/kg Se fed sows respectively.

In 1990th, twenty years later since the beginning of the Se study in sows, it was realised that organic Se was more effective than SS in sow nutrition. In particular, first-parity gilts were fed sodium selenite or organic selenium (Se-Yeast) at doses of 0.1 and 0.3 mg/kg, starting at approximately 60 days before breeding until weaning at 21 days. It was reported that the organic selenium source provided several advantages over the inorganic one (Mahan and Kim, 1996; Table 5.1 and Table 5.2):

- it was more effectively transferred via the placenta, resulting in higher Se content in loin and liver in the newborn piglet;
- it was more efficiently assimilated into colostrum and milk;
- it was more effective in maintaining the Se status of the developing piglet, resulting in higher loin Se concentration in piglets at weaning.

Table 5.1 Effect of selenium (Se) fed to reproducing gilts on Se concentration in gilt serum, milk and tissues (adapted from Mahan and Kim, 1996).

	Inorganic Se		**Organic Se**	
Item	**0.15**	**0.30**	**0.15**	**0.30**
Serum Se, mg/kg				
30 d postfeeding	0.193	0.195	0.197	0.199
60 d postfeeding	0.189	0.199	0.202	0.216
Breeding	0.180	0.175	0.185	0.209
35 d postcoitum	0.178	0.204	0.190	0.210
90 d postcoitum	0.163	0.174	0.167	0.183
Wean, 21 d	0.175	0.187	0.135	0.173
Milk Se, mg/kg				
0 d postpartum	0.129	0.110	0.132	0.116
7 d postpartum	0.038	0.048	0.054	0.081
14 d postpartum	0.035	0.047	0.043	0.078
21 d postpartum	0.041	0.044	0.051	0.090
Tissue Se, mg/kg				
Loin	0.093	0.115	0.131	0.224
Liver	0.464	0.422	0.367	0.528

Table 5.2. Effect of selenium (Se) fed to reproducing gilts on neonatal and weaning pig selenium measurements (adapted from Mahan and Kim, 1996).

	Inorganic Se, mg/kg		Se-yeast, mg/kg	
Item	**0.1**	**0.3**	**0.1**	**0.3**
Neonate				
Loin Se, mg/kg	0.039	0.055	0.068	0.085
Liver Se, mg/kg	0.236	0.250	0.283	0.310
Weaning, 21 d				
Serum Se, mg/kg	0.062	0.075	0.082	0.102
Loin Se, mg/kg	0.101	0.121	0.129	0.244
Liver Se, mg/kg	0.352	0.388	0.535	0.509

Several studies with sows have shown that increasing Se supplementation from either inorganic or organic source from 0.1 to 0.3 mg/kg generally resulted in increased Se concentration in milk, neonatal loin and weaning pig loin, but that the organic source produced a more marked improvement. Therefore, the higher milk Se concentration has been shown to improve the Se status in the nursery pig at weaning (Mahan and Kim, 1996) and this higher status may help reduce the occurrence and severity of post-weaning mortality frequently encountered in practice. The selenium and vitamin E content of milk also declines with advancing parity (Mahan, 1991a,b, 1994). This indicates that, in later parities, sows may need more vitamin E and Se to maintain their body reserves which can then be effectively transferred to their progeny, when necessary. Indeed, piglets born from older sows could be at greater risk of oxidative stress at birth and in their early post-natal development.

Mahan (2000) conducted a study with six dietary treatments in a 2×2 factorial arrangement with two additional treatments. Inorganic (sodium selenite) or organic (Se-Yeast) Se sources were added to the diet at 0.15 or 0.30 mg/kg Se. A non-Se-fortified corn-soybean meal basal diet served as a negative control, and a sixth group was fed 0.15 mg/kg Se from each source (inorganic and organic). A total of 43 sows were fed their treatment diets at 2.2 kg/day from 6 days pre-partum to parturition and throughout a 14-d lactation period. The major results can be summarised as follows:

- It was concluded that dietary Se supplementation is an important means to maintain antioxidant defences of the sow. For example, when the basal, non-supplemented diet was fed, sow serum GPx activity declined from 6 days pre-partum and remained low throughout lactation (Mahan, 2000). The inclusion of selenium into the sows' diet caused an increase in sow serum Se concentration and serum GPx activity at both 7- and 14 days post-partum.
- It was shown that colostrum Se concentration was much higher than that in the milk (Mahan, 2000). In general, the Se concentration of colostrum has been found to be markedly higher than that of following milk in several species (Falk *et al.*, 2019; Meyer *et al.*, 2011; Peters *et al.*, 2010; Salman *et al.*, 2013; Slavik *et al.*, 2008).

- It was confirmed that colostrum is an important source of selenium for newly born piglets. However, selenium transfer to colostrum was minimal when sodium selenite was provided in the diet of the sow. Indeed, the short-term feeding of selenite at 0.15 or 0.30 mg/kg Se did not influence the Se content of colostrum (Mahan, 2000). On the other hand, the inclusion of organic selenium in the sows' diet significantly increased Se content of colostrum.
- The milk Se content at days 7 and 14 of lactation was 2.5 to 3 times higher when the organic Se was provided in the diet, compared with the inorganic source (Figure 5.1).
- The low efficiency of the transfer of Se from sodium selenite to colostrum and milk was confirmed by using a combination of inorganic and organic Se, each at 0.15 mg/kg Se. When this combination was fed, colostrum and milk Se content were similar to those of sows fed 0.15 mg/kg Se from the organic source alone.
- Piglet serum GPx activity at days 7 or 14 of age was not affected by dietary Se level nor Se source fed to sows, but serum Se increased with increasing dietary Se level, particularly when sows were fed organic Se.
- With an assumed daily milk production in sows of 10 kg/d, approximately 30% of the inorganic and 80% of the organic dietary Se would be excreted with the milk (Mahan, 2000).

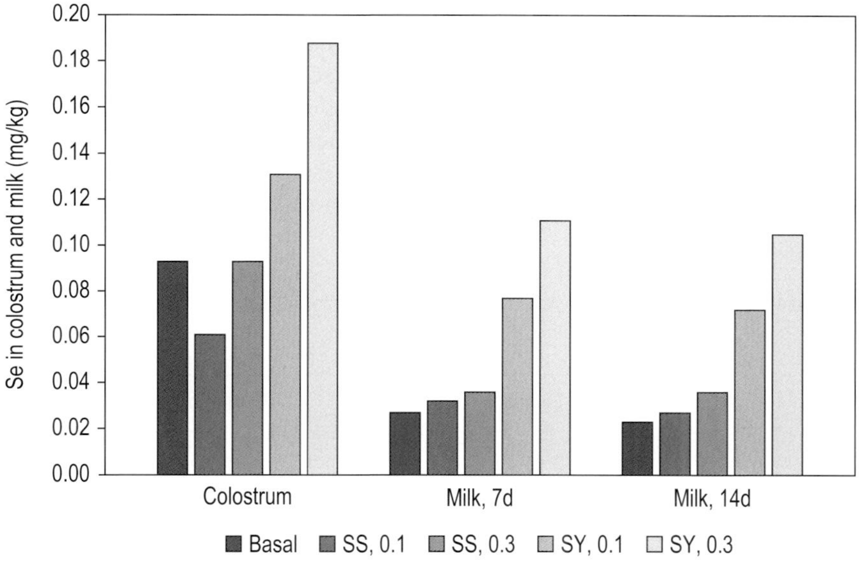

Figure 5.1. Selenium in sow's colostrum and milk depending on its dietary supplementation (adapted from Mahan, 2000).

These results demonstrate that organic Se increased milk Se more than did inorganic Se and increased the suckling piglets' serum Se content. Inorganic Se affected the sow serum GPx activity more than organic Se, but the latter was biologically more available to be incorporated into colostrum and milk. Indeed, selenium is essential in the most critical periods in the development of the pig: gestation, parturition and lactation, the periods immediately following birth and post weaning. It is also important for pig growth and meat quality as well as for the maintenance of semen quality in boar. Piglets are born with low antioxidant defences and their postnatal development and health depends on antioxidants transferred via colostrum and milk.

The most comprehensive study of the organic Se in sow nutrition over 4 parities was conducted by Mahan and Peters (2004). Their results showed that Se from organic sources (Se-Yeast) was more effectively transferred to colostrum, milk and sow hair, compared with inorganic Se (Table 5.3).

At 0.3 mg/kg dietary supplementation, Se levels in the liver, loin (Figure 5.2) and pancreas of the sows was substantially higher when organic selenium was provided in the diet.

Since Se in the loin represents major Se storage in the form of SeMet and therefore, increased Se concentration in muscles of sows due to organic Se in the diet could be a great advantage in terms of increasing resistance of sows to various stresses. The detailed mechanisms of SeMet usage from muscles reserves is described in Chapter 3 (Surai *et al.*, 2021). Indeed, The main form of Se in muscles of animals fed on grain-based diet is SeMet. For example, the Se-amino acids accounted for 91 (±8%) of the total selenium (mean of 95 samples of seven tissues analysed over a period of 18 months; Bierla *et al.*, 2008) with SeMet comprising more than 60% of total Se. When high doses of organic Se supplementation were used more than 95% of the

Table 5.3. Effect of dietary selenium (Se) on sow and litter Se status, mg/l/mg/kg (adapted from Mahan and Peters, 2004).

	Basal	Inorganic Se		Organic Se	
Item	**0**	**0.15**	**0.30**	**0.15**	**0.30**
Sow serum Se, 70 d	0.114	0.143	0.135	0.146	0.169
Sow serum Se, 110 d	0.088	0.125	0.124	0.130	0.159
Sow tissue Se					
Liver	0.298	0.450	0.571	0.583	0.781
Kidney	2.23	2.68	2.94	2.40	2.50
Pancreas	0.300	0.334	0.369	0.487	0.510
Neonate pig Se					
Liver	0.197	0.238	0.279	0.348	0.424
Litter, serum Se, weaning	0.056	0.067	0.062	0.078	0.092

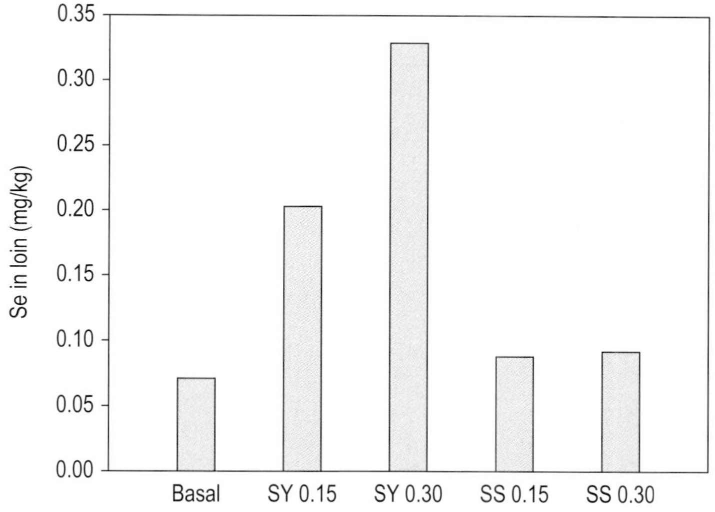

Figure 5.2. Selenium (Se) in sow's loin depending on its dietary supplementation (adapted from Mahan and Peters, 2004).

Se in chicken breast and leg muscles was found in the form of SeMet. It is known that animals cannot produce SeMet, it must come with the food. This means that only organic Se, in the form of SeMet or its analogue can substantially increase the Se concentration in the muscles in the chicken, pig or cattle.

Similarly, in neonate pig liver and loin (Figure 5.3), the Se concentration was double that of piglets from sows supplemented with sodium selenite. Again, increased Se concentration in muscles of neonate pig could be considered as an important investment by sows in future piglet resistance to stresses as discussed above.

When amount of Se per whole newly born piglet was calculated, it was very clearly shown that SS at 0.15 mg/kg supplementation to the sow was not transferred to the progeny piglets. For example, it was 0.075 mg/piglet in non-supplemented sows vs 0.079 mg/piglet after 0.15 mg/kg supplementation in the form of SS, while it was 0.123 mg/piglet (increased by 64%) after organic Se supplementation at the same dose. When organic Se supplementation of the sow was increased to 0.3 mg/kg, the amount of Se in the whole body of the newly born piglet more than doubled and reached 0.200 mg/piglet, while in SS sows at the same dose it was only 0.100 mg/piglet (Figure 5.4). Indeed, these data confirmed that SS is very poorly transferred via placenta (Mahan and Peters, 2004).

This phenomenon of organic Se transfer via placenta has a general biological validity, since the enhanced Se status of calves born to dams that received organic Se supplements during the latter stages of pregnancy has been reported previously (Gunter *et al.*, 2003; Guyot *et al.*, 2007; Pehrson *et al.*, 1999) and is most likely attributable to SeMet transfer from the dam to the calf. An increasing body of evidence clearly shows that

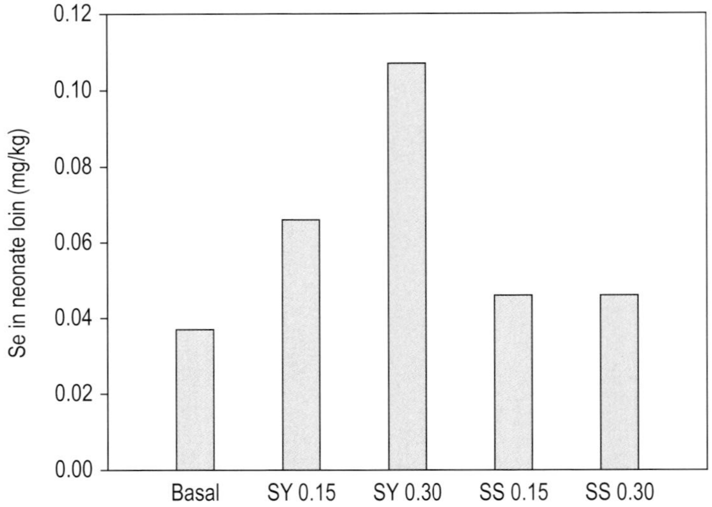

Figure 5.3. Selenium (Se) in neonatal piglet loin depending on its dietary supplementation (adapted from Mahan and Peters, 2004).

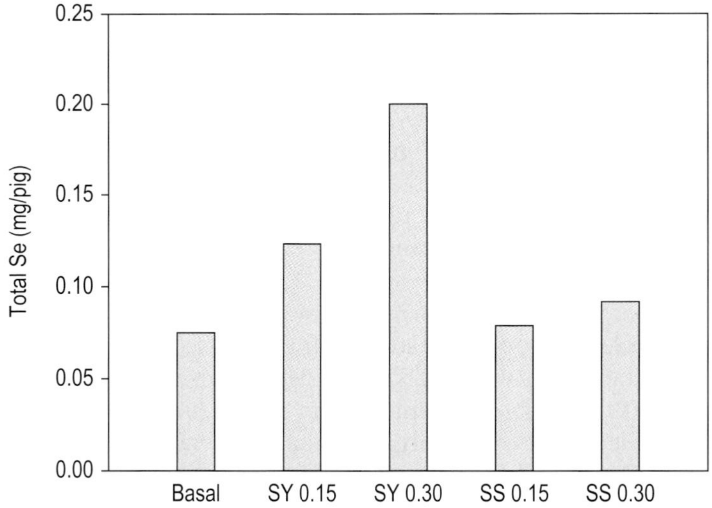

Figure 5.4. Maternal selenium (Se) transfer to the progeny piglets (adapted from Mahan and Peters, 2004).

the efficiency of Se transfer to the new-born animals depends on the form of Se in the maternal diet. However, molecular mechanisms of SeMet transfer via placenta are not well understood. For example, selenoprotein-based transfer mechanisms have been described in the mouse placenta, and they are responsible for an adequate Se supply to the developing foetus, especially when maternal Se supply is poor (Burk *et al.*, 2013). Interestingly, the transport of SeMet across the placenta is unregulated and has been reported to be dependent upon maternal SeMet supply (Burk *et al.*, 2013). It seems likely that similar Se transfer mechanisms exists in the placenta of various mammalian species and that the enhanced Se status of piglets born from Se-Yeast supplemented sows may have been a consequence of an effective placental SeMet transfer. Subsequent studies by Yoon and McMillan (2006), Quesnel *et al.* (2008), Svoboda *et al.* (2008), Peters *et al.*, (2010), Gelderman and Clapper, 2013 and Ma *et al.* (2014) have confirmed the superiority of organic sources of Se over inorganic sources in terms of Se status of sows or progeny piglets.

There is a great body of evidence indicating that replacement of SS by organic Se in cow diets leads to a significant improvement in the Se content of milk and colostrum (Figure 5.5 and 5.6).

It is well established that as the parity of the sow increases, milk selenium concentration decreases when the diet is supplemented with sodium selenite (Mahan and Peters, 2004) for the following reasons: sow selenium tissue reserves are depleted with advancing parity; tissue reserves are not readily mobilised; mammary tissue of older females may not incorporate or transfer dietary inorganic selenium or tissue Se reserves into milk as effectively as young sows. In general, a relationship between dietary Se and its concentration in milk is well researched in dairy cows. Based on

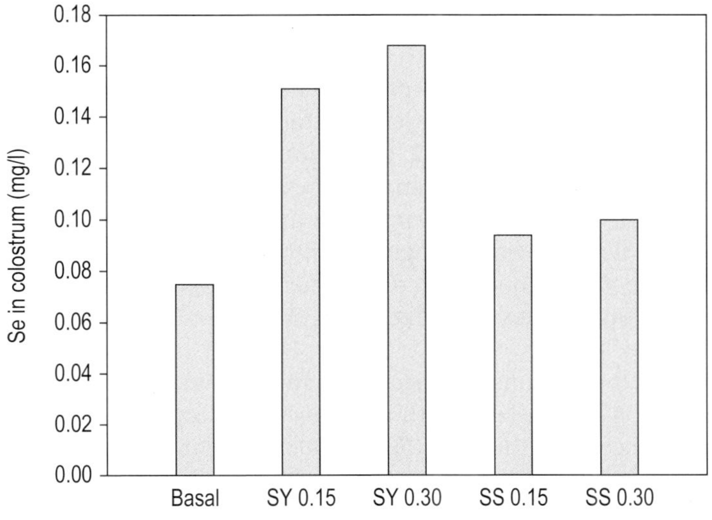

Figure 5.5. Selenium (Se) in sow's colostrum depending on its dietary supplementation (adapted from Mahan and Peters, 2004).

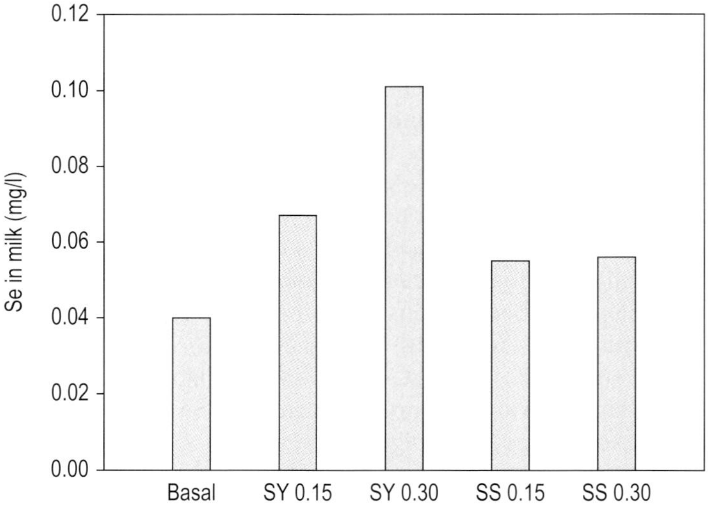

Figure 5.6. Selenium (Se) in sow's milk depending on its dietary supplementation (adapted from Mahan and Peters, 2004).

a meta-analysis of 42 studies investigating the effects of oral Se supplementation on milk Se concentration in cattle, it was concluded that cows supplemented with organic Se had significantly higher milk Se concentrations in comparison to those supplemented with inorganic forms of Se (Ceballos *et al.*, 2009). When comparing SS with Se-Yeast at 0.3 mg/kg, it was shown that the SeMet concentration in milk of Se-Yeast-supplemented animals increased from 61 to 111 ng Se/g, accounting for 44% of total Se, whilst the proportion of total Se comprised as SeMet in SS supplemented animals declined and was markedly lower (36 vs 111 ng Se/g) (Phipps *et al.*, 2008). In cows fed Se-Yeast, the efficiency of Se transfer from feed to milk has been reported to range from 9.9 to 12.5%, compared with 2.4-4.1% for those supplemented with SS (Givens *et al.*, 2004). A meta-analysis of milk Se data from 11 studies indicated that responses to Se-Yeast were four-fold greater than those of SS. A similar finding was found following the analysis of data from 130 individual cows looking at the efficiency of Se transfer from feed to milk (Juniper and Bertin, 2011). Furthermore, in Se-Yeast fed cows, the amounts of Se secreted daily into milk and apparently retained in tissues was shown to increase linearly with average daily intake of Se, total Se excretion was shown to be 66%, Se secretion in milk accounted for 17%, and Se retained in tissues accounted for 17% of total Se intake (Walker *et al.*, 2010].

It seems likely that a better response in terms of Se in the sow's milk to the organic Se dietary supplementation is associated with non-specific incorporation of SeMet into the milk proteins in place of methionine (Surai, 2006), similar as it was suggested for meat (Schrauzer and Surai, 2009) or eggs (Surai and Fisinin, 2014) as observed in dairy cows (Muniz-Naveiro *et al.*, 2007). It is interesting to note that, independently on the Se sources in the cow diet, in milk, the main increase in Se concentration was in the form of SeMet. Replacing SS with SY at 0.14 mg/kg increased the SeMet concentration of milk from 36 to 111 ng Se/g dried sample and its concentration increased further to 157 ng Se/g dried sample (by 4.4-fold) as the inclusion rate of SY increased further (0.29 mg/kg) to provide total amount Se in the diet 0.45 mg Se/kg (Phipps *et al.*, 2008). In 1990s it was shown that Se in plasma is represented mainly (~50%,) by SelP (Hill *et al.*, 1996; Whanger *et al.*, 1991) and GPx3 (up to 30%; Avissar *et al.*, 1989; Milner *et al.*, 1987; Whanger *et al.*, 1991). Recently, it was shown that SeCys concentration in the cow's plasma is about 3-fold higher than that of SeMet and this increased to 5-fold in SS animals (Juniper *et al.*, 2019). The authors also showed increased SeMet concentration in the plasma due to OH-SeMet dietary supplementation. Furthermore, occurrence and concentrations of SelP, SeAlb and SeMet have been described in plasma, colostrum and milk from sows (Falk *et al.*, 2019). The authors calculated that SelP, SeAlb and SeMet comprised about 57% (at farrowing) and 62% (at weaning) of total Se in the sows' plasma. However, the results of the trial with sodium selenite and pure SeMet in the sow's diets were not consistent. Total plasma Se was shown to be increased from farrowing to weaning but there was no significant difference for this parameter between groups during the trial. In fact, total Se and SeP levels in colostrum were increased due to replacement of SS by SeMet, however there was no dose response in this parameter, total Se and SeP concentrations in colostrum was not different between SS 0.4 vs 0.6 and SeMet 0.26 vs 0.43. SeMet concentration in the colostrum was quite low and it did not depend on dietary Se form or concentration used. Only SeAlb in colostrum of sows fed on SeMet (0.43 mg/kg) showed significant increase in comparison to SeMet 0.26. Replacement of SS with SeMet was shown to increase total Se and SeMet concentration in sow's milk with dose-dependent response (Falk *et al.*, 2019). Milk levels of SelP and SeAlb and thus the total Se concentration in milk decreased as lactation progressed possibly by active reduction of their transport and it seems likely that the sow's mammary gland actively regulates the transport of Se (Falk *et al.*, 2019). There were several points in the trial design which could affect the results. Firstly, before the experimental period sows were fed on the diet containing organic Se in the form of SeYeast at 0.15 mg/kg. This could affect Se reserves in the body and without a washing out period it would be difficult to take into account this variable. Secondly, a substantial (about 10%) increase in feed intake due to SeMet supplementation could affect final results. Thirdly, the dosage of SS (0.4 and 0.6 mg/kg) and SeMet (0.26 and 0.43 mg/kg) were quite different to conduct proper comparison between different Se forms.

In general, there is a substantial evidence that in cows differences in milk Se responses between comparable doses of organic and inorganic Se supplements are a consequence of SeMet incorporation (Calamari *et al.*, 2010; Phipps *et al.*, 2008). Similarly, studies in poultry and pigs have shown that increases in tissue Se as a response to dietary OH-SeMet supplementation are primarily a consequence of the incorporation of SeMet (Briens *et al.*, 2014; Jlali *et al.*, 2014). Unexpectedly, Se speciation analysis of colostrum from the recent study with cows indicated that the marked increase in the total Se concentration in colostrum due to OH-SeMet supplementation was attributable to SeCys rather than SeMet (Juniper *et al.*, 2019). Interestingly, the authors showed that the difference between SeCys and SeMet concentration in the colostrum is much less pronounced in comparison to plasma. It is interesting to note that sodium selenite fed to sows had some detrimental effects on piglets. In fact the percentage of piglet with spray legs (Figure 5.7) and stillborn piglets (Figure 5.8) was increased by selenite supplementation of the maternal diet. In contrast, in the same conditions, organic selenium had protective effects (Mahan and Peters, 2004). It could well be that prooxidant properties of selenite are responsible for those detrimental changes in sow's progeny.

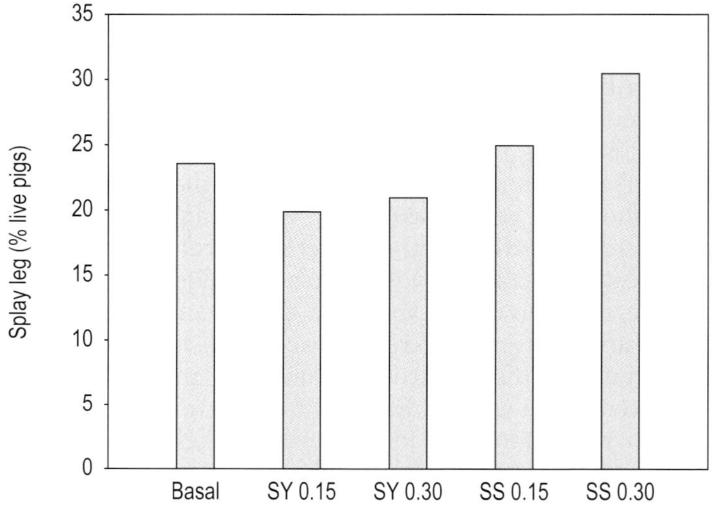

Figure 5.7. Effects of sow dietary selenium source and levels on litter splay legs (slight + severe) incidence (adapted from Mahan and Peters, 2004).

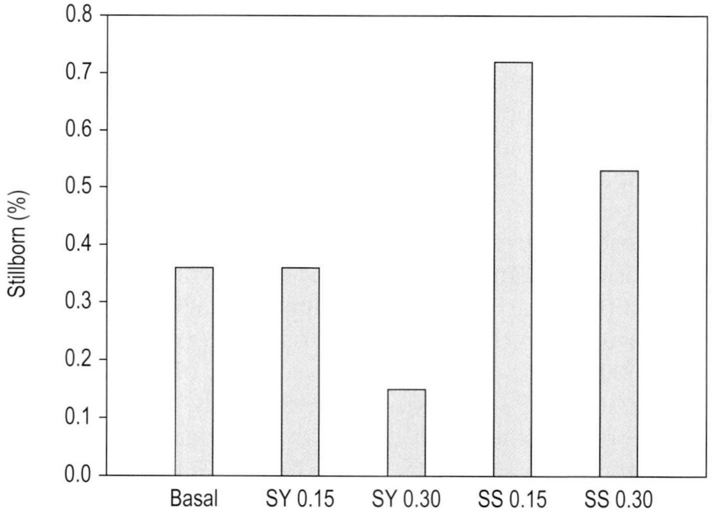

Figure 5.8. Effects of sow dietary selenium source and levels on stillborn piglets incidence (adapted from Mahan and Peters, 2004).

Interestingly, in a recent trial with sows fed on the basal diet containing 0.13 mg/kg feed-derived Se, dietary supplementation of SS at 0.3 mg/kg during gestation period was shown to have a tendency in increasing percentage of still-born piglets (from 6.95% up to 12.64%), while supplementing the same diet with OH-SeMet (*HMSeBA*, 2-hydroxy-4-methylselenobutanoic acid) significantly ($P<0.05$) decreased the percentage of still-born piglets down to 0.47% (Figure 5.9).

Indeed, the progeny from sows between their 4th and 7th parity are often deficient in selenium and mortality rates are increased. Furthermore, the higher the level of productivity, the greater the loss of Se from the sows' body reserves (Close, 2003). In fact, replacing sodium selenite with organic selenium in the sows' diet increases Se transfer across the placenta into the developing foetuses and increases Se concentration in both colostrum and milk. It could well be, that SeMet reserves in sow's body are depleted and this caused Se declining in milk with advanced parity. Therefore, even high inclusion rate of selenite cannot provide a form of selenium which is transferred to colostrum and milk and the only solution for this problem is to use organic selenium, providing SeMet which is effectively transferred to the colostrum and milk.

The next step in the studying effects of organic Se in sow reproduction was associated with usage of pure SeMet as a source of organic Se in the sow diet. For example, Hu *et al.* (2011) conducted an experiment to study the effects of different selenium sources on productive performance, serum and milk Se concentrations, and antioxidant status of sows. A total of 12 pregnant sows were randomly divided into two groups receiving the same basal gestation and lactation diets containing 0.042 mg Se/kg, supplemented with 0.3 mg Se/kg SS or SeMet, respectively. The feeding trial lasted for 60 days, with 32 and

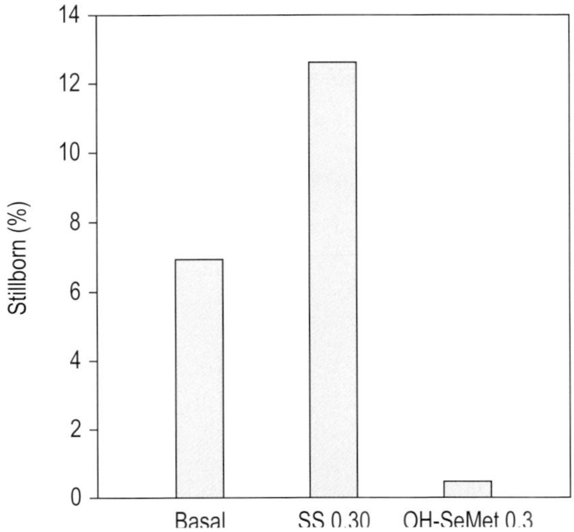

Figure 5.9. Effects of sow dietary selenium source and levels on stillborn piglets incidence (adapted from Mou *et al.*, 2020).

28 days for gestation and lactation period, respectively. They showed that replacement of SS by SeMet in the maternal diet improved Se concentration and antioxidant status of sows, thus maintain maternal health and increase productive performance after Se was transferred to its offspring. Indeed, the Se concentration in the serum, colostrum, and milk of sows were significantly increased in the SeMet-treated group. Furthermore, dietary SeMet (vs SS) significantly improved the antioxidant status of sows. In fact, there was an increased total antioxidant capability (T-AOC) and decreased MDA level in the serum of sows fed SeMet in comparison to SS-fed sows (Table 5.4).

Furthermore, in the same SeMet-fed group of sows significantly increased T-AOC, GPx, SOD and GSH, and decreased MDA level in the colostrum and milk were shown. It could well be that improved antioxidant defences of sows could be crucial in stress conditions of parturition and weaning, most sensitive periods of sow's life. Compared with SS, maternal SeMet intake significantly increased the weaning litter weight and average weight of progeny piglets. Continuing a previous work the efficiency of maternal SeMet intake on growth performance, Se distribution, and antioxidant status of offspring piglets was investigated by Zhan *et al.* (2011) in the same experimental design as in the previous study presented by Hu *et al.* (2011).

Table 5.4. Effects of different selenium source on serum antioxidant profile in sow (adapted from Hu *et al.*, 2011).[1]

	Sodium selenite	Selenomethionine
T-AOC, U/ml	2.90	4.03*
GPx, U/ml	1,423	1,446
SOD, U/ml	83.6	91.4
GSH, mg/l	4.92	5.32
MDA, nmol/ml	4.18	2.27*

[1] GPx = glutathione peroxidase; GSH = glutathione; MDA = malondialdehyde; SOD = superoxide dismutase; T-AOC = total antioxidant capacity.

Table 5.5. Effect of different maternal Se sources on Se concentration in colostrum and milk (adapted from Zhan *et al.*, 2011).[1]

	Sodium selenite	Selenomethionine
Colostrum, mg/l	0.17	0.19**
Milk, mg/l		
7 days	0.15	0.18**
14 days	0.15	0.16*
21 days	0.14	0.15*
28 days	0.11	0.13**

[1] * $P<0.05$; ** $P<0.01$.

A total of 12 pregnant sows were randomly divided into two groups and fed the same basal gestation and lactation diets containing 0.04 mg Se/kg, supplemented with 0.3 mg Se/kg sodium selenite (Control group) and SeMet (Experimental group), respectively (Table 5.6; Table 5.6).

The feeding trial lasted for 60 days, with 32 and 28 days for gestation and lactation period, respectively. The Se concentration in the colostrum and milk (at 7, 14, 21 and 28 days) and tissue Se content of piglets were significantly higher in the SeMet-treated group. Indeed, SeMet was shown to be more effective in increasing the Se content in serum, liver, kidney, pancreas, muscle, thymus and thyroid gland in weaned pigs at 28 days. It is necessary to note that SeMet supplementation of the maternal diet significantly improved not only GPx activity in the liver, kidney, pancreas, muscle, and serum, but also positively affected other important elements of the antioxidant system of the piglets. Indeed, there was a significantly improved total antioxidant activity in the liver, kidney, pancreas, muscle, and serum as well as SOD (pancreas, muscle, and serum) and GSH (kidney, pancreas, muscle, and serum) in the experimental piglets (Table 5.7). This was associated with decreased MDA content in liver, kidney, and pancreas of the piglets. Effects of dietary Se on a variety of indexes of antioxidant protection showed above, including SOD activity and GSH content in various pig tissues, reflect important interactions inside the antioxidant network. It could be speculated that such transcription factors as Nrf2 and NF-kB could mediate such interactions. In fact, recently mammalian selenoprotein thioredoxin reductase 1 has been shown to be a potent regulator of Nrf2 (Cebula *et al.*, 2015) and Se is found to downregulate NF-κB with decreasing synthesis of pro-inflammatory cytokines in cell culture (Zhang *et al.*, 2014).

Table 5.6. Effect of different maternal Se sources on Se concentration in serum and organs of the offspring piglets (adapted from Zhan *et al.*, 2011).[1]

	Sodium selenite	Selenomethionine
Serum, µg/ml	0.49	0.62**
Liver, µg/g	0.46	0.70**
Kidney, µg/g	2.33	2.49**
Pancreas, µg/g	0.35	0.54**
Muscles, µg/g	0.17	0.23**
Thymus, µg/g	0.15	0.16*
Thyroid gland, µg/g	0.13	0.15*

[1] * $P<0.05$; ** $P<0.01$.

Table 5.7 Effect of different maternal Se source in sow's diet on the antioxidant defences in the tissues of the offspring piglets (adapted from Zhan *et al.*, 2011).[1,2]

Tissue	Treatment	T-AOC, U/g	GPx, U/g	SOD, U/g	GSH, μg/g	MDA, nmol/g
Liver	SS	12.1	8.0	99.9	74.7	14.9
	SM	16.9*	9.1*	98.3	83.4	13.7*
Muscles	SS	24.3	5.9	98.7	40.8	16.3
	SM	28.9*	7.2*	115.8**	49.7*	14.9

[1] GPx = glutathione peroxidase; GSH = glutathione; MDA = malondialdehyde; SM = selenomethionine; SOD = superoxide dismutase; SS sodium selenite; T-AOC = antioxidant activity.
[2] * = P>0.05; ** = P<0.01.

The authors also found a positive effect of maternal SeMet on thyroid metabolism and digestive enzyme activity in the progeny piglets. Indeed, the increased triiodothyronine (T3) and decreased thyroxine (T4) concentration in the piglet serum due to SeMet maternal supplementation were shown. This could have a positive effect on protein synthesis and energy production in the experimental group. Compared with sodium selenite, maternal SeMet intake significantly increased protease, amylase, and lipase activities in the pancreatic tissue of its offspring. Probably improved thyroid status and digestive enzyme activities were responsible for a significant increase in the daily weight gain of piglet from birth to weaning (Zhan *et al.*, 2011). It seems likely, that effect of organic Se in the maternal diet on the growth and development of suckling piglets would depend on the dietary background Se level and the extent of the Se deficiency of the sows. Taking into account data showing crucial role of selenoproteins in cellular redox homeostasis and increased selenoprotein expression in response to various stresses (Surai, 2006) it is clear that an importance of sow and piglet antioxidant defences and optimal Se nutrition of sows are difficult to overestimate.

Another important point for sow Se nutrition is its transfer to the foetus and maintaining antioxidant defences in early embryo development. For example, the effect of Se as SS or SY on antioxidant status, hormonal profile, reproductive performance, and embryo development in first-parity gilts was evaluated (Fortier *et al.*, 2012). Forty-nine gilts were allocated to 1 of the 3 dietary treatments starting at first pubertal oestrus and lasting up to 30 d after AI: control (CONT): basal diet (BD, Se = 0.2 mg/kg); SS: (BD + 0.3 mg/kg of SS), and SY: (BD + 0.3 mg/kg of SY). It was shown, that the uterine transfer of Se to embryos was improved with SY as compared with SS, and this was concomitant with an enhanced development of embryos.

In particular, in SY gilts, the Se content of the whole litter and of individual embryos was 63 and 52% greater, respectively, than in SS gilts (Figure 5.10). Furthermore, the length, weight, and protein content of embryos were greater in SY than in SS gilts. The authors suggested that dietary organic Se would be critical for an efficient Se transfer to embryos and for their subsequent development in early gestation. It is interesting to note that there were 96 differentially expressed genes in the expanded blastocysts (embryos at day 5) of gilts fed organic (Se-Yeast) or inorganic (SS) selenium at 0.3 mg/kg. Therefore, there was an important response in terms of differentially expressed (DE) genes, but no specific biological process was identified (Dalto *et al.*, 2015).

It was shown that the Se concentration in foetal body and liver were greater from gilts fed SeYeast than from gilts fed SS, reflecting greater placental Se transfer from dam to offspring as a result feeding SeYeast compared with SS. Generally, Se concentration in the foetuses from gilts fed Se-Yeast were greater than those fed the inorganic diet (0.95 vs 0.75 mg/kg). Furthermore, foetal Se concentrations generally were elevated in response to the organic selenium ($P<0.01$), except of absence of statistical difference at d 73 (Figure 5.11; Ma et al., 2014).

Furthermore, piglets from sows fed Se-Yeast had a greater Se content at birth in both body and liver and the Se advantages (about 40%) persisted through weaning (Ma et al., 2014).

A comprehensive study was conducted with 234 multiparous sows in a 2×2 factorial arrangement with sodium selenite or Se-enriched yeast at 0.30 mg Se/kg diet each as the Se sources and vitamin E at 30 or 90 IU/kg (Chen *et al.*, 2016a). Sows fed organic Se had a trend to produced more live pigs ($P<0.10$), had greater litter birth weight (Figure 5.12; $P<0.05$; Figure 5.13) and weaned more pigs (Figure 5.13; $P=0.051$) in comparison to sows fed inorganic Se.

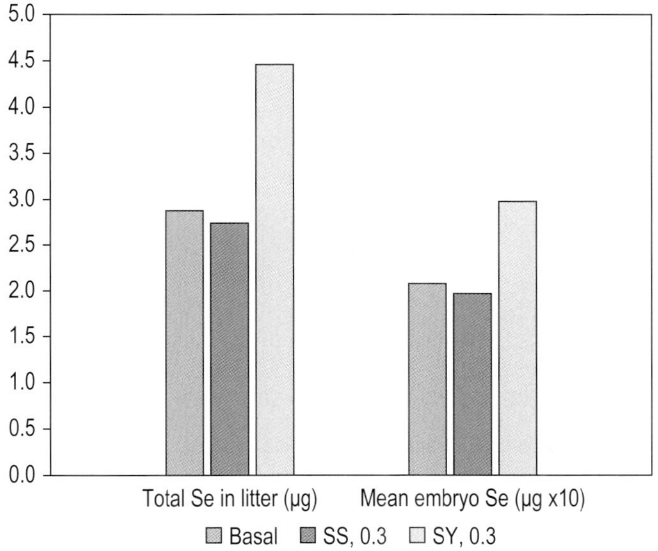

Figure 5.10. Selenium (Se) transfer to embryos on d 30 of gestation (adapted from Fortier *et al.*, 2012).

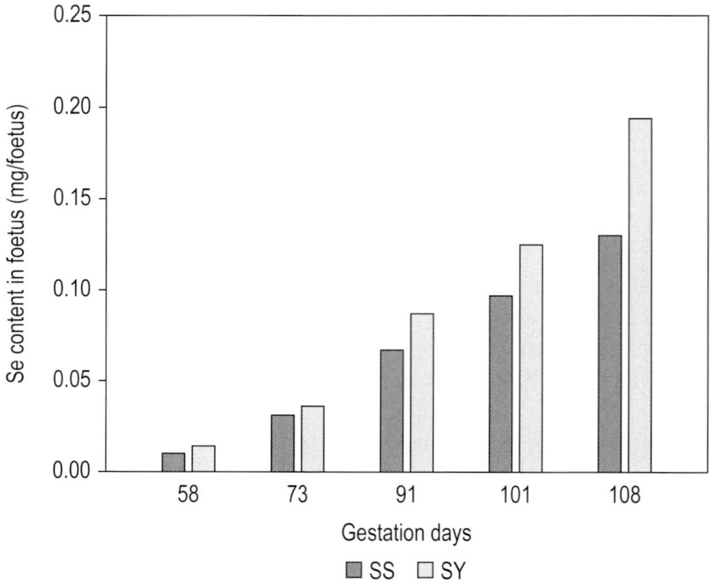

Figure 5.11. Selenium (Se) content in foetus at different days of gestation, mg/foetus (adapted from Ma *et al.*, 2014).

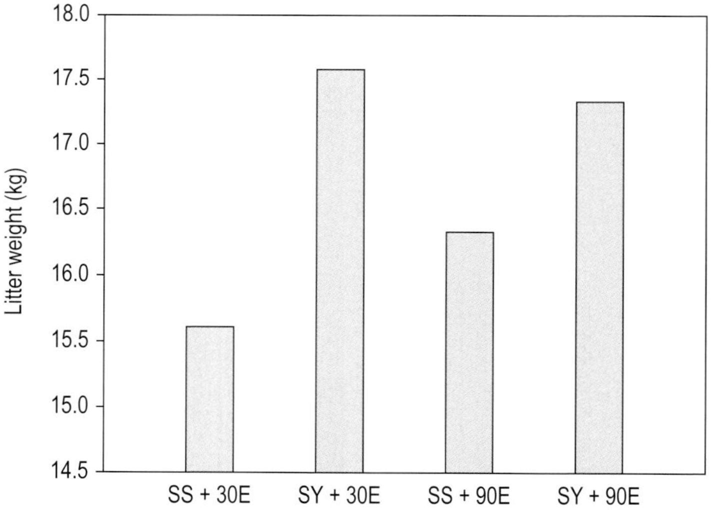

Figure 5.12. Effect of dietary selenium and vitamin E level on litter weight at parturition, kg (adapted from Chen *et al.*, 2016a).

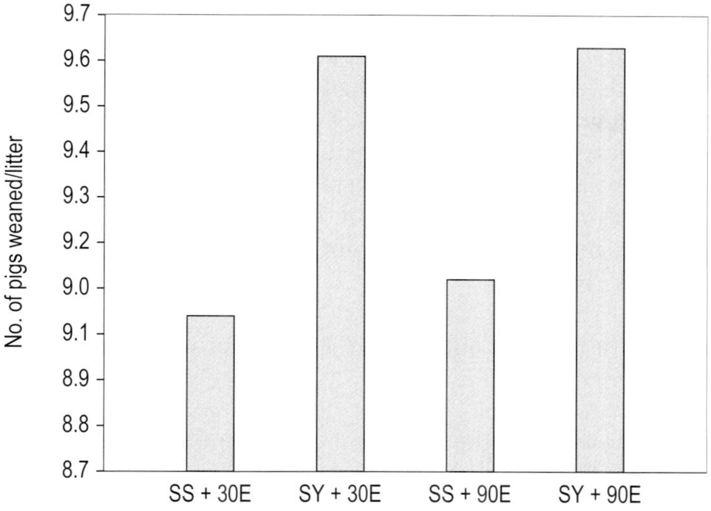

Figure 5.13. Effect of dietary selenium and vitamin E on number of pigs weaned per litter (adapted from Chen *et al.*, 2016a).

Recently, it has been shown that supplementing of the basal diet of sows containing 0.13 mg/kg feed-derived Se with SS or OH-SeMet at 0.3 mg/kg during gestation can improve the number of piglets born alive per sow. Indeed, while SS supplementation had only a trend in improving this parameter (14.3 vs 13.0), dietary supplementation of the organic Se in the form of OH-SeMet (*HMSeBA*, 2-hydroxy-4-methylselenobutanoic acid) was shown to significantly increase the number of piglets born alive (15.92 vs 13.0, $P<0.05$, Figure 5.14; Mou *et al.*, 2020).

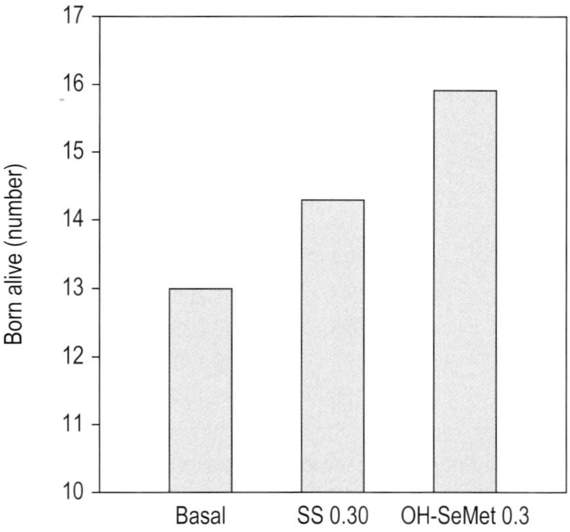

Figure 5.14. Effect of dietary selenium on number of pigs born alive (adapted from Mou *et al.*, 2020).

Organic Se in sow's diet significantly improved their antioxidant defences manifested by improved serum T-AOC, SOD activity, GPx activity, GSH content and decreased MDA at various periods of ontogenesis. (Chen *et al.*, 2016a). There was also an improvement in antioxidant status of the milk, but physiological importance of this parameter for sow's or progeny health is not obvious. The same experiment continues and at farrowing and at weaning, 8 piglets per treatment (1 piglet/litter) were selected and biochemical analyses related to antioxidant defences were assessed (Chen *et al.*, 2016b). The newly-born and weaned piglets from sows fed organic Se were shown to have improved Se status (Table 5.8).

Organic Se was shown to improve antioxidant defences of newly born and weaned piglets (Table 5.9).

Interestingly, when sows were fed the treatment diets from breeding until weaning, sows fed organic Se were shown to have more weaned piglets with similar individual pig weight at weaning compared with sows fed inorganic Se (Chen *et al.*, 2016b). It seems likely that different Se enrichment levels and Se sources in the sow's diet could potentially affect progeny piglet immune health, especially during the peri-weaning phase. For example, higher levels of IgG were observed in piglets from sows that were supplemented with OH-SeMet compared to offspring from the Se-deficient control and SS sows (Li *et al.*, 2019). Therefore, maternal supplementation with OH-SeMet is more effective than SS for improving piglet Se status, and promoting passive immunity, an essential mechanism for protecting newborn piglets against microbial infection (Li *et al.*, 2019). Since the porcine placenta acts as a strong barrier for the transport of maternal immunoglobulins to the foetus, the piglet depends on Igs passive transfer via colostrum and optimal Se status could be beneficial for this. In particular, lambs of Se-supplemented ewes had increased absorption of IgG (Rock *et al.*, 2001) and improved serum IgG (Hammer *et al.*, 2011). Furthermore, Se addition to colostrum was shown to increase immunoglobulin G absorption by newborn calves (Kamada *et al.*, 2007).

Table 5.8. Effect of Se in sow diet on Se status of newly born/weaned piglets, µg/g (adapted from Chen *et al.*, 2016b).[1,2]

Se sources	SS		SY	
Vit. E, mg/kg	E 30	E 90	E 30	E 90
Loin	0.045/0.081	0.045/0.077	0.082/0.099	0.059/0.092
Liver	0.255/0.264	0.195/0.247	0.205/0.290	0.298/0.278
Kidney	0.521/0.902	0.485/0.825	0.626/0.836	0.799/0.918
Pancreas	0.118/0.180	0.113/0.177	0.130/0.211	0.174/0.211
Thymus	0.085/0.164	0.098/0.156	0.137/0.178	0.161/0.206

[1] In newly born piglets organic Se significantly ($P<0.05$) increased Se level in kidney and had a tendency ($P<0.1$) increasing it in loin; in weaned piglets OS significantly ($P<0.05$) increased Se level in loin, pancreas and thymus in comparison to SS group.
[2] SS = sodium selenite; SY = Se-yeast.

Table 5.9. Effect of selenium (Se) source and vitamin E supplementation level (30 or 90 IU/kg) in sow diet on the antioxidant status in liver of the newborn/weaned piglets (adapted from Chen *et al.*, 2016b).[1]

Se sources	SS		SY	
Vitamin E suppl., mg/kg	30	90	30	90
GPx, U/ml	82.3/77.9	95.8/83.6	78.7/79.3	83.2/83.0
Total AOA in serum[2], U/ml	2.85/1.29	3.17/1.71	3.12/1.76	3.58/1.90
SOD, U/ml	39.3/60.5	42.0/62.8	39.5/64.1	43.5/70.5
GSH[2], mg/l	14.6/12.6	15.0/13.4	16.1/13.3	20.3/16.4
MDA, mmol/ml	2.03/0.61	2.03/0.65	1.78/0.66	2.18/0.71

[1] GPx = glutathione peroxidase; GSH = glutathione; MDA = malondialdehyde; SOD = superoxide dismutase; SS = sodium selenite; SY = Se-yeast; T-AOC = total antioxidant capacity.
[2] In newly born and weaned piglets organic Se significantly ($P<0.05$) increased total antioxidant activity and GSH in comparison to SS; the same parameters were also increased by increasing vitamin E supplementation.

Recently, it was shown that maternal supplementation of OH-SeMet during pregnancy improved antioxidant capacities and reduced inflammation level in placenta, and foetus. In fact, it has been shown that in comparison to the control group of sows fed on the diet containing 0.13 mg/kg feed derived Se, supplementing 0.3 mg/kg Se in the form of SeMet was associated with increased serum concentrations of total selenium and selenoprotein P in both sows and piglets. Furthermore, activities of antioxidant enzymes in the serum sows, umbilical cord and piglets, placentas and piglets' liver were also induced by dietary OH-SeMet supplementation in comparison to the control group, while MDA concentration was significantly ($P<0.05$) decreased in the blood of sows, umbilical cord and newborn piglets (Mou *et al.*, 2020). Importantly, maternal OH-SeMet intake during gestation was shown to up-regulate antioxidant-related selenoprotein gene expression in the placenta (GPx2, GPx3, $P<0.05$) and in the liver of newborn piglets (GPx1, GPx2, GPx3, TXNRD2, $P<0.05$). Moreover, sows and newborn piglets in the sodium selenite and OH-SeMet groups had a lower serum interleukin-6 ($P<0.05$) concentration, and placentas in the OH-SeMet group had lower IL-1β, IL-6 and IL-8 gene expression ($P<0.05$) (Mou *et al.*, 2020).

In a recent study by Chen *et al.* (2019) it was shown that increasing organic selenium supply (from 0.3 up to 1.2 mg/kg) for heat-stressed or actively cooled sows showed a tendency ($P=0.08$) in improving piglet preweaning survival, colostrum and milk composition, as well as maternal selenium, several parameters of plasma antioxidant status (GPx and MDA) and immunoglobulin transfer irrespective of the climatic conditions. The additional beneficial effect of very high Se doses (1.2 mg/kg) needs more careful elucidation, with several doses between 0.3 and 1.2 mg/kg tested, before a conclusion about an optimal Se dose for sows can be drawn. Indeed, increased organic Se supplementation in various stress conditions could be justified, but more research in this field is required. Furthermore, it would be important to find a proper combination of organic Se with other antioxidants in the feed. For example, a combined supplementation of Se (1.15 mg/kg as Se-Yeast), VE (44 IU/kg), plus Cr and betaine was shown to reduce mobilisation of body reserve of lactating sows in summer (Liu *et al.,* 2017). However, in such experiments it is difficult to answer a question about beneficial input of each individual compound in the mixture.

Analysis of the aforementioned studies indicates that compared with inorganic Se, organic sources of Se appear to better meet the needs of the modern, hyper-prolific sow for the following reasons (Figure 5.15):

- higher placental transfer of Se to the developing foetus;
- higher mammary transfer of Se to colostrum and milk;
- enhanced Se status of the piglet at birth and at weaning, as well as the sow;
- improved antioxidant status of the sows and newly born/weaned piglets;
- improved immunoglobulin transfer to the progeny piglets.

It should be mentioned, that most of aforementioned studies devoted to the maternal effect of Se in sow reproduction did not show any specific commercially-relevant benefit of replacement of traditional SS by organic Se sources in terms of productive or reproductive performances of sows or their progeny. From one hand, the number of animals used in aforementioned studies was too low to pick up possible small differences in sow reproductive performance or in growth and development of the progeny pigs. From the other hand, amount of stress imposed to sows and piglets in the experimental conditions with small number of animals probably was too low to compromise antioxidant defences and to increase a need for antioxidant defences, in particular highly available selenium. However, biochemical changes indicating improvement of antioxidant defences in sows and progeny piglets due to organic Se supplementation could be considered as a safety margin or 'insurance policy' in stressful conditions of commercial pig production or in areas with extremely low Se status in the feed. For example, in the two studies from China with background Se at 0.04 mg/kg, replacement of SS by SeMet was associated with a significant increase of the weaning litter weight and average weight of progeny piglets (Hu *et al.,* 2011), which probably, was responsible for an increase in the daily weight gain of piglet from birth to weaning (Zhan *et al.,* 2011). Similarly, in two more Chinese studies with comparatively low background Se level (0.05-0.07 mg/kg; Chen *et al.,* 2016a,b) with a bigger number of animals (55-60 sows pre group) showed a significant improvement in the litter weight at parturition and pigs weaned per litter as a result of replacing sodium selenite by Se-yeast at 0.3 mg/kg supplementation. Earlier, sow reproduction parameters, pre-weaning mortality and performance of piglets after weaning were evaluated and showed that substitution of SS by Se-Yeast in diets fed commercial sows resulted in more piglets weaned (9.61 vs 9.26; $P<0.001$) and higher litter weaning weight (53.15 vs 51.07 kg; $P<0.01$, Gourley *et al.,* 2005). Interestingly, the most recent study on Se effects during sow gestation clearly showed that replacement of sodium selenite with OH-SeMet (*HMSeBA,* 2-hydroxy-4-methylselenobutanoic acid) significantly decreased percentage of stillborn piglets and had a trend in increasing the number of piglets born alive (Mou *et al.,* 2020). Clearly, there is a need for a well-designed commercial large trials with sows and their progeny to prove commercially-relevant benefits of organic Se in sow nutrition. The summary of beneficial effects of organic Se in the sow's diet is shown in Table 5.10.

Figure 5.15. Selenium for sows – maternal effect.

Table 5.10. Advances of organic selenium (Se) in sows nutrition (adapted from Surai and Fisinin, 2016).[1]

Parameter	Se in the basal diet/ supplementation[2], mg/kg	Effect of organic vs inorganic Se	References
Sow			
Se status and antioxidant status of sows	0.042/0.3	Improved	Hu *et al.*, 2011
Se concentration in sow's whole blood	0.057-0.085/0.3-0.38/0.182/0.3	Improved	Gelderman and Clapper, 2013; Svoboda *et al.*, 2008
Se levels in the liver, loin and pancreas of the sows	0.055-0.085/0.15 or 0.3	Improved	Mahan and Peters, 2004
Se concentration in sow's serum	0.042/0.3	Improved	Hu *et al.*, 2011
Se content in colostrum and milk	0.04-0.23/0.15-0.38	Improved	Hu *et al.*, 2011; Mahan, 2000; Mahan and Kim, 1996; Mahan and Peters, 2004; Peters *et al.*, 2010; Quesnel *et al.*, 2008; Svoboda *et al.*, 2008; Yoon and McMillan, 2006; Zhan *et al.*, 2011
Antioxidant status of colostrum and milk	0.042/0.3	Improved	Hu *et al.*, 2011
Uterine transfer of Se	0.2/0.3	Improved	Fortier *et al.*, 2012
Development of pig embryos	0.2/0.3	Enhanced	Fortier *et al.*, 2012
Se concentration in foetus	0.25-0.27/0.3	Increased	Ma *et al.*, 2014
AO defences in the serum (T-AOC, SOD activity, GPx activity, GSH content)	0.25-0.27/0.3	Increased	Chen *et al.*, 2016a
AO defences in the serum (SOD activity at P110 and L0 and GPx activity at P60 and P110)	0.13/0.3	Increased	Mou *et al.*, 2020
TrxR in placenta	0.13/0.3	Increased	Mou *et al.*, 2020
IL8 in placenta	0.13/0.3	Decreased	Mou *et al.*, 2020
GPx2 and GPx3 in placenta	0.13/0.3	Increased	Mou *et al.*, 2020
Litter birth weight	0.25-0.27/0.3	Increased	Chen *et al.*, 2016a
Number of piglets weaned	0.25-0.27/0.3	Increased	Chen *et al.*, 2016a
Litter birth weight	0.05-0.07/0.3	Increased	Chen *et al.*, 2016a
Stillborn rate, %	0.13/0.3	Decreased	Mou *et al.*, 2020

Table 5.10. Continued.

Parameter	Se in the basal diet/ supplementation², mg/kg	Effect of organic vs inorganic Se	References
Newly-born piglets			
Se content in loin and liver in the newborn piglets	n.a./0.1 or 0.3/0.057-0.085/0.3-0.38	Improved	Chen *et al.*, 2016a; Ma *et al.*, 2014; Mahan and Kim, 1996; Svoboda *et al.*, 2008
Se content in heart in the newborn piglets	0.057-0.085/ 0.3-0.38	Improved	Chen *et al.*, 2016a; Svoboda *et al.*, 2008
Se content in kidney, pancreas and thymus in the newborn piglets	0.05-0.07/0.3	Increased	Chen *et al.*, 2016b
Se concentration in serum in the newborn piglets	0.23/0.3	Improved	Yoon and McMillan, 2006
Se in foetal body, foetal liver and maternal liver	n.a/0.3	Improved	Ma *et al.*, 2014
Se levels in the liver and loin of the neonate pigs	0.055-0.085/0.15 or 0.3	Improved	Mahan and Peters, 2004
Se levels in the liver and loin of the neonate pigs	0.055-0.085/0.15 or 0.3	Improved	Mahan and Peters, 2004
Se transfer via placenta and Se amount in a total newly born piglet	0.055-0.085/0.15 or 0.3	Improved	Mahan and Peters, 2004
Liver AO and GSH concentration	0.05-0.07/0.3	Increased	Chen *et al.*, 2016b
Serum GPx activity	0.13/0.3	Increased	Mou *et al.*, 2020
GPx3 in the liver	0.13/0.3	Increased	Mou *et al.*, 2020
Serum MDA	0.13/0.3	Decreased	Mou *et al.*, 2020
Passive transfer of maternal IgG to piglets on ppd2	0.056-0.077/0.3	Increased	Li *et al.*, 2019
Weaned piglets			
Se concentration in the whole blood of piglets at the age of 3 and 14 days	0.057-0.085/0.3-0.38	Improved	Svoboda *et al.*, 2008
Se concentration in piglet loin at weaning	n.a/0.1 or 0.3/0.056-0.077/0.3	Improved	Li *et al.*, 2019; Mahan and Kim, 1996
Se concentration in piglet plasma and liver at weaning, d21	0.077/0.3	Improved	Li *et al.*, 2019
Se concentration in piglet whole blood at weaning	0.05-0.07/0.3	Improved	Quesnel *et al.*, 2008
Se retention in body and liver of progeny piglets during 21-d period	n.a./0.3	Improved	Ma *et al.*, 2014

Table 5.10. Continued.

Parameter	Se in the basal diet/ supplementation[2], mg/kg	Effect of organic vs inorganic Se	References
Se content in serum, liver, kidney, pancreas, muscle, thymus and thyroid gland in weaned pigs at d 28	0.04/0.3	Improved	Zhan *et al.*, 2011
GPx activity in serum, liver, kidney, pancreas and muscle in weaned pigs at d 28	0.04/0.3	Improved	Zhan *et al.*, 2011
Total AOA in liver, kidney, pancreas, muscle and serum in weaned pigs at d 28	0.04/0.3	Improved	Zhan *et al.*, 2011
SOD activity in pancreas, muscle and serum in weaned pigs at d 28	0.04/0.3	Improved	Zhan *et al.*, 2011
MDA levels in liver, kidney and pancreas in weaned pigs at d28	0.04/0.3	Decreased	Zhan *et al.*, 2011
triiodothyronine (T3) concentration in the piglet serum at d 28	0.04/0.3	Increased	Zhan *et al.*, 2011
thyroxine (T4) concentration in the piglet serum at d 28	0.04/0.3	Decreased	Zhan *et al.*, 2011
Protease, amylase, and lipase activities in the pancreatic tissue of offspring piglets at d 28	0.04/0.3	Increased	Zhan *et al.*, 2011
weaning litter weight and average weight of progeny piglets	0.042/0.3	Increased	Hu *et al.*, 2011
daily weight gain of piglet from birth to weaning	0.04/0.3	Increased	Zhan *et al.*, 2011
Toxicity	n.a/0.3; 3; 7 or 10	Less toxic	Kim and Mahan, 2001a,b
Se excretion	n.a/0.1; 0.3 or 0.5	Decreased	Mahan and Parrett, 1996

[1] AO = antioxidant; AOA = antioxidant activity; GPx = glutathione peroxidase; GSH = glutathione; MDA = malondialdehyde; SOD = superoxide dismutase; SS = sodium selenite; SY = Se-yeast; T-AOC = total antioxidant capacity.

[2] n.a. = not available.

It is generally accepted that Se participates in various physiological functions as an integral part of a range of selenoproteins. The selenoprotein family includes 25 porcine selenoproteins and expression (mRNA, protein, and/or enzyme activity) of 2/3 of them in various tissues of pigs was shown to be affected by dietary Se intakes (Chen *et al.*, 2018). Expression of individual eukaryotic selenoproteins are shown to be characterised by high tissue specificity, depends on Se availability, can be regulated by hormones, and if compromised contributes to various pathological conditions (Surai, 2006, 2018). Furthermore, almost half of selenoproteins are involved in redox signalling and antioxidant defences (Surai, 2018). The aforementioned data indicate that main attention has been paid to GPx as an important element indicating Se status and potentially antioxidant defences. However, current evidence suggests that other Se-proteins could be even more important than GPx for Se action and that optimal levels may depend upon the form of Se being ingested (Ferguson *et al.*, 2012; Surai, 2018). Indeed, protective roles of selenoproteins in pig nutrition and reproduction is related to necessity for an effective antioxidant defence to prevent negative consequences of over-production of free radicals during gestation, births and weaning (Close *et al.*, 2008; Surai, 2006; Surai and Fisinin, 2016).

5.5 Conclusions

From the data presented above it is clear that oxidative stress in commercial sows is an important risk factor associated with decreased productive and reproductive performance. Taking into account some important features of antioxidant system of the sow (low vitamin E transfer via placenta and decreased vitamin C synthesis) at late gestation and parturition, selenium is shown to be an important dietary element in sow nutrition. It helps improving the antioxidant defences of sows and offspring piglets by converting to SeCys in selenoprotein synthesis. For the last two decades it has been clearly shown that organic selenium has many important advantages in sow nutrition in comparison to traditional sodium selenite. The aforementioned benefits of organic selenium for sow include: better Se status of sows, especially in the case of advanced parities; improved antioxidant defences of sows; increased Se concentration in colostrum and milk and improved antioxidant status of colostrum and milk; improved Se transfer via placenta; improved Se status of foetus and development of pig embryos. It was proven that replacement of sodium selenite by organic Se in the sow's diet improved Se status of newly-born piglets characterised by increased Se concentrations in tissues and whole body. In particular, increased Se concentration in piglet muscles could be considered as an important storage form of Se to be used in stress conditions. Furthermore, organic Se in sow's diet (vs sodium selenite) provides better Se status and antioxidant status of weaning piglets, as well as improves thyroid metabolism and increases activities of major digestive enzymes in piglet pancreas at time of weaning. In most of papers, Se-yeast was successfully used as a source of organic selenium. It seems likely that SeMet, the main active compound of the Se-Yeast, comprises 60-70% of total selenium. Furthermore, correct Se speciation in Se-Yeast is still a difficult task. A wide usage of pure SeMet in the animal industry would be restricted by problems with its stability. Recent results indicated that a new source of organic selenium in the form of 2-hydroxy-4-methylselenobutanoic acid supplied in the same dose as Se-Yeast in the chicken and pig diets could provide additional benefit in terms of Se reserves in the muscles as SeMet but also higher levels of SeCys supporting enhanced levels of antioxidant selenoproteins. One can thus expect higher Se levels in sow tissues as well as transfer to the progeny via placenta.

It should be mentioned that organic Se was effective in positive changes in various antioxidant-related biochemical parameters in sows and progeny piglets. However, most of the commercially-relevant parameters (growth rate, mortality, FCR, etc.) were not affected. Only in two most recent publications where sodium selenite was replaced by pure organic Se source in the sows diet, there were significant improvements in weaning litter weight and average weight of progeny piglets and daily weight gain of piglet from birth to weaning. In fact, most of university trials were conducted with comparatively low number of sows and level of stress was likely potentially lower than one can expect in large commercial pig units. Therefore, an optimal Se form and level in the sow's diet potentially can be translated into better antioxidant protection in stress conditions of commercial pig production and maintain sow reproductive performance and newly-born piglet viability at high level. Indeed, additional well-designed organic Se trials with big numbers of sows in commercial conditions are needed to explore a full potential of organic Se in sow nutrition.

Acknowledgements

Peter F. Surai is supported by a grant of the Government of Russian Federation (Contract No. 14.W03.31.0013).

References

Al-Gubory, K.H., Fowler, P.A. and Garrel, C., 2010. The roles of cellular reactive oxygen species, oxidative stress and antioxidants in pregnancy outcomes. The International Journal of Biochemistry & Cell Biology 42: 1634-1650.

Allen, R.G. and Venkatraj, V.S., 1992. Oxidants and antioxidants in development and differentiation. Journal of Nutrition 122: 631-635.

Avissar,N., Whitin, J., Allen, P., Palmer, I. and Cohen, H., 1989. Antihuman plasma glutathione peroxidase antibodies: immunologic investigations to determine plasma glutathione peroxidase protein and selenium content in plasma. Blood 73: 318-323.

Berchieri-Ronchi, C.B., Kim, S.W., Zhao, Y., Correa, C.R., Yeum, K.J. and Ferreira, A.L., 2011. Oxidative stress status of highly prolific sows during gestation and lactation. Animal 5: 1774-1779.

Bierla, K., Dernovics, M., Vacchina, V., Szpunar, J., Bertin, G. and Lobinski, R., 2008. Determination of selenocysteine and selenomethionine in edible animal tissues by 2D size-exclusion reversed-phase HPLC-ICP MS following carbamidomethylation and proteolytic extraction. Analytical and Bioanalytical Chemistry 390: 1789-1798.

Blodgett, D.J., Schurig, G.G., Kornegay, E.T., Meldrum, J.B. and Bonnette, E.D., 1989. Failure of an enhanced dietary selenium concentration to stimulate humoral immunity in gestating swine. Nutrition Reports International 40: 543-550.

Briens, M., Mercier, Y., Rouffineau, F., Mercerand, F. and Geraert, P.A., 2014. 2-hydroxy-4-methyl-selenobutanoic acid induces additional tissue selenium enrichment in broiler chickens compared with other selenium sources. Poultry Science 93: 85-93.

Buonocore, G., Perrone, S. and Bracci, R., 2001. Free radicals and brain damage in the newborn. Biology of Neonate 79: 180-186.

Burk, R.F., Olson, G.E., Hill, K.E., Winfrey, V.P., Motley, A.K. and Kurokawa, S., 2013. Maternal-fetal transfer of selenium in the mouse. FASEB Journal 27: 3249-3256.

Burton, G.J. and Jauniaux, E., 2011. Oxidative stress. Best Practise & Research. Clinical Obstetrics & Gynaecology 25: 287-299.

Calamari, L., Abeni, F. and Bertin, G., 2010. Metabolic and hematological profiles in mature horses supplemented with different selenium sources and doses. Journal of Animal Science 88: 650-659.

Ceballos, A., Sánchez, J., Stryhn, H., Montgomery, J.B., Barkema, H.W. and Wichtel, J.J., 2009. Meta-analysis of the effect of oral selenium supplementation on milk selenium concentration in cattle. Journal of Dairy Science 92: 324-342.

Cebula, M., Schmidt, E.E. and Arnér, E.S., 2015. TrxR1 as a potent regulator of the Nrf2-Keap1 response system. Antioxidants & Redox Signaling 23: 823-853.

Chavez, E.R., 1983. Supplemental value of ascorbic acid during late gestation on piglet survival and early growth. Canadian Journal of Animal Science 63: 683-687.

Chen, J., Han, J.H., Guan, W.T., Chen, F., Wang, C.X., Zhang, Y.Z., Lv, Y.T. and Lin, G., 2016a. Selenium and vitamin E in sow diets: I. Effect on antioxidant status and reproductive performance in multiparous sows. Animal Feed Science and Technology 221: 111-123.

Chen, J., Han, J.H., Guan, W.T., Chen, F., Wang, C.X., Zhang, Y.Z., Lv, Y.T. and Lin, G., 2016b. Selenium and vitamin E in sow diets: II. Effect on selenium status and antioxidant status of the progeny. Animal Feed Science and Technology 221: 101-110.

Chen, J., Zhang, F., Guan, W., Song, H., Tian, M., Cheng, L., Shi, K., Song, J., Chen, F., Zhang, S., Yang, F., Ren, C. and Zhang, Y., 2019. Increasing selenium supply for heat-stressed or actively cooled sows improves piglet preweaning survival, colostrum and milk composition, as well as maternal selenium, antioxidant status and immunoglobulin transfer. Journal of Trace Elements in Medicine and Biology 52: 89-99.

Chen, X.D., Zhao, Z.P., Zhou, J.C. and Lei, X.G., 2018. Evolution, regulation, and function of porcine selenogenome. Free Radical Biology & Medicine 127: 116-123.

Ching, S., 2000. Ontogeny of liver L-gulonolactone oxidase activity and tissue ascorbic acid concentration in fetal, nursing, and weaned pigs, and factors influencing its synthesis. Ph.D. thesis. Ohio State University, Columbus, OH, USA.

Ching, S., Mahan, D.C. and Dabrowski, K., 2001b. Liver L-gulonolactone oxidase activity and tissue ascorbic acid concentrations in nursing pigs and the effect of various weaning ages. Journal of Nutrition 131: 2002-2006.

Ching, S., Mahan, D.C., Ottobre, J.S. and Dabrowski, K., 2001a. Ascorbic acid synthesis in fetal and neonatal pigs and in pregnant and postpartum sows. Journal of Nutrition 131: 1997-2001.

Cindrova-Davies, T., Yung, H.W., Johns, J., Spasic-Boskovic, O., Korolchuk, S., Jauniaux, E., Burton, G.J. and Charnock-Jones, D.S., 2007. Oxidative stress, gene expression, and protein changes induced in the human placenta during labor. American Journal of Pathology 171: 1168-1179.

Close, W.H., 2003. Organic selenium may improve sow, piglet performance. Feedstuffs 75: 8.

Close, W.H., Surai, P.F. and Taylor-Pickard, J.A., 2008. Selenium in pig nutrition. In: Surai, P.F. and Taylor-Pickard, J. (eds) Current advances in selenium research and applications. Wageningen Academic Publishers, Wageningen, the Netherlands, pp.263-313.

Cools, A., Maes, D., Papadopoulos, G., Vandermeiren, J.A., Meyer, E., Demeyere, K., De Smet, S. and Janssens, G.P., 2011. Dose-response effect of fish oil substitution in parturition feed on erythrocyte membrane characteristics and sow performance. Journal of Animal Physiology and Animal Nutrition 95: 125-136.

Dalto, B.D., Tsoi, S., Audet, I., Dyck, M.K., Foxcroft, G.R. and Matte, J.J., 2015. Gene expression of porcine blastocysts from gilts fed organic or inorganic selenium and pyridoxine. Reproduction 149: 31-42.

Falk, M., Lebed, P., Bernhoft, A., Framstad, T., Kristoffersen, A.B., Salbu, B. and Oropeza-Moe, M., 2019. Effects of sodium selenite and L-selenomethionine on feed intake, clinically relevant blood parameters and selenium species in plasma, colostrum and milk from high-yielding sows. Journal of Trace Elements in Medicine and Biology 52: 176-185.

Fantel, A.G., 1996. Reactive oxygen species in developmental toxicity: review and hypothesis. Teratology 53: 196-217.

Ferguson, L.R., Karunasinghe, N., Zhu, S. and Wang, A.H., 2012. Selenium and its' role in the maintenance of genomic stability. Mutation Research 733: 100-110.

Fortier, M.E., Audet, I., Giguère, A., Laforest, J.P., Bilodeau, J.F., Quesnel, H. and Matte, J.J., 2012. Effect of dietary organic and inorganic selenium on antioxidant status, embryo development, and reproductive performance in hyperovulatory first-parity gilts. Journal of Animal Science 90: 231-240.

Gaykwad, C.K., De, U.K., Jadhav, S.E., Chethan, G.E., Akhilesh, Sahoo, N.R., Mondal, D.B., Gaur, G.K., Verma, M.R. and Chaudhuri, P., 2019. Adding α-tocopherol-selenium and ascorbic acid to periparturient sow diets influences hemogram, lipid profile, leptin, oxidant/antioxidant imbalance, performance and neonatal piglet mortality. Research in Veterinary Science 125: 360-369.

Gelderman, A. and Clapper, J., 2013. Effects of inorganic or organic selenium on immunoglobulins in swine. Journal of Animal Science and Biotechnology 4: 47.

Givens, D.I., Allison, R., Cottrill, B. and Blake, J.S., 2004. Enhancing the seleniumcontent of bovinemilk through alteration of the formand concentration of seleniumin the diet of the dairy cow. Journal of the Science of Food and Agriculture 84: 811-817.

Gourley, G.G., Lampe, J.F., Sparks, J.C. and Stumpf, T.T., 2005. Piglet survivability and performance: Sel-Plex® versus sodium selenite in sow and nursery diets. In: Lyons, T.P. and Jacques, K.A. (eds) Nutritional biotechnology in the feed and food industries. Nottingham University Press, Nottingham, UK, pp.181-188.

Gu, X.L., Li, H., Song, Z.H., Ding, Y.N., He, X. and Fan, Z.Y., 2019. Effects of isomaltooligosaccharide and *Bacillus* supplementation on sow performance, serum metabolites, and serum and placental oxidative status. Animal Reproduction Science 207: 52-60.

Gunter, S.A., Beck, P.A. and Phillips, J.K., 2003. Effects of supplementary selenium source on the performance and blood measurements in beef cows and their calves. Journal of Animal Science 81: 856-864.

Guyot, H., Spring, P., Andrieu, S. and Rollin, F., 2007. Comparative responses to sodium selenite and organic selenium in Belgian Blue cows and calves. Livestock Science 111: 258-263.

Hakansson, J., Hakkarainen, J. and Lundeheim, N., 2001. Variation in vitamin E, glutathione peroxidase and retinol concentrations in blood plasma of primiparous sows and their piglets, and in vitamin E, selenium and retinol contents in sows' milk. Acta Agriculture Scandinavica 51A: 224-234.

Hammer, C., Thorson, J., Meyer, A., Redmer, D., Luther, J., Neville, T., Reed, J., Reynolds, L., Caton, J. and Vonnahme, K., 2011. Effects of maternal selenium supply and plane of nutrition during gestation on passive transfer of immunity and health in neonatal lambs. Journal of Animal Science 89: 3690-3698.

Hidiroglou, M. and Batra, T.R., 1995. Concentrations of vitamin C in milk of sows and plasma of piglets. Canadian Journal of Animal Science 75: 275-277.

Hidiroglou, M., Farnworth, E., and Butler, G., 1993. Vitamin E and fat supplementation of sows and the effect on tissue vitamin E concentrations in their progeny. Reproduction Nutrition Development 33: 557-565.

Hill, K.E., Xia, Y., Åkesson, B., Boeglin, M.E. and Burk, R.F., 1996. Selenoprotein P concentration in plasma is an index of selenium status in selenium-deficient and selenium-supplemented Chinese subjects. Journal of Nutrition 126: 138-145.

Hostetler, C.E. and Kincaid, R.L., 2004a. Gestational changes in concentrations of selenium and zinc in the porcine fetus and the effects of maternal intake of selenium. Biological Trace Element Research 97: 57-70.

Hostetler, C.E. and Kincaid, R.L., 2004b. Maternal selenium deficiency increases hydrogen peroxide and total lipid peroxides in porcine fetal liver. Biological Trace Element Research 97: 43-56.

Hu, H., Wang, M., Zhan, X., Li, X. and Zhao, R., 2011. Effect of different selenium sources on productive performance, serum and milk Se concentrations, and antioxidant status of sows. Biological Trace Element Research 142: 471-480.

Hu, L., Che, L., Wu, C., Curtasu, M.V., Wu, F., Fang, Z., Lin, Y., Xu, S., Feng, B., Li, J., Zhuo, Y., Theil, P.K. and Wu, D., 2019. Metabolomic profiling reveals the difference on reproductive performance between high and low lactational weight loss sows. Metabolites 9: 12.

Jlali, M., Briens, M., Rouffineau, F., Geraert, P.A. and Mercier, Y., 2014. Evaluation of the efficacy of 2-hydroxy-4-methylselenobutanoic acid on growth performance and tissue selenium retention in growing pigs. Journal of Animal Science 92: 182-188.

Juniper, D.T. and Bertin, G., 2011. Distribution of total selenium and selenium species within the tissues and products of food producing animals offered diets containing either selenoyeasts or sodium selenite. In Proceedings of the TEMA14, Enshi, China, 19-24 September 2011.

Juniper, D.T., Rymer, C. and Briens, M., 2019. Bioefficacy of hydroxy-selenomethionine as a selenium supplement in pregnant dairy heifers and on the selenium status of their calves. Journal of Dairy Science 102: 7000-7010.

Kamada, H., Nonaka, I., Ueda, Y. and Murai, M., 2007. Selenium addition to colostrum increases immunoglobulin G absorption by newborn calves. Journal of Dairy Science 90: 5665-5670.

Kim, Y.Y. and Mahan, D.C., 2001a. Prolonged feeding of high dietary levels of organic and inorganic selenium to gilts from 25 kg body weight through one parity. Journal of Animal Science 79: 956-966.

Kim, Y.Y. and Mahan, D.C., 2001b. Comparative effects of high dietary levels of organic and inorganic selenium on selenium toxicity of growing-finishing pigs. Journal of Animal Science 79: 942-948.

Lauridsen, C., Engel, H., Jensen, S., Craig, A.M. and Traber, M.G., 2002. Lactating sows and suckling piglets preferentially incorporate RRR over all-rac-α-tocopherol into milk plasma and tissue. Journal of Nutrition 132: 1258-1264.

Li, Y., Liu, H., Zhang, L., Yang, Y., Lin, Y., Zhuo, Y., Fang, Z., Che, L., Feng, B., Xu, S., Li, J. and Wu, D., 2019. Maternal dietary fiber composition during gestation induces changes in offspring antioxidative capacity, inflammatory response, and gut microbiota in a sow model. International Journal of Molecular Sciences 21: 1.

Lipko-Przybylska, J. and Kankofer, M., 2012. Antioxidant defence of colostrum and milk in consecutive lactations in sows. Irish Veterinary Journal 65: 1.

Liu, F., Cottrell, J.J., Collins, C.L., Henman, D.J., O'Halloran, K.S.B. and Dunshea, F.R., 2017. Supplementation of selenium, vitamin E, chromium and betaine above recommended levels improves lactating performance of sows over summer. Tropical Animal Health and Production 49: 1461-1469.

Loudenslager, M.J., Ku, P.K., Whetter, P.A., Ullrey, D.E., Whitehair, C.K., Stowe, H.D. and Miller, E.R., 1986. Importance of diet of dam and colostrum to the biological antioxidant status and parenteral iron tolerance of the pig. Journal of Animal Science 63: 1905-1914.

Luo, Z.C., Fraser, W.D., Julien, P., Deal, C.L., Audibert, F., Smith, G.N., Xiong, X. and Walker, M., 2006. Tracing the origins of 'fetal origins' of adult diseases: programming by oxidative stress? Medical Hypotheses 66: 38-44.

Ma, Y.L., Lindemann, M.D., Pierce, J.L., Unrine, J.M. and Cromwell, G.L., 2014. Effect of inorganic or organic selenium supplementation on reproductive performance and tissue trace mineral concentrations in gravid first-parity gilts, fetuses, and nursing piglets. Journal of Animal Science 92: 5540-5550.

Mack, L.A., Lay, D.C. Jr, Eicher, S.D., Johnson, A.K., Richert, B.T. and Pajor, E.A., 2014. Growth and reproductive development of male piglets are more vulnerable to midgestation maternal stress than that of female piglets. Journal of Animal Science 92: 530-548.

Mahan, D.C., 1991a. Assessment of the influence of dietary vitamin E on sows and offspring in three parities: reproductive performance, tissue tocopherol, and effects on progeny. Journal of Animal Science 69: 2904-2917.

Mahan, D.C., 1991b. Vitamin E and selenium in swine nutrition. In: Miller E.R., Ullrey D.E. and Lewis A.J. (eds) Swine nutrition. Butterworth-Heineman, Boston, MA, USA, pp. 193-214.

Mahan, D.C., 1994. Effects of dietary vitamin E on sow reproductive performance over a five-parity period. Journal of Animal Science 72: 2870-2879.

Mahan, D.C., 2000. Effect of organic and inorganic selenium sources and levels on sow colostrum and milk selenium content. Journal of Animal Science 78: 100-105.

Mahan, D.C. and Kim, Y.Y., 1996. Effect of inorganic or organic selenium at two dietary levels on reproductive performance and tissue selenium concentrations in first-parity gilts and their progeny. Journal of Animal Science 74: 2711-2718.

Mahan, D.C. and Parrett, N.A., 1996. Evaluating the efficacy of selenium-enriched yeast and sodium selenite on tissue selenium retention and serum glutathione peroxidase activity in grower and finisher swine. Journal of Animal Science 74: 2967-2974.

Mahan, D.C. and Peters, J.C., 2004. Long-term effects of dietary organic and inorganic selenium sources and levels on reproducing sows and their progeny. Journal of Animal Science 82: 1343-1358.

Mahan, D.C. and Vallet, J.L., 1997. Vitamin and mineral transfer during fetal development and the early postnatal period in pigs. Journal of Animal Science 75: 2731-2738.

Mahan, D.C., Ching, S. and Dabrowski, K., 2004. Developmental aspects and factors influencing the synthesis and status of ascorbic Acid in the pig. Annual Review of Nutrition 24: 79-103.

Mahan, D.C., Moxon, A.L. and Cline, J.H., 1975. Efficacy of supplemental selenium in reproductive diets on sow and progeny serum and tissue selenium values. Journal of Animal Science 40: 624-631.

Mahan, D.C., Moxon, A.L. and Hubbard, M., 1977. Efficacy of inorganic selenium supplementation to sow diets on resulting carry-over to their progeny. Journal of Animal Science 45: 738-746.

Mahan, D.C., Penhale, L.H., Cline, J.H., Moxon, A.L., Fetter, A.W. and Yarrington, J.T., 1974. Efficacy of supplemental selenium in reproductive diets on sow and progeny performance. Journal of Animal Science 39: 536-543.

Mahan, D.C., Peters, J.C. and Hill, G.M., 2007. Are antioxidants associated with pig and sow mortalities? Swine Nutrition Conference Proceedings. Indianapolis, IN, USA, September 6, pp.13-21.

Meyer, A.M., Reed, J.J., Neville, T.L., Thorson, J.F., Maddock-Carlin, K.R., Taylor, J.B., Reynolds, L.P., Redmer, D.A., Luther, J.S., Hammer, C.J., Vonnahme, K.A. and Caton, J.S., 2011. Nutritional plane and selenium supply during gestation affect yield and nutrient composition of colostrum and milk in primiparous ewes. Journal of Animal Science 89: 1627-1639.

Milner, J.A., Sherman, L. and Picciano, M.F., 1987. Distribution of selenium in human milk. American Journal of Clinical Nutrition 45: 617-624.

Mou, D., Ding, D., Yan, H., Qin, B., Dong, Y., Li, Z., Che, L., Fang, Z., Xu, S., Lin, Y., Zhuo, Y., Li, C., Huang C, Zou Y., Li, L., Briens, M., Wu, D. and Feng, B., 2020. Maternal supplementation of organic selenium during gestation improves sows and offspring antioxidant capacity, inflammatory status, and promotes embryo survival. Food and Function 11: 7748-7761. https://doi.org/10.1039/D0FO00832J

Muñiz-Naveiro, O., Domínguez-González, R., Bermejo-Barrera, A., Bermejo-Barrera, P., Cocho, J.A. and Fraga, J.M., 2007. Selenium speciation in cow milk obtained after supplementation with different selenium forms to the cow feed using liquid chromatography coupled with hydride generation-atomic fluorescence spectrometry. Talanta 71: 1587-1593.

Myatt, L. and Cui, X., 2004. Oxidative stress in the placenta. Histochemistry and Cell Biology 122: 369-382.

Pallardo, F.V., Sastre, J., Asensi, M., Rodrigo, F., Estrela, J.M. and Viña, J., 1991. Physiological changes in glutathione metabolism in foetal and newborn rat liver. The Biochemical Journal 274: 891-893.

Pehrson, B., Holmgren, N. and Trafikowska, U., 2001. The influence of parenterally administered alpha-tocopheryl acetate to sows on the vitamin E status of the sows and suckling piglets and piglets after weaning. Journal of Veterinary Medicine. A, Physiology, Pathology, Clinical Medicine 48: 569-575.

Pehrson, B., Ortman, K., Madjid, N. and Trafikowska, U., 1999. The influence of dietary selenium as selenium yeast or sodium selenite on the concentration of selenium in the milk of Suckler cows and on the selenium status of their calves. Journal of Animal Science 77: 3371-3376.

Peters, J.C., Mahan, D.C., Wiseman, T.G. and Fastinger, N.D., 2010. Effect of dietary organic and inorganic micromineral source and level on sow body, liver, colostrum, mature milk, and progeny mineral compositions over six parities. Journal of Animal Science 88: 626-637.

Phipps, R.H., Grandison, A.S., Jones, A.K., Juniper, D.T., Ramos-Morales, E. and Bertin, G., 2008. Selenium supplementation of lactating dairy cows: effects on milk production and total selenium content and speciation in blood, milk and cheese. Animal 2: 1610-1618.

Pinelli-Saavedra, A. and Scaife, J.R., 2005. Pre and postnatal transfer of vitamin E to piglets in sows supplemented with vitamin E and vitamin C. Livestock Production Science 97: 231-240.

Pinelli-Saavedra, A., Calderón de la Barca, A.M., Hernández, J., Valenzuela, R. and Scaife, J.R., 2008. Effect of supplementing sows' feed with alpha-tocopherol acetate and vitamin C on transfer of alpha-tocopherol to piglet tissues, colostrum, and milk: aspects of immune status of piglets. Research in Veterinary Science 85: 92-100.

Quesnel, H., Renaudin, A., Le Floc'h, N., Jondreville, C., Père, M.C., Taylor-Pickard, J.A. and Le Dividich, J., 2008. Effect of organic and inorganic selenium sources in sow diets on colostrum production and piglet response to a poor sanitary environment after weaning. Animal 2: 859-866.

Rizzo, A., Roscino, M.T., Binetti, F. and Sciorsci, R.L., 2012. Roles of reactive oxygen species in female reproduction. Reproduction in Domestic Animals 47: 344-352.

Rock, M., Kincaid, R. and Carstens, G., 2001. Effects of prenatal source and level of dietary selenium on passive immunity and thermometabolism of newborn lambs. Small Ruminant Research 40: 129-138.

Salman, S., Dinse, D., Khol-Parisini, A., Schafft, H., Lahrssen-Wiederholt, M., Schreiner, M., Scharek-Tedin, L. and Zentek, J., 2013. Colostrum and milk selenium, antioxidative capacity and immune status of dairy cows fed sodium selenite or selenium yeast. Archives of Animal Nutrition 67: 48-61.
Sastre, J., Asensi, M., Rodrigo, F., Pallardó, F.V., Vento, M. and Viña, J., 1994. Antioxidant administration to the mother prevents oxidative stress associated with birth in the neonatal rat. Life Sciences 54: 2055-2059.
Saugstad, O.D., 1990. Oxygen toxicity in the neonatal period. Acta Paediatrica Scandinavica 79: 881-892.
Saugstad, O.D., 2003. Oxygen toxicity at birth: the pieces are put together. Pediatric Research 54: 789-783.
Saugstad, O.D., 2005. Oxidative stress in the newborn. A 30-year perspective. Biology of Neonate 88: 228-236.
Schrauzer, G.N., and Surai, P.F., 2009. Selenium in human and animal nutrition: resolved and unresolved issues. A partly historical treatise in commemoration of the fiftieth anniversary of the discovery of the biological essentiality of selenium, dedicated to the memory of Klaus Schwarz (1914-1978) on the occasion of the thirtieth anniversary of his death. Critical Reviews in Biotechnology 29: 2-9.
Shen, Y., Wan, H., Zhu, J., Fang, Z., Che, L., Xu, S., Lin, Y., Li. J. and Wu, D., 2015. Fish oil and olive oil supplementation in late pregnancy and lactation differentially affect oxidative stress and inflammation in sows and piglets. Lipids 50: 647-658.
Slavik, P., Illek, J., Brix, M., Hlavicova, J., Rajmon, R. and Jilek, F., 2008. Influence of organic versus inorganic dietary selenium supplementation on the concentration of selenium in colostrum, milk and blood of beef cows. Acta Veterinaria Scandinavica 50: 43.
Su, G., Zhao, J., Luo, G., Xuan, Y., Fang, Z., Lin, Y., Xu S, Wu, D., He, J. and Che, L., 2017. Effects of oil quality and antioxidant supplementation on sow performance, milk composition and oxidative status in serum and placenta. Lipids in Health and Disease 16: 107.
Surai, P.F., 2006. Selenium in nutrition and health. Nottingham University Press; Nottingham, UK.
Surai, P.F., 2018. Selenium in poultry nutrition and health. Wageningen Academic Publishers, Wageningen, the Netherland, 430 pp.
Surai, P.F and Fisinin, V.I., 2014. Selenium in poultry breeder nutrition: an update. Animal Feed Science and Technology 191: 1-15.
Surai, P.F. and Fisinin, V.I., 2016. Selenium in sow nutrition. Animal Feed Science and Technology 211: 18-30.
Surai, P.F., Kochish, I.I. and Fisinin, V.I., 2021. Selenium in feed: organic selenium concept development. Chapter 3. In: Surai, P.F. (ed.) Selenium in pig nutrition and health. Wageningen Academic Publishers, Wageningen, the Netherlands, pp. 61-110. https://doi.org/10.3920/978-90-8686-912-1_3
Surai, P.F., Kochish, I.I., Fisinin, V.I. and Juniper, D.T., 2019b. Revisiting oxidative stress and the use of organic selenium in dairy cow nutrition. Animals 9: 7.
Surai, P.F., Kochish, I.I., Fisinin, V.I. and Kidd, M.T., 2019a. Antioxidant defence systems and oxidative stress in poultry biology: an update. Antioxidants 8: 7.
Svoboda, M., Ficek, R. and Drabek, J., 2008. Efficacy of organic selenium from se-enriched yeast on selenium transfer from sows to piglets. Acta Veterinaria Scandinavica 77: 515-521.
Tanghe, S., Missotten, J., Raes, K. and De Smet, S., 2015. The effect of different concentrations of linseed oil or fish oil in the maternal diet on the fatty acid composition and oxidative status of sows and piglets. Journal of Animal Physiology and Animal Nutrition 99: 938-949.
Tian, L., Huang, J., Wen, A. and Yan, P., 2020. Impaired mitochondrial function results from oxidative stress in the full-term placenta of sows with excessive back-fat. Animals 10: 2.

Vergauwen, H., Degroote, J., Prims, S., Wang, W., Fransen, E., De Smet, S., Casteleyn, C., Van Cruchten, S., Michiels, J. and Van Ginneken, C., 2017. Artificial rearing influences the morphology, permeability and redox state of the gastrointestinal tract of low and normal birth weight piglets. Journal of Animal Science and Biotechnology 8: 30.

Walker, G.P., Dunshea, F.R., Heard, J.W., Stockdale, C.R. and Doyle, P.T., 2010. Output of selenium in milk, urine, and feces is proportional to selenium intake in dairy cows fed a total mixed ration supplemented with selenium yeast. Journal of Dairy Science 93: 4644-4650.

Wang, Y.S., Zhou, P., Liu, H., Li, S., Zhao, Y., Deng, K., Cao, D.D., Che, L.Q., Fang, Z.F., Xu, S.Y., Lin, Y., Feng, B., Li. J. and Wu, D., 2016. Effects of inulin supplementation in low- or high-fat diets on reproductive performance of sows and antioxidant defence capacity in sows and offspring. Reproduction of Domestic Animals 51: 492-500.

Wei, J.Y., Wang, J., Liu, W., Zhang, K.Z. and Sun, P., 2019. Short communication: Effects of different selenium supplements on rumen fermentation and apparent nutrient and selenium digestibility of mid-lactation dairy cows. Journal of Dairy Science 102: 3131-3135.

Wen, L.X., Yuan, H. and Xie, L., 2009. Sow reproductive stress and its syndrome. Animal Husbandry and Feed Science 1: 20-24.

Whanger, P.D., Butler, J.A. and Deagen, J.T., 1991. Determination of the distribution of selenium between selenoprotein P, glutathione peroxidase and albumin in plasma. FASEB Journal 75: 5.

Wisdom, S.J., Wilson, R., McKillop, J.H. and Walker, J.J., 1991. Antioxidant systems in normal pregnancy and in pregnancy-inducted hypertension. American Journal of Obstetrics and Gynecology 165: 1701-1704.

Yen, J.T., and Pond, W.G., 1983. Response of swine to periparturient vitamin C supplementation. Journal of Animal Science 56: 621-624.

Yin, J., Ren, W., Liu, G., Duan, J., Yang, G., Wu, L., Li, T. and Yin, Y., 2013. Birth oxidative stress and the development of an antioxidant system in newborn piglets. Free Radical Research 47: 1027-1035.

Yoon, I. and McMillan, E., 2006. Comparative effects of organic and inorganic selenium on selenium transfer from sows to nursing pigs. Journal of Animal Science 84: 1729-1733.

Zhan, X., Qie, Y., Wang, M., Li, X. and Zhao, R., 2011. Selenomethionine: an effective selenium source for sow to improve Se distribution, antioxidant status, and growth performance of pig offspring. Biological Trace Element Research 142: 481-491.

Zhang, W., Zhang, R., Wang, T., Jiang, H., Guo, M., Zhou, E., Sun, Y., Yang, Z., Xu, S., Cao, Y. and Zhang, N., 2014. Selenium inhibits LPS-induced pro-inflammatory gene expression by modulating MAPK and NF-κB signaling pathways in mouse mammary epithelial cells in primary culture. Inflammation 37: 478-485.

Zhao, Y. and Kim, S.W., 2019. Oxidative stress status and reproductive performance of sows during gestation and lactation under different thermal environments. Asian-Australasian Journal of Animal Sciences 33(5): 722-731. https://doi.org/10.5713/ajas.19.0334

Zhao, Y., Flowers, W.L., Saraiva, A., Yeum, K.J. and Kim, S.W., 2013. Effect of social ranks and gestation housing systems on oxidative stress status, reproductive performance, and immune status of sows. Journal of Animal Science 91: 5848-5858.

Zhao, Y., Flowers, W.L., Saraiva, A., Yeum, K-J. and Kim, S.W., 2011. Effect of heat stress on oxidative stress status and reproductive performance of sows. Journal of Animal Science 89: 108.

Zhou, Y., Xu, T., Wu, Y., Wei, H. and Peng, J., 2019. Oxidative stress and inflammation in sows with excess backfat: up-regulated cytokine expression and elevated oxidative stress biomarkers in placenta. Animals 9: 10.

Zou, T., Yu, B., Yu, J., Mao, X., Zheng, P., He, J., Huang, Z., Liu, Y. and Chen, D., 2016. Moderately decreased maternal dietary energy intake during pregnancy reduces fetal skeletal muscle mitochondrial biogenesis in the pigs. Genes & Nutrition 11: 19.

Chapter 6
Selenium in boar nutrition

P.F. Surai[1-7] and V.I. Fisinin[8]
[1]Vitagene and Health Research Centre, Bristol, BS4 2RS, United Kingdom; [2]Department of Hygiene and Poultry Sciences, Moscow State Academy of Veterinary Medicine and Biotechnology Named After K.I. Skryabin, 109472 Moscow, Russia; [3]Department of Microbiology and Biochemistry, Faculty of Veterinary Medicine, Trakia University, 6000 Stara Zagora, Bulgaria; [4]Department of Animal Nutrition, Faculty of Agricultural and Environmental Sciences, Szent Istvan University, 2103 Gödöllo, Hungary; [5]Saint-Petersburg State Academy of Veterinary Medicine, 196084 St. Petersburg, Russia; [6]Sumy National Agrarian University, 40000 Sumy, Ukraine; [7]Odessa National Academy of Food Technologies, 65039 Odessa, Ukraine; [8]All-Russian Institute of Poultry Husbandry, 141311 Sergiev Posad, Russia; psurai@feedfood.co.uk

Abstract

Selenium plays an important role in boar nutrition. In particular high levels of polyunsaturated fatty acids in boar testes and spermatozoa predispose them to be very susceptible to oxidative stress and lipid peroxidation. The requirement of boars for selenium varies depending on many environmental and other conditions and in general is considered to be 0.30 mg/kg feed and meeting boar Se requirements is an important challenge for pig nutritionists. In many countries there are legal limits as to how much Se may be included into the diet and this restricts flexibility in terms of addressing the Se needs of the developing and reproducing pigs. The analysis of data of various boar trials with different Se sources indicates, that in some cases when background Se levels were low, there were advantages of Se dietary supplementation. However, only an optimal Se status of animals is associated with the best antioxidant protection and could have positive effects on boar semen production and its quality. Nevertheless, in many cases background Se levels were not determined and therefore it is difficult to judge if the basic diets were deficient in Se. It can also be suggested that, because of higher efficacy of assimilation from the diet, and possibilities of building Se reserves in the body, organic selenium in the form of SeMet provided by a range of products is an important source of Se to better meet the needs of modern pig genotypes. In particular, organic Se in the form of OH-SeMet was shown to be an effective form of organic Se for pigs, poultry and ruminants.

Keywords: selenium, boar, semen, oxidative stress

DOI 10.3920/978-90-8686-912-1_6, © Wageningen Academic Publishers 2021

6.1 Introduction

Mammalian spermatozoa are unique in structure and chemical composition and are characterised by high proportions of polyunsaturated fatty acids (PUFAs) in the phospholipid fraction of their membranes (Cerolini *et al.*, 2000; Surai, 2002, 2006). This feature of these highly specialised cells is a reflection of the specific needs of their membranes for high levels of fluidity and flexibility, which are necessary for sperm motility and fusion with the egg. This functional advantage conferred by PUFAs is, however, associated with disadvantages in terms of the susceptibility of sperm to free radical attack and lipid peroxidation. Therefore, antioxidant protection is a vital element in maintaining sperm membrane integrity, motility and fertilising ability. It has been suggested (Surai, 2006; Surai *et al.*, 2001) that natural antioxidants (vitamin E, ascorbic acid and glutathione) together with antioxidant enzymes (superoxide dismutase and glutathione peroxidase) build an integrated antioxidant system in mammalian and avian semen capable of protecting it against free radicals and toxic products of their metabolism. The delicate balance between free radical production and antioxidant defence is considered to be an important determinant of boar semen quality and in particular its fertilising ability. In this respect, dietary Se is an important modulator of semen quality, including the antioxidant systems. The aim of the review is to update existing knowledge related to physiological roles of Se in boar nutrition and reproduction with a special emphasis to antioxidant defences in boar semen.

6.2 Selenium and male fertility

The essentiality of selenium for male fertility was shown in the early 1980s. This conclusion was based on the results of a range of different experiments with mammals which can be summarised as follows. In mild deficiency, Se is preferentially retained in rat testes and mammalian semen is considered to contain the highest selenium concentration of all other body tissues. In particular, in human, a significant positive correlation in the selenium concentration was demonstrated between the different reproductive organs with the testis having the highest concentrations of this element. After ^{75}Se intravenous injection, the highest levels of ^{75}Se were found in the kidney followed by seminal vesicles and testicles. Progressive selenium deficiency was associated with morphological alterations of spermatids and spermatozoa with subsequent complete disappearance of mature germinal cell (Figure 6.1). Impaired spermatogenesis due to Se deficiency has been reported in several animal species including pigs. Specific structural role of phospholipid hydroperoxide glutathione peroxidase (PH-GPx) in spermatozoa was shown (for review and references see Ahsan *et al.,* 2014 and Surai, 2006).

It is generally accepted that Se participates in various physiological functions as an integral part of a range of selenoproteins. The porcine selenoprotein family includes at least 25 eukaryotic proteins. Expression of individual eukaryotic selenoproteins is characterised by high tissue specificity, depends on Se availability, can be regulated by hormones, and if compromised contributes to various pathological conditions

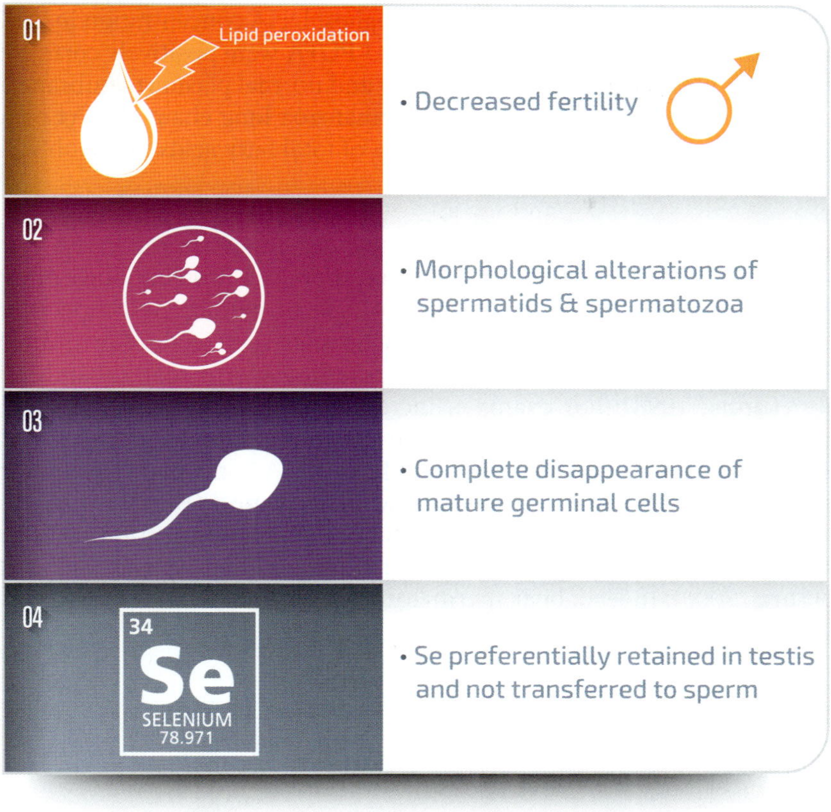

Figure 6.1. Selenium for boars – semen quality and selenium deficiency.

(Chen *et al.,* 2018; Lu *et al.,* 2019; Surai, 2006). Protective roles of selenoproteins in pig reproduction is related to necessity for an effective antioxidant defence to prevent negative consequences of over-production of free radicals during semen collection, manipulation (e.g. dilution, storage, etc.) and artificial insemination.

6.3 Fatty acid composition of mammalian semen

Lipids are important constituents of the mammalian semen. They serve as structural compounds of the spermatozoa membranes, are precursors of different biologically active compounds (eicosanoids) and can be used for energy production. It is widely accepted that docosahexaenoic fatty acid (DHA) is the most important spermatozoan PUFA in mammals, including man, bull, monkey, ram and boar (Cerolini *et al.,* 2000; Liu *et al.,* 2015; Poulos *et al.,* 1973; Surai, 2006). The importance of C_{22} polyunsaturates in relation to male fertility has been shown in humans where the amount of DHA in spermatozoa is positively correlated with sperm motility (Conquer *et al.,* 1999) and with the normal morphology of sperm cells (Lenzi *et al.,* 2000b). Therefore, the best morphological pattern corresponded to the highest DHA concentration in the human semen. Similarly, in boar semen DHA and n-3 PUFA were positively correlated with sperm motility, viability, normal morphology and normal plasma membrane (Am-in *et al.*, 2011). In general, in mammalian spermatozoa long chain PUFAs containing 20-22 carbon atoms comprise more than 50% of total fatty acids in the phospholipid fraction (Figure 6.2).

The biological reason for these species-specific differences in the PUFA profiles of spermatozoa is not clear at present. However, there is a growing body of evidence indicating that the fatty acid composition of sperm membranes, especially levels of PUFA, determine their biophysical characteristics such as fluidity and flexibility as appropriate for their specific functions, including sperm motility and fertilising capacity (Ladha, 1998). For example, increased PUFA concentrations in human spermatozoa were associated with increased sperm membrane fluidity (Comhaire *et al.*, 2000). The very high proportion of long chain PUFA in the mammalian, including boar, spermatozoa predisposes them to lipid peroxidation (Brouwers and Gadella, 2003) and it seems reasonable to suggest that antioxidant protection plays a crucial role in the maintenance of spermatozoan membrane integrity and their fertilising ability (Surai, 2006). Indeed, boar spermatozoa are rich in PUFAs and are vulnerable to lipid peroxidation.

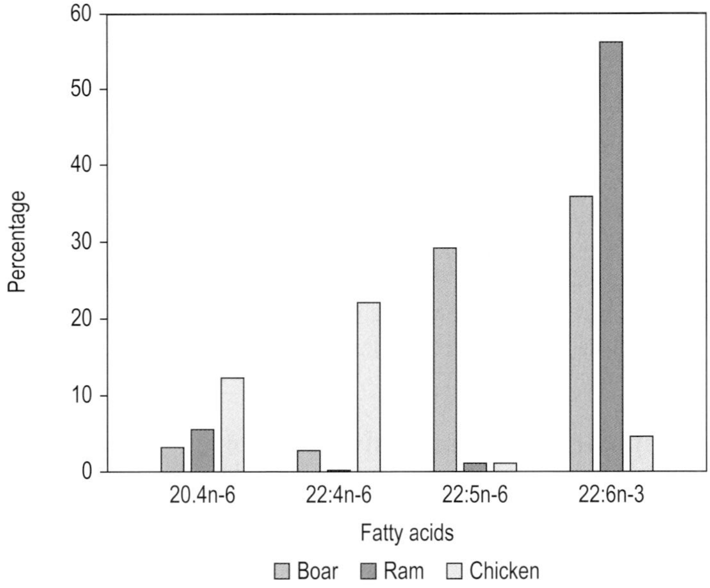

Figure 6.2. Polyunsaturated fatty acids in spermatozoa phospholipids, % (adapted from Surai, 2002).

Table 6.1. Fatty acid composition (g/100 g of fatty acids) of spermatozoa of boars fed diets differing in the ratio of n6/n-3 for 8 weeks (adapted from Liu *et al.*, 2015).

Fatty acids[1]	Ratio		
	n-6/n-3 (initial)	n-6/n-3 (6.6)	n-6/n-3 (2.2)
Myristic 14:0	9.53	9.49	8.65
Palmitic 16:0	25.90	22.72	224.42
Stearic 18:0	10.98	10.37	11.00
Oleic 18:1	2.04	1.20	1.15
Linoleic 18:2n-6	2.27	1.92	1.68
Arachidonic 20:4n-6	2.20	2.34	2.01
DPA 22:5n-6	10.85	9.32	8.65
DHA 22:6n-3	15.13	29.43	29.27
Total n-6	16.26	14.42	13.29
Total n-3	15.35	30.97	30.48
PUFA	31.40	47.29	44.74

[1] DHA = docosahexaenoic acid; DPA = docosapentaenoic acid; PUFA = polyunsaturated fatty acid.

It should be mentioned that the fatty acid profile of boar semen depends on the diet composition (Table 6.1) and enrichment of spermatozoa with DHA was considered to be an important approach to improve semen quality. In particular, feeding tuna oil to boars was shown to increase the proportion of spermatozoa with progressive motility and with a normal acrosome score and reduced the proportion of spermatozoa with abnormal morphologies (Rooke *et al.*, 2001). At the same time, the proportions (g per 100 g total fatty acids) of 22:6(n-3) in sperm phospholipid fatty acids were increased by 24% (from 34.5 to 42.9 g) by feeding tuna oil and 22:5(n-6) decreased from 29.8 to 17.9 g (Rooke *et al.*, 2001). The proportion of polyunsaturates, mainly 22:6n-3, decreased with a complementary increase in the content of the saturates, mainly 18:0. The inclusion of α-tocopherol into the diluent was effective in totally preventing the significant decrease of 22:6n-3 observed in sperm phospholipid in the control samples during the storage period.

6.4 Lipid peroxidation in semen

Toxicity of oxygen free radicals to the motility of human spermatozoa was observed more than 75 years ago (McLeod, 1943) and the toxic effect of H_2O_2 on chicken semen was shown by Wales *et al.* (1959). However, major attention to this subject came in 1970s. In fact, Jones and Mann published a milestone work in relation to lipid peroxidation in ram spermatozoa in 1973. These authors showed that ram spermatozoa produced, aerobically, an organic peroxide which can be determined quantitatively by the thiobarbituric acid reaction (Jones and Mann, 1973). They also proposed that lipid peroxidation may cause significant decline in respiratory rate and motility of spermatozoa. Lipid peroxidation irreversibly abolished the fructolytic and respiratory activity of spermatozoa. Moreover, the susceptibility of spermatozoa to peroxidation was increased in cells damaged prior to incubation (Jones and Mann, 1977a). Incubation of ram semen with peroxidised DHA caused the release of intracellular enzymes: lactate dehydrogenase, glutamicoxalacetic transaminase (GOT), hyaluronidase, acid phosphatase and β-N-acetylglucosaminidase, into the surrounding medium, in a similar manner to enzymatic release observed after cold shock (Jones and Mann, 1977b).

Peroxidised PUFAs added to washed human sperm suspension immobilised the spermatozoa rapidly and permanently (Jones *et al.*, 1978). Similarly, the toxicity of unsaturated fatty acids towards ram spermatozoa was shown to be directly related to their degree of peroxidation (Jones and Mann, 1977b). Therefore lipid peroxidation can be considered as one of the biochemical causes of sperm senescence under storage conditions *in vitro* (Jones and Mann, 1976, 1977a,b) and loss of motility in human spermatozoa submitted to aerobic incubation appears to be the consequence of enhanced lipid peroxidation (Engel *et al.*, 1999). In mammalian spermatozoa, 22:6n-3 and 20:4n-6 were predominantly oxidised during sperm storage (Griveau *et al.*, 1995; Jones and Mann, 1976). Thus, for human spermatozoa, peroxidation causes a dramatic decline of ethanolamine plasmalogen (from 10.1 to 4.3%) and PE (from 29.7 to 11,0%) while total amount of phospholipids decreased by about 30% (Jones *et al.*, 1979). Using radio-labelled fatty acids it was also demonstrated that PUFAs were predominantly incorporated into the plasmalogen fraction of phospholipid of human spermatozoa, while the peroxidation of PUFA from PE appeared to be very rapid (Alvarez and Storey, 1995). In this experiment, the level of arachidonic acid in phospholipids decreased from 3.9 to 2.6% and DHA from 47.5 to 27.2%. Peroxidation caused a loss of approximately 50% in the content of 20:4n-6 and 22:6n-3 fatty acids (Jones and Mann, 1976). Similarly, during peroxidation, phospholipids in human spermatozoan lost more than 65% of their DHA (Jones *et al.*, 1978). MDA concentrations in asthenospermia and oligoasthenospermia were significantly higher than in normospermia and a negative correlation between MDA concentration and sperm motility was observed (Huang *et al.*, 2000).

Therefore, the aforementioned publications clearly showed that lipid peroxidation:
- takes place in mammalian spermatozoa;
- caused decline in motility of spermatozoa;
- irreversibly abolished the fructolytic and respiratory activity of spermatozoa;
- increased release of intracellular enzymes from spermatozoa into medium;
- is the major biochemical cause of sperm senescence under storage conditions *in vitro*;
- caused predominant oxidation of 22:6n-3 and 20:4n-6 fatty acids.

Furthermore, those authors also showed that the susceptibility of spermatozoa to peroxidation was increased in cells damaged prior to incubation and that peroxidised PUFAs added to a washed sperm suspension immobilised the spermatozoa rapidly and permanently. Those publications presented results obtained with ram and human semen. However results on lipid peroxidation in other mammalian species have also been published, including boar (Castellano *et al.*, 2010; Cerolini *et al.*, 2000; Satorre *et al.*, 2012). Furthermore, lipid peroxidation in semen has been studied further in detail and several comprehensive reviews have discussed recent findings (Chen *et al.*, 2013; Naher *et al.*, 2013). The conclusion is that lipid peroxidation in mammalian semen is considered to be one of the most important factors causing infertility in man as well as causing decreased sperm quality during the storage of semen from farm animals.

Therefore, the mechanisms by which R OS disrupt sperm function probably involve the peroxidation of PUFA in the sperm plasma membrane. For example, it has been shown that in human spermatozoa, lipid peroxidation damages the cell plasma membrane, leading to loss of cytoplasmic components and hence to cell death – a process that is considered to play an important role in the pathophysiology of male infertility (Aitken *et al.*, 1993). Similarly, Breininger *et al.* (2005) showed a negative association between TBARS and sperm motility of boars ($r=-0.86$, $P<0.05$). Furthermore, increased level of MDA in stored boar semen was associated with a rapid loss of motility and membrane integrity (Kumaresan *et al.*, 2009). The reduction in motility might have been due to a ROS-induced lesion in ATP utilisation or in the contractile apparatus of the flagellum. It was shown that lipid peroxidation and the amount of ROS in the sperm after cryopreservation are similar between boars classified as good or bad freezers (Gómez-Fernández *et al.*, 2013). It is necessary to note that spermatozoa from individual boars can respond in a boar-dependent manner to different semen-processing techniques, including lipid peroxidation and DNA-fragmentation (Parrilla *et al.*, 2012).

Boar sperm subjected to cooling or freezing undergoes many stresses that may damage membrane structure, function and integrity, impairing motility, mitochondrial function, membrane potential and fertility (Radomil *et al.*, 2011). Since boar spermatozoa are rich in PUFAs, they are sensitive to lipid peroxidation. Indeed, adding various antioxidants to the boar semen diluent could have a protective effect during sperm storage and cryopreservation. For example, the freezing extenders with the presence of alginate led to higher SOD and GPx activities and lower MDA levels, in comparison to the control and improved frozen-thawed boar spermatozoa motility, functional integrity and antioxidative capacity (Hu *et al.*, 2014). These findings are especially important, since prolonged semen storage in liquid nitrogen induced a marked reduction in post-thaw sperm motility, mitochondrial function and plasma membrane integrity in most of the boars (Fraser *et al.*, 2014). In the experiment, post-thaw boar semen exhibited a marked decrease in osmotic resistance of the sperm acrosomal membrane with a significant increase in the sperm cryo-susceptibility to induced lipid peroxidation and DNA fragmentation. Antioxidant-prooxidant balance in boar semen is an important regulator of many physiological processes, including sperm attachment to oviductal epithelial cells, capacitation and hyperactivation, binding to the oocyte zona pellucida, acrosome reaction, penetration of the zona, and fusing with and penetrate the oolemma (Bailey *et al.*, 2000). Therefore, impaired sperm membrane function due to lipid and protein oxidation inevitably affects aforementioned processes and decreasing chances of successfully fertilising an oocyte *in vivo*. Therefore, most of the studies on mechanisms and consequences of lipid peroxidation have been associated with human spermatozoa with much less emphasis on boar semen. However, detrimental consequences of lipid peroxidation on boar semen quality are clearly shown. Molecular mechanisms and physiological and pathophysiological consequences of protein oxidation in boar testes and semen await investigations.

6.5 Selenium in boar semen

The essentiality of selenium for mail fertility was shown in the early 1980s (Behne *et al.*, 1982; Calvin *et al.*, 1987; Hansen and Deguchi 1996; Wu *et al.*, 1979). This conclusion was based on the results of a range of different experiments with mammals showing that:

- In mild deficiency, Se is preferentially retained in rat testes (Behne *et al.*, 1982).
- Mammalian semen is considered to contain the highest selenium concentration of all other body tissues (Behne *et al.*, 1986).
- In the human, a significant positive correlation in the selenium concentration was demonstrated between the different reproductive organs with the testis having the highest concentrations of this element (Oldereid *et al.*, 1998).
- After ^{75}Se intravenous injection, the highest levels of ^{75}Se were found in the kidney followed by seminal vesicles and testicles (Vanha-Perttula and Remes,1990). The seminal vesicle secretion was particularly rich in ^{75}Se and its fractionation resembled that of the seminal plasma.
- Alterations of spermatids and spermatozoa due to Se deficiency (Calvin *et al.*, 1987) with subsequent complete disappearance of mature germinal cells (Behne *et al.*, 1996).
- In selenium-deficient mice, the proportion of abnormal sperm ranged from 6.8 to 49.6%, while in the control group it was only 4.0-15.0%. The most frequently occurring abnormalities in sperm shape were found in the sperm head. However, there was also a tendency of increasing abnormalities in other spermatozoa regions, including neck, mid-piece and tail (Watanabe and Endo, 1991).
- In human semen, selenium was found mainly (more than 85%) in the seminal plasma and sperm motility was maximal when semen Se levels were between 50 and 69 ng/ml (Bleau, 1984).
- Se concentration in semen of various species vary considerable (Table 6.2).

Table 6.2 Sperm and seminal plasma selenium (Se) content and glutathione peroxidase (GPx) activity (adapted from Saaranen *et al.*, 1989).

Species	Sperm Se, µg/g dry weight	ng/10^9 sperm	Seminal plasma Se, µg/l	GPx in seminal plasma, U/mg prot.
Boar	1.2	71.4	71.0	0
Bull	3.6	91.1	492.8	409.6
Ram	4.1	503.4	102	21.0
Stallion	3.8	-	34.0	0
Human	1.8	34.4	33.4	15.6

It is well known that Se is involved in a regulation of animal reproduction. Furthermore, low Se in the diet had a detrimental effect on semen quality. For example, a prolonged Se deficiency in boars was shown to result in low sperm concentrations, reduced sperm motility, and sperm with a high incidence of cytoplasmic droplets (Liu *et al.*, 1982), while diets supplemented with Se from sodium selenite increased the number of spermatozoa ejaculated (Segerson *et al.*, 1981).

An extensive work on effects of selenium on boar semen quality has been conducted at the Columbus State University. For example, the effect of the prolonged feeding low dietary levels of Se on the semen production, quality and fertilising ability was studied. For this purpose, a low Se diet consisting of torula yeast and either dextrose or corn starch as the major basal ingredients and containing 0.063 mg/kg Se was fed to boars from weaning through sexual maturity (Marin-Guzman *et al.*, 1997). An important role of Se in the morphology of the sperm tail, and possibly in its metabolism, in boars has been suggested (Marin-Guzman, 1990). Furthermore, comparatively high Se concentrations in boar testes and sperm, even after fed low Se diet for a long period of time, confirmed an important functional role for Se in the sperm cell. Low Se diet was associated with decline in semen quality over the 16-wk collection period time, while sperm motility and the percentage of normal sperm were improved as a result of dietary Se supplementation. Furthermore, a lower fertilisation rate and fewer accessory sperm in the oocyte were observed as a result of Se inadequacy/deficiency in boars (Marin-Guzman *et al.*, 1997). Therefore, boars fed the nonfortified Se diets had sperm with lower motility and a higher percentage of sperm cells with bent and shoehook tails, while Se-supplementation was shown to improve sperm motility as evidenced by 3-fold increase in the percentage of normal sperm in comparison to Se-deficient boars (Mahan *et al.*, 2002; Marin-Guzman *et al.*, 1997; Table 6.3).

In continuation of the research in the same department, Marin-Guzman *et al.* (2000b) showed that Se is involved in a regulation of spermatozoa maturation in the epididymis. They used 10 mature boars (n=2 to 3/treatment group) which were fed from weaning to 18 months of age diets fortified with two levels of supplemental Se (0 or 0.5 mg/kg).

The low-Se diet caused changes in spermatozoa: the mitochondria in the tail midpiece were more oval with wider gaps between organelles and the plasma membrane connection to the tail midpiece was not tightly bound as when boars were fed Se. Furthermore, sperm ATP concentration was decreased (Figure 6.3) and percentage of immature spermatozoa with cytoplasmic droplets increased when boars were fed the low-Se diet (Marin-Guzman *et al.*, 2000b).

It seems likely that Se has a role in establishing the number of boar spermatozoal reserves and Sertoli cells. For example, when boar's diet was supplemented with Se 0 or 0.5 mg/kg for 18 months, testicular sperm reserves were higher in boars fed on the high Se diet (Marin-Guzman *et al.*, 2000a). In addition, the boars fed dietary Se had also a greater number of Sertoli cells and round spermatids at 6.2 month of age and by 18 month of age they also had more secondary spermatocytes. From the results

Table 6.3. Effect of selenium (Se) on boar (adapted from Mahan *et al.*, 2002; Marin-Guzman *et al.*, 1997).[1]

Item	Dietary selenium supplementation	
	0.0	0.5
Liver Se, mg/kg	0.54	1.15
Liver GPx, U/g	2.3	13.9
Testis Se, mg/kg	0.30	0.80
Testis GPx, U/g	1.19	1.24
Semen		
Volume, ml	160	163
Sperm concentration, no. $\times 10^9$	290	253
Total sperm, no. $\times 10^9$	43.5	43.5
GPx, U/ml	33	71
Se, mg/kg	0.031	0.134
Seminal plasma		
Se, mg/kg	0.024	0.060
GPx, U/ml	12.3	37.7
Sperm		
Se, mg/kg	0.42	0.94
GPx, U/g	579	977
Sperm production/g testis, no $\times 10^6$		
5.4 mo	39.4	50.7
6.2 mo	65.9	73.0
9 mo	64.0	89.8
18 mo	92.4	163.8
Semen quality		
ATP concentration, nmoles ATP/10^6 spermatozoa	1.15	1.55
Sperm motility, %	60.4	87.9
Normal sperm, %	24.2	61.9
Fertilisation rate, % of eggs	73	99
Accessory sperm, no./oocyte	14	60

[1] ATP = adenosine triphosphate; GPx = glutathione peroxidase.

presented above, it is obvious that Se may have several functional roles in the testes development and sperm function, including a structural role in the development of the spermatozoal midpiece and the Sertoli cells as well as being a component of GPx in the sperm. In the case of Se deficiency, a reduction in the percentage of normal spermatozoa is observed accompanied by a reduction in sperm motility. In particular, Se supplementation of the male diet is needed to maintain sperm membrane integrity during *in vitro* sperm manipulation including artificial insemination (Surai, 2006). It should be mentioned that effect of dietary Se on males depends on ingredient-derived background Se. Only in the case of dietary Se deficiency/inadequacy one can expect a beneficial effect of Se dietary supplementation. Therefore, there is a range of data

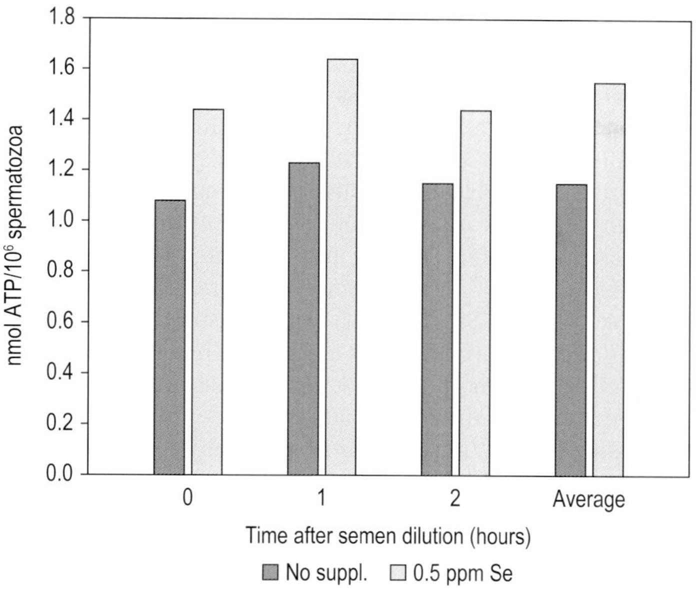

Figure 6.3. Effect of dietary selenium (Se) on boar spermatozoal ATP concentration at different times after semen dilution (Marin-Guzman *et al.* (2000b).

obtained in various experiments in which the addition of inorganic Se to the diets of rams, boars, and dairy or beef bulls did not affect their reproduction including semen quality (Bartle *et al.*, 1980; Buchanan-Smith *et al.*, 1969; Henson *et al.*, 1983; Segerson and Johnson, 1980; Segerson *et al.*, 1981; Smith *et al.*, 1979). It could well be that the animals in the aforementioned trials were adequate in Se and that is why an additional Se supplementation was not effective.

6.6 Antioxidant systems of boar semen

6.6.1 Enzymatic antioxidant defences

As mentioned in the Chapter 1 (Surai, 2021), during evolution living organisms have developed specific antioxidant protective mechanisms to deal with reactive oxygen species (ROS) and there are three major levels of antioxidant defence in various biological systems, including boar semen. Indeed, the first level of antioxidant defence of boar semen consists of SOD, CAT and GPx (Kowalowka *et al.*, 2008; Koziorowska-Gilun *et al.*, 2011; Strzezek *et al.*, 1999). Initially, purification and characterisation of SOD from supernatant fractions prepared from cold-shocked boar spermatozoa showed it to be cyanide-sensitive, to have a molecular weight of 31,000 and to contain Cu and Zn. In fact, a secretory extracellular form of SOD (EC-SOD) was purified to homogeneity from boar seminal plasma (Kowalowka *et al.*, 2008). The molecular properties and specificity of SOD confirmed that the purified enzyme is an extracellular form of Cu/Zn-superoxide dismutase occurring in boar seminal plasma. The antigenic determinants of SOD isolated from boar seminal plasma and spermatozoa were similar to each other (Orzołek *et al.*, 2013). Boar seminal plasma contains also catalase and glutathione (Koziorowska-Gilun *et al.*, 2011). However, it appears that boar semen is comparatively low in catalase, particularly when contrasted to catalase concentrations in blood and protective roles of catalase (functioning mainly when H_2O_2 concentration is largely above physiological levels) in antioxidant defence of mammalian semen is questionable (Vernet *et al.*, 2004). It should be noted, that peroxisomes that contain catalase are eliminated from germ cells during spermatogenesis (Nenicu *et al.*, 2007). Data concerning antioxidant enzymes in boar semen are variable. For example, SOD activity was measured in boar spermatozoa but GPx activity was not detected (Cerolini *et al.*, 2001). In contrast, both enzymatic activities were measured in seminal plasma. A large variability in SOD activity was found among the spermatozoa of different boars, with activities ranging from 167.7 to 926.6 and from 3.32 to 13.8 IU/mg protein in spermatozoa and seminal plasma, respectively.

Antioxidants secreted by the reproductive tract protect spermatozoa against the toxic effects of ROS after ejaculation. It was shown that the scavenging potential of the seminal plasma is dependent on the contributions of different antioxidants, originating in various fluids of boar reproductive tract (Koziorowska-Gilun *et al.*, 2011). The authors showed that the cauda epididymidal spermatozoa exhibited high SOD activity and relatively low activity of PH-GPx. The relative amounts of GPx, GR and GST activities in the cauda epididymidal spermatozoa were negligible, whereas CAT activity was undetectable. Greater SOD activity was found in the fluids of the cauda epididymis and prostate gland. Furthermore, the prostate gland fluid appeared to be the main source of CAT activity in the seminal plasma, whereas the highest level of GPx activity was derived from the cauda epididymidal fluid (Koziorowska-Gilun *et al.*, 2011). Since the concentration of catalase in spermatozoa and seminal plasma is low, glutathione and GPx are the main agents that can remove the hydrogen peroxide generated (Surai, 2006). A relatively high GSH level occurs in mouse spermatozoa, while only traces of GSH were found in boar spermatozoa (Luberda, 2005). Furthermore, in boar spermatozoa the activity of GPx and glutathione reductase was also low or undetectable (Li, 1975). However, GSH is present in boar seminal plasma in a significant amount (Strzezek, 2002; Strzezek *et al.*, 1999). It was speculated that boar spermatozoa are poorly adapted to counteract the toxic effects of induced ROS (Strzezek *et al.*, 2005) because of comparatively low levels of SOD and the lack of GPx in the seminal plasma. The author suggested that main compounds responsible for AO potential of the seminal plasma are GSH, L-ergothioneine and L-ascorbic acid. However, boar seminal plasma contains only moderate concentration of ascorbic acid (Audet *et al.*, 2004).

Since hydrogen peroxide, produced in the SOD driving superoxide radical detoxification and lipid peroxides are toxic for the spermatozoa (Wright *et al.*, 2014; Walczak-Jedrzejowska *et al.*, 2013), GPx plays an important role in protecting cell membrane lipid from peroxidation, thus maintaining the integrity of the cell (Flohe and Zimmermann, 1970). In fact, GPx in the sperm is considered to be the main enzyme, which removes peroxides and thereby protects cells against damage caused by free radicals and the products of lipid peroxidation *in vivo* (Griveau *et al.,* 1995).

In mammals 8 forms of GPx have been characterised and five of those are Se-dependent (for review, see Brigelius-Flohé and Maiorino, 2013), including a classical cytosolic glutathione peroxidase (cGPx or GPx1), gastrointestinal glutathione peroxidase (GI-GPx, GPx2), plasma glutathione peroxidase (pGPx, GPx3), phospholipid glutathione peroxidase (GPx4, PH-GPx) and so called GPx6, a close homologue to GPx3, which is a selenoprotein in humans but non-Se-GPx in rodents and other species (for review see Surai, 2006). Other non-Se-GPx includes GPx-5, an epididymis-specific Cys-GPx in mice, rats, pigs, monkey and humans showing a homology to GPx3; GPx7, a non-Se homolog to GPx4 and GPx8, a recently described membrane protein of the endoplasmic reticulum (Brigelius-Flohé and Maiorino, 2013). It should be mentioned that both Se-dependent and Se-independent GPx are working in concert providing antioxidant protection in various tissues. In particular, it has been suggested that the Se-independent GPx5 could function as a back-up system for Se-dependent GPx (Vernet *et al.,* 1999). For example, following dietary Se deprivation it was shown that the epididymis is still efficiently protected against increasing peroxidative conditions. In fact, the caput epididymis of selenium-deficient animals showed a limited production of lipid peroxides, a total GPx activity which was not dramatically affected by the shortage in selenium availability and an increase in GPx5 mRNA and protein levels was observed (Vernet *et al.,* 1999).

In 1985 Ursini and co-workers reported that a specific form of GPx, which used a phosphatidyl choline hydroperoxide as a substrate (PH-GPx), was Se-dependent (Ursini *et al.,* 1985). They showed that the enzyme was a monomer of 23 kDa. It contained one g-atom Se/22,000 g protein. Se was found there in the selenol form. The kinetic data were compatible with 'classical' glutathione peroxidase. The second-order rate constants (K_1) for the reaction of the enzyme with the hydroperoxide substrates indicated that, while H_2O_2 is reduced faster by the cGPx, linoleic acid hydroperoxide is reduced faster by PH-GPx. The authors suggested that this enzyme was active at the interface of the membrane and the aqueous phase of the cell. PH-GPx is distinguished from classical GPx as it is active in monomeric form and has a different amino acid composition (Sunde, 1993).

In fact, there are three isoforms of GPx4, namely, a cytosolic (cGPx4), a mitochondrial (mGPx4) and sperm nuclear GPx4 (snGPx4). It has been proven that cGPx4 is ubiquitously distributed in cells, while mGPx4 and snGPx4 are mainly expressed in testis with only marginal amounts in other tissues (Brigelius-Flohé and Maiorino, 2013). It is synthesised as a long form (L-form; 23 kDa) and a short form (S-form, 20 kDa) from mRNA that is transcribed from two initiation sites in exon Ia of PH-GPx genomic DNA (Imai and Nakagawa, 2003). S-form PH-GPx is the non-mitochondrial PH-GPx and L-form PH-GPx is the mitochondrial PH-GPx. The third form of PH-GPx, a 34 kDa selenoprotein, was detected in rat sperm nuclei and called sperm nuclei GPx (snGPx), which acts as a protamine thiol peroxidase and form disulphide cross-links among these proteins, thus stabilising and protecting DNA (Schneider *et al.,*

2009). The PH-GPx is unique in its capability of reducing ester lipid hydroperoxides even if they are incorporated in biomembranes or lipoproteins. For other members of GPx family, preliminary release of peroxides from the membrane by such enzymes as phospholipase C is an essential part of detoxification. It is well known that PH-GPx is widely expressed in normal tissue, and especially high in testis (Guerriero *et al.*, 2014), where it has an important role in spermatogenesis and sperm function and is under gonadotropin control. In this organ a relevant PH-GPx activity is strongly linked to mitochondria of cells undergoing differentiation to spermatozoa. The most extraordinary discovery about PH-GPx is related to its polymerisation and conversion from active enzyme to the structural protein. In fact, PH-GPx protein was identified as the major constituent of the keratin-like material that embeds the helix of mitochondria in midpiece of spermatozoa (Foresta *et al.*, 2002; Ursini *et al.*, 1999). Indeed failure of the expression of mitochondrial PH-GPx in spermatozoa is considered to be one of the causes of oligoasthenospermia in infertile men (Imai *et al.*, 2001). PH-GPx has the strong binding capacity to the sperm cell tails and to the sperm heads. The PH-GPx activities in tissues of rats fed the Se-deficient basal amino acid diet were 41, 50, 26 and 25% of the Se-adequate PH-GPx activities in liver, heart, kidney, and lung respectively (Lei *et al.*, 1995). The authors showed that testis had a 15-fold higher PH-GPx activity than liver and kidney, and a 25-fold higher PH-GPx activity than heart and lung. Furthermore, it was shown that PH-GPx mRNA levels were not affected by Se deficiency. Therefore, the pivotal link between Se, sperm quality and male fertility is PH-GPx, the enzyme responsible for the production of the correct architecture of the midpiece of spermatozoa.

Initially, it was shown that GPx activity in boar seminal plasma was very low (Saaranen *et al.*, 1989). However, in later publications GPx activity was successfully determined in boar seminal plasma (Kolodziej and Jacyno, 1995; Koziorowska-Gilun *et al.*, 2011; Marin-Guzman *et al.*, 1997). It was shown that in boar seminal plasma Se-dependent form of GPx comprised from 80.7 up to 90.8% total enzymatic activity (Cerolini *et al.*, 2001). It seems likely that relationship between GPx activity and boar semen quality depends on many factors and is not always straightforward. The levels of GPx activity detected in boar seminal plasma from normal ejaculates were nearly three times as high as GPx activity levels in spermatozoa (Jelezarsky *et al.*, 2008). In fact, the authors confirmed the presence of GPx with MW 20 kDa in boar seminal vesicles, prostate, bulbourethral glands, and spermatozoa, but not in seminal plasma. Furthermore, no immune reaction against GPx was present in boar accessory sex gland secretions and seminal plasma. PH-GPx in boar semen was shown to be affected by Se status (Martins *et al.*, 2015) and organic Se in the boar diet positively affected PH-GPx gene expression in boar testis (Speight *et al.*, 2012b). Aforementioned data clearly indicate that forms and protective functions of GPx in boar semen need further investigation. It is important to mention that only optimal Se status provides an effective antioxidant protections, since activities of antioxidant enzymes (GPx and catalase) decreased in livers of animals (mice) fed the marginal or excess dose of Se as compared to those in the Se-adequate group (Zhang *et al.*, 2013).

Aforementioned data indicate that main attention has been paid to GPx as an important element indicating Se status and potentially antioxidant defences. However, current evidence suggests that other Se-proteins could be even more important than GPx for Se action and that optimal levels may depend upon the form of Se being ingested (Ferguson *et al.*, 2012). Recently, the key role of glutathione S-transferase (GSTs) in preserving mitochondrial function and maintaining plasma membrane stability in boar semen has been shown (Llavanera *et al.*, 2020). In particular, the inhibition of GSTs reduces boar sperm quality and functionality parameters during their storage at 17 °C. Paraoxonases (PONs), an enzyme family participating in antioxidant defences (Shokri *et al.*, 2020), were identified in boar semen: PON-1 in seminal plasma and PON-2 in spermatozoa (Barranco *et al.*, 2015a). Seminal plasma PON-1 activity was sown to be positively correlated with the percentage of spermatozoa with rapid and progressive movement and negatively correlated with intracellular ROS in semen samples after 72 h of liquid storage (Barranco *et al.*, 2015b). Interestingly, PON-1 directly involved in boar sperm cryotolerance (Li *et al.*, 2018).

6.6.2 Non-enzymatic antioxidant defences

Vitamin E is the main chain-breaking feed-derived antioxidant in the biological membranes, including spermatozoa. However, information on vitamin E concentrations in sperm is quite limited, although some evidence exists that in mammalian spermatozoa the vitamin E levels is generally higher than that in avian species. For example, in boar semen, the α-tocopherol levels were found to be in the range of 365-497 ng/ml, which corresponds to about 1.6 μg/10^9 cells (Marin-Guzman *et al.*, 1997). Vitamin E supplementation (220 mg/kg) was shown to increase vitamin E in boar semen 5-fold (from 10 up to 53 ng/ml) in comparison to unsupplemented boars, while vitamin E was not detected in the seminal plasma even after its high dose dietary supplementation (Audet *et al.*, 2004; Marin-Guzman *et al.*, 1997) and its concentration in boar sperm comprised less than 10% of the plasma vitamin E level (Audet *et al.*, 2009b). At the same time vitamin E concentration in the tissues also increased: from 506 up to 1,745 ng/g in the liver and from 431 up to 1,061 ng/g in the testes (Marin-Guzman *et al.*, 1997). In sperm obtained from the caput epididymis of rats, vitamin E concentration was observed to be 3.56 μg/10^9 while in cauda epididymis spermatozoa its concentration was 0.94 μg/10^9 (Lenzi *et al.*, 2000b; Tramer *et al.*, 1998). Interestingly, vitamin E is distributed in various parts of rat spermatozoa and cauda spermatozoa contain only about 25% of vitamin E in comparison to the caput sperm cells (Table 6.4; Lenzi *et al.*, 2000b).

A wide range of α-tocopherol concentrations was detected in human semen. The first evidence for the presence of α-tocopherol in human semen was reported in 1985 by a group of Ukrainian scientists (Panchenko *et al.*, 1985). Indeed, vitamin E concentration was reported in human semen to be 160-290 ng/ml (normospermia), 120-160 ng/ml (astenospermia) and 54-120 ng/ml (oligozoospermia; Panchenko *et al.*, 1985). Later, in human seminal plasma vitamin E concentration was shown to be about 236.5±16-245.1±19 ng/ml and vitamin E consumption (100 mg, 3 times per day for 3 months) only slightly increased vitamin E in seminal plasma (up to 331.1±33.1 ng/ml; Moilanen *et al.*,

Table 6.4. Vitamin E distribution in rat spermatozoa (adapted from Lenzi *et al.*, 2000b).

	α-tocopherol
Whole epididymis spermatozoa, nmol/10^9 cells	3.42
Heads, nmol/10^9 heads	0.92
Tails, nmol/10^9 tails	1.32
Plasma and acrosomal membranes, nmol/g protein	351.1
Caput epididymis spermatozoa, nmol/10^9 cells	8.26
Cauda epididymis spermatozoa, nmol/10^9 cells	2.18

1993). Higher vitamin E concentrations were reported in the seminal plasma of fertile men (720.35 ng/ml; Benedetti *et al.*, 2012) and infertile men (628.8 ng/ml; Benedetti *et al.*, 2012). Alpha-tocopherol concentration was shown to greatly vary in human semen (between 87 and 2,173 ng/ml) and spermatozoa (10-245 ng/10^8 spermatozoa corresponding to 85±51 ng/10^8 spermatozoa (Therond *et al.*, 1996) and the authors suggested vitamin E to play an important role in maintaining the morphofunctional status of human spermatozoa. Moderate levels of vitamin E in the human semen (145.6-195.1 ng/ml) was reported by Zerbinati *et al.* (2020). Different tocopherols were identified in rabbit spermatozoa: α-tocopherol and δ-tocopherol were present in similar amount, corresponding to 117 and 118 ng/10^9 cells, respectively (Gliozzi *et al.*, 2009). In general, α-tocopherol in rabbit semen varied from 86.1 up to 387.6 ng/ml (Castelini *et al.*, 2007) and it was dependent on dietary vitamin E provision (Castelinni *et al.*, 2002). Rabit spermatozoa were shown to contain the largest proportion (46%) of total vitamin E, followed by semen droplets (39.7%), and seminal plasma (14.3%; Zaniboni *et al.*, 2004). In turkey spermatozoa vitamin E concentration was reported to be 58 ng/10^9 spermatozoa when males fed a diet supplemented with 60 mg/kg vitamin E and α-tocopherol concentration in turkey spermatozoa more than doubled reaching 121 ng/10^9 spermatozoa as a result of increasing vitamin E supplementation to 120 mg/kg (Zaniboni and Cerolini, 2009). The same authors reported increase in vitamin E concentration in turkey semen from 35.94 ng/10^9 spermatozoa up to 89.25 ng/10^9 spermatozoa due to increased vitamin E supplementation (Zaniboni *et al.*, 2006).

The associations between sperm vitamin E content and physiological and biochemical characteristics of spermatozoa have so far received only limited attention. Dietary supplementation of vitamin E (400 mg/kg) was shown to prevent sperm lipid peroxidation and resulted in higher total motility and sperm progressive motility in fresh and liquid stored (at 17 °C) boar semen (Liu *et al.*, 2017). Positive effects of vitamin E were also observed in other mammalian species. Thus, the α-tocopherol content in ram semen with 60-70% motile spermatozoa was higher than that observed in semen with only 30-40% motile spermatozoa (Kaludin *et al.*, 1989). In spite of a range of tocopherols present in the feed, only α-tocopherol was detected in the ram seminal plasma and spermatozoa (Kaludin and Georgiev, 1984). Significant increase in semen volume, motility and concentration and significant decrease in percentage of abnormal spermatozoa were detected in rams injected with vitamin E + selenium combination in comparison to the control group (Ozer Kaya *et al.*, 2020). A combination of increased levels of vitamin E and ascorbic acid in the rabbit diet was associated with significantly improved spermatozoa viability, the kinetics of spermatozoa movement and fertilising ability (Castellini *et al.*, 2000). Organic Se and vitamin E supplementation of dogs for 60 days was shown to enhance the antioxidant status of spermatozoa and improves the quality of the semen in males with lowered fertility (Domosławska *et al.*, 2018). Similarly, poor semen quality in dogs with low seminal plasma SOD can be also improved by vitamin E treatment (Kawakami *et al.*, 2016).

It is somehow surprising that vitamin E discovered in 1922 as a 'vitamin of reproduction' by Evance and Bishop was first identified and chemically analysed in semen only 60 years later. It seems likely that comparatively low vitamin E levels in semen and comparatively low sensitivity of analytical techniques used until 1980s would be a reason for such a delay in vitamin E identification in the semen. Indeed, the case of vitamin E role elucidation in poultry males, where it was first identified (about 4 years earlier than that in mammalian (e.g. human) species), could be considered as an important example of better understanding an important role of vitamin E in male reproduction. In fact, vitamin E was first detected in turkey semen in 1981 (Surai, 1981), and most α-tocopherol (85%) was shown to be located in the cells with only a very small proportion of this vitamin found in the seminal plasma (Surai, 1981, 1989a, 1992; Surai and Ionov, 1992). A similar distribution of vitamin E was described for chicken semen with about 88% of the semen's vitamin E to be located in the spermatozoa (Surai *et al.*, 1998). These data were confirmed by Blesbois *et al.* (1993) reporting vitamin E presence in both the cockerel's spermatozoa and seminal plasma. In particular, fresh ejaculates were shown to contained 1.6 μg vitamin E/ml with an average spermatozoa concentration to be 6.5×10^9 spermatozoa per ml. Furthermore, the mean concentration (± SD) was 0.25 ± 0.07 μg/10^9 cells in the spermatozoa and 0.074 ± 0.048 μg/10^9 spermatozoa in the homologous seminal plasma (Blesbois *et al.*, 1993). It is important to mention that there are species-specific differences in vitamin E concentration in avian semen. For example, in duck spermatozoa, vitamin E concentration was shown to be about 3 times lower compared to chickens (Surai *et al.*, 2000b).

Similar to other species, in duck semen, most vitamin E was associated with spermatozoa. Indeed, it is also noticeable that vitamin E concentration in duck sperm only reached 48.1 ng/10^9 cells (Surai *et al.*, 2000b), which was comparable to that in turkey spermatozoa (35.9-89.3 ng/10^9 spermatozoa (Zaniboni *et al.*, 2006), but it was somehow lower compared to that observed in the fowl (182.5 ng/10^9; Surai *et al.*, 2000b; 74.6-174.1 ng/10^9; Surai *et al.*, 1997). Similar to the other avian species semen, in duck semen, most vitamin E was also located in spermatozoa (Surai *et al.*, 2000b). Vitamin E concentration in spermatozoa depends on its dietary provision. For example, increase in vitamin E supplementation of the turkey diet (from 20 to 80 mg/kg) was shown to significantly elevate α-tocopherol concentration in spermatozoa (Surai, 1983a,b; Surai and Ionov, 1992). On the whole, depending on dietary supplementation, vitamin E concentration in chicken semen varied more than 2-fold: from 0.46 μg/ml (12 mg/kg feed-derived α-tocopherol, without vitamin E supplementation) up to 1.04 μg/ml (12 mg/kg feed-derived α-tocopherol + vitamin E dietary supplementation at 200 mg/kg) (Surai *et al.*, 1997). Similarly, increased vitamin E supplementation (150 vs 25 mg/kg) of quail male diet caused a significant increase in α-tocopherol concentration in semen associated with a significant (>3-fold) reduction of lipid peroxidation and improvement of semen quality as indicated by increased semen volume and motility, decreased percentage of dead, and abnormal spermatozoa (Golzar Adabi *et al.*, 2011).

Increasing the α-tocopherol content of the spermatozoa membrane was shown to make spermatozoa more resistant to the 'unnatural' stresses incurred during artificial insemination, short-term sperm storage, and cryopreservation (Surai, 1983b, 1991, 1992, 2002; Surai *et al.*, 2019b). Generally, vitamin E is implicated as an important natural stabiliser of spermatozoa membranes. To validate this hypothesis, 2 main approaches were used. First, the release of aspartate aminotransferase (AST) from spermatozoa during sperm manipulation was used as a marker of sperm membrane integrity. In particular, during *in vitro* sperm storage, the AST activity increased in the medium/diluent and simultaneously decreased in the spermatozoa (Surai, 1989b; Surai and Ionov, 1981). Similar changes in AST activity were also observed in chicken spermatozoa during a freeze-thaw procedure (Matsumoto *et al.*, 1985), and a highly significant (r=0.99) correlation between AST activity in seminal plasma and the percentage of dead spermatozoa was shown (Bilgili *et al.*, 1985).

The second approach employed low concentrations of detergent (Triton X-100) in the sperm storage medium/diluent to induce sperm membrane damage. This treatment was shown to significantly increase the release of AST from spermatozoa (Surai, 1989b). Thus, AST leakage from spermatozoa during sperm storage *in vitro* and sperm treatment by low concentrations of a detergent were used to confirm a membrane stabilising effect of vitamin E. Therefore, stabilising effect of increased vitamin E concentration in turkey spermatozoa was shown (Surai, 1982) with increased semen resistance to lipid peroxidation, measures as MDA formation during semen storage (Surai, 1984; Surai *et al.*, 1997). Increased vitamin. E concentration in turkey spermatozoa was associated with improved motility, viability and fertilising ability after artificial insemination (Surai, 1983a; 1991; Surai *et al.*, 2019b). Improved antioxidant defences of turkey semen were also shown to be beneficial during sperm storage and cryopreservation (Surai, 1989a) with vitamin E incorporated into spermatozoa membranes to be much more effective than added directly into the diluent (Surai, 1991). However, it should be mentioned that 'increased' vitamin E supplementation of the turkey male in our experiments conducted in 1980-1990s described above (60-80 mg/kg), nowadays become industrial standards and the control group in those experiments supplemented with 20 mg/kg vitamin E could be called a negative control.

Increase in the α-tocopherol concentration of semen by dietary manipulation did result in a significant reduction in the susceptibility of the semen to lipid peroxidation. In fact, the susceptibility of semen to peroxidation showed a very high negative correlation (r=-0.998) with the α-tocopherol content of the semen. Furthermore, the resistance of testes homogenates to *in vitro* peroxidation also increased due to the dietary supplementation with α-tocopherol (Surai *et al.,* 1998). In a similar manner, vitamin E supplementation (200 mg/kg vs no supplementation) reduced the lipid peroxidation of rooster sperm (Safari *et al.,* 2018). Furthermore, during semen storage, the susceptibility of spermatozoa to lipid peroxidation significantly increased, as a result of initiation of spontaneous lipid peroxidation (Cecil and Bakst, 1993). In such conditions, the protective effect of vitamin E in the spermatozoa has been clearly demonstrated (Surai *et al.,* 1998; Breque *et al.,* 2003). It seems likely that only 'physiological' levels of vitamin E in the chicken diet could be beneficial for semen quality and protective effects of vitamin E have some species-specific differences. For example, the α-tocopherol concentration in chicken semen increased significantly due to high (100 to 20,000 mg/kg) vitamin E supplementation, whereas lipid peroxidation in the semen samples substantially decreased due to increased vitamin E supplementation. However, the reproductive performance of cockerels was negatively influenced by very high (1,000 to 20,000 mg/kg) doses of vitamin E (Danikowski *et al.,* 2002). The authors suggested several possible explanations for the negative effect of high vitamin E supplementation on the cockerel reproduction, including pro-oxidant properties of high vitamin dosages; however, decreased lipid peroxidation in semen samples contradicts this suggestion.

Negative effects of high vitamin E dosage on vitamin D metabolism and membrane structure were also considered by the authors. It seems likely that vitamin E excess can negatively affect cell signalling (Zingg, 2019) in testicles leading to deterioration in semen quality. Furthermore, effects of vitamin E access on expression of various transcription factors, including Nrf2 and NF-κB, and gene expression await further investigation. Furthermore, lipid manipulation and various stress factors can substantially increase semen susceptibility to lipid peroxidation and protective effect of vitamin E would be much more pronounced.

In the early 1970, it was shown that inclusion of a high level (10%) of safflower oil into the cockerel diet without vitamin E supplementation decreased fertility down to 11%, while dietary vitamin E provision (32.4 mg/kg) was associated with fertility maintenance at 69% (Tinsley *et al.,* 1971). In the case of lipid manipulation of the cockerel semen by increasing proportion on n-3 PUFA, dietary vitamin E supplementation at 200 mg/kg was shown to be effective in decreasing lipid peroxidation and improving sperm membrane integrity, viability, and motility by enhancing antioxidant status (Safari *et al.,* 2018). The diet supplemented with tuna oil and 40 mg vitamin E/kg diet markedly depleted vitamin E from the tissues of the birds and decreased the concentration of vitamin E in the semen associated with increased semen susceptibility to lipid peroxidation. Interestingly, these negative effects of omega-3 PUFA were largely prevented by the higher (200 mg/kg) vitamin E supplementation (Surai *et al.,* 2000a).

Therefore, increased vitamin E concentration in chicken semen was associated with a significant (>3-fold) decrease in lipid peroxidation, which was initially elevated by inclusion of high levels of PUFA. The best sperm quality was found in cockerels fed on the control diet rich in omega-6 PUFA supplemented with vitamin E at 200 mg/kg, while enrichment of the diet with omega-3 PUFA increased vitamin E requirement to 300 mg/kg (Cerolini *et al.*, 2006). Increased vitamin E supplementation (300 vs 100 mg/kg) was associated with improved semen antioxidant defences as indicated by decreased lipid peroxidation, increased spermatozoa concentration, and motility. Similarly, chicken sperm concentration was decreased in the fish oil supplemented cockerels fed on the diet containing moderate level (40 mg/kg) of vitamin E and such an effect was prevented by increasing the vitamin E to 200 mg/kg (Cerolini *et al.*, 2005). Replacing omega-6 fatty acids in the turkey male diet by omega-3 fatty acids simultaneously with increased (120 vs 60 mg/kg) vitamin E supplementation was associated with improved semen antioxidant defences, as indicated by a significant (more than double) increase in vitamin E concentration and decrease in semen lipid peroxidation.

The changes in antioxidant defences were also associated with improved semen quality (Zaniboni and Cerolini, 2009; Zaniboni *et al.*, 2006). Under heat stress conditions (daily temperature 33 to 36 °C and relative humidity 60 to 70%), inclusion of high level of vitamin E (200 mg/kg) in the diets containing 6 mg/kg feed-derived vitamin E enhanced the semen quality parameters, including the spermatozoa count and motility, and reduced the percentage of dead spermatozoa. Dietary vitamin E also increased GPx activity, total antioxidant potential, and decreased lipid peroxidation in seminal plasma (Ebeid, 2012).

Observations that addition of antioxidants to boar semen improves sperm quality provide indirect evidence for the damaging effects of ROS on sperm function. In our previous study it was shown that boar sperm viability progressively decreased during storage and vitamin E inclusion into the diluent significantly inhibited this trend (Cerolini *et al.*, 2000). In particular vitamin E inclusion decreased significantly peroxidation (TBARS production of spermatozoa). Spermatozoa stored in the treatment diluent became rapidly enriched in α-tocopherol (aT) with a concomitant decrease of vitamin E content in the medium. The inclusion of aT into the diluent was effective in totally preventing the significant decrease of 22:6n3 observed in sperm phospholipid in the control samples during the storage period. Similarly, the addition of vitamin E to diluents was able to increase sperm's resistance to lipid peroxidation in boars (Breininger *et al.*, 2005; Johannisson *et al.*, 2003). Indeed, the addition of 400 μg/ml of vitamin E reduce s the MDA production in boar semen stored at 15 °C for 72 h (Mendez *et al.*, 2013). Furthermore, removal of low-molecular weight components from seminal plasma, together with the addition 200 μM of α-tocopherol, was shown to prevent the lipid peroxidation, DNA fragmentation and preserve mitochondrial activity leading to better semen quality (higher progressive motility, a higher proportion of morphologically normal spermatozoa and a significantly lower level of acrosomal reacted spermatozoa) compared to non-dialyzed semen samples after 72 h of storage (Zakošek Pipan *et al.*, 2017).

In fact, α-tocopherol, supplemented at 200 μM concentration in boar semen extender during cryopreservation was shown to have a positive effect on post-thawed sperm survivability (Jeong *et al.*, 2009). Similarly, exogenous Trolox (the water-soluble vitamin E analogue) was shown to positively affect post-thaw sperm viability, including spermatozoa motility and mitochondrial membrane potential (Pena *et al.*, 2003). Interestingly, the protective effect of vitamin E was shown to be dependent of the portion of the boar ejaculate considered (Pena *et al.*, 2004). Protective effects, including reduction of lipid peroxidation, of vitamin E in the diluent for boar semen cryopreservation was comparable with those for GSH and butylated hydroxytoluene (Baishya *et al.*, 2018).

In general, protective effect of inclusion of vitamin E into the semen diluent under various stress conditions was clearly shown with different animal species and human (Adami *et al.*, 2018). Indeed, vitamin E and GSH supplementation at 3 mM and 1 mM, respectively, was shown to protect the Beetal buck spermatozoa from oxidative damage during preserving the semen samples at 4 °C for up to 72 h (Sarangi *et al.*, 2017). Increased dietary vitamin E supplementation was also shown to improve back semen quality (Majid *et al.*, 2015). Water soluble vitamin E analogue (Trolox, 200 μM) was shown to improve rabbit sperm quality (the motility, intact acrosome, membrane integrity and mitochondrial potentials) during the process of cryopreservation (Zhu *et al.*, 2015). Vitamin E was also shown to be an effective protectant during the cooling process of equine semen at 5 °C for 72 h as evident by increased total spermatozoa motility and decreased lipid peroxidation (Nogueira *et al.*, 2015).

In humans, the in-vitro addition of α-tocopherol in provoked oxidative stress has been reported to inhibit the lipid peroxidation of sperm membranes and to enhance the ability of spermatozoa to fuse with zona-free hamster oocytes (Aitken and Clarkson, 1988; Aitken *et al*, 1989) while it did not counteract a free radical-induced loss of motility (Baker *et al*, 1996). A recent systematic review and meta-analysis confirmed positive effects of *in vitro* adding various antioxidants, including vitamin E, during the freeze thaw process on prevention of the oxidative stress and protecting progressive sperm motility, viability and reduction in sperm DNA fragmentation (Bahmyari *et al.*, 2020). Indeed, supplementing the cryomedium with vitamin E was shown to improved DNA integrity (Kalthur *et al.*, 2011). Similarly, vitamin E increased post-thaw sperm motility, however, it did not change the sperm viability and DNA integrity (Taylor *et al.*, 2009). Semen cryopreservation is considered to increase the production of reactive oxygen species leading to damages to sperm membranes. Addition of GSH to the thawing media resulted in a lower number of capacitated viable spermatozoa, a decrease in the number of spermatozoa with changes in the sulfhydryl groups in membrane proteins, a reduction of the ROS generation, a lower chromatin condensation, and a higher penetration ability of oocytes *in vitro* and a higher proportion of decondensated sperm heads (Gadea *et al.*, 2005).

Boar sperm motility, viability and AR increased significantly in SS, SeMet and Vitamin E (5.5 ug/l and 1.0 mM, respectively) treated semen, compared with the control (Tareq *et al.*, 2009). Furthermore, treatment of the sperm with SeMet and SeMet + Vit-E in the presence of 300 μM ammonia also resulted in a significant increase of the same semen quality parameters. Clearly, there is a need for further elucidation of the composition and efficacy of antioxidant defence system in boar semen.

Although excessive production of ROS are usually detrimental to spermatozoa, there is a growing body of evidence indicating that low levels of ROS are involved in the physiological control of a range of mammalian sperm functions (Surai, 2006). Therefore, in various attempt to decrease lipid and protein oxidation in boar semen during storage by adding various antioxidants into the medium (diluents) one should be very careful because ROS mediate capacitation of boar spermatozoa by various signalling pathways, such as the ERK pathway, which phosphorylates different proteins depending on their molecular mass (Awda and Buhr, 2010; Awda *et al.*, 2009). In fact, the authors demonstrated that there was a time-dependent, significant increase in the percentage of acrosome reactions seen in boar sperm exposed to a ROS-generating system and then incubated in capacitating medium. In general, low concentrations of ROS can induce hyperactivation, capacitation, sperm-oocyte fusion and acrosomal loss and high ROS concentrations inhibit sperm-oocyte fusion, decrease motility and damage DNA (for review see Surai, 2006).

Enhancement of the antioxidant capacity of semen by nutritional means could present a major opportunity for improving male fertility. The beneficial consequences of effective protection against lipid peroxidation are likely to result from two related mechanisms. Firstly, defence against peroxidative damage is essential to prevent any reduction in functionally important C20-22 polyunsaturated fatty acid levels of spermatozoan phospholipids and to maintain the structural integrity of the spermatozoa. Secondly, minimisation of lipid peroxidation will prevent accumulation of the toxic products of peroxidation. Additional possibilities may also be envisaged for the use of antioxidants in improving the viability of semen during cryopreservation.

Taking into account data showing that selenoproteins are essential in mammals because of their crucial role in cellular redox homeostasis and indicating that increased selenoprotein expression in response to stress conditions is an important evolutionary mechanism of adaptive ability of animals and human (Surai, 2006) it is clear that an importance of semen antioxidant defences and optimal Se nutrition of boars is difficult to overestimate.

6.7 Practical aspects of boar selenium nutrition

As indicated in previous chapters, a choice of Se sources in the male diet is of great importance. Recently it has been shown that Se is assimilated from organic sources much more efficiently compared to commonly used selenite (Surai, 2006, 2018; Surai *et al.*, 2018). This can be translated into higher Se accumulation in the animal tissues and building selenium reserves, which can be effectively used in stress conditions. Data on the effect of different forms of Se on boar reproduction are summarised in Table 6.5.

Table 6.5. Effect of selenium (Se) on boar reproduction.[1]

Experimental design	Effects	References
12 boars divided into 3 groups: CON, 0.3 mg/kg selenite; INO, 0.5 mg/kg selenite and ORG, 0.5 mg/kg Se-Yeast for 10 weeks, semen storage – 72 h; No data on Se in a basal diet	No effects on semen quality, ORG increased PH-GPx activity, but did not improve chilled semen viability	Martins *et al.*, 2015
18 boars divided into 3 groups: BD, no Se suppl., BD + 0.3 mg/kg selenite and BD + 0.3 mg/kg Se-Yeast from day 21 up to day 383; No data on Se in basal diet	No effects on semen quality, Se increased in plasma independently of Se form, Se; level and GPx activity in semen were not affected	Lovercamp *et al.*, 2013
Crossbred boars were weaned at 28 d of age and randomly assigned to 1 of 3 dietary treatments: basal (BD) diets containing 0.034 mg/kg Se, BD + 0.3 mg/kg selenite (SS) or BD + 0.3 Se-Yeast (SY) and used in 3 experiments (n=10 boars/dietary treatment).	SY and SS did not affect semen quality (fresh or extended) and its fertilising ability, while SS decreased percentage of sperm with abnormal head in comparison to BD; numerically (but not significantly, P=0.11) Se-yeast tended to improve fertility. The negative effects of day of semen collection on semen quality were least pronounced in boars fed diets supplemented with Se-Yeast	Speight *et al.*, 2012a
The same treatment design as Speight *et al.*, 2012a; boars were grown until BW 137 kg.	No effect of Se on reproduction organs; SY was more effective in Se accumulation in tissues; SY positively affected PH-GPx gene expression in testis	Speinght *et al.*, 2012b
60 boars were randomly allocated at Day (D) 0 into 2 groups. Group A received the ration supplemented with SS (0.4 mg/kg) whereas Group B fed the same diet supplemented with SY (0.4 mg/kg). The sperm was investigated during 4 months (D0, D30, D60, D75, D90, D105 and D120); No data on Se in a basal diet	Changing from inorganic Se to organic Se in the diet of boars increased sperm concentration but reduced some motility parameters and resistance to oxidative stress.	Lopez *et al.*, 2010
28 boars were divided to 4 experimental groups. Each group was supplemented with SY or SS (0.3 and 0.6 mg Se/kg). The duration of the experiment was 18 weeks. No data on Se in a basal diet	Dietary supplements with 0.6 mg organic Se/kg of feed mixture improve antioxidant potential of breeding boars ejaculate: increased Se level and GPx activity, GSH/GSSH ratio and total AO activity	Horky *et al.*, 2012
10 boars divided into 2 groups fed a diet supplemented with 0.45 mg/kg Se in the form SS or SY (0.15SS + 0.3SY) for 84 days	No effect of SY on semen quality	Tongchalam *et al.*, 2012
25 boars were fed diets without Se supplementation (12) or supplemented with 0.5 mg/kg Se (13) as SS from 5.4 to 18 mo. of age; Basal Se level 0.06 mg/kg	By 18 mo of age the Se-fed boars had higher numbers of sperm reserves, had more Sertoli cells, more secondary spermatocytes and more round spermatids.	Marin-Guzman *et al.*, 2000b

Table 6.5. Continued.

Experimental design	Effects	References
6 boars divided into 2 groups. Basal diet contained 0.063 mg/kg Se was fed to boars without Se supplementation (contr.) or supplemented with -0.5 mg/kg in the form of SS (Exp.) from weaning and the experiment: from 9 mo of age for 16 weeks; Se in BD- 0.067 mg/kg	Boars fed the nonfortified Se diet had sperm with lower motilities and a higher percentage of sperm cells with bent and shoehook tails and characterised by a lower fertilisation rate of oocytes with fewer accessory sperm penetrating the zona pellucida.	Marin-Guzman *et al.*, 1997
The experimental design is the same as in Martin-Guzman, 2000	When the low-Se diet was fed the mitochondria in the tail midpiece were more oval with wider gaps between organelles and decreased level of ATP was observed as well as increased number of immature spermatozoa with cytoplasmic droplets. The plasma membrane connection to the tail midpiece was not tightly bound as when boars were fed Se.	Marin-Guzman *et al.*, 2000a

[1] ATP = adenosine triphosphate; BD = basal diet; BW = body weight; CON = control; GSH = glutathione; GSSH = oxidised glutathione; INO = inorganic Se; ORG = organic Se; PH-GPx = phospholipid hydroperoxide glutathione peroxidase; SS = sodium selenite; SY = Se-yeast.

When 10 boars were divided into 2 groups and fed a diet supplemented with 0.45 mg/kg Se in the form SS or the same amount Se as a mixture of 0.15 mg/kg as SS plus 0.3 mg/kg Se as SY for 84 days there was no effect of additional organic Se on boar semen quality (Thongchalam *et al.*, 2012). Dietary supplements with 0.6 mg organic Se/kg of feed mixture improved antioxidant potential of breeding boars ejaculate: increased Se level, GPx activity, GSH/GSSH ratio and total AO activity (Horky *et al.*, 2012). In an experimental setup 60 boars were randomly allocated at Day (D) 0 into 2 groups. Group A received the ration supplemented with selenium as SS (0.4 mg/kg) whereas Group B fed the same diet supplemented with selenium as SY (0.4 mg/ kg) and semen quality was assessed during 4 months (D0, D30, D60, D75, D90, D105 and D120) (Lopez *et al.*, 2010). It was shown that changing from inorganic Se to organic Se in the diet of boars significantly increased Se concentration in the spermatozoa (Figure 6.4) and sperm concentration (434 vs 514 $\times 10^6$ sperm/ml, $P<0.05$).

In another experiment crossbred boars were weaned at 28 d of age and randomly assigned to 1 of 3 dietary treatments: basal diets (BD) containing 0.034 mg/kg Se, BD + 0.3 mg/kg Se as SS or BD + 0.3 mg/kg Se as SY and used in 3 experiments (n=10 boars/dietary treatment) (Speight *et al.*, 2012a). SY and SS did not affect semen quality (fresh or extended) and its fertilising ability, while SS decreased percentage of sperm with abnormal head in comparison to BD; numerically (but not significantly, $P=0.11$) Se-yeast tended to improve fertility (70.7% for SY vs 60.9% for BD and 58.5%

for SS; Figure 6.5). It could well be that with bigger number of animals and some commercial stress conditions this trend of fertility improvement due to organic Se supplementation could be translated in a significant difference.

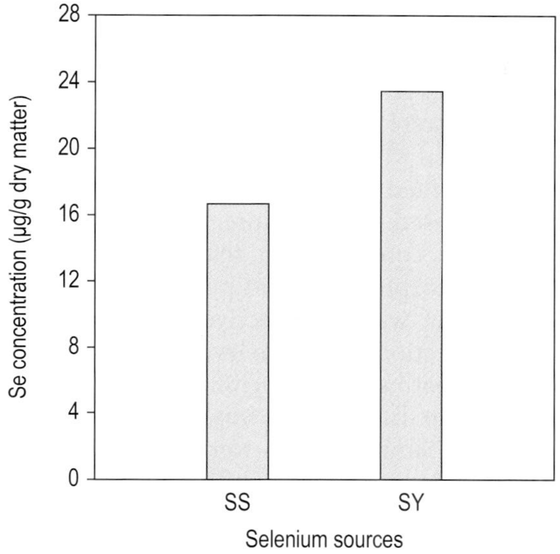

Figure 6.4. Selenium (Se) concentration in boar spermatozoa depending on its dietary supplementation (adapted from Lopez *et al.*, 2010).

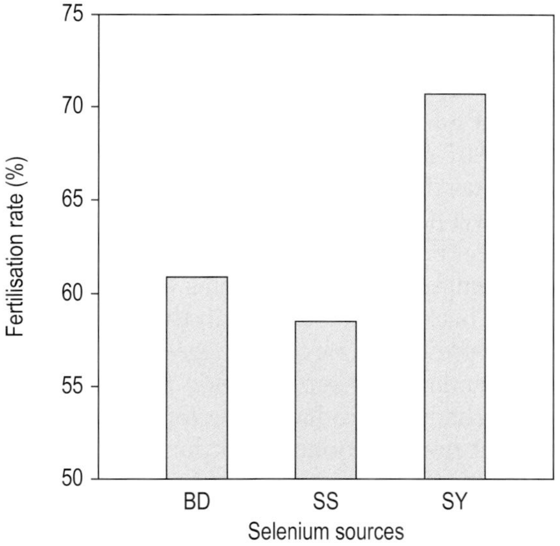

Figure 6.5. Effect of dietary selenium on fertilisation rate of boar spermatozoa (adapted from Speight *et al.*, 2012a).

The negative effects of day of semen collection on semen quality were least pronounced in boars fed diets supplemented with Se-Yeast. Furthermore, Se-Yeast was more effective than SS in Se accumulation in tissues and positively affected PH-GPx gene expression in boar testis (Speight *et al.*, 2012).

In a study 18 boars were divided into 3 groups, including a control group fed a non-supplemented basal diet and two experimental groups fed the basal diet supplemented with 0.3 mg/kg selenium in either an organic (Se-Yeast) or inorganic form (selenite; (Lovercamp *et al.*, 2013). Boars were fed dietary treatments from weaning at 21 d of age until the study was terminated when they were 383 d of age. It was demonstrated that boars fed the Se-supplemented diet had increased plasma levels of selenium independently on the Se form used. Furthermore, dietary treatment did not affect semen quality including volume, concentration, total sperm in the ejaculate, sperm motility, progressive motility, morphology, lipid peroxidation. It is also interesting to note that Se supplementation was not effective in improvement of quality of extended semen stored post collection. Selenium levels in the semen and GPx activity were not affected by dietary treatments. Unfortunately, the authors did not show the background level of Se in the diet and it is impossible to determine if the basic diet was adequate in selenium. Similar results were obtained when 12 boars were divided into 3 groups: CON, fed on the diet supplemented with Se at 0.3 mg/kg as SS; INO, supplemented with Se at 0.5 mg/kg as SS and ORG, supplemented with Se at 0.5 mg/kg as Se-Yeast, for 10 weeks (Martins *et al.*, 2015). Replacing sodium selenite with Se-yeast at 0.5 mg/kg dietary supplementation was associated with a significant increase in PH-GPx activity, but did not improve chilled semen viability in 72 h. In fact, abnormal spermatozoa head percentage and proximal droplet percentage significantly decreased, but abnormal tail percentage significantly increased due to Se-Yeast dietary supplementation.

Aforementioned results indicate that effects of dietary Se on boar reproduction depend on the basic level of Se in the diet. In the case when Se level in the diet is comparatively low (0.06-0.07 mg/kg; (Marin-Guzman *et al.*, 1997, 2000b), there were characteristic detrimental changes in testes structure. In another study, having even lower background Se level (0.034 mg/kg, Speight *et al.*, 2012a), an additional Se supplementation did not affect boar semen quality. Unfortunately, in recent studies (Horky *et al.*, 2012; Lovercamp *et al.*, 2013; Martins *et al.*, 2015; Thongchalam *et al.*, 2012) information about background Se levels in the basic diets is not available. However, based on the results presented one could suggest that the background dietary Se level was adequate to maintain boar reproduction and as a result an additional Se supplementation did not change reproductive performance and semen quality. It should be also mentioned that aforementioned trials done with boars were performed on very low number of animals (the maximum number is 10 boars per of replicate). Clearly, taking into account the large variability in reproductive performances between individuals males, there is a need to perform trials using a large number of boars to be sure of the conclusions. The general relationship between selenium and male fertility is shown in Figure 6.6.

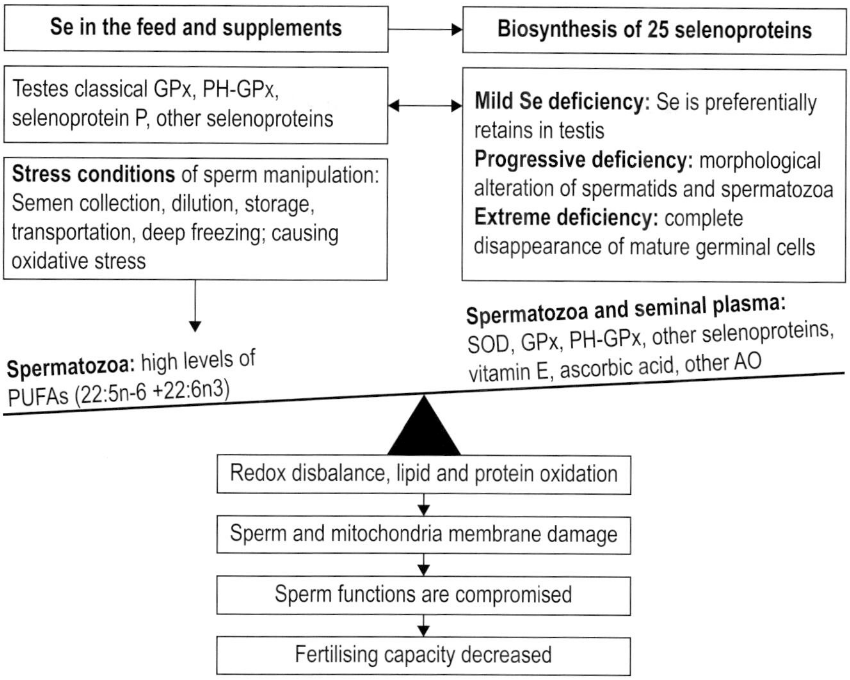

Figure 6.6. Selenium (Se) and male fertility (adapted from Surai, 2006).

Artificial insemination (AI) is a major pig breeding technique commonly performed through the use of liquid semen diluted with a proper extender and stored at 15-20 °C, usually for 1 to 5 days (Johnson *et al.,* 2000), however, AI-fertility rates using such sperm usually decline within 72 h of its storage (Waberski *et al.,* 2011). This decrease in fertilising ability of stored boar semen is associated with disturbances of semen redox balance, an increased of oxidative stress (OS) and lipid peroxidation with following reduction in sperm motility, viability, and plasma membrane stability (Kumaresan, *et al.,* 2009).

Generally speaking, the main advantage of organic Se in boar nutrition, similar to other animal species, including poultry, is coming from Se reserves accumulated in tissues, mainly in muscles, in the form of SeMet (Surai, 2006; Surai and Fisinin, 2016; Surai *et al.,* 2018). In fact, protective effect of organic selenium is more pronounced under stressful conditions. For example, in a study conducted by Speight *et al.* (2012a) in the case of intensive semen collection there was a positive effect of organic selenium, while in other cases of normal (non-intensive) semen collection from boar organic Se was not different from selenite. Therefore, Se reserves in the body (mainly in the muscles) built in the form of SeMet non-specifically incorporated into the proteins in place of Met could be considered as an important element in increasing adaptive ability of animals (boars) to various stresses. This could increase their reproductive performance in stressful commercial conditions. The benefit of organic selenium in boar diets lies in its efficient absorption, transport and accumulation of body reserves. This results in improved antioxidant status of the testes and semen. As the levels of major natural antioxidants (vitamin E, ascorbic acid and carotenoids) in boar semen are comparatively low (Audet *et al.*, 2004, 2009a,b), the antioxidant enzymes become a critical arm of antioxidant defence. Therefore, enhanced GPx activity in tissues and semen as a result of organic selenium supplementation of the boar diet may have a positive impact on semen quality in stressful commercial conditions of pig production. Furthermore, recent data indicate that the antioxidant/prooxidant balance in semen is an important element in maintaining membrane integrity and function including sperm viability and fertilising capacity (Majzoub *et al.*, 2020; Martins and Agarwal, 2019). Therefore, the antioxidant system is a crucial element in male reproduction and dietary selenium has a unique role via antioxidant mechanisms. Since Se levels in European feedstuffs is substantially lower than in the USA (Surai, 2006) one could expect better response to dietary Se supplementation than can be seen in the USA. The current recommendations on the Se requirements of sexually active boars are 0.30 mg/kg (NRC, 2012), but in commercial conditions this could be increased depending on the level of stress. Indeed, the Se requirements of pigs, including boars, depends on their living conditions, especially on the amount of stress caused by environmental, managemental and technological factors, as well as levels of other antioxidants (vitamin E) and pro-oxidants (unsaturated fat) in the diet. Under experimental laboratory conditions, stressors are generally lower than in commercial pig producing systems and low Se doses could be adequate to maintain optimal selenoprotein expression and activity (Surai, 2006, 2018).

It could be suggested that boars at big scale commercial farms located in Europe and other areas with low Se dietary background level and having high challenges would need increased antioxidant defences and most likely would benefit from additional dietary Se supplementation in the organic form (Figure 6.7). Indeed, as it is mentioned above, building Se reserves in the body which can be used in stress conditions for additional synthesis of various selenoproteins is the main advantage of organic Se in boar nutrition (Surai, 2006, 2018; Surai *et al.*, 2018). It should be also mentioned that in the last decade, there has been significant progress in characterising selenoproteins and understanding their physiological functions. Furthermore all the antioxidants in the body working cooperatively building an integrated antioxidant system which is regulated at the vitagene level (Calabrese *et al.*, 2012; Surai, 2018; Surai *et al.*, 2019a) and thus, it is extremely challenging to measure the effect on semen quality of any single antioxidant alone.

6.8 Conclusions

From the information presented, it is clear that selenium plays an important role in boar nutrition. The requirement of boars for selenium varies depending on many environmental and other conditions and in general is considered to be 0.30 mg/kg feed. It seems likely that reproducing sows and boars are especially sensitive to Se deficiency, and to meeting their requirements is an important challenge for pig nutritionists. In

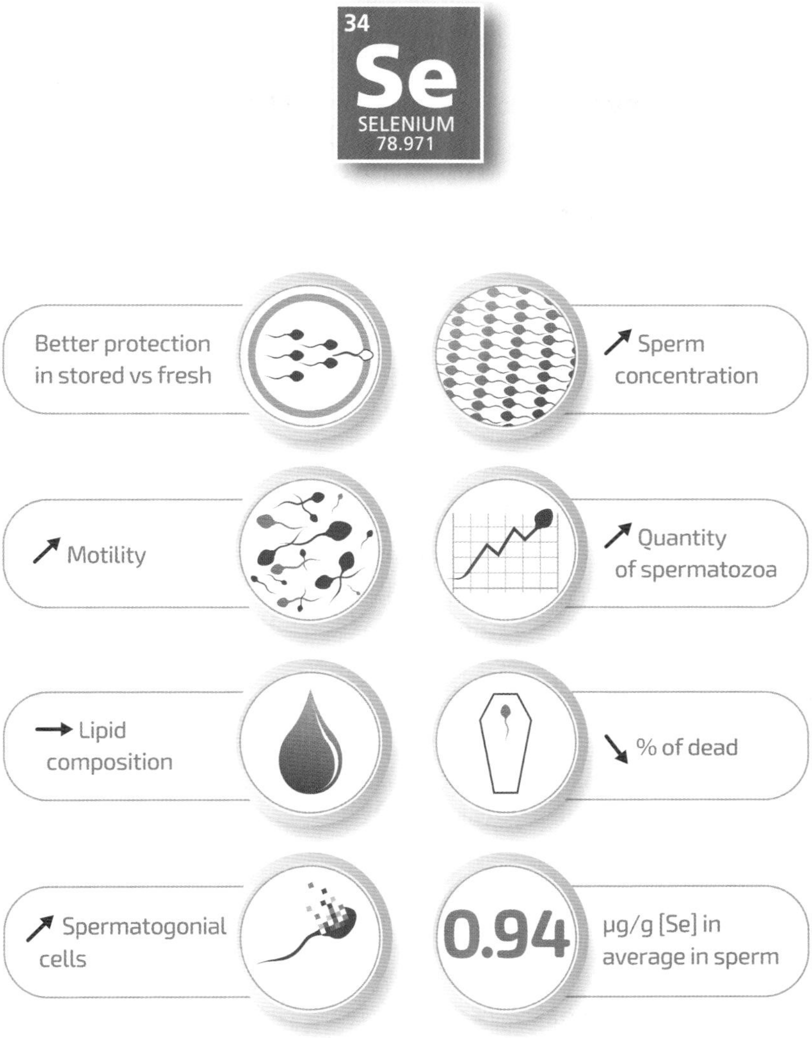

Figure 6.7. Selenium for boars – semen quality and selenium supplementation.

fact, in many countries there are legal limits as to how much Se may be included into the diet and this restricts flexibility in terms of addressing the Se needs of the developing and reproducing pigs. The analysis of data of various boar trials with different Se sources indicates, that in some cases when background Se levels were low, there were advantages of Se dietary supplementation. It is necessary to take into account that only an optimal Se status of animals is associated with the best antioxidant protection and could have positive effects on boar semen production and its quality. However, in many cases background Se levels were not determined and therefore it is difficult to judge if the basic diets were deficient in Se. It can also be suggested that, because of higher efficacy of assimilation from the diet, and possibilities of building Se reserves in the body, organic selenium in the form of SeMet provided by a range of products, including Se-Yeast and SeMet preparations is an important source of Se to better meet the needs of modern pig genotypes. Indeed, organic Se in the form of OH-SeMet was shown to be an effective form of organic Se for pigs (Jlali *et al.*, 2014; Li *et al.*, 2020), poultry (Briens *et al.*, 2013; 2014; Jlali *et al.*, 2013; Zhao *et al.*, 2017) and ruminants (Juniper *et al.*, 2019; Sun *et al.*, 2017, 2019; Wei *et al.*, 2019).

Acknowledgements

Peter F. Surai is supported by a grant of the Government of Russian Federation (Contract No. 14.W03.31.0013).

References

Adami, L.N.G., Belardin, L.B., Lima, B.T., Jeremias, J.T., Antoniassi, M.P., Okada, F.K. and Bertolla, R.P., 2018. Effect of *in vitro* vitamin E (alpha-tocopherol) supplementation in human spermatozoon submitted to oxidative stress. Andrologia 50(4): e12859.

Ahsan, U., Kamran, Z., Raza, I., Ahmad, S., Babar, W., Riaz, M.H. and Iqbal, Z., 2014. Role of selenium in male reproduction – a review. Animal Reproduction Science 146: 55-62.

Aitken, R.J. and Clarkson, J.S., 1988. Significance of reactive oxygen species and antioxidants in defining the efficacy of sperm preparation techniques. Journal of Andrology 9: 367-376.

Aitken, R.J., Clarkson, J.S. and Fishel, S., 1989. Generation of reactive oxygen species, lipid peroxidation, and human sperm function. Biology of Reproduction 41: 183-197.

Aitken, R.J., Harkiss, D. and Buckingham, D., 1993. Relationship between iron-catalysed lipid peroxidation potential and human sperm function. Reproduction 98: 257-265.

Alvarez, J.G. and Storey, B.T., 1995. Differential incorporation of fatty acids into and peroxidative loss of fatty acids from phospholipids of human spermatozoa. Molecular Reproduction and Development 42: 334-346.

Am-in, N., Kirkwood, R.N., Techakumphu, M. and Tantasuparuk, W., 2011. Lipid profiles of sperm and seminal plasma from boars having normal or low sperm motility. Theriogenology 75: 897-903.

Audet, I., Bérubé, N., Bailey, J.L., Laforest, J.P. and Matte, J.J., 2009a. Effects of dietary vitamin supplementation and semen collection frequency on reproductive performance and semen quality in boars. Journal of Animal Science 87: 1960-1970.

Audet, I., Bérubé, N., Bailey, J.L., Laforest, J.P., Quesnel, H. and Matte, J.J., 2009b. Effects of dietary vitamin supplementation and semen collection frequency on hormonal profile during ejaculation in the boar. Theriogenology 71: 334-341.

Audet, I., Laforest, J.P., Martineau, G.P. and Matte, J.J., 2004. Effect of vitamin supplements on some aspects of performance, vitamin status, and semen quality in boars. Journal of Animal Science, 82: 626-633.

Awda, B.J. and Buhr, M.M., 2010. Extracellular signal-regulated kinases (ERKs) pathway and reactive oxygen species regulate tyrosine phosphorylation in capacitating boar spermatozoa. Biology of Reproduction 83: 750-758.

Awda, B.J., Mackenzie-Bell, M. and Buhr, M.M., 2009. Reactive oxygen species and boar sperm function. Biology of Reproduction 81: 553-561.

Bahmyari, R., Zare, M., Sharma, R., Agarwal, A. and Halvaei, I., 2020. The efficacy of antioxidants in sperm parameters and production of reactive oxygen species levels during the freeze-thaw process: A systematic review and meta-analysis. Andrologia 52: e13514.

Bailey, J.L., Blodeau, J.F. and Coemier, N., 2000. Semen cryopreservation in domestic animals: a damaging and capacitating phenomenon minireview. Journal of Andrology 21: 1-7.

Baishya, S.K., Kumar Biswas, R., Govindasamy, K., Deka, B.C., Sinha, S. and Singh, M., 2018. Effect of reduced glutathione, water soluble vitamin E analogue and butylated hydroxytoluene on the post thaw characteristics of boar spermatozoa. CryoLetters 39: 227-234.

Baker, H.W., Brindle, J., Irvine, D.S. and Aitken, R.J., 1996. Protective effect of antioxidants on the impairment of sperm motility by activated polymorphonuclear leukocytes. Fertility and Sterility 65: 411-419.

Barranco, I., Roca, J., Tvarijonaviciute, A., Rubér, M., Vicente-Carrillo, A., Atikuzzaman, M., Ceron, J.J., Martinez, E.A. and Rodriguez-Martinez, H., 2015a. Measurement of activity and concentration of paraoxonase 1 (PON-1) in seminal plasma and identification of PON-2 in the sperm of boar ejaculates. Molecular Reproduction and Development 82: 58-65.

Barranco, I., Tvarijonaviciute, A., Perez-Patiño, C., Alkmin, D.V., Ceron, J.J., Martinez, E.A., Rodriguez-Martinez, H. and Roca, J., 2015b. The activity of paraoxonase type 1 (PON-1) in boar seminal plasma and its relationship with sperm quality, functionality, and *in vivo* fertility. Andrology 3: 315-320.

Bartle, J.L., Senger, P.L. and Hillers, J.K., 1980. Influence of injected selenium in dairy bulls on blood and semen selenium, glutathione peroxidase and seminal quality. Biology of Reproduction 23: 1007-1013.

Behne, D., Duk, M. and Elger, W., 1986. Selenium content and glutathione peroxidase activity in the testis of the maturing rat. The Journal of Nutrition 116: 1442-1447.

Behne, D., Hofer, T., von Berswordt-Wallrabe, R. and Elger, W., 1982. Selenium in the testis of the rat: studies on its regulation and its importance for the organism. The Journal of Nutrition 112: 1682-1687.

Behne, D., Weiler, H. and Kyriakopoulos, A., 1996. Effects of selenium deficiency on testicular morphology and function in rats. Journal of Reproduction and Fertility 106: 291-297.

Benedetti, S., Tagliamonte, M.C., Catalani, S., Primiterra, M., Canestrari, F., De Stefani, S., Palini, S. and Bulletti, C., 2012. Differences in blood and semen oxidative status in fertile and infertile men, and their relationship with sperm quality. Reproductive Biomedicine Online, 25: 300-306.

Bilgili, S.F., Renden, J.A. and Sexton, K.J., 1985. Fluorometry of poultry semen: its application in the determination of viability, enzyme leakage, and fertility. Poultry Science 64: 1227-1230.

Bleau, G., Lemarbre, J., Faucher, G., Roberts, K.D. and Chapdelaine, A., 1984. Semen selenium and human fertility. Fertility and Sterility 42: 890-894.

Blesbois, E., Grasseau, I. and Blum, J.C., 1993. Effects of vitamin E on fowl semen storage at 4 degrees C. Theriogenology 39: 771-779.

Breininger, E., Beorlegui, N.B., O'Flaherty, C.M. and Beconi, M.T., 2005. Alpha-tocopherol improves biochemical and dynamic parameters in cryopreserved boar semen. Theriogenology 63: 2126-2135.

Bréque, C., Surai, P. and Brillard, J.P., 2003. Roles of antioxidants on prolonged storage of avian spermatozoa *in vivo* and *in vitro*. Molecular Reproduction and Development 66: 314-323.

Briens, M., Mercier, Y., Rouffineau, F., Mercerand, F. and Geraert, P.A., 2014. 2-hydroxy-4-methylselenobutanoic acid induces additional tissue selenium enrichment in broiler chickens compared with other selenium sources. Poultry Science 93: 85-93.

Briens, M., Mercier, Y., Rouffineau, F., Vacchina, V. and Geraert, P.A., 2013. Comparative study of a new organic selenium source v. seleno-yeast and mineral selenium sources on muscle selenium enrichment and selenium digestibility in broiler chickens. British Journal of Nutrition 110: 617-624.

Brigelius-Flohé, R. and Maiorino, M., 2013. Glutathione peroxidases. Biochimica et Biophysica Acta (BBA)-General Subjects 1830: 3289-3303.

Brouwers, J.F. and Gadella, B.M., 2003. *In situ* detection and localization of lipid peroxidation in individual bovine sperm cells. Free Radical Biology and Medicine 35: 1382-1391.

Buchanan-Smith, J.G., Nelson, E.C., Osburn, B.I., Wells, M.E. and Tillman, A.D., 1969. Effects of vitamin E and selenium deficiencies in sheep fed a purified diet during growth and reproduction. Journal of Animal Science 29: 808-815.

Calabrese, V., Cornelius, C., Dinkova-Kostova, A.T., Iavicoli, I., Di Paola, R., Koverech, A., Cuzzocrea, S., Rizzarelli, E. and Calabrese, E.J., 2012. Cellular stress responses, hormetic phytochemicals and vitagenes in aging and longevity. Biochimica et Biophysica Acta (BBA)-Molecular Basis of Disease 1822: 753-783.

Calvin, H.I., Grosshans, K., Musicant-Shikora, S.R. and Turner, S.I., 1987. A developmental study of rat sperm and testis selenoproteins. Journal of Reproduction and Fertility 81: 1-11.

Castellano, C.A., Audet, I., Bailey, J.L., Laforest, J.P. and Matte, J.J., 2010. Dietary omega-3 fatty acids (fish oils) have limited effects on boar semen stored at 17 C or cryopreserved. Theriogenology 74: 1482-1490.

Castellini, C., Lattaioli, P., Bernardini, M. and Dal Bosco, A., 2000. Effect of dietary alpha-tocopheryl acetate and ascorbic acid on rabbit semen stored at 5 degrees C. Theriogenology 54: 523-533.

Castellini, C., Lattaioli, P., Bosco, A.D. and Beghelli, D., 2002. Effect of supranutritional level of dietary alpha-tocopheryl acetate and selenium on rabbit semen. Theriogenology 58: 1723-1732.

Castellini, C., Mourvaki, E., Dal Bosco, A. and Galli, F., 2007. Vitamin E biochemistry and function: a case study in male rabbit. Reproduction in Domestic Animals 42: 248-256.

Cecil, H. C. and Bakst, M.R., 1993. *In vitro* lipid peroxidation of turkey spermatozoa. Poultry Science 72: 1370-1378.

Cerolini, S., Maldjian, A., Pizzi, F. and Gliozzi, T.M., 2001. Changes in sperm quality and lipid composition during cryopreservation of boar semen. Reproduction 121: 395-401.

Cerolini, S., Maldjian, A., Surai, P. and Noble, R., 2000. Viability, susceptibility to peroxidation and fatty acid composition of boar semen during liquid storage. Animal Reproduction Science 58: 99-111.

Cerolini, S., Surai, P.F., Speake, B.K. and Sparks, N.H.C., 2005. Dietary fish and evening primrose oil with vitamin E effects on semen variables in cockerels. British Poultry Science 46: 214-222.

Cerolini, S., Zaniboni, L., Maldjian, A. and Gliozzi, T., 2006. Effect of docosahexaenoic acid and α-tocopherol enrichment in chicken sperm on semen quality, sperm lipid composition and susceptibility to peroxidation. Theriogenology 66: 877-886.

Chen, S.J., Allam, J.P., Duan, Y.G. and Haidl, G., 2013. Influence of reactive oxygen species on human sperm functions and fertilizing capacity including therapeutical approaches. Archives of Gynecology and Obstetrics 288: 191-199.

Chen, X.D., Zhao, Z.P., Zhou, J.C. and Lei, X.G., 2018. Evolution, regulation, and function of porcine selenogenome. Free Radical Biology and Medicine 127: 116-123.

Comhaire, F.H., Christophe, A.B., Zalata, A.A., Dhooge, W.S., Mahmoud, A.M.A. and Depuydt, C.E., 2000. The effects of combined conventional treatment, oral antioxidants and essential fatty acids on sperm biology in subfertile men. Prostaglandins, Leukotrienes and Essential Fatty Acids 63: 159-165.

Conquer, J.A., Martin, J.B., Tummon, I., Watson, L. and Tekpetey, F., 1999. Fatty acid analysis of blood serum, seminal plasma, and spermatozoa of normozoospermic vs. asthernozoospermic males. Lipids 34: 793-799.

Danikowski, S., Sallmann, H.P., Halle, I. and Flachowsky, G., 2002. Influence of high levels of vitamin E on semen parameters of cocks. Journal of Animal Physiology and Animal Nutrition 86: 376-382.

Domosławska, A., Zdunczyk, S., Franczyk, M., Kankofe,r M. and Janowski, T., 2018. Selenium and vitamin E supplementation enhances the antioxidant status of spermatozoa and improves semen quality in male dogs with lowered fertility. Andrologia 50: e13023.

Ebeid, T.A., 2012. Vitamin E and organic selenium enhances the antioxidative status and quality of chicken semen under high ambient temperature. British Poultry Science 53: 708-714.

Engel, S., Schreiner, T. and Petzoldt, R., 1999. Lipid peroxidation in human spermatozoa and maintenance of progressive sperm motility. Andrologia 31: 17-22.

Ferguson, L.R., Karunasinghe, N., Zhu, S. and Wang, A.H., 2012. Selenium and its' role in the maintenance of genomic stability. Mutation Research 733: 100-110.

Flohé, L. and Zimmermann, R., 1970. The role of GSH peroxidase in protecting the membrane of rat liver mitochondria. Biochimica et Biophysica Acta (BBA)-Bioenergetics 223: 210-213.

Foresta, C., Flohé, L., Garolla, A., Roveri, A., Ursini, F. and Maiorino, M., 2002. Male fertility is linked to the selenoprotein phospholipid hydroperoxide glutathione peroxidase. Biology of Reproduction 67: 967-971.

Fraser, L., Strzeżek, J. and Kordan, W., 2014. Post-thaw sperm characteristics following long-term storage of boar semen in liquid nitrogen. Animal Reproduction Science 147: 119-127.

Gadea, J., Gumbao, D., Matás, C. and Romar, R., 2005. Supplementation of the thawing media with reduced glutathione improves function and the *in vitro* fertilizing ability of boar spermatozoa after cryopreservation. Journal of Andrology 26: 749-756.

Gliozzi, T.M., Zaniboni, L., Maldjian, A., Luzi, F., Maertens, L. and Cerolini, S., 2009. Quality and lipid composition of spermatozoa in rabbits fed DHA and vitamin E rich diets. Theriogenology 71: 910-919.

Golzar Adabi, S., Cooper, R.G., Kamali, M.A. and Hajbabaei, A., 2011. The influence of inclusions of vitamin E and corn oil on semen traits of Japanese quail (*Coturnix coturnix japonica*). Animal Reproduction Science 123: 119-125.

Gómez-Fernández, J., Gómez-Izquierdo, E., Tomás, C., Moce, E. and De Mercado, E., 2013. Is sperm freezability related to the post-thaw lipid peroxidation and the formation of reactive oxygen species in boars? Reproduction in Domestic Animals 48: 177-182.

Griveau, J.F., Dumont, E., Renard, P., Callegari, J.P. and Le Lannou, D., 1995. Reactive oxygen species, lipid peroxidation and enzymatic defence systems in human spermatozoa. Reproduction 103: 17-26.

Guerriero, G., Trocchia, S., Abdel-Gawad, F.K. and Ciarcia, G., 2014. Roles of reactive oxygen species in the spermatogenesis regulation. Frontiers in Endocrinology: 5: 56.

Hansen, J.C. and Deguchi, Y., 1996. Selenium and fertility in animals and man – a review. Acta Veterinaria Scandinavica 37: 19-30.

Henson, M.C., Kattesh, H.G., Hitchcock, J.P. and Kincaid, S.A., 1983. The effects of dietary selenium on growth and selected reproductive parameters in young boars. Animal Science 37: 401-407.

Horky, P., Jancikova, P., Sochor, J., Hynek, D., Chavis, G.J., Ruttkay-Nedecky, B., Cernei, N., Zitka, O., Zeman, L., Adam, V. and Kizek, R., 2012. Effect of organic and inorganic form of selenium on antioxidant status of breeding boars ejaculate revealed by electrochemistry. International Journal of Electrochemical Science 7: 9643-9657.

Hu, J., Geng, G., Li, Q., Sun, X., Cao, H. and Liu, Y., 2014. Effects of alginate on frozen-thawed boar spermatozoa quality, lipid peroxidation and antioxidant enzymes activities. Animal Reproduction Science 147: 112-118.

Huang, Y.L., Tseng, W.C., Cheng, S.Y. and Lin, T.H., 2000. Trace elements and lipid peroxidation in human seminal plasma. Biological Trace Element Research 76: 207-215.

Imai, H. and Nakagawa, Y., 2003. Biological significance of phospholipid hydroperoxide glutathione peroxidase (PHGPx, GPx4) in mammalian cells. Free Radical Biology and Medicine 34: 145-169.

Imai, H., Suzuki, K., Ishizaka, K., Ichinose, S., Oshima, H., Okayasu, I., Emoto, K., Umeda, M. and Nakagawa, Y., 2001. Failure of the expression of phospholipid hydroperoxide glutathione peroxidase in the spermatozoa of human infertile males. Biology of Reproduction 64: 674-683.

Jelezarsky, L., Vaisberg, C., Chaushev, T. and Sapundjiev, E., 2008. Localization and characterization of glutathione peroxidase (GPx) in boar accessory sex glands, seminal plasma, and spermatozoa and activity of GPx in boar semen. Theriogenology 69: 139-145.

Jeong, Y.J., Kim, M.K., Song, H.J., Kang, E.J., Ock, S.A., Kumar, B.M., Balasubramanian, S. and Rho, G.J., 2009. Effect of α-tocopherol supplementation during boar semen cryopreservation on sperm characteristics and expression of apoptosis related genes. Cryobiology 58: 181-189.

Jlali, M., Briens, M., Rouffineau, F., Geraert, P.A. and Mercier, Y., 2014. Evaluation of the efficacy of 2-hydroxy-4-methylselenobutanoic acid on growth performance and tissue selenium retention in growing pigs. Journal of Animal Science 92: 182-188.

Jlali, M., Briens, M., Rouffineau, F., Mercerand, F., Geraert, P.A. and Mercier, Y., 2013. Effect of 2-hydroxy-4-methylselenobutanoic acid as a dietary selenium supplement to improve the selenium concentration of table eggs. Journal of Animal Science 91: 1745-1752.

Johannisson, A., Wallgren, M. and Rodriguez-Martinez, H., 2003. Antioxidant supplementation *in vitro* improves boar sperm motility and mitochondrial membrane potential after cryopreservation of different fractions of the ejaculate. Animal Reproduction Science 78: 85-98.

Johnson, L.A., Weitze, K.F., Fiser, P. and Maxwell, W.M.C., 2000. Storage of boar semen. Animal Reproduction Science 62: 143-172.

Jones, R. and Mann, T., 1973. Lipid peroxidation in spermatozoa. Proceedings of the Royal Society 184: 103-107.

Jones, R. and Mann, T., 1977b. Damage to ram spermatozoa by peroxidation of endogenous phospholipids. Journal of Reproduction and Fertility 50: 261-268.

Jones, R. and Mann, T., 1977a. Toxicity of exogenous fatty acid peroxides towards spermatozoa. Journal of Reproduction and Fertility 50: 255-260.

Jones, R. and Mann. T., 1976. Lipid peroxides in spermatozoa; formation, role of plasmalogen, and physiological significance. Proceedings of the Royal Society 193: 317-333.

Jones, R., Mann, T. and Sherins, R.J., 1978. Adverse effects of peroxidized lipid on human spermatozoa. Proceedings of the Royal Society 201: 413-417.

Jones, R., Mann, T. and Sherins, R.J., 1979. Peroxidative breakdown of phospholipids in human spermatozoa, spermicidal properties of fatty acid peroxides, and protective action of seminal plasma. Fertility and Steriliy 31: 531-537.

Juniper, D.T., Rymer, C. and Briens, M., 2019. Bioefficacy of hydroxy-selenomethionine as a selenium supplement in pregnant dairy heifers and on the selenium status of their calves. Journal of Dairy Science 102: 7000-7010.

Kalthur, G., Raj, S., Thiyagarajan, A., Kumar, S., Kumar, P. and Adiga, S.K., 2011. Vitamin E supplementation in semen-freezing medium improves the motility and protects sperm from freeze-thaw-induced DNA damage. Fertility and Sterility 95: 1149-1151.

Kaludin, I. and Georgiev, G.T., 1984. Vŭrkhu absorbtsiiata i natrupvaneto na tokoferoli v organizma na koch [Tocopherol absorption and accumulation in the body of rams]. Veterinarno-meditsinski nauki 21: 94-97.

Kaludin, I., Ivanov, I.G. and Dimitrova I., 1989. Influence of selenium and methionine on the tocopherol content of ram semen. Veterinarna Sbirka 87: 50-52.

Kawakami, E., Kobayashi, M., Hori, T. and Kaneda, T., 2016. Therapeutic effects of vitamin E supplementation in 4 dogs with poor semen quality and low superoxide dismutase activity in seminal plasma. Journal of Veterinary Medical Science 77: 1711-1714.

Kolodziej, A.Q. and Jacyno, E., 1995. Effect of selenium and vitamin E supplementation on reproductive performance of young boars. Archives of Animal Breeding 48: 68-75.

Kowalowka, M., Wysocki, P., Fraser, L. and Strzezek, J., 2008. Extracellular superoxide dismutase of boar seminal plasma. Reproduction in Domestic Animals 43: 490-496.

Koziorowska-Gilun, M., Koziorowski, M., Fraser, L. and Strzeżek, J., 2011. Antioxidant defence system of boar cauda epididymidal spermatozoa and reproductive tract fluids. Reproduction in Domestic Animals 46: 527-533.

Kumaresan, A., Kadirvel, G., Bujarbaruah, K.M., Bardoloi, R.K., Das, A., Kumar, S. and Naskar, S., 2009. Preservation of boar semen at 18 °C induces lipid peroxidation and apoptosis like changes in spermatozoa. Animal Reproduction Science 110: 162-171.

Ladha, S., 1998. Lipid heterogeneity and membrane fluidity in a highly polarized cell, the mammalian spermatozoon. The Journal of Membrane Biology 165: 1-10.

Lei, X.G., Evenson, J.K., Thompson, K.M. and Sunde, R.A., 1995. Glutathione peroxidase and phospholipid hydroperoxide glutathione peroxidase are differentially regulated in rats by dietary selenium. The Journal of Nutrition 125: 1438-1446.

Lenzi, A., Gandini, L., Maresca, V., Rago, R., Sgro, P., Dondero, F. and Picardo, M., 2000a. Fatty acid composition of spermatozoa and immature germ cells. Molecular Human Reproduction 6: 226-231.

Lenzi, A., Gandini, L., Picardo, M., Tramer, F., Sandri, G. and Panfili, E., 2000b. Lipoperoxidation damage of spermatozoa polyunsaturated fatty acids (PUFA): scavenger mechanisms and possible scavenger therapies. Frontiers in Bioscience 5: 1-15.

Li, J., Barranco, I., Tvarijonaviciute, A., Molina, M.F., Martinez, E.A., Rodriguez-Martinez, H., Parrilla, I. and Roca, J., 2018. Seminal plasma antioxidants are directly involved in boar sperm cryotolerance. Theriogenology 107: 27-35.

Li, K.T., 1975. The glutathione and thiol content of mammalian spermatozoa and seminal plasma Biology of Reproduction 12: 641-646.

Li, N.Y., Sun, Z.J., Ansari, A.R., Cui, L., Hu, Y.F., Li, Z.W., Briens, M., Kai, L., Sun, L.H., Karrow, N.A. and Liu, H.Z., 2020. Impact of maternal selenium supplementation from late gestation and lactation on piglet immune function. Biological Trace Element Research 194: 159-167.

Liu, C.H., Chen, Y.M., Zhang, J.Z., Huang, M.Y., Su, Q., Lu, Z.H., Yin, R.X., Shao, G.Z., Feng, D. and Zheng, P.L., 1982. Preliminary studies on influence of selenium deficiency to the developments of genital organs and spermatogenesis of infancy boars. Acta Veterinaria et Zootechnica Sinica 13: 73-77.

Liu, Q., Zhou, Y.F., Duan, R.J., Wei, H.K., Jiang, S.W. and Peng, J., 2015. Effects of dietary n-6: n-3 fatty acid ratio and vitamin E on semen quality, fatty acid composition and antioxidant status in boars. Animal Reproduction Science 162: 11-19.

Liu, Q., Zhou, Y., Duan, R., Wei, H., Jiang, S. and Peng, J., 2017. Lower dietary n-6: n-3 ratio and high-dose vitamin E supplementation improve sperm morphology and oxidative stress in boars. Reproduction, Fertility and Development 29: 940-949.

Llavanera, M., Delgado-Bermúdez, A., Olives, S., Mateo-Otero, Y., Recuero, S., Bonet, S., Fernández-Fuertes, B., Yeste, M. and Barranco, I., 2020. Glutathione S-transferases play a crucial role in mitochondrial function, plasma membrane stability and oxidative regulation of mammalian sperm. Antioxidants 9: 2.

López, A., Rijsselaere, T., Van Soom, A., Leroy, J.L.M.R., De Clercq, J.B.P., Bols, P.E.J. and Maes, D., 2010. Effect of organic selenium in the diet on sperm quality of boars. Reproduction in Domestic Animals 45: e297-e305.

Lovercamp, K.W., Stewart, K.R., Lin, X. and Flowers, W.L., 2013. Effect of dietary selenium on boar sperm quality. Animal Reproduction Science 138: 268-275.

Lu, Z., Wang, P., Teng, T., Shi, B., Shan, A. and Lei, X.G., 2019. Effects of dietary selenium deficiency or excess on selenoprotein gene expression in the spleen tissue of pigs. Animals 9: 12.

Luberda, Z., 2005. The role of glutathione in mammalian gametes. Reproductive Biology 5: 5-17.

Mahan, D., Zawadzki, J. and Guerrero, R., 2002. Mineral metabolism and boar fertility: observations from Latin America to Europe. In: Lyons, T.P. and Jacques, K.A. (eds) Nutritional Biotechnology in the feed and food industries. Nottingham University Press, Nottingham, UK, pp. 411-418.

Majid, A., Qureshi, M.S. and Khan, R.U., 2015. *In vivo* adverse effects of alpha-tocopherol on the semen quality of male bucks. Journal of Animal Physiology and Animal Nutrition 99: 841-846.

Majzoub, A., Arafa, M., El Ansari, W., Mahdi, M., Agarwal, A., Al-Said, S. and Elbardisi, H., 2020. Correlation of oxidation reduction potential and total motile sperm count: its utility in the evaluation of male fertility potential. Asian Journal of Andrology 22: 317-322. https://doi.org/10.4103/aja.aja_75_19

Marin-Guzman, J., 1990. Studies evaluating dietary selenium and vitamin E on semen quality, *in vivo* oocyte fertilization, spermatozoal ultrastructure and testicular histology of boars. Ph.D. Thesis. Ohio State University, Columbus, OH, USA.

Marin-Guzman, J., Mahan, D.C. and Pate, J.L., 2000a. Effect of dietary selenium and vitamin E on spermatogenic development in boars. Journal of Animal Science 78: 1537-1543.

Marin-Guzman, J., Mahan, D.C. and Whitmoyer, R., 2000b. Effect of dietary selenium and vitamin E on the ultrastructure and ATP concentration of boar spermatozoa, and the efficacy of added sodium selenite in extended semen on sperm motility. Journal of Animal Science 78: 1544-1550.

Marin-Guzman, J., Mahan, D.C., Chung, Y.K., Pate, J.L. and Pope, W.F., 1997. Effects of dietary selenium and vitamin E on boar performance and tissue responses, semen quality, and subsequent fertilization rates in mature gilts. Journal of Animal Science 75: 2994-3003.

Martins, A.D. and Agarwal, A., 2019. Oxidation reduction potential: a new biomarker of male infertility. Panminerva Medica 61: 108-117.

Martins, S.M.M.K., De Andrade, A.F.C., Zaffalon, F.G., Bressan, F.F., Pugine, S.M.P., Melo, M.P.D., Chiaratti, M.R., Marino, C.T., Moretti, A.S. and Arruda, R.P.D., 2015. Organic selenium supplementation increases PHGPx but does not improve viability in chilled boar semen. Andrologia 47: 85-90.

Matsumoto, Y., Terada, T. and Tsutsumi, Y., 1985. Glutamic oxaloacetic transaminase released from chicken spermatozoa during freeze-thaw procedures. Poultry Science 64: 718-722.

McLeod, J., 1943. The role of oxygen in the metabolism and motility of human spermatozoa. American Journal of Physiology 138: 512-518.

Mendez, M.F., Zangeronimo, M.G., Rocha, L.G., Faria, B.G., Pereira, B.A., Fernandes, C.D., Chaves, B.R., Murgas, L.D. and Sousa, R.V., 2013. Effect of the addition of IGF-I and vitamin E to stored boar semen. Animal 7: 793-798.

Moilanen, J., Hovatta, O. and Lindroth, L., 1993. Vitamin E levels in seminal plasma can be elevated by oral administration of vitamin E in infertile men. International Journal of Andrology 16: 165-166.

Naher, Z.U., Ali, M., Biswas, S.K., Mollah, F.H., Fatima, P., Hossain, M.M. and Arslan, M.I., 2013. Effect of oxidative stress in male infertility. Mymensingh Medical Journal 22: 136-142.

National Research Council (NRC), 2012. Selenium. In: Nutrient requirements of swine. 11th ed. National Academies Press, Washington, DC, USA.

Nenicu, A., Lüers, G.H., Kovacs, W., Bergmann, M. and Baumgart-Vogt, E., 2007. Peroxisomes in human and mouse testis: differential expression of peroxisomal proteins in germ cells and distinct somatic cell types of the testis. Biology of Reproduction 77: 1060-1072.

Nogueira, B.G., Sampaio, B.F., Souza, M.I., Costa, E., Silva, E.V. and Zúccari, C.E., 2015. Coenzyme Q10 and α-tocopherol prevent the lipid peroxidation of cooled equine semen. Reproduction of Domestic Animals 50: 1003-1010.

Oldereid, N.B., Thomassen, Y. and Purvis, K., 1998. Selenium in human male reproductive organs. Human Reproduction 13: 2172-2176.

Orzołek, A., Wysocki, P., Strzeżek, J. and Kordan, W., 2013. Superoxide dismutase (SOD) in boar spermatozoa: Purification, biochemical properties and changes in activity during semen storage (16 C) in different extenders. Reproductive Biology 13: 34-40.

Ozer Kaya, S., Gur, S., Erisir, M., Kandemir, F.M., Benzer, F., Kaya, E., Turk, G. and Sonmez, M., 2020. Influence of vitamin E and vitamin E-selenium combination on arginase activity, nitric oxide level and some spermatological properties in ram semen. Reproduction in Domestic Animals 55: 162-169.

Panchenko N.I., Kochubei S.I., Surai P.F., and Yunda I.F., 1985. A method of determination of spermatozoa functions. USSR Patent SU1552343 (Priority: 9.08.1985; Published 23.03.1990).

Parrilla, I., Del Olmo, D., Sijses, L., Martinez-Alborcia, M.J., Cuello, C., Vazquez, J.M., Martinez, E.A. and Roca, J., 2012. Differences in the ability of spermatozoa from individual boar ejaculates to withstand different semen-processing techniques. Animal Reproduction Science 132: 66-73.

Pena, F.J., Johannisson, A., Wallgren, M. and Martinez, H.R., 2003. Antioxidant supplementation *in vitro* improves boar sperm motility and mitochondrial membrane potential after cryopreservation of different fractions of the ejaculate. Animal Reproduction Science 78: 85-98.

Peña, F.J., Johannisson, A., Wallgren, M. and Rodriguez Martinez, H., 2004. Antioxidant supplementation of boar spermatozoa from different fractions of the ejaculate improves cryopreservation: changes in sperm membrane lipid architecture. Zygote 12: 117-124.

Poulos, A., Darin-Bennett, A. and White, I.G., 1973. The phospholipid-bound fatty acids and aldehydes of mammalian spermatozoa. Comparative Biochemistry and Physiology Part B: Comparative Biochemistry 46: 541-549.

Radomil, L., Pettitt, M.J., Merkies, K.M., Hickey, K.D. and Buhr, M.M., 2011. Stress and dietary factors modify boar sperm for processing. Reproduction in Domestic Animals 46: 39-44.

Rooke, J.A., Shao, C.C. and Speake, B.K., 2001. Effects of feeding tuna oil on the lipid composition of pig spermatozoa and *in vitro* characteristics of semen. Reproduction 121: 315-322.

Saaranen, M., Suistomaa, U. and Vanha-Perttula, T., 1989. Semen selenium content and sperm mitochondrial volume in human and some animal species. Human Reproduction 4: 304-308.

Safari Asl, R., Shariatmadari, F., Sharafi, M., Karimi Torshizi, M.A. and Shahverdi, A., 2018. Improvements in semen quality, sperm fatty acids, and reproductive performance in aged Ross breeder roosters fed a diet supplemented with a moderate ratio of n-3: n-6 fatty acids. Poultry Science 97: 4113-4121.

Sarangi, A., Singh, P., Virmani, M., Yadav, A.S., Sahu, S., Ajithakumar, H.M., Kumari, A. and Rath, A.P., 2017. Effect of antioxidants supplementation on the quality of Beetal buck semen stored at 4 °C. Veterinary World 10: 1184-1188.

Satorre, M.M., Breininger, E. and Beconi, M.T., 2012. Cryopreservation with α-tocopherol and Sephadex filtration improved the quality of boar sperm. Theriogenology 78: 1548-1556.
Schneider, M., Forster, H., Boersma, A., Seiler, A., Wehnes, H., Sinowatz, F., Neumüller, C., Deutsch, M.J., Walch, A., Hrabé de Angelis, M. and Wurst, W., 2009. Mitochondrial glutathione peroxidase 4 disruption causes male infertility. The FASEB Journal 23: 3233-3242.
Segerson, E.C. and Johnson, B.H., 1980. Selenium and reproductive function in yearling Angus bulls. Journal of Animal Science 51: 395-401.
Segerson, E.C., Getz, W.R. and Johnson, B.H., 1981. Selenium and reproductive function in boars fed a low selenium diet. Journal of Animal Science 53: 1360-1367.
Shokri, Y., Variji, A., Nosrati, M., Khonakdar-Tarsi, A., Kianmehr, A., Kashi, Z., Bahar, A., Bagheri, A. and Mahrooz, A., 2020. Importance of paraoxonase 1 (PON1) as an antioxidant and antiatherogenic enzyme in the cardiovascular complications of type 2 diabetes: genotypic and phenotypic evaluation. Diabetes Research and Clinical Practice 161: 108067. https://doi.org/10.1016/j.diabres.2020.108067
Smith, D.G., Senger, P.L., McCutchan, J.F. and Landa, C.A., 1979. Selenium and glutathione peroxidase distribution in bovine semen and selenium-75 retention by the tissues of the reproductive tract in the bull. Biology of Reproduction 20: 377-383.
Speight, S.M., Estienne, M.J., Harper, A.F., Barb, C.R. and Pringle, T.D., 2012b. Effects of organic selenium supplementation on growth performance, carcass measurements, tissue selenium concentrations, characteristics of reproductive organs, and testis gene expression profiles in boars. Journal of Animal Science 90: 533-542.
Speight, S.M., Estienne, M.J., Harper, A.F., Crawford, R.J., Knight, J.W. and Whitaker, B.D., 2012a. Effects of dietary supplementation with an organic source of selenium on characteristics of semen quality and *in vitro* fertility in boars. Journal of Animal Science, 90: 761-770.
Strzeżek, J., 2002. Secretory activity of boar seminal vesicle glands. Reproductive Biology 2: 243-266.
Strzezek, J., Lapkiewicz, S. and Lecewicz, M., 1999. A note of antioxidant capacity of boar seminal plasma. Animal Science Papers and Reports 17: 181-188.
Strzezek, J., Wysocki, P., Kordan, W., Kuklinska, M., Mogielnicka, M., Soliwoda, D. and Fraser, L., 2005. Proteomics of boar seminal plasma – current studies and possibility of their application in biotechnology of animal reproduction. Reproductive Biology 5: 279-290.
Sun, L.L., Gao, S.T., Wang, K., Xu, J.C., Sanz-Fernandez, M.V., Baumgard, L.H. and Bu, D.P., 2019. Effects of source on bioavailability of selenium, antioxidant status, and performance in lactating dairy cows during oxidative stress-inducing conditions. Journal of Dairy Science 102: 311-319.
Sun, P., Wang, J., Liu, W., Bu, D.P., Liu, S.J. and Zhang, K.Z., 2017. Hydroxy-selenomethionine: a novel organic selenium source that improves antioxidant status and selenium concentrations in milk and plasma of mid-lactation dairy cows. Journal of Dairy Science 100: 9602-9610.
Sunde, R.A., 1993. Intracellular glutathione peroxidases – structure, regulation, and function. In: Burk, R.F. (ed.) Selenium in biology and human health. Springer-Verlag, New-York, NY, USA, pp. 45-77.
Surai, P.F., 1981. Fat-soluble vitamins in turkey sperm. Proceedings of 2nd Conference of Young Scientists, Zagorsk, USSR, pp 55-56.
Surai, P.F., 1982. The effect of high doses of vitamin E in turkey male diet on the GOT and LDH activity in stored sperm. Nauchno-Tekhnicheskii Byulletin of the Ukrainian Poultry Research Institute 12: 24-29.
Surai, P.F., 1983a. Biochemical and functional changes in turkey tissues and sperm as a function of the levels of vitamins E and A in feed. Cand. Biol. Sci. (PhD) Thesis. Ukrainian Poultry Research Institute, Borky, Ukraine.

Surai, P.F., 1983b. Stabilizing effect of vitamins A and E on turkey sperm membranes. Ptitsevodstvo (Kiev) 36: 49-52.
Surai, P.F., 1984. Lipid peroxidation in turkey semen. Ptitsevodstvo (Kiev) 37: 46-49.
Surai, P.F., 1989a. Relations between vitamin E concentration in poultry spermatozoa and some semen biochemical and physiological characteristics. Proceedings of the 8th International Symposium on Current Problems in Avian Genetics, Smolenice, Czechoslovakia, pp. 171-173.
Surai, P.F., 1989b. Detergent treatment of poultry spermatozoa: release of some enzymes. Proceedings of the 8th International Symposium on Current Problems in Avian Genetics, Smolenice, Czechoslovakia, pp. 174-175.
Surai, P.F., 1991. Nutritional and biochemical aspects of vitamins in poultry. DSc (Biol.) Thesis. Ukrainian Poultry Research Institute, Borky, Ukraine.
Surai, P.F., 1992. Vitamin E feeding of poultry males. Proceedings of the 19th World Poultry Congress, Amsterdam, the Netherlands 1: 575-577.
Surai, P.F., 2002. Natural antioxidants in avian nutrition and reproduction. Nottingham University Press, Nottingham, UK.
Surai, P.F., 2006. Selenium in nutrition and health. Nottingham University Press, Nottingham, UK.
Surai, P.F., 2018. Selenium in poultry nutrition and health. Wageningen Academic Publishers, Wageningen, the Netherland, 430 pp.
Surai, P.F., 2021. Antioxidant defence systems in animal health. Chapter 1. In: Surai, P.F. (ed.) Selenium in pig nutrition and health. Wageningen Academic Publishers, Wageningen, the Netherlands, pp. 17-38. https://doi.org/10.3920/978-90-8686-912-1_
Surai, P.F. and Fisinin, V.I., 2014. Selenium in sow nutrition. Animal Feed Science and Technology 211: 18-30.
Surai, P.F. and Ionov, I.A., 1981. Changes in GOT activity in turkey semen as a result of storage. Nauchno-Tekhnicheskii Byulletin of the Ukrainian Poultry Research Institute 11: 24-30.
Surai, P.F. and Ionov, I., 1992. Vitamin E in fowl sperm. Proceedings of the 12th International Congress on Animal Reproduction, The Hague, the Netherlands 1: 535-537.
Surai, P.F., Brillard, J-P., Speake, B.K., Blesbois, E., Seigneurin, F. and Sparks, N.H.C., 2000b. Phospholipid fatty acid composition, vitamin E content and susceptibility to lipid peroxidation of duck spermatozoa. Theriogenology 53: 1025-1039.
Surai, P.F., Cerolini, S. Wishart, G.J., Speake, B.K., Noble, R.C. and Sparks, N.H.C., 1998. Lipid and antioxidant composition of chicken semen and its susceptibility to peroxidation. Avian and Poultry Biology Reviews 9: 11-23.
Surai, P.F., Fujihara,N., Speake, B.K., Brillard, J-P., Wishart, G.J. and Sparks, N.H.C., 2001. Polyunsaturated fatty acids, lipid peroxidation and antioxidant protection in avian semen. Review. Asian Australasian Journal of Animal Sciences 7: 1024-1050.
Surai, P.F., Kochish, I.I., Fisinin, V.I. and Kidd, M.T., 2019a. Antioxidant defence systems and oxidative stress in poultry biology: an update. Antioxidants 8: 7.
Surai, P.F., Kochish, I.I., Fisinin, V.I. and Velichko, O.A., 2018. Selenium in poultry nutrition: from sodium selenite to organic selenium sources. The Journal of Poultry Science 55: 79-93.
Surai, P.F., Kochish, I.I., Romanov, M.N. and Griffin, D.K., 2019b. Nutritional modulation of the antioxidant capacities in poultry: the case of vitamin E. Poultry Science 98: 4030-4041.
Surai, P.F., Kutz, E., Wishart, G.J., Noble, N.C. and Speake, B.K., 1997. The relationship between the dietary provision of α-tocopherol and the concentration of this vitamin in the semen of chicken: effects on lipid composition and susceptibility to peroxidation. Reproduction 110: 47-51.

Surai, P.F., Noble, R.C., Sparks, N.H.C. and Speake, B.K., 2000a. Effect of long-term supplementation with arachidonic or docosahexaenoic acids on sperm production in the broiler chicken. Journal of reproduction and fertility 120: 257-264.

Tareq, K.M.A., Akter, Q.S., Takagi, Y., Hamano, K.I., Sawada, T. and Tsujii, H., 2009. Effect of selenium and vitamin E on acrosome reaction in porcine spermatozoa. Reproductive Medicine and Biology 28(2): 73-81.

Taylor, K., Roberts, P., Sanders, K. and Burton, P., 2009. Effect of antioxidant supplementation of cryopreservation medium on post-thaw integrity of human spermatozoa. Reproductive Biomedicine Online 18: 184-189.

Therond, P., Auger, J., Legrand, A. and Jouannet, P., 1996. α-tocopherol in human spermatozoa and seminal plasma: relationships with motility, antioxidant enzymes and leukocytes. MHR: Basic science of Reproductive Medicine 2: 739-744.

Thongchalam, K., Rukkwamsuk, T. and Chopmchai, S., 2012. Blood and semen selenium concentrations and semen quality in boars fed diets supplemented with organic or inorganic selenium. Journal of Animal and Veterinary Advances 11: 603-608.

Tinsley, I.J., Arscott, G.H. and Lowry, R.R., 1971. Fertility and testicular fatty acid composition in the chicken as influenced by vitamin E and ethoxyquin. Lipids 6: 657-660.

Tramer, F., Rocco, F., Micali, F., Sandri, G. and Panfili, E., 1998. Antioxidant systems in rat epididymal spermatozoa. Biology of Reproduction 59: 753-758.

Ursini, F., Heim, S., Kiess, M., Maiorino, M., Roveri, A., Wissing, J. and Flohé, L., 1999. Dual function of the selenoprotein PHGPx during sperm maturation. Science 285: 1393-1396.

Ursini, F., Maiorino, M. and Gregolin, C., 1985. The selenoenzyme phospholipid hydroperoxide glutathione peroxidase. Biochimica et Biophysica Acta (BBA)-General Subjects 839: 62-70.

Vanha-Perttula, T. and Remes, E., 1990. Incorporation of selenium-75 into seminal plasma and spermatozoa of the bull. Andrologia 22: 34-41.

Vernet, P., Aitken, R.J. and Drevet, J.R., 2004. Antioxidant strategies in the epididymis. Molecular and Cellular Endocrinology 216: 31-39.

Vernet, P., Rock, E., Mazur, A., Rayssiguier, Y., Dufaure, J.P. and Drevet, J.R., 1999. Selenium-independent epididymis-restricted glutathione peroxidase 5 protein (GPX5) can back up failing Se-dependent GPXs in mice subjected to selenium deficiency. Molecular Reproduction and Development 54: 362-370.

Waberski, D., Henning, H. and Petrunkina, A.M., 2011. Assessment of storage effects in liquid preserved boar semen. Reproduction in Domestic Animals 46: 45-48.

Walczak-Jedrzejowska, R., Wolski, J.K. and Slowikowska-Hilczer, J., 2013. The role of oxidative stress and antioxidants in male fertility. Central European Journal of Urology 66: 60-67.

Wales, R.G., White, I.G. and Lammond, D.R., 1959. The spermicidal activity of hydrogen peroxide *in vitro* and *in vivo*. Journal of Endocrinology 18: 236-244.

Watanabe, T. and Endo, A., 1991. Effects of selenium deficiency on sperm morphology and spermatocyte chromosomes in mice. Mutation Research 262: 93-99.

Wei, J.Y., Wang, J., Liu, W., Zhang, K.Z. and Sun, P., 2019. Short communication: effects of different selenium supplements on rumen fermentation and apparent nutrient and selenium digestibility of mid-lactation dairy cows. Journal of Dairy Science 102: 3131-3135.

Wright, C., Milne, S. and Leeson, H., 2014. Sperm DNA damage caused by oxidative stress: modifiable clinical, lifestyle and nutritional factors in male infertility. Reproductive Biomedicine Online 28: 684-703.

Wu, A.S., Oldfield, J.E., Shull, L.R. and Cheeke, P., 1979. Specific effect of selenium deficiency on rat sperm. Biology of Reproduction 20: 793-798.

Zakošek Pipan, M., Mrkun, J., Nemec Svete, A. and Zrimšek, P., 2017. Improvement of liquid stored boar semen quality by removing low molecular weight proteins and supplementation with α-tocopherol. Animal Reproduction Science 186: 52-61.

Zaniboni, L. and Cerolini, S., 2009. Liquid storage of turkey semen: changes in quality parameters, lipid composition and susceptibility to induced *in vitro* peroxidation in control, n-3 fatty acids and alpha-tocopherol rich spermatozoa. Animal Reproduction Science 112: 51-65.

Zaniboni, L., Gliozzi, T., Maldjian, A., Luzi, F. and Cerolini, S., 2004. Fatty acid and tocopherol composition of semen components in the rabbit. Proceedings of the 8th World Rabbit Congress, Puebla, Mexico, pp. 365-370.

Zaniboni, L., Rizzi, R. and Cerolini, S., 2006. Combined effect of DHA and alpha-tocopherol enrichment on sperm quality and fertility in the turkey. Theriogenology 65: 1813-1827.

Zerbinati, C., Caponecchia, L., Fiori, C., Sebastianelli, A., Salacone, P., Ciacciarelli, M. and Iuliano, L., 2020. Alpha- and gamma-tocopherol levels in human semen and their potential functional implications. Andrologia 52: e13543.

Zhang, Q., L. Chen, K. Guo, L. Zheng, B. Liu, W. Yu, C. Guo, Z. Liu, Y. Chen and Z. Tang, 2013. Effects of different selenium levels on gene expression of a subset of selenoproteins and antioxidative capacity in mice. Biological Trace Element Research 154: 255-261.

Zhao, L., Sun, L.H., Huang, J.Q., Briens, M., Qi, D.S., Xu, S.W. and Lei, X.G., 2017. A novel organic selenium compound exerts unique regulation of selenium speciation, selenogenome, and selenoproteins in broiler chicks. The Journal of Nutrition 147: 789-797.

Zhu, Z., Fan, X., Lv, Y., Zhang, N., Fan, C., Zhang, P. and Zeng, W., 2015. Vitamin E analogue improves rabbit sperm quality during the process of cryopreservation through its antioxidative action. PLoS One 10: e0145383.

Zingg, J.M., 2019. Vitamin E: regulatory role on signal transduction. IUBMB Life 71: 456-478.

Chapter 7
Selenium in piglet nutrition

P.F. Surai[1-7] *and I.I. Kochish*[2]
[1]*Vitagene and Health Research Centre, Bristol, BS4 2RS, United Kingdom;* [2]*Department of Hygiene and Poultry Sciences, Moscow State Academy of Veterinary Medicine and Biotechnology Named After K.I. Skryabin, 109472 Moscow, Russia;* [3]*Department of Microbiology and Biochemistry, Faculty of Veterinary Medicine, Trakia University, 6000 Stara Zagora, Bulgaria;* [4]*Department of Animal Nutrition, Faculty of Agricultural and Environmental Sciences, Szent Istvan University, 2103 Gödöllo, Hungary;* [5]*Saint-Petersburg State Academy of Veterinary Medicine, 196084 St. Petersburg, Russia;* [6]*Sumy National Agrarian University, 40000 Sumy, Ukraine;* [7]*Odessa National Academy of Food Technologies, 65039 Odessa, Ukraine; psurai@feedfood.co.uk*

Abstract

Selenium is shown to play an important role in pig nutrition. The requirement of pigs in selenium varies depending on many environmental and other conditions and in general is considered to be 0.15-0.30 mg/kg feed. The main advantage of organic Se in the diet of growing pigs are related to building Se reserves in the body. Indeed, in stress conditions, when requirement for antioxidants substantially increased, but feed consumption is usually decreased Se reserves, accumulated in tissues, could be a key for the resistance of pigs to detrimental effects of free radicals and toxic products of their metabolism. Furthermore, it was clearly shown that by using organic Se in the pig diet it is possible to improve antioxidant defences in pigs which potentially can improve adaptive abilities of animals to various stresses. Environmental concerns about selenium as a potential pollutant lead to legal restrictions of levels of Se, which can be included into animal diets. Therefore, better assimilation of selenium from organic sources could potentially help pig nutritionists to overcome consequences of such a limitation. When organic Se is used in the diet of the growing pigs, there is also an opportunity to produce Se pork which could help dealing with Se deficiencies in various parts of the world. Effect of antioxidants on pork quality is extensively studied and it seems likely that a combination of increased doses of vitamin E with organic selenium is an optimal solution for pig industry.

Keywords: selenium, growing pigs, SeMet, OH-SeMet, oxidative stress

7.1 Introduction

Pig production from weaning up to slaughter time is associated with a range of stresses and Se as an integral part of 25 selenoproteins is involved in the antioxidant defences and redox homeostasis maintenance in pig tissues. Similar to reproducing animals,

DOI 10.3920/978-90-8686-912-1_7, © Wageningen Academic Publishers 2021

in comparison to sodium selenite, organic Se in the diet of growing piglets is more effectively accumulated in tissues, especially in muscles. This builds Se reserves, which can be used in stress conditions when Se requirement increases while its supply could be inadequate. Therefore, main results related to Se nutrition of the growing piglets will be presented in this chapter with a specific emphasis to organic selenium as an effective form of this element in pig nutrition.

7.2 Stresses in weaning pigs

Weaning represents a significant challenge to young piglets and the success of the process determines future productive and reproductive characteristics of animals. The transition from a liquid diet to one that is usually based on dry feed ingredients is accompanied by major changes in digestive physiology, immune status and social and physical environment needs. Indeed, successful weaning is the key to effective growth, feed efficiency and the subsequent development of the pig. This is stressful time for pigs and multiple stressors at weaning include (Close and Cole, 2000):

- loss of maternal relationship;
- separation from littermates;
- movement to a new environment including new housing system with limited thermoregulation ability;
- the change of diet composition and transition from liquid feeding to dry feed with major changes in digestive physiology;
- mixing with other pigs and need for social adjustment;
- possible crowding.

This is a period of active pig growth and it is well accepted that growth rate of the pig is determined by a number of factors and selenium is involved in most of them (Huo *et al.*, 2003):

- **Genotype.** Genetic potential of pigs is constantly improving and therefore their ability to utilise nutrients is also improved. However, the price for this improvement is an increased sensitivity to various stresses and therefore antioxidant defences are an integral part of the body to prevent damaging effects of various stressors. Indeed, genotypes with a higher potential for lean tissue growth, sometimes are characterised by a lower feed intake, and therefore need more nutritional attention than other genotypes.
- **Nutrition.** To meet increased demands for nutrient of fast growing pigs is an important challenge for the nutritionist. Since digestive system of the piglet at birth is not mature and actively developing for the first few weeks, colostrum and milk are extremely important sources of essential nutrients. Antioxidants compounds such as vitamin E and selenium are transferred from mother to progeny providing essentials for building antioxidant defences of fast-growing piglet.
- **Environment.** Temperature is the most important environmental factor affecting growth and development of piglets. Indeed, thermoregulation of piglets is dependent on thyroid hormone function. Selenium as part of iodothyronine deiodinase is involved in activation of thyroid hormone and in thermoregulation

of the piglets. Since the fat reserves in the piglet body at birth is limited (about 1% of body mass), its efficient usage is a priority and aforementioned selenoenzymes are involved in this process. Furthermore, environmental pollutants such as ammonia, carbon monoxide, sulphated hydrogen cause oxidative stress in piglets and need for additional antioxidant protection, provided by selenoproteins is increased.

- **Health status.** Immune system of the newly born piglets is very weak and therefore they are extremely sensitive to various pathogens. Indeed, colostrum and milk provide immunological protection, but a building of immune system of piglets usually takes several weeks. During this period of time antioxidants plays an important immunoregulating role (see Chapter 6; Surai and Fisinin, 2021).
- **Management.** The decisions about time of weaning, choice of diets and supplements and some other management factors affect growth and development of piglets. In particular, a choice of feed supplements could have a positive or detrimental effect on growth performance of pigs. Indeed, the daily and long-term management decisions made by personnel will greatly affect the efficiency of pig production.

Young pigs enter the postweaning period with a variable Se reserves with following declining Se concentrations in pig serum and tissues. This period of postnatal development of pigs is critical in terms of Se deficiency development and still 1-2% death rate occurs in the nursery and the occurrence of the Se deficiency is still prevalent on most pig farms (Mahan, 1996). In general, Se deficiency can affect growth and development of growing pigs with the largest, fastest growing weanling pigs being more sensitive to sudden death. It has been shown that weaning stress can exacerbate Se deficiency if Se reserves in the body are not sufficient. In particular, Mahan (1991) suggested that the 21-day-old weaning pig fed a practical diet requires 0.5 mg/kg supplemental Se for about 2 weeks and it could be reduced to 0.35 mg/kg by 5 weeks postweaning.

7.3 Selenium nutrition of growing pigs

Se metabolism in growing pigs was extensively studied. Four-week-old weaned pigs were fed a corn-soybean meal diet supplemented with 0, 0.3, 0.5 or 1.0 mg/kg Se from sodium selenite for 5 weeks (Mahan, 1985). It was shown that faecal Se excretion increased with increasing dietary Se level. In contrast, urinary Se decreased during the postweaning periods for pigs fed the basal diet but increased linearly as dietary Se increased. It was reported that unsupplemented diet was deficient in Se, since there was a net loss of Se from the body during the initial 2 weeks postweaning, whereupon, it became positive. It was shown that Se retention in the postweaning pig increased in direct proportion to the amount consumed when diets contained up to 1.0 mg/kg Se. In general, apparent digestibility of Se was approximately 70%, whereas it ranged between 30 to 40% for pigs fed the basal diet during 5 weeks (Mahan, 1985). Balance experiments with Se in pigs showed the following:

- Low selenium supplementation (0.1 mg/kg) was associated with increased percent of Se retention in the body (from 45.5 to 61.7%, Groce *et al.*, 1971; Table 7.1).

- High selenium dietary supplementation (0.5 mg/kg) substantially decreased Se retention in the body (15.2%; Groce *et al.,* 1971).
- Inclusion in the grower pig diet of organic selenium in the form of selenised yeast was associated with a substantial increase in percent of retention of Se in the pig body (Mahan and Parret, 1996).

When diet for Crossbred barrows was supplemented with 0.1; 0.3 or 0.5 mg/kg selenium in the form of selenite or Se-Yeast, selenium retention increased as dietary Se levels increased, particularly when the Se-Yeast was provided (Mahan and Parrett, 1996; Table 7.2).

Table 7.1. Daily selenium (Se) balance in pigs supplemented with sodium selenite (adapted from Groce *et al.,* 1971).

Item	Basal	Basal+0.1 mg/kg Se	Basal+0.5 Se
Se, mg/kg (as fed)	0.042	0.147	0.543
Se intake, µg	18.9	66.2	244.4
Total Se excretion, µg	10.3	25.3	207.3
Faecal Se, µg	8.4	18.4	57.7
Urinary Se, µg	1.9	6.9	149.6
Se retention, µg	8.6	40.9	37.1
Percent retention	45.5	61.7	15.2

Table 7.2. Effect of dietary selenium on selenium level in tissues of grower and finisher pig, mg/kg (adapted from Mahan and Parrett, 1996).

Item	Basal	Inorganic selenium			Organic selenium		
Serum	**0**	**0.1**	**0.3**	**0.5**	**0.1**	**0.3**	**0.5**
60 kg body weight (feeding from 22.2 kg BW)							
Serum	0.056	0.142	0.162	0.162	0.115	0.159	0.170
Loin	0.114	0.167	0.170	0.189	0.216	0.300	0.362
Liver	0.229	0.615	0.682	0.756	0.533	0.679	0.820
Pancreas	0.421	0.526	0.519	0.494	0.532	0.645	0.604
Kidney	1.248	1.603	2.043	1.824	1.708	1.870	1.756
105 kg body weight (feeding from 65.8 kg BW)							
Serum	0.044	0.154	0.167	0.167	0.117	0.168	0.187
Loin	0.08	0.101	0.124	0.13	0.143	0.212	0.262
Liver	0.222	0.492	0.675	0.699	0.467	0.737	0.901
Pancreas	0.368	0.435	0.484	0.459	0.490	0.614	0.725
Kidney	1.402	2.599	3.512	3.738	2.140	2.990	3.040

Similar results were reported when selenite or organic selenium in the form of selenised yeast was fed to grower and finisher pig (Table 7.3; Mahan *et al.*, 1999).

Increasing selenite concentration in the diet from 0.05 to 0.3 mg/kg only slightly affected Se concentration in the loin and pancreas, but substantially increased Se concentration in the liver. In contrast, organic Se in the pig diet was much more effective in transferring to loin and pancreas. It is interesting to note, that as dietary Se levels increased, urinary Se excretion was substantially increased when pigs were fed sodium selenite, whereas excess of selenium in the form of Se-Yeast was mainly excreted via faecal route. Indeed, loin Se contents were higher in grower and finisher pigs as dietary Se levels increased when the Se-enriched yeast was fed.

In an experiment conducted in China effect of organic selenium in the form of Se-Yeast on growth performance and diarrhoea incidence of weaning piglets from days 21 through 42 was investigated (Huang *et al.*, 2004). A total of 108 piglets weaned at 21 days were randomly allotted to three dietary treatments, with three pens and 12 piglets in each pen. When dietary sodium selenite (0.2 mg/kg) was replaced by the same amount of organic selenium in the form of Se-Yeast the following advantages were seen:

- average daily gain significantly increased (290.9 vs 274.1 g/day);
- decreased incidence of diarrhoea (1.32% vs 1.72%);
- decreased the feed cost/kg weight gain by 11%.

Diets and pig tissues from 19 states of the USA were analysed for Se and a great variability of dietary Se, as well as tissue Se were shown (Table 7.4; Adapted from, Mahan *et al.*, 2005)

Table 7.3. Effect of dietary selenium on tissue selenium concentration, mg/kg (adapted from Mahan *et al.*, 1999).

Item	Basal	Inorganic selenium				Organic selenium			
	0	0.05	0.1	0.2	0.3	0.05	0.1	0.2	0.3
55 kg body weight									
Loin	0.087	0.091	0.103	0.113	0.120	0.113	0.143	0.184	0.206
Liver	0.195	0.425	0.464	0.526	0.522	0.319	0.398	0.577	0.513
Pancreas	0.314	0.327	0.360	0.400	0.368	0.335	0.396	0.529	0.529
Kidney	1.305	1.716	1.771	1.76	1.655	1.597	1.734	1.973	1.972
105 kg body weight									
Loin	0.085	0.106	0.114	0.118	0.124	0.134	0.17	0.249	0.332
Liver	0.258	0.452	0.436	0.478	0.534	0.469	0.516	0.627	0.717
Pancreas	0.290	0.355	0.391	0.367	0.383	0.395	0.426	0.483	0.573
Kidney	1.811	2.283	2.333	2.328	2.376	2.282	2.333	2.473	2.677

Table 7.4. Dietary Se and loin selenium (Se) in pigs from 19 different states of the USA (adapted from Mahan *et al.*, 2005).

State	Dietary Se (as fed basis), mg/kg	Selenite added, mg/kg	Loin Se (wet basis), mg/kg
South Dakota	0.651/0.493	0.30	0.495/0.386
Kansas	0.625	0.30	0.510
Oklahoma	0.583	0.20	0.368
Nebraska	0.555/0.330	0.30	0.362/0.313
Louisiana	0.462	0.3	0.250
Iowa	0.433/0.235	0.0	0.396/0.278
Arkansas	0.401/0.152	0.2	0.381/0.212
Texas	0.396	0.15	0.426
Indiana	0.363/0.052	0.15	0.110/0.063
Alabama	0.354	0.3	0.126
Ohio	0.352	0.30	0.155
Florida	0.341	0.3	0.155
Illinois	0.321/0.036	0.26	0.123/0.059
Wisconsin	0.304/0.178	0.20	0.256/0.125
Kentucky	0.304	0.30	0.094
Georgia	0.296	0.3	0.116
Michigan	0.295/0.040	0.3	0.135/0.052
Virginia	0.290/0.027	0.30	0.098/0.034
North Carolina	0.227	0.20	0.139

The results of the study showed a wide variation in the Se concentration in the diets, even though selenite was added, suggesting that the organic component was the major difference in Se concentrations among diets (Mahan *et al.*, 2005). It is interesting, that low Se levels in pig loin from Alabama, Illinois, Kentucky, Georgia, Michigan, Virginia and North Carolina are related to low Se level in the basic diet (in table: dietary Se – selenite added), while the highest Se in the loin in South Dakota and Kansas are related to extremely high Se levels in feed ingredients. Earlier publications on Se level in the pig diet without supplementation confirmed this conclusion: in Illinois, Michigan and Virginia Se level in the complete pig feed was reported to be 0.036, 0.040 and 0.027 mg/kg respectively (Ku *et al.*, 1972). The correlation of dietary Se to loin Se was high ($r=0.84$, $P<0.01$), reflecting high impact of feed-derived organic Se (SeMet) on the loin Se concentration (Mahan *et al.*, 2005). Recently the effects of adding organic or inorganic Se to corn and soybean meal based pig diets differing in their natural organic Se within the US regions have been studied (Mahan *et al.*, 2014). The corn Se content from the various states ranged from 0.026 to 0.283 mg Se/kg while the soybean meal Se content ranged from 0.086 to 0.798 mg Se/kg. Main results are shown in Table 7.5.

Table 7.5. Effect of selenium source and level on grower-finisher pig loin responses at each state, mg/kg (adapted from Mahan *et al.*, 2014).[1]

State	Basal	OS, 0.15	OS, 0.30	SS, 0.15	SS, 0.30
North Carolina	0.082	0.187	0.278	0.093	0.109
Kentucky	0.096	0.204	0.328	0.136	0.123
Ohio	0.119	0.294	0.450	0.148	0.173
Illinois	0.126	0.287	0.448	0.159	0.163
Georgia	0.130	0.230	0.355	0.138	0.141
Wisconsin	0.267	0.358	0.496	0.283	0.268
Texas	0.289	0.391	0.453	0.296	0.341
Nebraska	0.345	0.445	0.532	0.348	0.349
South Dakota	0.527	0.621	0.690	0.517	0.527
Average	0.221	0.335	00.448	0.235	0.244

[1] OS = organic Se; SS = sodium selenite.

Tissue and serum Se concentrations were shown to be higher ($P<0.01$) when supplemental organic Se was fed, whereas serum GPx was found to be greater ($P<0.01$) as Se level increased. There were linear increases ($P<0.01$) in loin and quadratic increases ($P<0.01$) in liver and hair Se concentrations as dietary Se level increased within each state (Mahan *et al.*, 2014).

A total of 108 castrates at average body weight (BW) of 60 kg were allotted to three treatments, each of which was replicated three times with 12 pigs per replicate (four per pen). The control groups received the basal diet containing 0.045 mg Se/kg. In experimental groups SS or SeMet at 0.3 mg Se/kg were added to the basal diet and the experiment lasted 40 d (Zhan *et al.*, 2007). SS was shown to significantly increase Se concentration in serum, muscle, liver, pancreas and kidney tissue, while SeMet increased it further in liver, muscle and pancreas showing significance (P<0.05) in comparison to the SS group. The feeding trial lasted for 42 days was conducted in China. A total of 216 crossbred weaning pigs (weaned on d 28, initial body weight 9.94±1.26 kg) were randomly allocated to 6 treatment groups with 6 replicates each containing six pigs. Basal diet contained 0.06 mg Se/kg without any additional Se supplement, SS group was supplemented with 0.3 mg/kg of Se from sodium selenite, and other groups were supplemented with 0.1, 0.3, 0.5, and 0.7 mg/kg of Se from DL-SeMet, respectively. The measured values of Se in the diets were shown to be 0.06, 0.35, 0.44, 0.51, 0.57 and 0.99 mg/kg, respectively (Cao *et al.*, 2014). The results of the trial in terms of Se in plasma were rather inconsistent. In particular, it was shown that SS or SeMet at 0.1 and 0.3 mg/kg did not affect Se concentration in the pig plasma on day 21, while pigs supplemented with 0.5 mg/kg SeMet showed significant higher plasma Se concentration compared with the BD, SS, and 0.1-0.3 mg/kg SeMet groups ($P<0.05$, Figure 7.1).

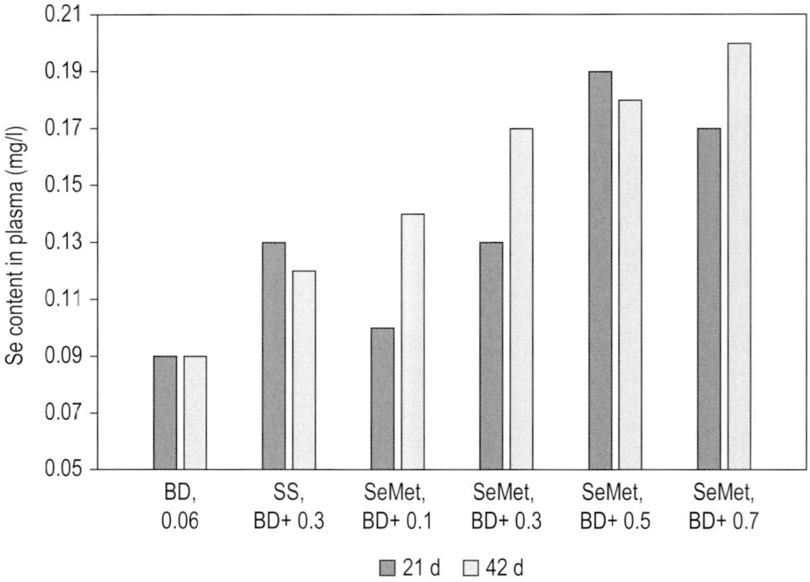

Figure 7.1. Effect of dietary selenium (Se) on its content in pig plasma (adapted from Cao *et al.*, 2014).

However, further increase SeMet supplementation (0.7 mg/kg) was not effective, since Se concentration in plasma was not different from pigs fed on SS or obtaining SeMet at 0.1 or 03 mg/kg. On d 42, in comparison to BD group plasma Se concentration significantly increased only when SeMet was supplemented at 0.3-0.7 mg/kg, while in comparison to SS group only highest (0.7 mg/kg) SeMet dietary supplementation was able to significantly increase Se concentration. (Cao *et al.*, 2014).

Similar inconsistency was seen in terms of Se supplementation on antioxidant defences of serum, liver or muscles (Table 7.6). First, Se supplementation did not affect concentration of final product of lipid peroxidation (MDA) in serum. Secondly, Se supplementation decreased MDA concentration in the liver and muscles, but effect SeMet was not dose dependent and was not different from that of SS. Thirdly, total AO activity of serum and tissues was not affected by Se supplementation. The exception is a group supplemented with 0.1 mg/kg SeMet (serum and liver), which has no proper explanation.

The response of GPx to Se supplementation was also variable: SS improve GPx activity only in the liver, while effects of SeMet was not dose dependent and achieved significance in comparison to BD only in few cases.

Recently, results of another trial with SeMet dietary supplementation in pigs conducted in Brazil has been published (Silva *et al.*, 2019). A total of 128 hybrid pigs with an average weight of 76 kg were randomly distributed in eight treatments and eight replicates according to body weight and fed experimental diets for 30 days. The experimental treatments included two Se levels from sodium selenite-SS (0.3 and 0.6 mg/kg), four Se

Table 7.6. Effect of dietary Se on antioxidant defences in pigs (adapted from Cao *et al.*, 2014).[1]

Serum	BD 0.06	BD + SS, 0.3	BD + DL-SeMet, 0.1	BD+ DL-SeMet, 0.3	BD+ DL-SeMet, 0.5	BD+ DL-SeMet, 0.7
T-AOC, U/ml	8.56	8.43	9.19	9.87	9.23	9.35
GPx, U/ml	787.5	774.4	837.7	892.1	865.2	859.4
MDA, nmol/ml	3.74	3.36	3.25	3.97	3.96	3.98
Liver						
T-AOC, U/ml	0.40	0.43	0.63	0.44	0.38	0.46
GPx, U/ml	54.50	76.40	81.61	59.15	61.02	66.02
MDA, nmol/ml	1.45	1.04	0.84	0.82	0.86	1.08
Muscle						
T-AOC, U/ml	0.85	0.77	0.83	0.73	0.74	0.80
GPx, U/ml	90.05	100.81	109.89	132.30	129.19	105.11
MDA, nmol/ml	0.16	0.10	0.12	0.11	0.11	0.11

[1] BD = basal diet; DL-SeMet = DL-selenomethionine; GPx = glutathione peroxidase; MDA = malondialdehyde; SS = sodium selenite; T-AOC = total antioxidant capacity.

levels from SeMet (0.3, 0.4, 0.5 and 0.6 mg/kg) and two combinations of SS with SeMet (SS 0.15 + SM 0.15 mg/kg and SS 0.3 + SM 0.3 mg/kg) providing 0.3 and 0.6 mg/kg Se in the diet respectively (Silva *et al.*, 2019). Se supplementation was shown to be not effective in improving major performance indices. The Se concentration in muscles in response to Se dietary supplementation is the major parameter reflecting quality and efficacy of organic Se supplementation. Indeed, SeMet is major Se form in muscles (Bierla *et al.*, 2008) and since it is not synthesised in animals, only SeMet from the diet can be incorporated non-specifically into proteins in place of Met (Surai, 2006, 2018) leading to increase Se concentration in muscles. The results of the aforementioned trial are rather inconsistent in terms of Se concentration in muscles in response to SeMet dietary supplementation. There are at least two possible reasons for such results. First, Se analysis in muscles is quite difficult procedure and, based on our own experience, direct usage of the hydride generation atomic absorption spectroscopy (as it was the case in the study) could give quite variable results because of low sensitivity of Se detection. Usually, a fluorometric detection combined with atomic absorption or ICP-MS could resolve the problem. Secondly, as mentioned in Chapter 3 (Surai *et al.*, 2021), SeMet as chemical is not stable during storage and it would be important to check the level of pure SeMet (not total Se) in the product before adding to the diet.

A total of 72 finishing pigs were divided into four groups, which received a BD (0.1 mg Se/kg) without Se supplementation or one supplemented with SS, Se-methylselenocysteine MeSeCys, or SeMet at 0.25 mg Se/kg for 60 days (Zhang *et al.*, 2020). Se supplementation did not affect pig growth performance. SeMet was shown to significantly increase Se concentrations in the heart, muscle, and liver in comparison to SS or MeSeCys supplemented pigs. Interestingly, both organic Se sources (SeMet and MeSeCys) were shown to significantly increase the immunoglobulin A (IgA), IgG, and IgM serum levels compared with BD and SS supplemented pigs (Zhang *et al.*, 2020). In a trial conducted in Norway twenty-four female finisher pigs, with an average body weight of 25.6 kg and divided into 4 groups were used. The trial diets were either the Se deficient, non-supplemented basal diet containing 0.05 mg Se/kg (BD) or the same diet supplemented with SS, SY or L-SeMet at 0.3 mg/kg and dietary content of Se was as follows: 0.05; 0.33; 0.32 and 0.32 mg/kg respectively (Falk *et al.*, 2018). The Se-concentrations in various skeletal muscles including M. longissimus dorsi (LD), M. semitendinosus, M. semimembranosus, M. biceps brachii, M. quadriceps femoris, and M. psoas major were shown to be significantly different between the groups, with highest concentrations in pigs fed SeMet, followed by those fed Se yeast, SS, and BD diet (Figure 7.2). These data clearly showed that dietary organic Se can build Se reserves in various muscles, which potentially can be used in stress condition, when feed consumption goes down but Se requirement increases (Surai, 2018).

Indeed, transcriptional analyses of LD showed that selenogenes *SelenoW* and *SelenoH* were higher expressed in pigs fed Se supplemented diets compared with BD (control). Furthermore, the expression of interferon gamma and cyclooxygenase 2 was lower in the Se-supplemented pigs versus control. Interestingly, in stress conditions (after LPS exposure) glutathione peroxidase 1 and *SelenoN* were more reduced in pigs fed SS compared with pigs fed organic Se (Falk *et al.*, 2018). This could reflect better adaptability of pigs fed diets supplemented with organic selenium. Indeed, most of studies devoted to a comparison of various Se sources were conducted in healthy pigs under physiological conditions. However, effect of Se provided by different Se sources on the antioxidant defences and growth of weanling pigs under stress conditions

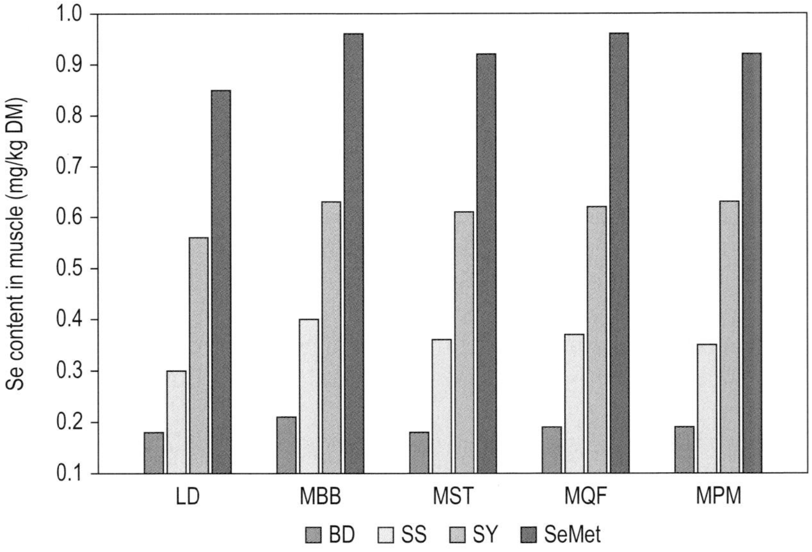

Figure 7.2. Selenium (Se) concentration in pig muscles collected after 64 days of trial duration (adapted from Falk *et al.*, 2018).

imposing oxidative insults is largely unknown. Therefore, organic Se concept based on assumption that SeMet accumulated in muscles could be an additional source of Se to upregulate various selenoproteins in stress conditions could predict and explain protective effects of organic Se in comparison to SS. Indeed, sufficient or even increased Se supply is needed to replenish cellular GPx and possibly other selenoproteins in oxidative stress situations. Therefore, the development and use of effective stress models could help distinguishing between efficacy of various Se sources. In particular, in addition to LPS treatment mentioned above, diquat toxicity associated with increased free radical production could be one of such models. In a recent study, plasma GPx activity was the highest in SY-consumed piglets post-diquat challenge (Doan *et al.*, 2020). Interesting, SY, but not SS or so called soy-chelated selenium, restored plasma GPx activity to the level of control group at 6 h, on d 2 and d 7 post diquat-injection and was able to restore hepatic GPx activity on d 7 post injection (Doan *et al.*, 2020).

In an experiment conducted in France 112 crossbred gilts with initial body weight of 26.73±3.15 kg were allotted to the 7 dietary treatments on the basis of weight and fed experimental diets for 32 days. Each group included 8 replicate pens of 2 pigs per pen. The dietary treatments were composed with the same basal diet (BD containing 0.11 mg Se/kg) and differed only in Se sources and levels: BD (control diet) without Se supplementation, and BD diet supplemented with 0.1 or 0.3 mg/kg of SS (SS-0.1 and SS-0.3), SY (SY- 0.1 and SY-0.3), or OH-SeMet (SO-0.1 and SO-0.3). Water and feed were provided ad libitum throughout the experimental period. Gilts were weighed individually at the beginning and at the end of experimental period and feed intake per pen was monitored weekly (Jlali et al., 2014). The obtained results indicated that Se dietary supplementation did not affect pig's performance: final BW and ADG did not differ among treatments, no difference in ADFI and G:F were found among the treatments groups during the study. The Se concentrations in plasma and liver are summarised in Table 7.7.

Table 7.7. Effect of dietary Se on serum and tissue Se concentrations, ng/g fresh weight (adapted from Jlali *et al.*, 2014).[1]

	BD 0.11	BD + SS, 0.1	BD + SS, 0.3	BD+ SY, 0.1	BD+ SY, 0.3	BD+ OH-SeMet, 0.1	BD+ OH-SeMet, 0.3
Plasma	91.24	120.43	120.68	117.46	142.12	128.10	160.66
Liver, mg/kg	0.34	0.44	0.48	0.46	0.65	0.56	0.74

[1] BD = basal diet; OH-SeMet = hydroxy-selenomethionine; SS = sodium selenite; SY = Se-yeast; T-AOC = total antioxidant capacity.

Pigs fed diets supplemented with Se were shown to have higher plasma and liver Se concentrations (P<0.05) than those fed BD diet without Se added. However, at low level of supplementation (0.1 mg Se/kg of diet) plasma Se concentration did not differ among pigs supplemented with inorganic Se (SS) and those supplemented with organic Se (SY and SO), while at 0.3 mg/kg organic Se for more effective than SS in increasing plasma Se levels. At 0.3 mg/kg Se supplementation both SY and SO were shown to increase Se concentration in the liver in comparison to piglets fed on SS. However, at low Se supplementation (0.1 mg/kg) SO was more effective that SY in increasing Se concentration in the liver. Interestingly, in tenderloin (psoas major) muscle, only dietary organic Se was able to increase Se concentration in comparison to BD fed piglets. compared with control diet: muscle Se concentration remains similar between pigs fed the BD diet and those fed diets supplemented with SS regardless of the dietary Se level. Furthermore, compared with pigs fed diets supplemented with Se from SY, pigs fed Se as OH-SeMet were shown to have higher muscle Se concentrations in psoas major muscle (P<0.05 Figure 7.3; Jlali et al., 2014).

Assessment of the relative bioavailability of Se from each Se sources in plasma, liver, and muscle are summarised in Table 7.7 show that compared with SS, SY and SO were more efficient (P<0.01) to improve Se concentrations in plasma, liver, and muscle. However, OH-SeMet was characterised by the highest bioavailability being more effective in enhancing Se concentrations (P<0.01) in plasma (+69.66%), liver (+40.52%), and muscle (+62.22%) than SY (Jlali et al., 2014).

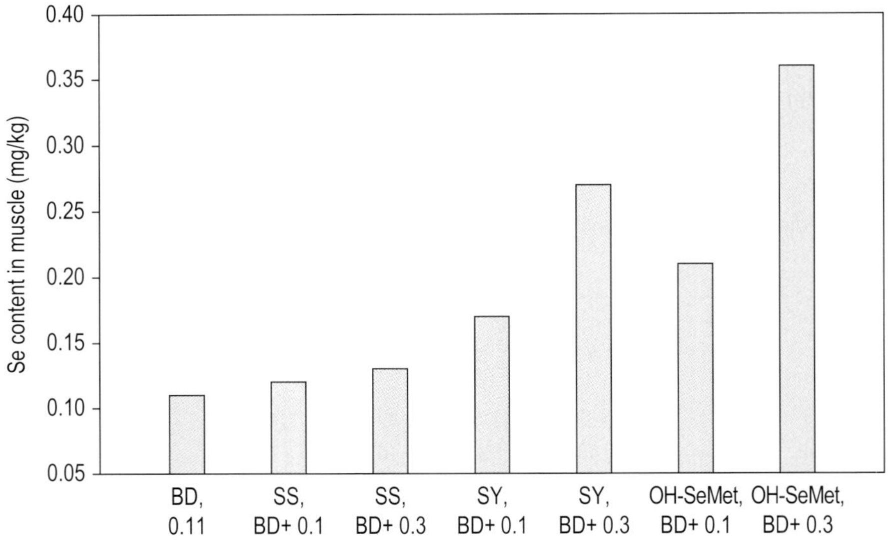

Figure 7.3. Selenium (Se) content in pig muscles, mg/kg (adapted from Jlali *et al.*, 2014).

Recently, a comprehensive study with growing piglets has been conducted to evaluate efficacy of various form of Se in pig nutrition (Chao *et al.*, 2019). A total of 252 crossbred piglets with initial body weight of 7.72±0.18 kg were assigned into seven dietary treatment groups with six replicates per treatment and six pigs per replicate based on their initial body weight and gender. The dietary treatments included BD-basal diet without Se supplementation containing 0.13 mg/kg Se (a negative control group), BD + 0.3 mg Se from Na_2SeO_3 per kg (a positive control group), and five experimental groups supplemented with OH-SeMet (HMSeBA; BD + 0.1, 0.2, 0.3, 0.4 and 0.5 mg Se per kg from OH-SeMet, respectively). After 6 days of adaptation on BD diet, the pigs were fed corresponding diets as described above for 28 d with *ad libitum* access to feed and water. Similar to results of many previous studies, the obtained results showed that dietary Se supplementation did not affect growth performance of piglets. However, OH-SeMet supplementation was shown to increase the Se concentrations in serum, liver, kidney and muscle compared with unsupplemented BD and BD+SS groups ($P<0.05$; Table 7.8; Chao *et al.*, 2019). In fact, this study clearly showed that organic Se in the form of OH-SeMet dose-dependently increased Se concentration in muscles (Figure 7.4), building Se reserves in the piglet body.

This reserve can be used in stress conditions to supply additional Se for selenoprotein synthesis (Surai, 2018; Surai *et al.*, 2018). As expected, adding SS into the piglet diet did not affect Se concentration in the muscles. Indeed, the main form of Se in muscles is SeMet which is non-specifically incorporated into the muscle replacing Met. Since SeMet cannot be synthesised in animal body, muscle Se is not responsive to inorganic Se in the pig diet. Additional benefit of organic Se in the pig diet is related to improvement of antioxidant defence mechanisms in the growing piglets.

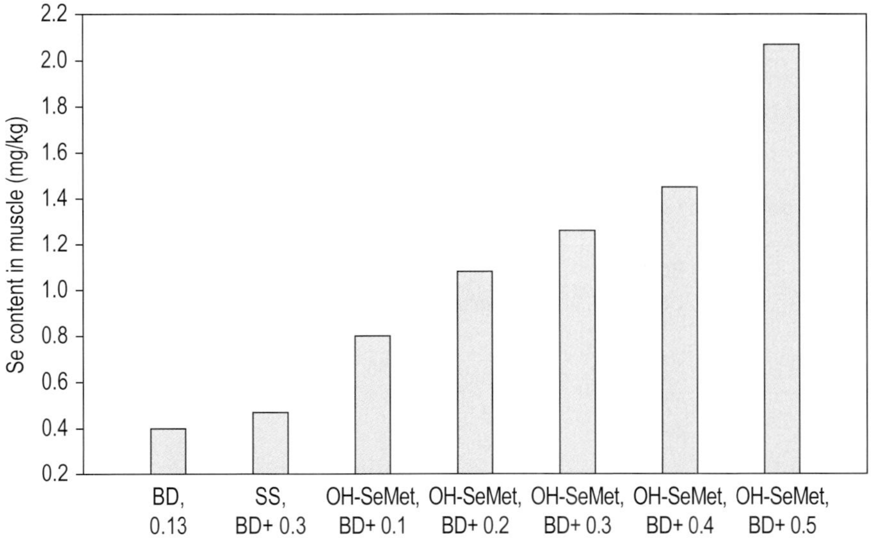

Figure 7.4. Selenium (Se) content in pig muscles, mg/kg (adapted from Chao *et al.*, 2019).

Table 7.8. Effect of dietary Se on serum and tissue Se concentrations (adapted from Chao *et al.*, 2019).[1]

	BD 0.13	BD + SS, 0.3	BD + OH-SeMet, 0.1	BD+ OH-SeMet, 0.2	BD+ OH-SeMet, 0.3	BD+ OH-SeMet, 0.4	BD+ OH-SeMet, 0.5
Serum, mg/l	0.07	0.09	0.102	0.13	0.15	0.17	0.16
Liver, mg/kg	1.07	1.45	1.53	1.89	2.27	2.51	3.09
Kidney, mg/kg	4.66	5.38	5.45	6.18	6.75	6.81	6.14

[1] BD = basal diet; OH-SeMet = hydroxy-selenomethionine; SS = sodium selenite.

Compared with BD group, supplementation with 0.2 and 0.4 mg Se from OH-SeMet was shown to increase serum total antioxidant capability (T-AOC) and addition of 0.4 and 0.5 mg Se in the form of OH-SeMet increased serum glutathione peroxidase (GPx) activities ($P<0.05$; Table 7.9; Chao *et al.*, 2019).

Improvement of Se status of the growing piglets by inclusion of organic Se in the form of OH-SeMet at 0.1, 0.2, 0.4, 0.5 mg Se was shown to increase GPx activities and decreased lipid peroxidation as evidenced by reduced malondialdehyde (MDA) contents in the liver. Furthermore, OH-SeMet supplementation at 0.3 mg Se/kg was found to increase T-AOC and GPx activities in the liver ($P<0.05$). Interestingly, compared with SS supplemented group, 0.3 mg Se from OH-SeMet was shown to increase serum superoxide dismutase (SOD) and hepatic T-AOC activities, and decreased the serum MDA level ($P<0.05$; Table 7.10; Chao *et al.*, 2019).

Therefore, the aforementioned data clearly showed that OH-SeMet is an effective form of Se in pig nutrition. It is well assimilated from the feed building Se reserves in the body. Those reserves could be considered as an 'insurance factor' in stress conditions helping to deal with overproduction of free radicals, maintaining redox balance, productive and reproductive performance of animals.

Table 7.9. Effect of dietary Se on serum antioxidant defences (adapted from Chao *et al.*, 2019).[1]

	BD 0.13	BD + SS, 0.3	BD + OH-SeMet, 0.1	BD+ OH-SeMet, 0.2	BD+ OH-SeMet, 0.3	BD+ OH-SeMet, 0.4	BD+ OH-SeMet, 0.5
T-AOC, U/ml	1.19	1.25	1.27	1.53	1.48	1.63	1.39
SOD, U/ml	51.0	49.6	55.0	51.7	54.3	55.6	55.2
GPx, U/ml	430	454	446	445	456	495	562
MDA, nmol/ml	2.75	3.00	2.53	2.50	2.31	2.46	2.57

[1] BD = basal diet; OH-SeMet = hydroxy-selenomethionine; SS = sodium selenite.

Table 7.10. Effect of dietary Se on liver antioxidant defences (adapted from Chao *et al.*, 2019).[1]

	BD 0.13	BD + SS, 0.3	BD + OH-SeMet, 0.1	BD+ OH-SeMet, 0.2	BD+ OH-SeMet, 0.3	BD+ OH-SeMet, 0.4	BD+ OH-SeMet, 0.5
T-AOC, U/ml	0.36	0.35	0.41	0.39	0.48	0.42	0.40
SOD, U/ml	78.7	89.2	89.2	84.6	88.4	92.7	83.9
GPx, U/ml	131	189	182	185	193	209	192
MDA, nmol/ml	3.40	2.44	2.52	2.55	2.77	2.64	2.45

[1] BD = basal diet; OH-SeMet = hydroxy-selenomethionine; SS = sodium selenite.

It should be mentioned that most research trials were conducted in physiological conditions, with healthy animals fed balanced diets. In such conditions Se requirement is quite low and very often can be met by feed ingredient-derived Se. Therefore, Se supplementation did not affect major productive or reproductive performance of animals. The comparisons between different Se sources are mainly related to efficiency of Se assimilation from the diet and its usage by animals for maintaining optimal Se status and maximum expression of various selenoproteins participating in regulation of various physiological functions with specific action as an essential part of the antioxidant defence mechanisms. Therefore, it is clear that organic Se in the form of SeMet is the best option for feed formulators to include it into the premixes and in comparison to traditional sodium selenite, there is a range of advantages of organic Se for growing pigs in stress conditions of commercial pig production (Figure 7.5).

7.4 Conclusions

From the data presented above it is clear that Se plays an important role in pig nutrition. The requirement of pigs in selenium varies depending on many environmental and other conditions and in general is considered to be 0.15-0.30 mg/kg feed. The main advantage of organic Se in the diet of growing pigs are related to building Se reserves in the body. Indeed, in stress conditions, when requirement for antioxidants substantially increased, but feed consumption is usually decreased Se reserves, accumulated in tissues, could be a key for the resistance of pigs to detrimental effects of free radicals and toxic products of their metabolism. Furthermore, the aforementioned data clearly showed that by using organic Se in the pig diet it is possible to improve antioxidant defences in pigs which potentially can improve adaptive abilities of animals to various stresses. Environmental concerns about selenium as a potential pollutants lead to legal restrictions of levels of Se, which can be included into animal diets. Therefore, better assimilation of selenium from organic sources could potentially help pig nutritionists to overcome consequences of such a limitation. When organic Se is used in the diet of the growing pigs, there is also an opportunity to produce Se pork which could help to deal with Se deficiencies in various parts of the world. Effect of antioxidants on pork quality is extensively studied and it seems likely that a combination of increased doses

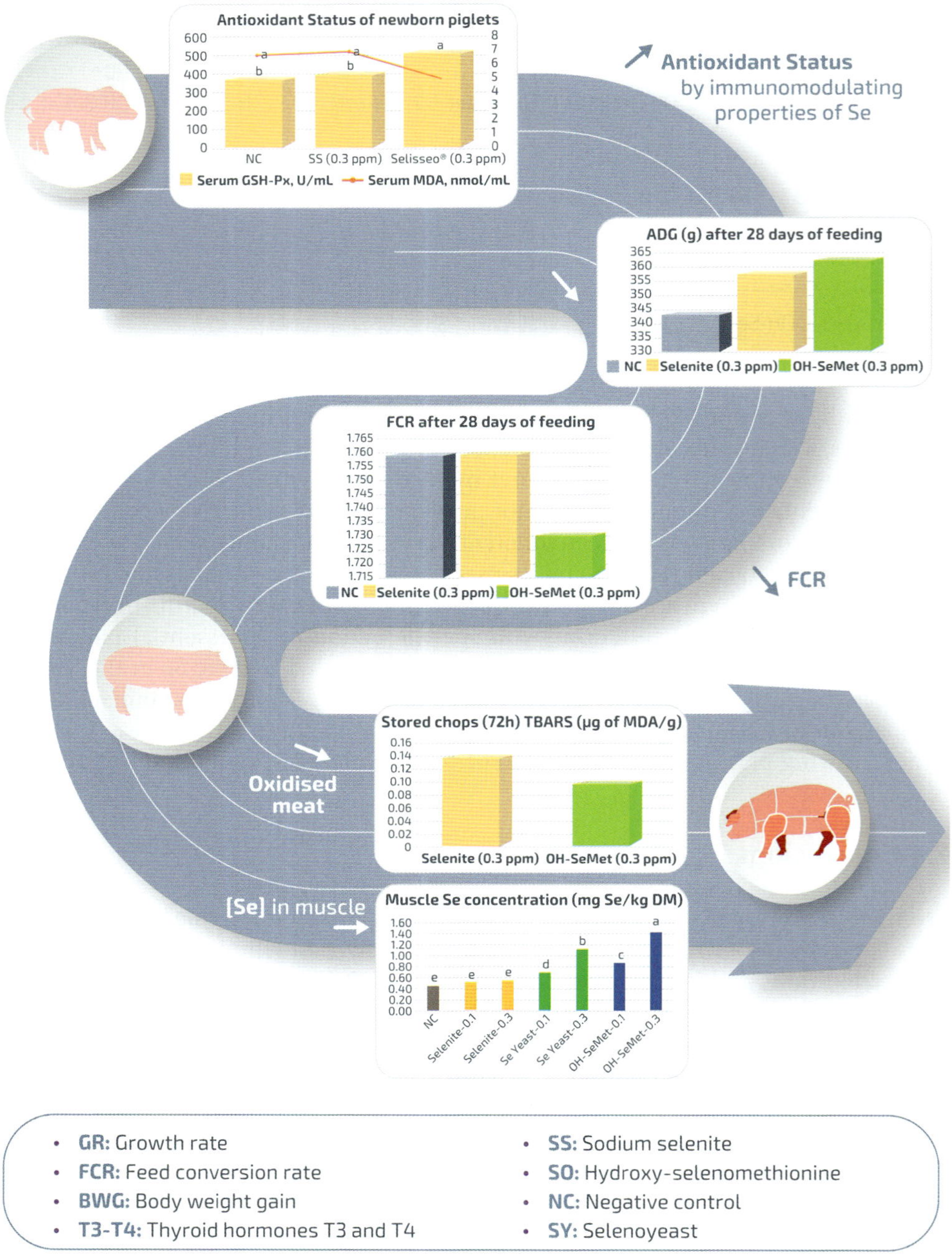

Figure 7.5. Selenium for growing pigs – pig growth and development.

of vitamin E with organic selenium is an optimal solution for pig industry (For details see Chapter 8; Mercier *et al.*, 2021).

Acknowledgements

Peter F. Surai and Ivan I. Kochish are supported by a grant of the Government of Russian Federation (Contract No. 14.W03.31.0013).

References

Bierla, K., Dernovics, M., Vacchina, V., Szpunar, J., Bertin, G. and Lobinski, R., 2008. Determination of selenocysteine and selenomethionine in edible animal tissues by 2D size-exclusion reversed-phase HPLC-ICP MS following carbamidomethylation and proteolytic extraction. Analytical and Bioanalytical Chemistry 390: 1789-1798.

Cao, J., Guo, F., Zhang, L., Dong, B. and Gong, L., 2014. Effects of dietary Selenomethionine supplementation on growth performance, antioxidant status, plasma selenium concentration, and immune function in weaning pigs. Journal of Animal Science and Biotechnology 5: 46.

Chao, Y., Yu, B., He, J., Huang, Z., Mao, X., Luo, J., Luo, Y., Zheng, P., Yu, J. and Chen, D., 2019. Effects of different levels of dietary hydroxy-analogue of selenomethionine on growth performance, selenium deposition and antioxidant status of weaned piglets. Archives of Animal Nutrition 73: 374-383.

Close, W.H. and Cole, D.J.A., 2000. Nutrition of sows and boars. Nottingham University Press, Nottingham, UK.

Doan, N., Liu, Y., Xiong, X., Kim, K., Wu, Z., Bravo, D.M., Blanchard, A. and Ji, P., 2020. Organic selenium supplement partially alleviated diquat-induced oxidative insults and hepatic metabolic stress in nursery pigs. British Journal of Nutrition 124(1): 23-33. https://doi.org/10.1017/S0007114520000689

Falk, M., Bernhoft, A., Framstad, T., Salbu, B., Wisløff, H., Kortner, T.M., Kristoffersen, A.B. and Oropeza-Moe, M., 2018. Effects of dietary sodium selenite and organic selenium sources on immune and inflammatory responses and selenium deposition in growing pigs. Journal of Trace Elements in Medicine and Biology 50: 527-536.

Groce, A.W., Miller, E.R., Keahey, K.K., Ullrey, D.E. and Ellis, D.J.,1971. Selenium supplementation of practical diets for growing-finishing swine. Journal of Animal Science 32: 905-911.

Huang, R., Zhu, W., Hang, S., Miao, C., Wu, Y., Liang, L., Xu, J, and Lu, X., 2004. Effect of Bio-Mos and Sel-Plex on growth performance and incidence of diarrhea of weaning piglets. Abstracts of Posters presented at Alltech's 20th Annual Symposium on Nutritional Biotechnology in the Feed and Food Industries, Lexington, KY, USA, p.49.

Huo, Y.J., Wang, T. and Xu, R.J., 2003. Nutrition and metabolism of neonatal pig. In: Xu, R.J. and Cranwell, P. (eds) Gastrointestinal physiology and nutrition of neonatal pigs. Nottingham University Press, Nottingham, UK, pp. 185-211.

Jlali, M., Briens, M., Rouffineau, F., Geraert, P.A. and Mercier, Y., 2014. Evaluation of the efficacy of 2-hydroxy-4-methylselenobutanoic acid on growth performance and tissue selenium retention in growing pigs. Journal of Animal Science 92: 182-188.

Ku, P.K., Ely, W.T., Groce, A.W. and Ullrey, D.E., 1972. Natural dietary selenium, α-tocopherol and effect on tissue selenium. Journal of Animal Science 34: 208-211.

Mahan, D.C., 1985. Effect of inorganic selenium supplementation on selenium retention in postweaning swine. Journal of Animal Science 61: 173-178.

Mahan, D.C., 1991. Vitamin E and selenium in swine nutrition. In: Miller E.R., Ullrey D.E. and Lewis A.J. (eds) Swine nutrition. Butterworth-Heineman, Boston, MA, USA, pp. 193-214.

Mahan, D.C., 1996. Are we still having vitamin E and selenium deficiencies in pigs? In: Illini PorkNet Papers, The Online Resource for the Pork Industry, pp. 1-21. Available at: https://tinyurl.com/4qkf9tkg.

Mahan, D.C. and Parrett, N.A., 1996. Evaluating the efficacy of selenium-enriched yeast and sodium selenite on tissue selenium retention and serum glutathione peroxidase activity in grower and finisher swine. Journal of Animal Science 74: 2967-2974.

Mahan, D.C., Azain, M., Crenshaw, T.D., Cromwell, G.L., Dove, C.R., Kim, S.W., Lindemann, M.D., Miller, P.S., Pettigrew, J.E., Stein, H.H. and Van Heugten, E., 2014. Supplementation of organic and inorganic selenium to diets using grains grown in various regions of the United States with differing natural Se concentrations and fed to grower-finisher swine. Journal of Animal Science 92: 4991-4997.

Mahan, D.C., Brendemuhl, J.H., Carter, S.D., Chiba, L.I., Crenshaw, T.D., Cromwell, G.L., Dove, C.R., Harper, A.F., Hill, G.M., Hollis, G.R., Kim, S.W., Lindemann, M.D., Maxwell, C.V., Miller, P.S., Nelssen, J.L., Richert, B.T., Southern, L.L., Stahly, T.S., Stein, H.H., Van Heugten, E. and Yen, J.T., 2005. Comparison of dietary selenium fed to grower-finisher pigs from various regions of the United States on resulting tissue Se and loin mineral concentrations. Journal of Animal Science 83: 852-857.

Mahan, D.C., Cline, T.R. and Richert, B., 1999. Effects of dietary levels of selenium enriched yeast and sodium selenite as selenium sources fed to growing-finishing pigs on performance, tissue selenium, serum glutathione peroxidase activity, carcass characteristics, and loin quality. Journal of Animal Science 77: 2172-2179.

Mercier, Y., Batonon-Alavo, D. and Geraert, P.A., 2021. Selenium and pork meat quality: muscle redox status, selenium, sulphur amino acids and beyond. Chapter 8. In: Surai, P.F. (ed.) Selenium in pig nutrition and health. Wageningen Academic Publishers, Wageningen, the Netherlands, pp. 235-260. https://doi.org/10.3920/978-90-8686-912-1_8

Silva, V.A., Bertechini, A.G., Clemente, A.H.S., De Freitas, L.F.V.B., Nogueira, B.R.F., De Oliveira, B.L. and Ramos, A.L.S., 2019. Different levels of selenomethionine on the meat quality and selenium deposition in tissue of finishing pigs. Journal of Animal Physiology and Animal Nutrition 103: 1866-1874.

Surai, P.F., 2006. Selenium in nutrition and health. Nottingham University Press, Nottingham, UK.

Surai, P.F., 2018. Selenium in poultry nutrition and health. Wageningen Academic Publishers, Wageningen, the Netherland, 430 pp.

Surai, P.F. and Fisinin, V.I., 2021. Selenium in boar nutrition. Chapter 6. In: Surai, P.F. (ed.) Selenium in pig nutrition and health. Wageningen Academic Publishers, Wageningen, the Netherlands, pp. 173-213. https://doi.org/10.3920/978-90-8686-912-1_6

Surai, P.F., Kochish, I.I. and Fisinin, V.I., 2021. Selenium in feed: organic selenium concept development. Chapter 3. In: Surai, P.F. (ed.) Selenium in pig nutrition and health. Wageningen Academic Publishers, Wageningen, the Netherlands, pp. 61-110. https://doi.org/10.3920/978-90-8686-912-1_3

Surai, P.F., Kochish, I.I., Fisinin, V.I. and Velichko, O.A., 2018. Selenium in poultry nutrition: from sodium selenite to organic selenium sources. The Journal of Poultry Science 55: 79-93.

Zhan, X., Wang, M., Zhao, R., Li, W. and Xu, Z., 2007. Effects of different selenium source on selenium distribution, loin quality and antioxidant status in finishing pigs. Animal Feed Science and Technology 132: 202-211.

Zhang, K., Zhao, Q., Zhan, T., Han, Y., Tang, C. and Zhang, J., 2020. Effect of different selenium sources on growth performance, tissue selenium content, meat quality, and selenoprotein gene expression in finishing pigs. Biological Trace Element Research 196: 463-471.

Chapter 8

Selenium and pork meat quality: muscle redox status, selenium, sulphur amino acids and beyond

Y. Mercier, D. Batonon-Alavo*[2] *and P.A. Geraert*[3]
[1]*Adisseo France SAS, CERN, Route de Malicorne, 03600 Commentry, France;* [2]*Adisseo USA Inc., 4400 North Point Pkwy, Alpharetta, GA 30022, USA;* [3]*Adisseo France SAS, 10 Place du Général de Gaulle, 92160 Antony, France; yves.mercier@adisseo.com*

Abstract

Selenium has long been known to be efficiently deposited in meat tissues. Moreover, increasing the selenium deposited in muscle tissues, protect the meat from oxidative degradation in particular lipid and protein oxidation. Indeed, the most impacting factor for consumers' buying decision is certainly meat colour which depends on myoglobin oxidation. Drip loss is also an important consumer issue and linked with the stability of the muscle cell membrane reinforcing the need for increased oxidative stability. Selenium, and particularly its organic form, has long been shown to be effective to improve such parameters. The relationship between methionine, cysteine, glutathione and selenoproteins is also extensively addressed in this chapter to give a better understanding on the nutritional solutions to improve meat quality and guarantee a better oxidative stability. Recently, meat defaults such as destructured hams appeared. While those defaults might have different origins, enhancing the oxidative stability of the muscle tissue is also of primary importance. Further research is needed to understand mechanisms of protective effects of selenium and selenoproteins on meat quality under various commercially challenging conditions.

Keywords: selenium, methionine, selenoprotein, meat quality, pork, destructured ham

8.1 Introduction

Meat quality must be considered as multifactorial. Hygiene and safety aspects along the food chain are not discussed and are anyway mandatory; technological aspects and further processing abilities are of high interest for slaughterhouses and for processors while nutritional benefits or simply organoleptic aspects are the most important aspects for final consumers. From fresh meat to salted delicatessens, muscles can have many possible fates. Beyond the hygienic aspects and the nutritive value, meat quality traits are based (1) on the visual aspects that can be assessed by the consumers at retail place such as colour, drip loss and (2) on the organoleptic aspects such as tenderness, juiciness, odours and flavour (Figure 8.1). Moreover, depending on the retail form of

Peter F. Surai (ed.) **Selenium in pig nutrition and health**
DOI 10.3920/978-90-8686-912-1_8, © Wageningen Academic Publishers 2021

the meat and the different packaging process, visual and organoleptic properties can be largely modified (Renerre, 1990).

Meat colour and its stability during the retail period is certainly the most impacting factor for consumers' buying decision. Meat colour is under the dependence of muscle pigments, mainly myoglobin with its content and its reduction/oxidation (redox) status. The colour criteria were largely studied in beef meat because of its high quantity of myoglobin and the rapid change of redox status that changes the colour from a bright red (highly appreciated) to brownish colour systematically rejected by the consumers. Since the colour stability problems are mainly linked to the myoglobin content, it is thus linked to the muscle type and the species that present different myoglobin levels (Surendranath and Poulson, 2013). Surface spectrometers allow to determine the three components of the meat colour, i.e. L*, a*, b* respectively for lightness, redness and yellowness, and its evolution during storage.

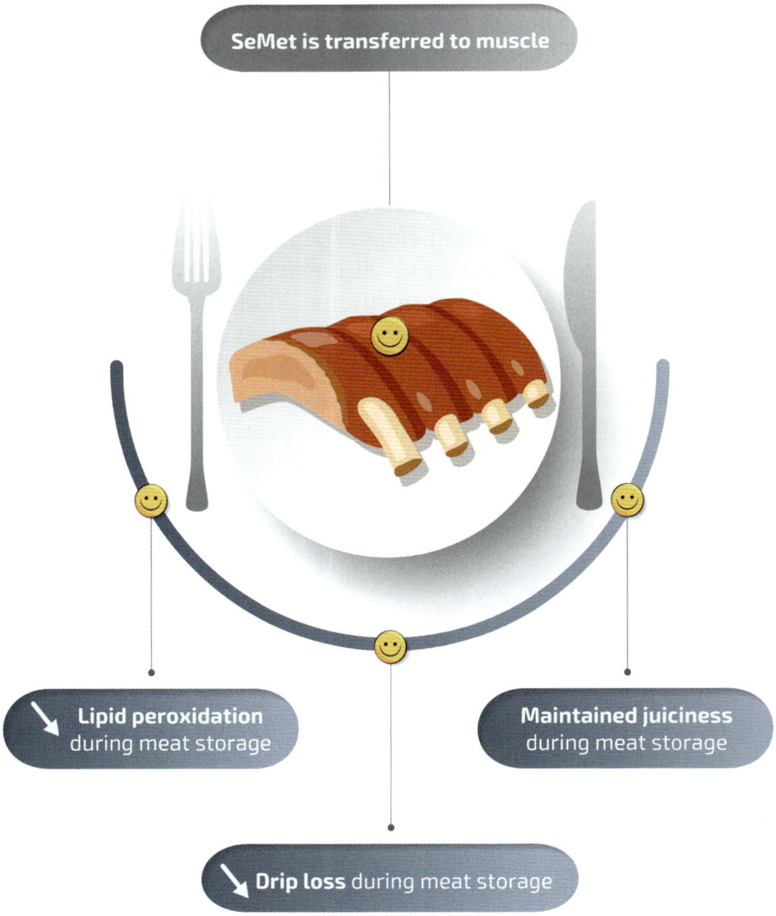

Figure 8.1. Selenium for pigs – meat production and quality.

Drip loss and, therefore, the amount of exudate in the retailing package can also impact the buying decision of consumers. If colour stability is certainly the first aspect for beef meat, drip loss is a bigger problem in pork and poultry meat (Pearce *et al.*, 2011). The assessment of this quality criteria is easy to measure as the difference of the weight of a piece of meat after a given period. This parameter might also have other applications since it is well correlated with water holding capacity (WHC) which is an indicator of the cooking losses, particularly important in case of further processing.

Among the organoleptic aspects, tenderness is an important factor for the determination of meat quality. This factor is linked to the muscle type but also to the maturation time allowing tenderisation processes. This quality trait is more relevant for beef and pork meat whereas it is not an issue for poultry meat or fish (Ouali *et al.*, 2006). Tenderness is under the influence of the collagen amount in the muscle that will determine the optimal cooking process: grilled when collagen content is low or boiled when collagen is high. However, tenderness is also under the dependency of the maturation process that involves intramuscular proteolytic systems (e.g. cathepsin, calpain, proteasome) which degrade myofibrillar structures during refrigerated storage and allow to decrease toughness of the meat.

8.2 Pork quality aspects and their relationship with redox status

Meat is a protein and lipid reach food susceptible to oxidative injuries during storage that would alter visual aspect by colour change (Guillen and Guzman, 1998) through myoglobin oxidation; odour and flavour by generating volatiles compounds from lipid oxidation (Ahn, 2002; Ahn and Kim, 1998; Ahn and Lee, 2002; Byrne *et al.*, 2002; Campo *et al.*, 2006; Dietze *et al.*, 2007). Among the alteration of the muscle component, oxidation of protein would also affect meat characteristics. Meat is consumed for the quality of its protein composition rich in essential and balanced amino acids and its high digestibility. However, the oxidation process by changing and altering the amino acids will affect the digestibility and nutritional property of the meat (Papuc *et al.*, 2017). Interestingly, within these four quality aspects of the meat, at least three of them depend on ultimate pH of the meat and are highly interdependent.

After slaughter and exsanguination, peripheral tissues like muscles are deprived of nutrients and oxygen that trigger this process. After death, muscle metabolism will be turned into anaerobic condition that leads to lactic acid formation as end product of the glycolysis (England *et al.*, 2013). This lactic acid accumulation, due to the lack of evacuation of the products by the blood flow, results in pH drop of the muscle. This pH drop is under the dependency of the glycogen reserves of the muscle, which is affected by many factors like feed withdrawal, muscular activity before slaughter, genetic traits. Under 'standard' conditions (if 'standard' can be applied to the diversity of practices), the pH drop of muscle starts a few minutes after the death of the animal until an ending point which is generally assessed 24 hours after slaughter and named: Ultimate pH (pHu).

Specific genetic alteration in pigs referred as Halothane gene and RN- gene have been largely documented for the impact on pH drop and the impacts on pork quality. The halothane gene leads to pale, soft and exudative meats generally referred as 'PSE meats'. A PSE meat is caused by an extensive protein denaturation due to the combination of a low pH and simultaneously high temperature early post-mortem (Briskey, 1964; Offer and Knight, 1988). This protein alteration results in high lightness values (L*) and poor ability of water holding capacity resulting in high drip loss. Moreover, the denaturation of the myofibrillar proteins results in a loss of structure of the post-mortem muscle which is considered as soft. This type of quality alteration has been widely studied in the 1990's and finally solved by eliminating the responsible genotype from the selection line to eliminate or at least minimise these issues (Rosenvold and Andersen, 2003). The RN- genotype has been identified as having low technological yield and the mutation took the name of Napole yield, based on the French name 'Rendement Napole' for RN. The effects of this RN- gene have been associated with high muscle glycogen stores and an extended pH decline post-mortem (Estrade *et al.,* 1993; Monin and Sellier, 1985) resulting in acidic meat (Naveau, 1986) that can be assimilated to PSE-like meat. This genotype has now been removed for the genetic selection schemes by the breeding companies. Moreover, these two genetic defaults were additive for colour and water holding capacity (Hamilton *et al.,* 2000; Le Roy *et al.,* 1999). Texture, colour and drip loss (or exudation) are linked to ultimate pH value. Indeed, low pH is associated with high Lightness (L* values), high exudation levels and softness. These kinds of meat are also described for low transformation rates during cooking process associated with low water holding capacity (Berri *et al.,* 2015); whereas, high pH values are associated with firm texture, low L* value and low exudation. Moreover, besides the organoleptic problem of a high pH, higher microbial proliferation and poor conservation are also observed. Another consequence of the pH drops, and the energy depletion of the muscle tissue is the muscle shortening resulting in myofibrillar contraction. During this process, some extracellular water will also be expulsed from the tissue, thus constituting the first weight loss of the muscle. This muscle shortening, also called rigor mortis, leads to the tougher status of the muscle that will be attenuated during the refrigerated storage and the maturation process. During this process of maturation, the endogenous proteolytic systems, including cathepsins, calpains, proteasomes and caspases, will breakdown myofibrillar proteins that will decrease hardness (Ouali *et al.,* 2006). This process duration is largely different between species and lasts for few hours in poultry until a few weeks for beef. Moreover, depending on the initial lipid content of the muscle, the enzymatic activity and the level of oxidation flavour develops during this phase.

The complexity and interactive aspect on meat quality traits are presented in Figure 8.2 (Geay *et al.,* 2002).

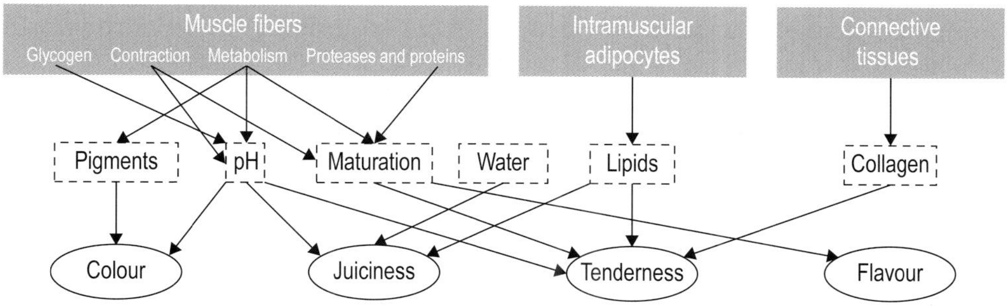

Figure 8.2. Factors affecting meat quality (adapted from Geay *et al.*, 2002).

8.2.1 Oxidative process in meat

Redox reactions are well known as a modulator of metabolic pathway as the oxidised or reduced status of a given molecule can be compared to an electric switch with two positions: 'On/Off' (Toledano *et al.*, 2007). It is also well known that cells are equipped with different antioxidant systems that can directly scavenge free radicals (e.g. super oxide dismutase (SOD) that detoxify O_2^{-*}-radicals), vitamins that can detoxify oxidised product (e.g. Vitamin E stops the elongation chain of lipid oxidation). However, all these scavenging systems need to be supplied directly (vitamins E and A) or need specific nutrients/minerals to be synthetised by the cell (Cu and Zn for SOD, Se for selenoproteins, sulphur amino acids for glutathione and taurine synthesis). This is currently admitted for living animals but when discussing redox processes in meat the concept appears differently since the building blocks needed to synthesise these components are not anymore supplied and because new synthesis of antioxidant enzyme is not possible either after death. Oxidative processes are reactions in which loss of electrons or hydrogen atoms or gain of oxygen occurs. In contrast, reduction processes are reactions in which gain of electrons or hydrogen atoms or loss of oxygen occurs. Because an electron donor does not exist without an electron acceptor, the oxidising agent and the reducing agent form a redox couple. This definition that's looks chemical is modulated when applicated to living organisms by changing the term 'oxidant' to Pro-Oxidant and the term Reductant to Antioxidant (Papuc *et al.*, 2017). Free radicals or Reactive species are the main pro-oxidative compounds in biological systems.

Reactive species contain unpaired electron that can be linked to oxygen atoms constituting a reactive oxygen species (ROS) or to nitrogen atom constituting reactive nitrogen species (RNS). Whatever the centred atom, these molecules are highly reactive and react quickly with neighbouring molecules or cellular constituent by extracting electron and hence oxidised the neighbour molecule. In meat (or postmortem muscle), the generation of Reactive Species, mainly ROS, is due to different phenomena such as mitochondrial swelling and breakage and change in oxygen consumption (Tang *et al.*, 2005). Also, Photosensitisers like porphyrins, riboflavin, and myoglobin which are present in the meat, can convert triplet oxygen into singlet oxygen through the energy supplied as light radiation (Afonso *et al.*, 1999; Min and Boff, 2002).

This singlet oxygen can react quickly with PUFA by forming hydroperoxide with conjugated or unconjugated double bonds. Min and Boff (2002) suggested that this reaction is accelerated by the degree of unsaturation of the lipid but not by the double bond position. This reactive species can also impact proteins and more preferably the amino acid with high electron density like cysteine, methionine, tryptophan, tyrosine and histidine (Davies, 2012). The superoxide anion (O_2^{-*}) is considered as the most reactive ROS. The genesis of this ROS can be due to conversion of xanthine dehydrogenase into xanthine oxidase or due to autoxidation of heme protein as main source in meat (Thomas, 1995; Baron and Andersen, 2002). The most important reaction of this superoxide anion is the dismutation into hydrogen peroxide. This reactive oxygen species which is not a free radical per se can have different fate in this reaction chain leading to new radical through different reaction such as Fenton reaction with transition metal or Haber-Weiss reaction with another superoxide anion. In meat the transition metal, realised from haemoprotein or metalloproteins, can be of high abundance and hence the formation of hydroxyl radical (OH*) through the Fenton reaction is commonly admitted as main source of OH*. Reaction chain from Superoxide to hydroxyl radical is summarised in the Figure 8.3.

The extended list of reactive species based on the unpaired electron can be continued by including RNS or even the lipid peroxide, ferrylmyoglobin, that are reactive forms of the cellular constituent (Papuc *et al.*, 2017 for review). However, what is important to consider is that all these reactive compounds have a high reactivity that would impair the muscle constituent and hence affect meat quality during storage. All these initiating elements would appear quickly after the death of the animal and the effect obtained in term of the quality degradation after these initial stages must be understood as a chain reaction leading to ultimate oxidative products mainly from lipid and protein degradation in meat.

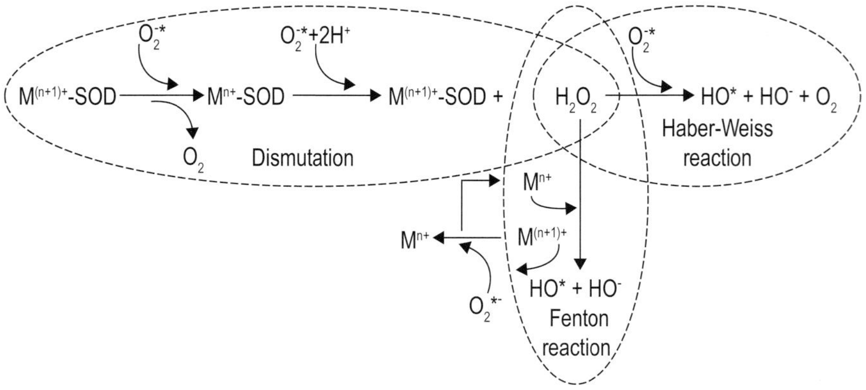

Figure 8.3. Superoxide anion precursor for hydrogen peroxide, hydroxyl radical and reduced transition metal (adapted from Papuc *et al.*, 2017).

Lipids and mainly membrane phospholipids which are more unsaturated are often considered as the first target of these prooxidant compounds. Indeed, the high tendency of unsaturated fatty acids, especially polyunsaturated fatty acids, to rapidly oxidise is important in further biochemical changes that have undesirable effects, such as rancidity and colour deterioration (Wood *et al.*, 2003). The unsaturated fatty acyl group of phospholipids from muscle fibre membranes (sarcolemma, mitochondria, and sarcoplasmic reticulum) is the first to be oxidised (Morrissey *et al.*, 1998; Shleikin and Medvedev, 2014). Figure 8.4 details the main events for lipid oxidation that can be divided into 3 main steps. The initiation phase corresponds to the initial event of an hydrogen atom extraction from a methylene group by a radical compound that will create a lipid radical (Alkyl radical: L*) where an unpaired electron that would result in relocation of double bounds links along the lipid chain. Under this form the lipid radical will be prone to interact with oxygen molecule (02) to form an hydroperoxyl radical (LOO*). This hydroperoxyl radical can react back with a neighbour PUFA by extracting an hydrogen and hence obtaining a new lipid radical (LOO* + LH → LOOH + L*). This reaction can be then extended as a snowball effect that would propagate the oxidation to the other PUFA in the membrane explaining the term of elongation of this second step. This reaction can occur 100 times before two L* combine and terminate the process (Estevez, 2015). The Hydroperoxyl Radical find the stability by this hydrogen extraction and enter in the termination steps by forming a lipid hydroperoxide (LOOH) that can be assimilated at the hydrogen peroxide describe above. Peroxides are commonly formed as primary lipid oxidised products and can be broken to form lower molecular weight volatile and non-volatile compounds which are considered as secondary oxidation products such as carbonyls, alcohols, hydrocarbons, and furans (Min and Ahn, 2005). Among these, aldehydes are one of most abundant products found in meat: hexanal, malondialdehyde (MDA) (Figure 8.3), and 4-hydroxy-2-trans-nonenal (Bekhit *et al.*, 2013, Ladikos and Lougovois, 1990; Min and Ahn, 2005; Papuc *et al.*, 2017). These oxidation products will obviously lead to meat quality alteration by increasing notably the off-flavour and off-odours at meat level. Also, due to membrane degradation during the elongation step and the breaking of the phospholipid structure this lipid oxidation is thought to increase drip loss of the meat by leaking the cellular membrane leading to intracellular liquid dripping. Moreover, many authors have highlighted the toxic effect of lipid oxidation products ingestion by human that would be an aggravating element in the atherosclerosis risk (Khan-Merchant *et al.*, 2002; Steinberg, 1997), leading to carcinogenic compounds (De Bont and Van Larebeke, 2004; Marquez-Ruiz *et al.*, 2008; Neiderhofer *et al.*, 2003; Pietras *et al.*, 2012).

Beyond the lipid oxidation, proteins which are largely represented in meat, can also be affected by the oxidative process of meat. It is admitted that the protein oxidation is occurring in a different kinetic way than lipids as the sensitivity and the end products and/or effect on protein backbone depending on the considered amino acids is different. For instance, sulphur containing amino acids are considered as the most susceptible to oxidation due to the high reactivity of sulphur.

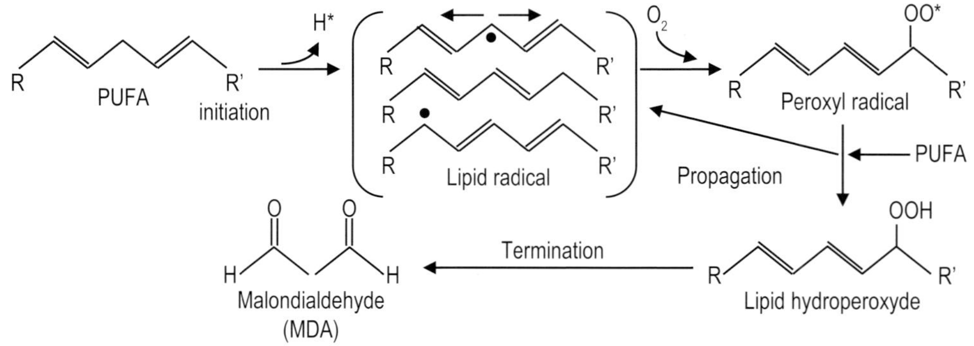

Figure 8.4. Different phases of the polyunsaturated fatty acid (PUFA) oxidative processes leading to propagation phase and then terminal products.

Cysteine residue that present a free thiol group (SH) is considered as the most susceptible to oxidation due to the high reactivity of the SH group. Free radical, extract the labile proton of the SH group leading to a thiyl group S* that would react quickly with other thiol to form disulphide bonds (S-S) leading to form link into the back bone of same protein, changing the conformation of this protein, or with a neighbouring protein leading to protein di-sulphur bonds (Turell *et al.*, 2008). This reaction of di-sulphur bonds forming intra or inter protein is accompanied with a loss of protein solubility and increase protein aggregation (Rowe *et al.*, 2004). Importantly, thyil group can react as well with reduced glutathione. Also, μ-calpain and m-Calpain, proteolytic enzyme systems involved in the muscle fibres degradation leading to improved tenderness during meat aging, contain His and Cys in the catalytic site. Therefore, the increased protein oxidation decreases the catalytic activity of these enzymes which limits degradation of cytoskeletal proteins that increase shrinking of the overall muscle cell, which affect the water holding capacity (Huff-Lonergan and Lonergan, 2005).

Similarly, methionine can be easily oxidised by reactive species into methionine sulfoxide (Met-SO) and methionine sulfone (Met-SOO). *In vivo*, these sulphur oxidised products can be reduced back by enzymatic systems like Thioredoxine reductase or Methionine sulfoxide reductase involving both NADPH (Levine *et al.*, 2000). Methionine oxidation and back reduction has been considered *in vivo* a way to protect protein from oxidation by playing the role of shell for the rest of the protein back bone (Levine *et al.*, 2000). It is interesting to notice methionine sulfoxide reductase and thioredoxin reductase, which are existing under different isoforms, are for some of them seleno-enzymes showing the close relationship between selenium and sulphur in the redox systems.

Tyrosine is also highly susceptible to oxidation due to the hydroxyl on the benzene group. The oxidation of tyrosine residue would lead to form di-tyrosine bridges intra or inter protein affecting the conformation of these proteins (Zhang *et al.*, 2013). Tryptophan residues can be oxidised into N-Formylkynurenine and kynurenine which are potentially toxic compounds (Friedman and Cuq, 1988). During chilled storage of meat and meat products, lysine residues form oxidation products, such as α-aminoadipic semialdehyde and α aminoadipic acid and Schiff bases (Utrera *et al.*, 2012). ROS can also abstract hydrogen atom from the carbon of amino acid residues resulting in the formation of carbon-centred radical. In the presence of oxygen these carbon-centred radicals can be turned into alkylperoxyl radical producing protein hydroperoxide. The decomposition of this hydroperoxide by transition metal will produce alkoxy radicals that can be subject to protein fragmentation (Davies, 2016). Protein backbone fragmentation generates carbonyl and imine derivatives.

It is also important to note that lipid and protein oxidation are obviously not separated events and that they can interact through because their high reactivity and will lead as well to carbonyl formation by condensation between products (Papuc *et al.*, 2017 for review).

8.2.2 Dietary selenium and meat quality

Meat consumption and its related quality aspects have evolved during the past decades. Indeed, consumers are more and more looking for good quality products instead of the quantity and would privilege products with some health benefits, also for meat products. In this view, many studies using selenium supplementations have been conducted with different objectives. Selenium per se is an indispensable micronutrient that can be deficient in some geographical area inducing health issues in the population. According to Jones *et al.* (2017) the soil selenium availability might be impacted by climate change leading to selenium deficiency especially in agricultural area by the end of the 21st century. Dietary selenium supply in farm animals is common practice and the commercial availability of organic forms of selenium have improved significantly the selenium deposition in animal products. In pig muscles many authors have reported the beneficial effect of using organic selenium forms instead of inorganic ones on the muscle selenium deposition (Bobcek *et al.*, 2004; Jlali *et al.*, 2014; Mahan and Parrett, 1996; Mahan *et al.*, 1999). Indeed, dietary Se concentration of 0.3 mg/kg led to a loin tissue Se concentration of 0.332 mg/kg with seleno-yeast (SY) as a source of Se whereas only 0.124 mg/kg was reached with sodium selenite (SS) as a source of Se (Mahan *et al.*, 1999). Both SeMet and SS treatments (0.3 mg/kg) increased Se concentration in serum, muscle and liver tissue. But, Se concentration was substantially higher in muscle and liver in SeMet treated group although pigs were only fed supplemented diets for 40 days (Zhan *et al.*, 2007). The literature is plenty of examples showing the higher deposition of Selenium in animal tissues by using organic Se for the selenomethionine deposition in proteins. However, and as the organic selenium source market has developed, comparative studies highlighting that a ranking between organic selenium sources in term of tissue selenium deposition can be established. For instance, muscle Se concentration was higher in SeY group than in Se-enriched algae *Chlorella* spp. group (Svoboda *et al.*, 2009). Today some purified organic selenium sources are commercially available and help to increase one step further the selenium supplementation in tissue and then its availability in meat. Jlali *et al.* (2014) compared the selenium deposition in various tissues of pigs from 25 to 50 kg body weight by using 0.1 and 0.3 mg/kg of different selenium sources from inorganic (Sodium selenite) and two organic sources (seleno-yeast and hydroxy-selenomethionine (HMSeBA)). Seleno yeasts are referred to contain more than 95% organic selenium under different forms that can varied between producers, strain and process. The results obtain in this study showed clearly that the pure source (HO-SeMet) allow higher selenium deposition in the different tissues assessed (plasma +69.66%, liver +40.52%, and muscle +62.22%) than SY (P<0.01).

With the development of omega-3 enriched meat products, combined antioxidant strategy is even more important. To improve the N-3/N-6 in the human foods, high dietary level of rapeseed oil in animal feed is often recommended to increase the n-3 muscle content, however it challenges pork meat stability due to oxidative processes (Lauridsen *et al.,* 1999; Morissey *et al.,* 1998) and the impact on human health. To counteract this issue of lower oxidative stability by increased n-3 fatty acids and more generally long chain PUFA muscle content, antioxidant strategies using Vit E and/or selenium have been developed (Morel *et al.,* 2013; Jiang *et al.,* 2017). A large range of studies have demonstrated the efficacy of dietary supplementation in organic selenium and vitamin E on enhancing the antioxidant capacities of the animals, reduction of drip loss, maintenance of meat colour over storage time and better shear-force (Chen *et al.,* 2019; Mahan *et al.,* 1999; Mateo *et al.,* 2007; Silva *et al.,* 2019). However, some authors did not report any positive effect on these parameters (Kawecka *et al.,* 2013; Lisiak *et al.,* 2014; Wolter *et al.,* 1999). Such discrepancies between the results reported with or without positive effects on drip loss, lipid oxidation (MDA or TBARS), etc. might be linked to the different environmental oxidative conditions faced by the animals. Indeed, the breeding conditions as well as the pre-slaughter conditions are not often well described, as well, all potential pro-oxidant situations (heat, dietary or sanitary stresses for instance) which could have further challenged the antioxidant potential defences of the animals. The dietary levels of Se supplementation might also be an issue. As the inorganic Se, sodium selenite, mostly used might have toxic effect, there is often a small range of dietary Se supplementation tested. In case of organic Se, high dose of supplementation largely beyond the usual levels can be tested and are showing technical benefits.

In a recent study, Calvo *et al.* (2016) compared dietary supplementation with selenium, organic and inorganic as well as vitamin E for 26 days before slaughter. While Se supplementation did not modify MDA production, GSSH concentration or colour parameters of meat, it significantly improves WHC (reduction of drip loss) and post-mortem muscle proteolytic activity. Vitamin E (100 mg/kg) was the most efficient treatment for stabilising colour, reducing MDA and GSSH production. The vitamin E supplementation was not as effective as organic Se administration to reduce drip loss, which could confirm the relation between muscle proteolysis and WHC.

Calvo *et al.* (2017) further showed that dietary organic selenium improved the lipid stability of meat by reducing TBARS concentration with time when compared with inorganic selenium. Dietary organic selenium also resulted in higher a*, b* values, and colour intensity in pig meat. Neither electric conductivity nor drip loss were affected by the selenium source. The organic Se supplementation group had lower TBARS in muscle samples after 7 days of refrigerated storage and higher a* values than the inorganic Se group. However, the effect of dietary selenium source on some meat quality characteristics was affected by muscle pH. Hence as the muscle pH increases, the drip loss decreases but this effect is more marked with the dietary organic Se enrichment. Muscle pH seems to modulate the action of selenium in pork, especially some meat characteristics such as drip loss. Se deficiency reduced the pH values of meat and increased drip loss, which was consistent with the negative correlation between drip loss and pH (Huff-Lonergan *et al.*, 2002). In comparison with Se deficiency, adequate Se supplementation (0.3 mg/kg Se as organic SeMet) only slightly decreased (Mahan *et al.*, 1999) or significantly decreased drip loss (Zhan *et al.*, 2007).

Among the 24-25 selenoproteins identified in mammals and birds, most exhibit redox regulation actions (Rederstorff *et al.*, 2006). Among those selenoproteins, the glutathione peroxidase (GPX) enzyme family has well been described with cellular GPX (GPX1), gastrointestinal epithelium-specific GPX (GPX2), plasma GPX (GPX3), phospholipid hydroperoxide GPX (GPX4) and sperm-bound GPX (GPX5) (Rederstorff *et al.*, 2006). GPX1 and GPX4 have been identified in pig muscle (Lei *et al.*, 1997). Although GPXs (GPX1 and/or GPX4) have been postulated to enhance meat oxidative stability by scavenging peroxides and decreasing lipid peroxidation, the molecular mechanism underlying this phenomenon has not been elucidated. Se deficiency is clearly impairing the level of GPX1 (Li *et al.*, 2011) and increasing the dietary Se supplementation enhanced the GPX1 in muscle, liver and other tissues. Drip loss is not only influenced by oxidative stability while TBARS is a more sensitive indicator of oxidative stability. Li *et al.* (2011) showed a high correlation coefficient for TBARS content and GPX1 activity (-0.82), which supported the view that the improvement of meat oxidative stability by Se was due to the protective effect of GPX1 against oxidative damage and lipid oxidation. However, this relationship was not observed in the liver as already reported by Juniper *et al.* (2009). These studies also suggest that, in addition to GPX1, other potential Se-dependent antioxidants exist in muscle. Indeed, another selenoprotein (SelW) is highly present in muscle and its expression can be further increased with Se supplementation. In contrast to other selenoproteins whose expression reaches maximum level at adequate Se level, SelW expression in muscle can further increase due to excessive Se intake (Li *et al.*, 2011). In addition, its antioxidant activity has further been reported (DaeWon *et al.*, 2004; Jeong, *et al.*, 2002). Using high level of organic Se supplementation (Se-yeast), Li *et al.* (2011) showed that among the 12 selenoproteins genes investigated in the muscle, only the expression of Sepw1 was influenced by dietary Se level. The Se deficiency induced decrease of Sepw1 mRNA while dietary supplementation significantly increases its expression. A few studies have also shown that the regulation of SelW expression is different from that of the other selenoproteins (Whanger, 2000) such as a clear response to high dietary Se supplementation (4 mg/kg) (Vendeland *et al.*, 1995). SelW can protect cells from oxidative stress, and it has been suggested to be a glutathione-dependent antioxidant (DaeWon *et al.*, 2004; Jeong *et al.*, 2002). The correlation coefficients between drip loss and GPX1 activity, Sepw1mRNA level, and TBARS were: -0.65, -0.90 and 0.59, respectively (Li *et al.*, 2011). The correlation and regression analyses of TBARS, GPX1 and Sepw1 suggest that the apparent effect of TBARS and GPX1 on drip loss depends on the expression of Sepw1 gene, and second, lipid peroxidation is actually not affected by GPX1 activity. The role of GPX1 in improving the oxidative stability of meat is thus not clear, but SelW appeared the key Se-dependent antioxidant that improved the oxidative stability. In conclusion, the antioxidant status of pig tissue was regulated by the dietary Se level, and the inclusion of organic Se in the pig diet increased the expression of the Sepw1 gene and improved meat quality through enhanced water-holding capacity (Li *et al.*, 2011).

8.2.3 Sulphur amino acids and meat quality

Methionine (Met) is an essential and first limiting amino acid in classical broiler diets and second one, after lysine, in pig diets (Baker, 2009). In animal nutrition, Met and Cysteine (Cys) are building blocks for tissue proteins and essential substrates for the synthesis of other substances of physiological importance. For example, Met participates in methyl group metabolism and synthesis of other sulphur amino acids (SAA), notably Cys. Cys is required for the synthesis of glutathione (GSH) and taurine, which are essential compounds for host defence against oxidative stress.

Although protein metabolism is regulated by numerous factors such as physiological status, genetic, environmental factors and hormones, amino acids are recognised to be essential in such regulation. Several authors have examined growth performance and muscle development as responses to Met doses. A deficiency of 16% in SAA applied for 17 weeks, from 12 to 90 kg BW, significantly reduced the final BW (-13%), average daily gain (-15%) and gain-to-feed ratio (-12%) for pigs (Conde-Aguilera *et al.*, 2014). A severe deficiency (28% below recommended SAA levels) applied on a shorter period for 10 days, from 42 to 52 days old, also reduced weaned piglets growth performance and decreased the fractional rate of protein synthesis in the Longissimus (Conde-Aguilera *et al.*, 2016). However, the response to Met supply could also vary between muscles. The deficiency in SAA in pigs did not affect the relative weights or protein synthesis rate in the Rhomboideus and Semitendinous whereas the longissimus was significantly affected (Conde-Aguilera *et al.*, 2016). This specific effect could be related to the nature of the muscle, fast glycolytic vs slow or fast oxidative. Fast-twitch glycolytic fibres which represent 80% of the total fibres of the Longissimus are more sensitive to a period of starvation, restricted feeding than slow-twitch oxidative fibres (Kelly and Goldspink, 1982; Kelly *et al.*, 1986).

Oxidative stress refers to the 'disturbance of the balance between pro-oxidative factors and antioxidants in favour of the former' (Sies, 1997, 2015). Muscle tissue from farm animals is highly susceptible to oxidative reactions due to the generation of ROS in mitochondria (Lykkesfeldt and Svendsen, 2007). The occurrence of oxidative stress is linked to assorted metabolic syndromes and myopathies with consequences on carcass conformation, composition and meat quality (Bekhit *et al.*, 2013; Estévez, 2015). The physiological processes aiming to maintain redox homeostasis in the living muscle are deregulated upon slaughter, leading to a pro-oxidative environment in which lipids and proteins are main targets for ROS (Bekhit *et al.*, 2013; Estévez *et al.*, 2020).

SAA requirements are reached in practical diets through the supplementation of either DL-Met, DL-2-hydroxy-4-methylthio-butanoic acid (OH-Met) or L-Met. The effects of SAA on redox activity in living animals have been extensively studied in poultry (Metayer *et al.*, 2008; Wang *et al.*, 2009; Wen *et al.*, 2017) contrarily to pigs where only a few studies exist (Conde-Aguilera *et al.*, 2016; Li *et al.*, 2017; Lebret *et al.*, 2018). Conde-Aguilera *et al.* (2016) demonstrated that SAA deficiency affects muscle properties in a muscle-dependent manner increasing the oxidative capacity of rhomboideus and reducing growth and glycolytic metabolism of longissimus. According to Castellano *et al.* (2015), a dietary Met deficiency (40%) during 10 days in post weaned piglets increased the levels of lipogenesis and lipolytic indicators in porcine adipose tissues. In addition, the antioxidant defences of fat tissues were drastically affected by the Met deficiency leading to a decreased GSH/GSSG ratio and a higher compensatory antioxidant enzyme activity (SOD, CAT and GPX) as compared to tissues from animals provided with an adequate Met diet.

Redox processes are different in meat than in the muscle. Because the building blocks needed to synthesise antioxidant components are no longer supplied and because the synthesis of antioxidant enzymes is impossible after the animal's death, strategies must be applied to reduce oxidative processes in muscle. According to Estevez *et al.* (2020), these actions may be destined to: (1) strengthen the endogenous antioxidant defences via dietary means and/or (2) apply technological strategies based on formulation with antioxidants and avoid oxygen and light exposure. Fortunately, the antioxidant protection provided by protein bound or free SAA, is not restricted to the living cell. Upon slaughter, such antioxidant mechanisms remain active and may contribute to protect muscle proteins and other muscle cell components against oxidative damage during subsequent storage and processing. Supplementation of SAA over nutritional requirements has been found to protect muscle lipids and proteins against oxidation, with subsequent benefits in terms of water-holding capacity, colour stability and sensory properties. Lebret *et al.* (2018) conducted a trial in growing-finishing pigs in which a short-term supply of high doses of Met as OH-Met (3 times: Met+ and 5 times the requirement: Met++ treatment) over 14 days before slaughtering improved the ultimate pH (pHu) and meat antioxidant potential. Moreover, growth performance or muscle protein content were not affected, but lipid content was reduced with the increase of Met. In addition, the highest dose of Met increased the level of GSH in the muscle, thus enhancing its capacity to cope with ROS (Figure 8.5). pHu was higher with the highest dose of Met, which supports the trend of reduction of drip losses and defrosting losses obtained with this dose of Met. Meat colour was less pronounced and evolved less during storage.

The benefits of Met supplementation on pork quality have also been shown on a longer period (from weaning to slaughter). In low-weight birth piglets, Li *et al.* (2017) demonstrated that the concentration of malondialdehyde (MDA) in pork was lower when the L-Met level was increased by 30%. The higher concentration of GSH in muscles from supplemented pigs was hypothesised as one major cause for their enhanced oxidative stability as compared to muscles from control pigs. Similarly, Yuan *et al.* (2017) showed a decrease of the drip loss (Figure 8.6) and the shear-force of longissimus muscle of pigs following an increased SAA consumption from weaning to slaughter. The increased glycolytic potential and decreased protein deposition might explain the decreased shear force, while the decreased drip loss was associated with down-regulation of genes (like fast glycolytic IIx) regulating fibre types. Moreover, fresh muscle from barrows fed OH-Met diets showed a faster pH drop in the 45 minutes post-mortem and a higher lightness value compared with the control and DL-Methionine treatments (Yuan *et al.*, 2017).

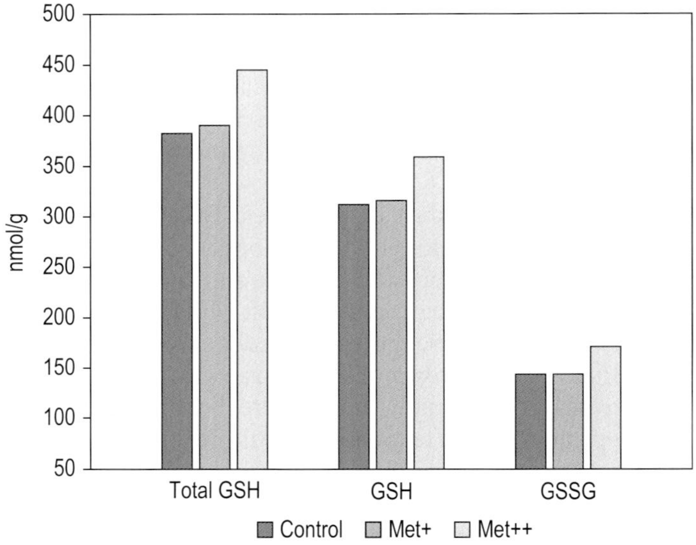

Figure 8.5. Non enzymatic antioxidants measured in the longissimus lumborum muscle of pigs fed with adequate methionine (Met) levels (Control) and Met levels three times (Met+) or five times (Met++) above growth requirements (adapted from Lebret *et al.* (2018). GSH = glutathione; GSSG = glutathione disulfide.

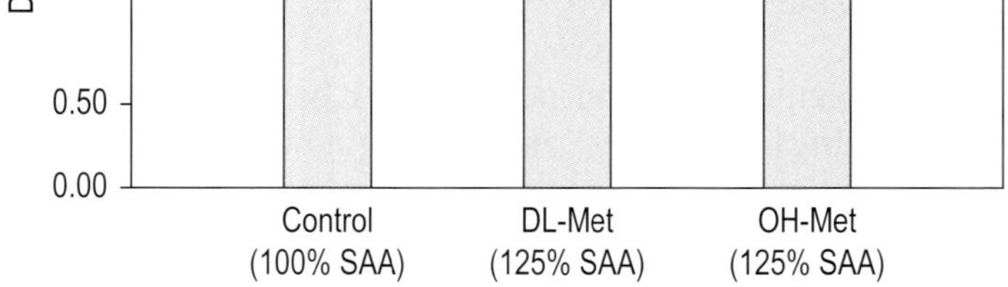

Figure 8.6. Drip loss of pigs fed with adequate sulphur amino acids levels (Control) and 25% above growth requirements, using either DL-Met or OH-Met (adapted from Yuan *et al.*, 2017). SAA = serum amyloid A.

8.2.4 Selenium and pork meat defaults

In pork, the muscle content has continued to progress over the last decade thanks to genetics, passing from 44-49 to 58-62% (Listrat *et al.*, 2015). However, as in poultry, this selection on growth and muscle development was accompanied by meat defaults that appeared already in the 1990's and changed the overall meat quality. Economic importance of those meat quality issues both for consumers and meat processors is more and more actively recognised worldwide nowadays (Semenova *et al.*, 2019).

The problem of a decrease in meat quality and prevalence of specific meat defaults associated with myopathy is under scientific investigation worldwide in pigs and poultry. Studies show that fast-growing animals, where lean mass has largely increased and fatness decreased, demonstrate high frequency of detection of myopathies, including those that are caused by stress, which are of great importance for meat quality. The study of myopathic conditions demonstrates that animals that can gain muscle mass quickly are prone to the emergence and development of the disease (Petracci and Cavani, 2012). It has largely been observed in broiler production with apparition of wooden breast, spaghetti meat, white stripping and other defaults and in pork, with PSE-like meat defaults and de-structured hams. These defaults are accompanied by microstructural changes in muscle tissue which inevitably leads to meat quality deterioration and loss of meat functional-technological and consumer properties.

Almost 30 years ago, meat defaults appear in the pork industry, impacting all levels of pork meat chain, from slaughter, cut-up to further processing and cure meats particularly with the de-structured hams. 10 to 20% of hams indeed exhibit indices of de-structuration (Minvielle *et al.*, 2005). This is even more obvious with the automatic high-speed cutting systems for ham slices which is becoming more and more important (Minvielle *et al.*, 2005). Heavy losses are indeed reported during ham slicing, due to the pommade aspect of the cut. Identifying the key factors influencing those aspects and the possibility to predict or anticipate their occurrence appeared a raising concern over the last years for the whole pork meat processing and cured meat industry (Vautier *et al.*, 2004). The occurrence of de-structuration indeed increases with greater lean mass and reduced fat contents (Minvielle *et al.*, 2005), is also accompanied by an alteration of colour, very pale and greyish hams with lower pHu (Vautier *et al.*, 2004).

Highly de-structured muscles lost their organised structure. They had higher glycolytic potential compared to unaffected muscles. Some histological similarities appeared between de-structured and PSE meats. Emergence of 'de-structured hams' led to significant losses in production of cooked hams (grey colour, the unbound and fragmented structure of the finished product, the presence of holes, bad consistency and slicing) (Minvielle *et al.*, 2001; Minvielle *et al.*, 2005). As for wooden breast and white stripping in broilers, a classification of de-structured hams has been proposed with 4 grades: (1) ham muscles with no de-structuration; (2) the defect is seen, superficial, located only in the semimembranosus muscle; (3) the semimembranosus muscle is strongly altered and the defect partly affects adjacent muscles; (4) ham muscles are fully de-structured (Minvielle *et al.*, 2001). Up to 17% of pork carcasses were affected in a field survey analysing more than 6,000 pork carcasses in France. Some features appeared related to the meat defect such as high proportion of type 2B fibres (glycolytics), meat pH below 5.6, less fat deposition, etc. but causes leading to that defects are not yet fully understood.

The conditions prior to slaughter also seem determinant for the apparition of those defaults. A field study testing different slaughter plants, concluded that there was a significant effect of the ambient temperature on the day of slaughter on the occurrence of the default. The proportion of hams affected increase by 12.8 points when temperature increase by more than 10 °C, the proportion of de-structured hams going from 0.5 to 7.5%. Such a trend is also in agreement with the effect of temperature on pHu and the relationship between pHu and de-structuration suggesting the importance of a better monitoring of the pre-slaughter conditions on muscle energy reserve and evolution of pHu (Vautier *et al.*, 2004). Another important factor is the duration of fast which will reduce the glycogen reserves and thus maintain higher pHu (Minvielle *et al.*, 2005). The chilling process is also an important determinant of the meat quality. A faster chilling rate could reduce drip loss by 21% and increase shear force after 2 days of maturation by 21%. Conversely, a slow chilling rate increased by more than 3 times the frequency of de-structured hams as well as reducing the occurrence of PSE-like zones while maintaining other meat quality parameters (Vautier *et al.*, 2008). Ultimate pH seems indeed to be a reliable predictor of these defects.

Pathogenesis of different myopathies, including nutritional, is characterised by a specific sequence. As muscle necrosis leads to damage and finally destruction of sarcolemma of fibres, normal ionic gradients cannot be maintained. Calcium ions, which are usually in lower concentration in cytoplasm than in the extracellular fluid, diffuse into a cell. Entry of calcium ions leads to activation of calcium-dependent proteases, such as calpains, and the following destruction of myofibrils and other structural elements of a muscle fibre. An increase in the concentration of calcium ions in sarcoplasm determines formation of areas of myofibril supercontraction in the structure of muscle fibres. A change in homeostasis results in destruction of cytoskeletal proteins, damage of the structure of cell organelles and their necrosis in addition to the development of contractures of myofibrillar structures (Semenova *et al.*, 2019).

However, it seems more and more obvious that selection for growth rate, muscle development and feed efficiency has challenged the energy metabolism of the cells: from reduction of blood supply at the muscle level, increase of fibre diameter, impaired mitochondria functioning, oxidative stress at the muscle cell levels. Moreover, all parameters increasing the stress exposure of the animals will further strengthen oxidative challenges.

Stress in farm animals is also an important cause of meat defaults such as PSE (pale, soft, exudative) and DFD (dark, firm and dry) meats as well as more recent defaults RSE (red, soft and exudative) and RFN (red, firm and non-exudative) (Semenova *et al.*, 2019). In pork production, the proportion of meat with reduced quality resulting from stress before slaughter ranges from 10 to 30%, and in several countries up to 60%. The porcine stress syndrome (PSS) is manifested as paralysis of the heart and necrosis of longissimus dorsi. In the USA, up to 8% of pork carcasses are rejected due to myopathy caused by PSS. One reason has been identified: the presence of a mutation in the ryr1 gene. This gene is responsible for synthesis of ryanodine receptor protein, which is found in the sarcoplasmic reticulum of a muscle fibre. The main function of this protein is regulation of the calcium ion concentration in cytoplasm. Even insignificant stress impact on an animal – carrier of this mutation leads to a sharp increase in the content of calcium ions in sarcoplasm. A rise in myoplasmic calcium is accompanied by an increase in heat production due to activation of phosphorylase and breakdown of ATP, glycogenolysis is enhanced with production of lactic acid, CO_2 and excess of heat. An ATP deficiency is created, which leads to a damage of the actin-myosin complex, emergence of super-contractions of myofibrillar structures of muscle fibres and fast muscle rigidity (Semenova *et al.*, 2019). Blood pH drops and metabolic acidosis develops. Mild forms of the disease are latent and are detected only during carcass cutting. Histologically, areas with hyaline degeneration and necrosis of muscle fibres are revealed. However, the genetic factor ryr1 is not the only responsible factor of quality issues and economic losses. In commercial breeding, slaughter and processing of pigs, the management before and during slaughter is important to limit the development of myopathies. Therefore, the development of pig husbandry shows that genetic knowledge is significant, but it does not make it possible to fully solve the problem of PSS. It was estimated that only 4% of low-quality meat was conditioned by genetics. Pork quality and yield are still influenced by multiple ante-mortem and post-mortem factors including transportation and pre-slaughter holding, methods of slaughter and carcass chilling. All this indicate an increased necessity to pay close attention to stability of pork quality and the ways of its improvement.

There are several explanations of the myopathy development that are not connected with genetics (Semenova *et al.*, 2019), namely:

- excessive hypertrophy of muscle fibres of the glycolytic type, immaturity of collagen and/or inadequate development of intramuscular connective tissue, disorders of blood supply of muscles;
- heat stress;
- high muscle temperatures due to struggle, stress, and high metabolic rate in muscles before slaughter;
- large muscle mass of different carcass parts, which is difficult to rapidly chill after slaughter.

Nevertheless, to prevent PSS, it is necessary to continue creating favourable conditions for animal holding, adhering to the rules of their transportation and the norms of humane slaughter.

When considering myopathies, there is a long history of nutritional origin linked with deficiency of selenium and vitamin E, as well as with an increased level of oxidative stress emerging (including hot conditions) and the high animal growth rate. Dietary Se deficiency in animal can cause various types of muscular dystrophy, such as exudative diathesis in chick, mulberry heart in pig, and white muscle disease in lamb and calf (Rederstorff *et al.*, 2006). Although there are some differences in the pathologies of these muscular diseases, they can be related to oxidative stress. The oxidative stress induced by Se deficiency and excess selenite metabolism can reduce the water-holding capacity of meat (Mahan *et al.*, 1999; Zhan *et al.*, 2007).

The most common nutrition deficiency that leads to nutritional myopathy is selenium deficiency. However, the status in vitamin E can also be a co-determinant of the occurrence of myopathy. Nutritional myopathy, like other types of myopathies, can be accompanied with necrosis of muscle fibres. Animals with high feed conversion efficiency and significant muscle mass show the highest sensitivity to myopathy and cardiomyopathy caused by selenium deficiency. The problem of such myopathies emerges in different parts of the world not only in animals but also in humans. Nutritional myopathy has long been considered a disease of young animals, especially very young. Fast postnatal growth may apparently predispose to the problems of nutrient deficiency, although nutritional muscle degeneration is also found in adult animals. At the same time, spontaneous nutritional myopathy can arise prenatally and muscle damages can be observed already at birth.

In general, an oxidative load or stress due to environment conditions (hot climates, pre-slaughter stress), highly concentrated and energetic diets, will challenge the antioxidant reserves such as selenium and vitamin E. With the same content of α-tocopherol, polyunsaturated fats and selenium in pig diets, accumulation of vitamin E and selenium in their organs and muscle tissue can thus be highly different. In case of low selenium content in a diet, an increased intake of vitamin E can retard the development of myopathy associated with selenium deficiency. Oxidative metabolism at the muscle cell level, cell membrane integrity, calcium transport membrane efficacy are often involved in myopathy occurrence. Vitamin E and selenium are important antagonists of free radicals. Some of free radicals are endogenous products of normal cell function, others are external and undesirable factors that take part in cell metabolism. They can be produced both in cells and outside cells as results of the oxidative metabolism, high growth and metabolic rates will increase ROS production, inflammation, sanitary challenges are also an important source of increasing ROS. One of the main sources of free radicals is a process of cell detoxification and microbial killing. However, free radicals can initiate cell damage causing peroxidation of membrane lipids and physico-chemical damage of protein molecules including mitochondria, endoplasmic reticulum and cytosol. Protection against an impact of free radicals is provided by the constant presence of antioxidant molecules, such as tocopherols, ascorbate, beta-carotene as well as by the presence of selenoproteins and selenocysteine-containing enzymes (glutathione peroxidase, thioredoxin reductase, etc.). Nutritional myopathy can thus be related to an insufficient antioxidant protection challenging the muscle cell membranes affecting their capacity to support differential ionic gradients. This initiates a sequence of events when calcium entry leads to hypercontraction of myofibrils and their necrosis.

In addition to drawbacks of a diet associated with deficiency of selenium and vitamin E, there are several factors that can also lead to the development of nutritional myopathy. Among them are feeding with rancid or oxidised fats, mycotoxins, trace element deficiencies, etc. Nutritional myopathy of pigs is largely observed in all countries with intensive breeding, lean animals and low feed conversion. In outbreaks of nutritional myopathy, both growing and adult pigs can be affected. In nutritional myopathy in pigs, mineralisation of damaged muscle fibres might not be that abundant and, even when it occurs and visible, it is difficult to find it in pale porcine muscles. This explains the fact that in nutritional myopathy of pigs, lesions in the heart and liver are found much more frequently; while still a low report of skeletal muscle lesions is mentioned. Nevertheless, lesions ranged from microscopic to pronounced macroscopic can be observed in muscles in experimental investigations of nutritional myopathy in pigs. In the severe form of the disease microscopic lesions of porcine muscles consist in multifocal polyphasic necrosis (Falk *et al.*, 2018). Mild changes were defined as the presence of individual, separated muscle fibres with increased volume and loss of striation or a very low number of degenerating muscle fibres, sometimes with mild infiltration of macrophages. Moderate changes were defined as multifocal degenerating or necrotic muscle fibres with or without infiltration of macrophages, while severe changes were defined as multifocal, relatively widespread degenerating or necrotic muscle fibres with or without infiltration of macrophages. Therefore, selenium and vitamin E play an important role in prophylaxis of nutritional myopathies. With that, selenium being a constituent of selenoproteins is a key factor as it affects endocrine, immune, inflammatory and reproductive processes. Glutathione peroxidases belonging to selenoproteins inactivate peroxides and thereby support physiological functions of muscle tissue. Recently, it has been found that levels of selenium and vitamin E intake that earlier were considered enough, do not exclude the development of nutritional myopathy in modern genotypes with increased metabolic efficiency. Moreover, considering the wide range of potential oxidative stress challenges faced by modern pigs it is key to reduce adverse effects on meat quality.

A direct effect of dietary selenium supplementation has not been shown on de-structured meats, particularly de-structured hams. Further studies are thus needed to investigate the potential nutritional solutions to reduce the occurrence of de-structured meats as it is under investigation nowadays in broiler production with the breast myopathies: wooden breast, spaghetti meat.

8.3 Conclusions

Pork quality aspects, as more generally meat quality aspects, are referring to oxidative processes and/or stressing conditions that can occurred during the life of the animal but also mainly after its death. As the post-mortem processes linked to meat stability and shelf life can almost be more related to chemical processes than biological processes and because after the death of the animal the only improvement can be done on further processing products, dietary antioxidant supply to the animal has to be particularly focused on. Animal protein is still increasing around the world and the animal protein demand remains high, the animal production need maintaining high level of production by improving genetic and ensuring optimal nutrition. The rearing condition while largely improved around the world are still facing challenging conditions due to climate and or other environment pressure factors. These, challenging conditions require maintaining a high and well balance redox status of the animal to cope with these conditions but also ensure good antioxidant reserves for meat life. Even if, during the post-mortem period basic biological functions are restricted or even stopped, the antioxidant cooperative systems are still valid and necessary to limit prooxidant processes. Dietary selenium supply is indubitably a good way to ensure and preserve as long as possible the pork quality by limiting the oxidative processes and hence keep the marketed meat in line with customers expectation during the retailing phase. However, this selenium supply should be understood and think in a good harmony with other antioxidant actors likes vitamins and sulphur products to ensure the teamwork. Meat quality default, for which it is somewhat difficult to demonstrate antioxidant effects would be certainly solved through genetic improvement. However, for the time being it was also highlighted that antioxidant depletion and/or challenging rearing conditions lead to worst situations in terms of appearance and suggest that improving living animal redox status can contribute to improvements. It was also raised in this chapter that new nutritional strategies using over-nutritional dosage of selenium and/or methionine allowed getting interesting improvement on meat quality. These emerging strategies need to be pursued and further tested to ensure new data and draw accurate nutritional recommendations that will be associated with meat quality aspects in the future.

References

Afonso, S.G., Enriquez de Salamanca, R. and Batlle, A.M.C, 1999. The photodynamic and non-photodynamic actions of porphyrins. Brazilian Journal of Medical and Biological Research 32: 255-266. https://doi.org/10.1590/S0100-879X1999000300002.

Ahn, D.U., 2002. Production of volatiles from amino acid homopolymers by irradiation. Journal of Food Science 67: 2565-2570.

Ahn, D.U. and Kim, S.M., 1998. Prooxidant effects of ferrous iron, hemoglobin, and ferritin in oil emulsion and cooked-meat homogenates are different from those in raw-meat homogenates. Poultry Science 77: 348-355.

Ahn, D.U. and Lee, E.J., 2002. Production of off-odor volatiles from liposome-containing amino acid homopolymers by irradiation. Journal of Food Science67: 2659-65.

Baker, D., 2009. Advances in protein-amino acid nutrition of poultry. Amino Acids 37(1): 29-41.

Baron, C.P. and Andersen, H.J., 2002. Myoglobin-induced lipid peroxidation – a review. Journal of Agricultural and Food Chemistry 50: 3887-3897.

Bekhit, A.E.-D.A., Hopkins, D.L., Fahri, F.T. and Ponnampalam, E.N., 2013. Oxidative processes in muscle systems and fresh meat: sources, markers, and remedies. Comprehensive Reviews in Food Science and Food Safety 12(5): 565-597. https://doi.org/10.1111/1541-4337.12027

Berri, C., 2015. La viande de volaille: des attentes pour la qualité qui se diversifient et des défauts spécifiques à corriger. Inra Productions Animales 28: 115-118.

Bobcek, B., Lahucky, R., Mrazova, J., Bobcek, R., Novotna, K. and Vasicek, D., 2004. Effects of dietary organic selenium supplementation on selenium content, antioxidative status of muscles and meat quality of pigs. Czech Journal of Animal Science 49(9): 411-417.

Briskey, E.J., 1964. Etiological status and associated studies of pale, soft, exudative porcine musculature. Advances in Food Research 13: 89-178.

Byrne, D.V., Bredie, W.L., Mottram, D.S. and Martens, M., 2002. Sensory and chemical investigations on the effect of oven cooking on warmed-over flavor development in chicken meat. Meat Science 61: 127-139.

Calvo, L., Toldrá, F., Aristoy, M.C., López-Bote, C.J. and Rey, A.I., 2016. Effect of dietary organic selenium on muscle proteolytic activity and water-holding capacity in pork. Meat Science 121: 1-11.

Calvo, L., Toldrá, F., Rodríguez, A., López-Bote, C. and Rey, A.I., 2017. Effect of dietary selenium source (organic vs. mineral) and muscle pH on meat quality characteristics of pigs. Food Science & Nutrition 5(1): 94-102.

Campo, M.M., Nute, G.R., Hughes, S.I., Enser, M., Wood, J.D. and Richardson, R.I., 2006. Flavour perception of oxidation in beef. Meat Science 72: 303-311.

Castellano, R., Perruchot, M.H., Conde-Aguilera, J.A., Van Milgen, J., Collin, A., Tesseraud, S., Mercier, Y. and Gondret, F., 2015. A methionine deficient diet enhances adipose tissue lipid metabolism and alters anti-oxidant pathways in young growing pigs. Plos One 10(7): e0130514. https://doi.org/10.1371/journal.pone.0130514

Chen, J., Tiana, M., Guana, W., Wena, T., Yang, F., Chen, F., Zhanga, S., Songa, J., Ren, C., Zhang, Y. and Songa, H., 2019. Increasing selenium supplementation to a moderately-reduced energy and protein diet improves antioxidant status and meat quality without affecting growth performance in finishing pigs. Journal of Trace Elements in Medicine and Biology 56: 38-45.

Conde-Aguilera, J., Lefaucheur, L., Tesseraud, S., Mercier, Y., Le Floc'h, N. and Van Milgen, J., 2016. Skeletal muscles respond differently when piglets are offered a diet 30% deficient in total sulfur amino acid for 10 days. European Journal of Nutrition 55: 117-126.

Conde-Aguilera, J.A., Cobo-Ortega, C., Mercier, Y., Tesseraud, S. and Van Milgen, J., 2014. The amino acid composition of tissue protein is affected by the total sulfur amino acid supply in growing pigs. Animal 8(03): 401-409.

DaeWon, J., EunHae, K., TaeSoo, K., YounWook, C., Hyun, K. and IckYoung, K., 2004. Different distributions of selenoprotein W and thioredoxin during postnatal brain development and embryogenesis. Molecules and Cells 17: 156-159.

Davies, M.J., 2012. Free radicals, oxidants and protein damage. Australian Biochemitry 43: 8-12.

Davies, M.J., 2016. Protein oxidation and peroxidation. Biochemistry Journal 473(7): 805-825.

De Bont, R. and Van Larebeke, N., 2004. Endogenous DNA damage in humans: a review of quantitative data. Mutagenesis 19: 169-185.

Dietze, K., Lammers, M. and Ternes, W., 2007. Effect of natural additive on warmed-over flavor of dark turkey meat analyzed with dynamic headspace GC/MS. Fleischwirtsch International 23(5): 67-70.

England, E.M., Scheffler, T.L., Kasten, S.C., Matarneh, S.K. and Gerard, D.E., 2013. Exploring the unknowns involved in the transformation of muscle to meat. Meat Science 95: 837-843.

Estévez, M., 2015. Oxidative damage to poultry: from farm to fork. Poultry Science 94(6): 1368-1378. https://doi.org/10.3382/ps/pev094

Estévez, M., Geraert, P.A., Liu, R., Delgado, J., Mercier, Y. and Zhang, W., 2020. Sulphur amino acids, muscle redox status and meat quality: more than building blocks – invited review. Meat Science 163: 108087. https://doi.org/10.1016/j.meatsci.2020.108087

Estrade, M., Vignon, X. and Monin, G., 1993. Effect of the RN-gene on ultrastructure and protein fractions in pig muscle. Meat Science 35: 313-319.

Falk, M., Bernhoft, A., Framstad, T., Salbu, B., Wisløff, H., Kortner, T., Kristoffersen, A.B. and Oropeza-Moe, M., 2018. Effects of dietary sodium selenite and organic selenium sources on immune and inflammatory responses and selenium deposition in growing pigs. Journal of Trace Elements in Medicine and Biology 50: 527-536.

Friedman, M. and Cuq, J.L., 1988. Chemistry, analysis, nutritional value, and toxicology of tryptophan in food – a review. Journal of Agricultural and Food Chemistry 36(5): 1079-1093.

Geay, Y., Bauchard, D., Hocquette, J.F. and Culioli, J., 2002. Valeur diététique et qualités sensorielle des viandes de ruminants. Incidence de l'alimentation des animaux. Inra Productions Animales 15: 37-52.

Guillen S.R. and Guzman C.M., 1998. The thiobarbibutric acid (TBA) reaction in food: a review. Critical Reviews in Food Science and Nutrition 38(4): 315-330.

Hamilton, D.N., Ellis, M., Miller, K.D., McKeith, F.K. and Parrett, D.F., 2000. The effect of the Halothane and Rendement Napole genes on carcass and meat quality characteristics of pigs. Journal of Animal Science 78: 2862-2867.

Huff-Lonergan, E. and Lonergan, S.M., 2005. Mechanisms of water-holding capacity of meat: the role of postmortem biochemical and structural changes. Meat Science 71: 194-204.

Huff-Lonergan, E., Baas, T.J., Malek, M., Dekkers, J.C., Prusa, K. and Rothschild, M.F., 2002. Correlations among selected pork quality traits. Journal of Animal Science 80: 617-627.

Jeong, D., Kim, T.S., Chung, Y.W., Lee, B.J. and Kim, I.Y., 2002. Selenoprotein W is a glutathione-dependent antioxidant *in vivo*. FEBS Letters 517: 225228.

Jiang, J., Tang, X., Xue, Y., Lin, G. and Xiong, Y.L., 2017. Dietary linseed oil supplemented with organic selenium improved the fatty acid nutritional profile, muscular selenium deposition, water retention, and tenderness of fresh pork. Meat Science 131: 99-106. https://doi.org/10.1016/j.meatsci.2017.03.014

Jlali, M., Briens, M., Rouffineau, F. Geraert, P.A. and Mercier, Y., 2014. Evaluation of the efficacy of 2-hydroxy-4-methylselenobutanoic acid on growth performance and tissue selenium retention in growing pigs. Journal of Animal Science 92(1): 182-188. https://doi.org/10.2527/jas.2013-6783

Jones, G.D., Droz, B., Greve, P., Gottschalk, P., Poffet, D., McGrath, S.P., Seneviratne, S.I., Smith, P. and Winkel, L.H.E., 2017. Selenium deficiency risk predicted to increase under future climate change. PNAS 114(11): 2848-2853.

Juniper, D.T., Phipps, R.H., Ramos-Morales, E. and Bertin, G., 2009. Effects of dietary supplementation with selenium enriched yeast or sodium selenite on selenium tissue distribution and meat quality in lambs. Animal Feed Science and Technology 149: 228-239.

Kawęcka, M., Jacyno, E., Matysiak, B., Kołodziej-Skalska, A. and Pietruszka, A., 2013. Effects of selenium and vitamin E supplementation on selenium distribution and meat quality of pigs. Acta Agriculturae Scand Section A 63(4): 194-200.

Kelly, F.J. and Goldspink, D.F., 1982. The differing responses of four muscle types to dexamethasone treatment in the rat. Biochemical Journal 208(1): 147-151. https://doi.org/10.1042/bj2080147

Kelly, F.J., McGrath, J.A. Goldspink, D.F. and Cullen, M.J., 1986. A morphological/biochemical study on the actions of corticosteroids on rat skeletal muscle. Muscle Nerve 9(1): 1-10. https://doi.org/10.1002/mus.880090102

Khan-Merchant, N., Penumetcha, M., Meilhac, O. and Parthasarathy, S., 2002. Oxidized fatty acids promote atherosclerosis only in the presence of dietary cholesterol in low-density lipoprotein receptor knockout mice. Journal of Nutrition 132(11): 3256-3262.

Ladikos, D. and Lougovois, V., 1990. Lipid oxidation in muscle foods: a review. Food Chemistry 35: 295-314.
Lauridsen, C., Nielsen, J.H., Henckel, P. and Sorensen, M.T. 1999. Antioxidative and oxidative status in muscles of pigs fed rapeseed oil, vitamin E, and copper. Journal of Animal Science 77: 105-115.
Le Roy, P., Moreno, C., Elsen, J.M., Caritez, J.C., Billon, Y., Lagant, H., Talmant, A., Vernin, P., Amigues, Y., Sellier, P. and Monin, G., 1999. Interactive effects of the HAL and RN major genes on carcass quality traits in pigs: preliminary results. In Proceeding of 50th Annual meeting of the European Association for Animal Production. Zurich, Switzerland.
Lebret, B., Batonon-Alavo, D.I., Perruchot, M.H., Mercier, Y. and Gondret, F., 2018. Improving pork quality traits by a short-term dietary hydroxy methionine supplementation at levels above growth requirements in finisher pigs. Meat Science 145: 230-237.
Lei, X.G., Ross, D.A. and Roneker, K.R., 1997. Comparison of age-related differences in expression of phospholipid hydroperoxide glutathione peroxidase mRNA and activity in various tissues of pigs. Comparative biochemistry and physiology. Part B, Biochemistry & Molecular Biology 117: 109-114.
Levine, R.L., Moskovitz, J. and Stadtman, E.R., 2000. Oxidation of methionine in proteins: roles in antioxidant defense and cellular regulation. IUBMB Life 50(4-5): 301-307.
Li, J.G., Zhou, J.C., Zhao, H., Lei, X.G., Xia, X.J., Gao, G. and Wang, K.N., 2011. Enhanced water-holding capacity of meat was associated with increased Sepw1 gene expression in pigs fed selenium-enriched yeast. Meat Science 87: 95-100.
Li, Y., Zhang, H., Chen, Y.P., Ying, Z.X., Su, W.P., Zhang, L.L. and Wang, T., 2017. Effects of dietary l-methionine supplementation on the growth performance, carcass characteristics, meat quality, and muscular antioxidant capacity and myogenic gene expression in low birth weight pigs. Journal of Animal Science 95(9): 3972-3983.
Lisiak, D., Janiszewski, P., Blicharski, T., Borzuta1, K., Grześkowiak, E., Lisiak, B., Powałowski, K., Samardakiewicz, L., Batorska, M., Skrzymowska, K. and Hammermeister, A., 2014. Effect of selenium supplementation in pig feed on slaughter value and physicochemical and sensory characteristics of meat. Annals of Animal Science 14(1): 213-222.
Listrat, A., Lebret, B., Louveau, I., Astruc, T., Bonnet, M., Lefaucheur, L. and Bugeon, J., 2015. Comment la structure et la composition du muscle déterminent la qualité des viandes ou chairs? INRA Productions Animales 28(2): 125-136.
Lykkesfeldt, J. and Svendsen, O., 2007. Oxidants and antioxidants in disease: oxidative stress in farm animals. The Veterinary Journal 173(3): 502-511. https://doi.org/10.1016/j.tvjl.2006.06.005
Mahan, D.C. and Parrett, N.A., 1996. Evaluating the efficacy of selenium-enriched yeast and sodium selenite on tissue selenium retention and serum glutathione peroxidase activity in grower and finisher swine. Journal of Animal Science 74(12): 2967-2974.
Mahan, D.C., Cline, T.R. and Richert, B., 1999. Effects of dietary levels of selenium-enriched yeast and sodium selenite as selenium sources fed to growing-finishing pigs on performance, tissue selenium, serum glutathione peroxidase activity, carcass characteristics, and loin quality. Journal of Animal Science 77(8): 2172-2179.
Marquez-Ruiz, G., Garcia-Martinez, M.C. and Holgado, F., 2008. Changes and effects of dietary oxidized lipids in the gastrointestinal tract. Lipids Insights 2: 11-19.
Mateo, R.D., Spallholz, J.E., Elder, R., Yoon, I. and Kim, S.W., 2007. Efficacy of dietary selenium sources on growth and carcass characteristics of growing-finishing pigs fed diets containing high endogenous selenium. Journal of Animal Science 85: 1177-1183.

Metayer, S., Seiliez, I., Collin, A., Duchene, S., Mercier, Y., Géraert, P.A. and Tesseraud, S., 2008. Mechanisms through which sulfur amino acids control protein metabolism and oxidative status. The Journal of Nutritional Biochemistry 19(4): 207-215. https://doi.org/10.1016/j.jnutbio.2007.05.006

Min, B. and Ahn, D.U. Mechanism of lipid peroxidation in meat and meat products – a review. Food Science and Biotechnology 14(1): 152-163.

Min, D.B. and Boff, J.M., 2002. Chemistry and reaction of singlet oxygen in foods. Comprehensive Reviews in Food Science and Food Safety 1: 60-72.

Minvielle, B., Boulard, J., Vautier, A. and Houix, Y., 2005. Viandes déstructurées: effets combinés des durées de transport et d'attente sur la fréquence d'apparition du défaut. Viandes et Produits Carnés Vol 24(4): 126-130.

Minvielle, B., Houix, Y. and Lebret, B., 2001. Viandes déstructurées: Facteurs de risque. Caractérisation colorimétrique, biochimique et histologique. TechniPorc 24(4): 21-28.

Monin, G. and Sellier, P., 1985. Pork of low technological quality with a normal rate of muscle pH fall in the immediate post-mortem period: the case of the Hampshire breed. Meat Science 13: 49-63.

Morel, P.C.H., Leong, J., Nuijten, W.G.M., Purchas, R.W. and Wilkinson, B.H.P., 2013. Effect of lipid type on growth performance, meat quality and the content of long chain n-3 fatty acids in pork meat. Meat Science 95(2): 151-159.

Morrissey, P.A., Sheehy, P.J.A., Galvin, K. and Kerry, J.P., 1998. Lipid stability in meat and meat products. Meat Science 49: 73-86.

Naveau, J., 1986. The genetic determination of meat quality in pigs. Heritability of the technological yield of meat prepared by the "Napole" method. Journées de la Recherche Porcine en France 18: 265-276.

Niederhofer, L.J., Daniels, J.S., Rauzer, C.A., Greene, R.E. and Marnett, L.J., 2003. Malonaldehyde, a product of lipid peroxidation is mutagenic in human cells. Journal of Biological Chemistry 278: 31426-31433.

Offer, G. and Knight, P., 1988. The structural basis of water-holding in meat. Part 2: drip losses. In: Lawrie, R. (ed.) Development in meat science (Vol. 4). Elsevier Applied Science, London, UK, pp. 172-243.

Ouali, A., Herrera-Mendez, C.H., Coulis, G., Becila, S., Boudjellal, A., Aubry, L. and Sentendreu, M.A., 2006. Revisiting the conversion of muscle into meat and the underlying mechanisms. Meat Science 74: 44-58.

Papuc, C., Goran, G.V., Predescu, C.N. and Nicorescu, V., 2017. Mechanisms of oxidative processes in meat and toxicity induced by postprandial degradation products: a review. Comprehensive Reviews in Food Science and Food Safety 16(1): 96-123. https://doi.org/10.1111/1541-4337.12241

Pearce, K.L., Rosenvold, K., Andersen, H.J. and Hopkins D.L., 2011. Water distribution and mobility in meat during the conversion of muscle to meat and ageing and the impacts on fresh meat quality attributes – a review. Meat Science 89: 111-124.

Petracci, M. and Cavani, C., 2012. Muscle growth and poultry meat quality issues. Nutrients 4(1): 1-12.

Pietras, M.P., Pieszka, M., Bederska-Łojewska, D. and Orczewska-Dudek, S., 2012. Cholesterol and lipid peroxides in animal products and health implications – a review. Annals of Animal Science 12(1): 25-52.

Rederstorff, M., Krol, A. and Lescure, A., 2006. Understanding the importance of selenium and selenoproteins in muscle function. Cellular and Molecular Life Sciences 63: 52-59.

Renerre, M., 1990. Factors involved in the discoloration of beef meat. International Journal of Food Science Technology 25: 613-630.

Rosenvold, K. and Andersen, H.J., 2003. The significance of preslaughter stress and diet on colour and colour stability of pork. Meat Science 63(2): 199-209.

Rowe, L., Maddock, K., Lonergan, S.M. and Huff-Lonergan, E., 2004. Influence of early post-mortem protein oxidation on beef quality. Animal Science 82: 785-93.

Semenova, A.A., Kuznetsova, T.G., Nasonova, V.V., Nekrasov, R.V. and Bogolyubova, N.V., 2019 Myopathy as a destabilizing factor of meat quality formation. Theory and Practice of Meat Processing 4(3): 24-31.

Shleikin, A.G. and Medvedev, Y.V., 2014. Role of peroxidation and heme catalysis in coloration of raw meat. Acta Scientiarum Polonorum Technologia Alimentaria 13(2): 123-127.

Sies, H., 1997. Oxidative stress: oxidants and antioxidants. Experimental Physiology 82(2): 291-295. https://doi.org/10.1113/expphysiol.1997.sp004024

Sies, H., 2015. Oxidative stress: a concept in redox biology and medicine. Redox Biology 4(0): 180-183. https://doi.org/10.1016/j.redox.2015.01.002

Silva, V.A., Bertechini, A.G., Clemente, A.H.S., Villas Boas de Freitas, L.F., Nogueira, B.R.F., Lopes de Oliveira, B., De Lemos Souza Ramos, A., 2019. Different levels of selenomethionine on the meat quality and selenium deposition in tissue of finishing pigs. Journal of Animal Physiology and Animal Nutrition 103: 1866-1874.

Steinberg, D., 1997. Oxidative modification of LDL and atherosclerosis. Circulation 95: 1062-1071.

Surendranath, P. and Poulson, J., 2013. Myoglobin chemistry and meat color. Annual Review of Food Science and Technology 4: 79-99.

Svoboda, M., Kotrbáček, V., Ficek, R. and Drábek, J., 2009. Effect of organic selenium from Se-enriched alga (*Chlorella* spp.) on selenium transfer from sows to their progeny. Acta Veterinaria Brno 78: 373-377.

Tang, J., Faustman, C., Mancini, R.A., Seyfert, M. and Hunt, M.C., 2005. Mitochondrial reduction of 420 metmyoglobin: dependence on the electron transport chain. Journal of Agricultural and Food Chemistry 53: 5449-5455.

Thomas, M.J., 1995. The role of free radicals and antioxidants: how do we know that they are working? Critical Reviews in Food Science and Nutrition 35: 21-39.

Toledano, M.B., Kumar, C., Le Moan, N., Spector, D. and Tacnet, F., 2007. The system biology of thiol redox system in *Escherichia coli* and yeast: differential functions in oxidative stress, iron metabolism and DNA synthesis. FEBS letters 581: 3598-3607.

Turell, L., Botti, H., Carballal, S., Ferrer-Sueta, G., Souza, J.M., Duran, R., Freeman, B.A., Radi, R., Alvarez, B., 2008. Reactivity of sulfenic acid in human serum albumin. Biochemistry 47: 358-367.

Utrera, M., Rodriguez-Carpena, J.G., Morcuende, D. and Estevez, M., 2012. Formation of lysine-derived oxidation products and loss of tryptophan during processing of porcine patties with added avocado byproducts. Journal of Agricultural and Food Chemistry 60: 3917-3926.

Vautier, A., Boulard, J., Houix, Y. and Minvielle, B., 2008. Prediction level of meat quality criteria on 'PSE-like zones' defect of pork's ham. 54th International Congress of Meat Science and Technology, Cape Town, South Africa. Available at: https://tinyurl.com/yn7qb2ax.

Vautier, A., Minvielle, B., Boulard, J., Bouyssiere, M. and Houix, Y., 2004 Viandes destructurées. Effets du système d'abattage et des conditions météorologiques. TechniPorc 27(3): 10-23.

Vendeland, S.C., Beilstein, M.A., Yeh, J., Ream, W. and Whanger, P.D., 1995. Rat skeletal muscle selenoprotein W: cDNA clone and mRNA modulation by dietary selenium. Proceedings of the National Academy of Sciences 92: 8749-8753.

Wang, Z.G., Pan, X.J., Peng, Z.Q., Zhao, R.Q. and Zhou, G.H., 2009. Methionine and selenium yeast supplementation of the maternal diets affects color, water-holding capacity, and oxidative stability of their male offspring meat at the early stage. Poultry Science 88(5): 1096-1101. https://doi.org/10.3382/ps.2008-00207

Wen, C., Jiang, X.Y., Ding, L.R., Wang, T. and Zhou, Y.M., 2017. Effects of dietary methionine on growth performance, meat quality and oxidative status of breast muscle in fast- and slow-growing broilers. Poultry Science 96(6): 1707-1714. https://doi.org/10.3382/ps/pew432

Whanger, P.D., 2000. Selenoprotein W: a review. Cellular and Molecular Life Sciences (CMLS) 57: 1846-1852.

Wolter, B., Ellis, M., Mc Keith, F.K., Miller, K.D. and Mahan, D.C., 1999. Influence of dietary selenium source on growth performance, and carcass and meat quality characteristics in pigs. Canadian Journal of Animal Science 79: 119-121.

Wood, J.D., Richardson, R.I., Nute, G.R., Fisher, A.V., Campo, M.M., Kasapidou, E., Sheard, P.R. and Enser, M., 2003. Effects of fatty acids on meat quality: a review. Meat Science 66: 21-32.

Yuan, P., Mercier, Y., Rouffineau, F., Batonon-Alavo, D.I. and Fang, Z.F., 2017. Increased inclusion of sulfur amino acids in growing-finishing pig diets improved meat quality. In: Troy, D., McDonnell, C., Hinds, L. and Kerry, J. (eds) 63rd International Congress of Meat Science and Technology – nurturing locally, growing globally. Wageningen Academic Publishers, Wageningen, the Netherlands, pp. 200-201.

Zhan, X.A., Wang, M., Zhao, R.Q., Li, W.F. and Xu, Z.R., 2007. Effects of different selenium source on selenium distribution, loin quality and antioxidant status in finishing pigs. Animal Feed Science and Technology 132: 202-211.

Zhang, W., Xiao, S. and Ahn, D.U., 2013. Protein oxidation: basic principles and implication for meat quality. Critical Reviews in Food Science and Nutrition 53: 1191-201.

Chapter 9
Selenium and immunity

P.F. Surai[1-7]
[1]*Vitagene and Health Research Centre, Bristol, BS4 2RS, United Kingdom;* [2]*Department of Hygiene and Poultry Sciences, Moscow State Academy of Veterinary Medicine and Biotechnology Named After K.I. Skryabin, 109472 Moscow, Russia;* [3]*Department of Microbiology and Biochemistry, Faculty of Veterinary Medicine, Trakia University, 6000 Stara Zagora, Bulgaria;* [4]*Department of Animal Nutrition, Faculty of Agricultural and Environmental Sciences, Szent Istvan University, 2103 Gödöllo, Hungary;* [5]*Saint-Petersburg State Academy of Veterinary Medicine, 196084 St. Petersburg, Russia;* [6]*Sumy National Agrarian University, 40000 Sumy, Ukraine;* [7]*Odessa National Academy of Food Technologies, 65039 Odessa, Ukraine; psurai@feedfood.co.uk*

Abstract

Selenium affects all components of the immune system, including the development and expression of nonspecific, humoral, and cell-mediated responses. In general, a deficiency in Se appears to result in immunosuppression, whereas optimal Se supplementation appears to result in augmentation and/or restoration of immunologic functions. Since free radicals can be damaging to intestinal structures, antioxidant functions of selenium are responsible for prevention damages to intestinal lymphoid structure as well as damages to intestinal enterocyte membranes. This could be especially important in relation to gut immunosuppression caused by toxins/mycotoxins, nutritional deficiencies and infectious agents. Se is proven to have an important role in resolution of inflammation in various tissues, including gut, via affecting redox balance and expression of various transcription factors including NF-κB and PPARγ. When considering immuno-facilitating properties of selenium, it is necessary to take into account several points, including interactions between various antioxidants and prooxidants, possible increased Se requirement for immunomodulation under stress conditions, Se source with organic selenium to be at an advantage because it is better assimilated from feed and better accumulated in tissues, etc. As a result of antioxidant (selenium) deficiency increased oxidative stress of a host can lead to increased virus mutation rate and change in a viral pathogen resulting in emerging viral pathogens with new pathogenic properties. This possibility was not fully exploited in animal production but seems important to study more extensively. There is a need for further research related to molecular mechanisms of protective effects of various antioxidants, including selenoproteins, on immunity under various commercially-relevant stress conditions.

Keywords: selenium, immunity, immunomodulation, oxidative stress, pigs

DOI 10.3920/978-90-8686-912-1_9, © Wageningen Academic Publishers 2021

9.1 Introduction

Animal defence against various diseases depends on the efficacy of the immune system responsible for elimination of foreign substances (e.g. parasites, bacteria, moulds, yeast, fungi, viruses and various macromolecules) or the creation of specific inhospitable conditions within the host for a wide range of pathogens. This protective capacity is based on the effective immune system which is considered to be an important determinant of animal health and wellbeing. In that sense, a remarkable ability of components of the immune system to distinguish between self and non-self is a great achievement of animal evolution.

Commercial pig production is based on balanced feed, meeting requirements in major nutrients and optimised environmental conditions. However, it is very difficult to avoid various stresses which are responsible for immunosuppression and increased susceptibility to various diseases and as a result decreased productive and reproductive performance of pigs. For example, mycotoxins are among major immunosuppressive agents in pig diet. In such situations immunomodulating properties of certain macro- and micronutrients are of great importance for the pig industry. In fact, almost all nutrients in the diet play a crucial role in maintaining an 'optimal' immune response, and both insufficient and excessive intakes can have negative consequences on the immune status and susceptibility to a variety of pathogens. Indeed, immunocompetence can be presented as a 'violin music'. One can have a best violoncellist in the world with a Stradivari violin in the hands, however, before the violin is properly tuned there is no music, just noise. That is exactly what happened with immunocompetence when the immune system is overreacted. This is associated with allergy and other immune conditions. Furthermore, to maintain such an increased immune reactivity important nutrients are used. On the other hand, an underreacting immune system is also a problem, since it is not able to adequately protect the body from invaders, including microbes, viruses, etc. Only well-tuned immune system can do a great job of the optimal protection without big losses in productive and reproductive performance of pigs. Information has been actively accumulated for the last 20 years indicating that selenium is among major immunomodulating agents and its requirement for such action could be higher than that for animal growth and development (Surai, 2006, 2018). In fact, selenium has been shown to have immunomodulatory effects in a variety of species when administered in quantities in excess of established dietary requirements. Therefore, selenium as an essential component of selenocysteine-containing proteins is involved in most aspects of cell biochemistry and function and immune cell activity is also Se-dependent and Se deficiency attenuates the host immune response, thereby increasing the risk of bacterial and viral infections.

9.2 Immune system and its evaluation

There are two major types of immune function in pigs: natural (innate) and acquired (adaptive) immunity (Figure 9.1).

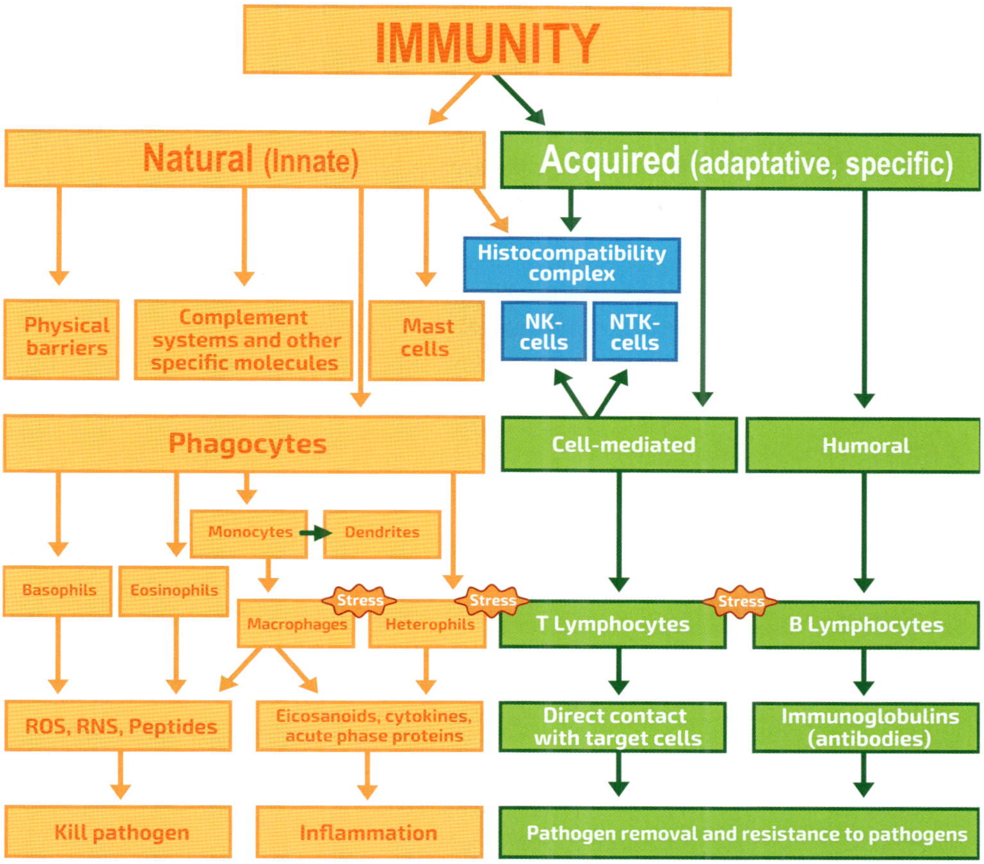

Figure 9.1. General scheme of the immune system (adapted from Surai 2002, 2006, 2018).

9.2.1 Natural immunity

Natural immunity, called the innate immune system, includes physical barriers (e.g. skin, mucus coat of the GI tract), specific molecules (e.g. agglutinins, precipitins, defensins, acute-phase proteins, lysozyme, etc.), innate immune cells, including dendritic cells (DC), macrophages, neutrophils, mast cells, natural killer (NK) cells, and natural killer T (NKT) cells, represent the first line of defence against pathogens and foreign agents (Table 9.1).

It should be mentioned that major immune cells are derived from blood stem cells (Figure 9.2).

Table 9.1. Key elements of the immune system (adapted from Surai, 2006, 2018).

Cells	**Significance**
Monocytes, macrophages	Phagocytosis, synthesis of interleukins (IL) IL1, IL6, IL8 and other substances
Neutrophils	Phagocytosis of bacteria, viruses and toxins
Eosinophils	Destruction of parasites
Basophils	Initiation of inflammatory processes
Dendritic cells	Antigen processing, lymphocyte activation, cytokine secretion
Mast cells	Release of inflammatory mediators
B cells (B lymphocytes)	Production of plasma cells (immunoglobulins), antigen-specific; 10% of total lymphocytes
T helper cells (Helper T lymphocytes)	Antigen-specific, produce cytokines: IL2, IL3, IL4, IL5, IL9 and IL10; 55% of total lymphocytes
Cytotoxic T cells (T lymphocytes)	Destruction of tumour cells and virus-infected cells; antigen-specific, 25% of total lymphocytes
Suppresser T cells (T lymphocytes)	Inhibition of immune reactions (development of autoimmune diseases)
Natural killer cells	Destruction of tumour cells and virus-infected cells, 10% of total lymphocytes
Macromolecules	
Immunoglobulins	Binding foreign cells and proteins; promotion of their ingestion by phagocytes
Interferons (IFN-α; IFN-β;IFN-γ)	Activation of macrophages (γ-interferon); inhibition of viral replication
Complement system: a set of over 20 soluble glycoproteins	Destruction of foreign cells
Interleukins	Regulation of specific types of leukocytes
Leukotrienes	Promotion of inflammatory process
Lysozymes	Dissolution of bacterial membranes
Collectines – a group of carbohydrate-binding proteins	Act as opsonins in non-adaptive immune response to pathogen
Acute phase proteins – a group of plasma proteins produced in the liver in response to microbial stimulus	Maximise activation of the complement system

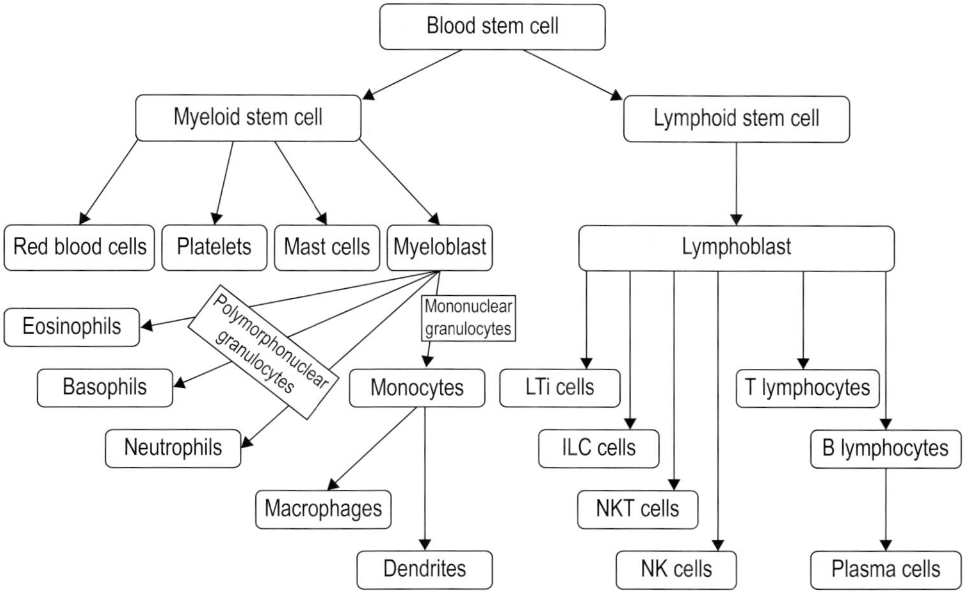

Figure 9.2. Immune cell formation scheme.

9.2.2 PAMPS, DAMPs and immune receptors

In the case of invasion by a pathogen (bacteria, viruses, parasites, fungi, etc.) pathogen- or damage-associated molecular patterns (PAMPs and DAMPs) present on the cells are recognised by the cells of the innate immune system through receptors called pattern recognition receptors (PRRs) that are specialised in their recognition being a key element of the immune system (Korver, 2012; Mair *et al.*, 2014). PRRs are shown to be mainly expressed by antigen presenting cells including dendritic cells and macrophages, but they can also be found in other immune and non-immune cells. These receptors are shown to be strategically localised in the immune cell: at the cell surface to recognise extracellular pathogens such as bacteria or fungi, in the endosomes to sense intracellular invaders such as viruses and they also can be found in the cytoplasm. In general, PRRs can be divided into four families: Toll-like receptors (TLR), nucleotide-binding oligomerisation domain-like receptors (NLR), C-type lectin receptors and RIG-1 like receptors and DNA receptors (cytosolic sensors for DNA) with TLR being most studied among them (Kumar *et al.*, 2011). In fact, TLRs are a family of conserved pattern recognition receptors that recognise specific microbial-associated molecular patterns and helping the cell distinguish between self and non-self materials (Coffey and Werling, 2011). There are 12 functional TLRs in mice and ten in humans and pigs (Table 9.2) and they are type I membrane glycoproteins that recognise extracellular or endosomal/endolysosomal PAMPs via leucine rich repeats (Mair *et al.*, 2014).

In fact, TLRs provide a link between innate and adaptive immunity. Signalling by TLRs involves five adaptor proteins known as MyD88, MAL, TRIF, TRAM and SARM. Recently additional functions for MyD88 apart from NF-B activation, including activation of the transcription factors IRF1, IRF5 and IRF7, and also a role outside the TLRs in interferon-γ signalling insights have been revealed (O'Neill and Bowie, 2007).

Furthermore, MAL and TRAM are shown to be bridging adaptors, with MAL recruiting MyD88 to TLR2 and TLR4, and TRAM recruiting TRIF to TLR4 to allow for IRF3 activation. Finally, the fifth adaptor, SARM, has been revealed, to negatively regulate TRIF (O'Neill and Bowie, 2007). The presence of TLR ligands as adjuvants in conjunction with a vaccine is shown to increase the efficacy and response to the immunisation with a particular antigen (Coffey and Werling, 2011). Indeed, similar to innate immune cells, B cells express TLRs and B cell-intrinsic TLRs appear to cooperate with adaptive Ig receptors to achieve rapid humoral immunity and to preserve long-term humoral memory (Cerutti *et al.*, 2011). In fact, multiple innate immune receptors including TLR2, TLR4, TLR5, TLR9, NOD1/2, and NLRP3 have been implicated in the recognition of metabolic stress and initiation of inflammatory responses in various tissues (Table 9.3; Jin *et al.*, 2013).

Table 9.2. Main Toll-like receptors (TLRs) in pigs (adapted from Kumar *et al.*, 2011; Mair *et al.*, 2014).[1]

TLRs	Main ligand	Signalling pathway	Main immune response	Main target pathogen
TLR 1	Triacylglyceride	MyD88-NF-κB	Proinflammatory cytokines	Gram-positive bacteria & fungi
TLR 2	Peptidoglycans Zymosan	MyD88-NF-κB	Proinflammatory cytokines	Gram-positive bacteria & fungi
TLR 3	dsRNA	TRIF-IRF3/7	Type-1 IFN	RNA viruses
TLR 4	LPS	MyD88-NF-κB MyD88-IRF3/7	Proinflammatory cytokines type-1 IFN	Gram-negative bacteria
TLR 5	Flagellin	MyD88-NF-κB	Proinflammatory cytokines	Flagellated bacteria
TLR 6	Diacylglyceride	MyD88-NF-κB	Proinflammatory cytokines	Gram-positive bacteria & fungi
TLR 7	ssRNA	MyD88-NF-κB MyD88-IRF7	Proinflammatory cytokines type-1 IFN	Endosomal bacteria & viruses
TLR 8	ssRNA	TRIF-IRF3/7	Proinflammatory cytokines	RNA viruses
TLR 9	Unmethylated CpG DNA	MyD88-NF-κB MyD88-IRF7	Proinflammatory cytokines type-1 IFN	Endosomal bacteria & viruses
TLR 10	*Listeria monocytogenes*	MyD88-NF-κB	Proinflammatory cytokines	*Listeria monocytogenes*

[1] CpG DNA = shorthand for 5'–C–phosphate–G–3', that is, cytosine and guanine separated by only one phosphate group; phosphate links any two nucleosides together in DNA; dsRNA = double-stranded RNA; IFN = interferon; LPS = lipopolysaccharide; Proinfl.= proinflammatory; ssRNA = single-stranded RNA.

Table 9.3. TLR4 endogenous and exogenous ligands (adapted from Garcia *et al.*, 2020).[1,2]

Endogenous		
	beta-defensin (mDF2β)	epithelial cells and keratinocytes
	biglycan	extracellular matrix (ECM)
	chaperone GP96 (HSP90B1)	various types of cells
	chaperone Hsp22 (HspB8)	various types of cells
	chaperone Hsp60	various types of cells
	chaperone Hsp70	various types of cells
	FNIII FDA	endothelial cells
	heme	hepatocytes and erythrocytes
	heparan sulphate	various types of cells
	hyaluronan	plasma membrane-ECM
Exogenous		
Bacteria	(mono- or di-phosphoryl) lipid A	gram negative bacteria
	lipopolysaccharide (LPS)	gram negative bacteria
	LPS-Bq*	*Bartonella quintana*
	LPS-Rs*	*Rhodobacter sphaeroides*
	lipooligosaccharide (LOS)	*Neisseria meningitides*
	seeligeriolysin O (LSO)	*Listeria seeligeri*
	pneumolysin	*Streptococcus pneumoniae*
Fungus	mannan	*Candida albicans*
	glucuronoxylomannan	*Cryptococcus neoformans*
Protozoa	glycoinositolphospholipid (GIPL)	*Trypanosoma cruzi*
	P8 PGLC	*Leishmania pifanoi*
Virus	protein F	respiratory syncytial virus
	envelope proteins	mouse mammary tumor virus
Plants	*Artocarpus integrifolia* lectin (AIL)*	*Artocarpus integrifolia*
	morphine-3-glucuronide (M3G)	*Papaver somniferum*
	plant agglutinin (PNA)	*Arachis hypogaea*
	soybean agglutinin (SBA)	*Glycine max*
	taxol	*Taxus brevifolia*

[1] FNIII EDA = alternatively spliced type III repeat extra domain A of cellular finronectin; Gp96 = glycoprotein 96; Hsp = heat shock protein; mDF2β = marine beta-defensin 2; p8 PGLC = p8 proteoglycolipid complex.
[2] TLR4 antagonists are marked with an asterisk.

9.2.3 The complement system

The complement system, a chief component of the innate immunity, was identified more than 100 years ago as a heat-labile factor in blood serum that kills bacteria. Complement is a complex, self-regulatory system of approximately 35 proteins that usually designated either as complement components (C) or complement factors which are designed to attack and kill extracellular bacteria in body fluids (Vorup-Jensen and Boesen, 2011). Complement acts via three main pathways, called classical, lectin and alternative one. These pathways are initiated by sensory molecules recognising unwanted material amenable to clearance. There is a series of proteolytic activation steps that generate effector molecules of complement, such as potent anaphylatoxin C5a and fragments of C3 that are strong opsonins facilitating phagocytosis. The final step of the complement action includes formation of the membrane attack complex, which is able to lyse Gram negative bacteria and to damage eukaryotic cells (Blom, 2017). Therefore, microbial pathogens are attacked and eliminated by a pore-forming proteins specially adapted for their targets. These specific proteins are able to inflict physical damage thereby damaging permeability barriers and structural protection (McCormack *et al.*, 2013) with the following penetration by additional effectors to complete target destruction. In fact, the complement system is highly inducible with a liberation of both antimicrobial elements (e.g. the C9 membrane attack complex) and elements that attract and modulate the cellular immune system (Juul-Madsen *et al.*, 2008). Indeed, the combination of physical and chemical attack is highly efficient allowing animal survival despite continuous attacks by various microbes being huge in number and possessing the ability to rapidly replicate and mutate (McCormack *et al.*, 2013). The complement system is considered to be a phylogenetically ancient component of innate immunity, but it seems likely that it also plays an important role in adaptive immunity by acting as a bridge between both systems (Borceux *et al.*, 2020).

Therefore, phagocytes such as neutrophils, dendritic cells, and macrophages are known to kill pathogens through the release of toxic chemicals such as ROS and RNS and through degranulation. The phagocytes also process and present antigen to cells of the acquired immune system. The humoral factors of innate responses such as the complement, acute phase proteins, defensins, cathelicidins, pentraxins, heat shock proteins and the release of histamine and heparin from mast cells, are shown to play a significant role in bacterial infections, but are less understood in relation to viral infection, especially in livestock species, including pigs (Toka and Golde, 2013).

9.2.4 Macrophages

Macrophages perform a range of functions, including phagocytosis of foreign particles, destruction of bacterial or tumour cells, secretion of prostaglandins and cytokines (Table 9.4) and as a result regulating activity of lymphocytes and other macrophages (Qureshi, 1998).

Table 9.4. Macromolecules produced in immune cells (adapted from Surai, 2006, 2018).[1]

Name	Site of production	Principal effects
IL-1	MPH, epithelial cells	Stimulation of T- and B-cell proliferation and production of inflammation-modifying proteins in the liver; induction of acute phase proteins; increase of effector cell access; tissue destruction
IL-2	T-helper cells, Mast cells	Proliferation of T- and NK cells, increase in NK cell activity; activation of production of cytokines
IL-3	T-helper cells	Induction of proliferation and differentiation of stem cells in bone marrow
IL-4	T-helper cells, mast cells	Stimulation of T- and B-cell proliferation and IgG and IgE synthesis in plasma cells; increase in macrophage cytotoxicity; induction of Th2 and inhibition of Th1 responses
IL-5	T-helper cells, mast cells	Stimulation of B-cell proliferation and activation and IgA synthesis in plasma cells stimulation of eosinophil production and accumulation
IL-6	MPH, T-helper cells, fibroblasts, endothelial cells, mast cells	Induction of differentiation of B-cells and production of inflammation-modifying proteins in the liver; lymphocyte activation; fever
IL-7	Thymic epithelial cells, bone marrow cells	Stimulation of pre-T and pre-B-cell proliferation
IL-8	Monocytes, MPH, fibroblasts, keratinocytes	Promotion of neutrophil activation, adhesion to endothelial cells and migration to inflamed site
IL-9	T-helper cells	Stimulation of mast cell proliferation
IL-10	T-helper cells, MPH	Inhibition of production of IFN-γ and some ILs; B-cell activation
IL-11	Mesenchymal cells	Stimulation of megakaryocyte production
IL-12	B cells, MPH	Stimulation of production of T-helper cells type 1 and IFN-γ synthesis by T cells and NK cells
IFN-γ	T cells, NK cells	MPH activation; induction of Th1 and inhibition of Th2 responses
TNF-α	MPH, T cells	Activation of vascular endothelium; fever; shock; mobilisation of metabolites
Platelet-activating factor	MPH	Promotion of neutrophil migration and eosinophil mobilisation
Fibroblast growth factor	MPH	Induction of division of connective tissue cells at sites of injury
Complement	MPH	Lysis of foreign cells
α-l-protease inhibitor	MPH	Inhibition of proteases and breakdown of healthy tissues
α-2-macroglobin	MPH	Inhibition of proteases and breakdown of healthy tissues
Leukotrienes	MPH	Promotion of leukocyte accumulation at inflammatory foci

Table 9.4. Continued.

Name	Site of production	Principal effects
Granulocyte colony-stimulating factor	MPH	Increase of the production of granulocytes (neutrophils) in the bone marrow
Granulocyte-macrophage colony-stimulating factor	MPH	Increase of granulocyte and monocyte production

[1] IFN = interferons; IL = interleukin; MPH = macrophages; NK = natural killer; TNF = tumour necrosis factor.

Macrophages are known to differentiate from hemopoietic progenitors after cell migration to the different compartments of the body under tight control by various adhesion molecules including integrins, immunoglobulin-superfamily members, selectins, and others. In the tissues, macrophages are shown to adapt to local microenvironment and develop different phenotypes. In fact, differentiated residential macrophages can comprise 10-15% of the total cells in tissue (Mair *et al.,* 2014). Macrophages express MHC Class I, MHC Class II and costimulatory molecules. Pig macrophages play a central role in body defence against microbial infections and they express PRR including the extensively studied TLR (Mair *et al.,* 2014; Wu and Kaiser, 2011). The immune system utilises a range of cytokines to deliver signals in the immune microenvironment (Delgoffe *et al.,* 2011). In fact, phagocytosis is the major mechanism by which microbes are removed from the body and is especially important for defence against extracellular microbes. As a result of stimulation (for example by microbes) monocytes differentiate into macrophages that are more powerful in mediating host defence (Klasing, 1998a). Macrophages are equipped with a range of receptor systems including the scavenger receptors, complement receptors, Fc receptors, C-type lectins and mannose receptors mediating opsonic and non-opsonic recognition (Kaspers *et al.,* 2008; Mair *et al.,* 2014). The phagocytic process includes several stages (Lydyard *et al.,* 2000):
- Movement of phagocyte towards the microbe using chemotactic signals.
- Physical contact of the micro-organisms with macrophage and attachment to the phagocyte surface with a nonspecific or receptor-mediated binding.
- Endocytosis (engulfment) of microbe resulting in a phagosome by invagination of surface membrane.
- Fusion of the phagosome with a lysosome.
- Killing microbes by bombarding them with oxidants (superoxide and hydroxyl radicals, hydrogen peroxide, nitric oxide, hypochlorous acid, etc.).

Macrophages are also considered to be important antigen presenting cells having regulatory and effector functions in the specific immune response and tissue homeostasis maintenance (Crisci *et al.,* 2019). Therefore, via antigen uptake and presentation as well as by the production of inflammatory cytokines macrophages are believed to provide a first line of defence against pathogens and their life cycle can be divided in four different stages: differentiation, priming, activation, and resolution (Gordon and Martinez, 2010). In general ROS, RNS and eicosanoids (e.g. leukotrienes and prostaglandins) have received substantial attention as major metabolites produced by macrophages (Dietert and Golemboski, 1998) (Figure 9.3). Because of this powerful weapon, macrophages bind, internalise, and degrade foreign antigens (e.g. bacteria) quite quickly. It takes only 15 minutes for chicken macrophages to kill more than 80% of the internalised *Salmonella* (Qureshi *et al.*, 1998). Therefore, natural immunity works rapidly and gives rise to the acute inflammatory response. They contain various substances (including enzymes producing free radicals and small peptides with an antibiotic activity) involved in microbial killing. They have also receptors for chemoattractive factors released from microbes. In addition to ROS, RNS and eicosanoids mentioned above, macrophages also synthesise and secrete a great number of such communicational molecules as cytokines, including the pro-inflammatory cytokines called interleukin (IL)-1, IL-6 and tumour necrosis factor (TNF)-α. They also produce cytokine inhibitors, endocrine hormones, neurotransmitters (Klasing, 1998a) which regulate specific immunity (initiating and directing the immune and inflammatory responses) and many other related physiological responses. Therefore, these cells are important amplifiers of the immune response, both by cytokine production and by serving to present parasite derived peptides to T cells. "Classical activation" of macrophages via IFN-γ (produced mainly by activated Th1 cells or NK cells) is associated with a pro-inflammatory, antimicrobial activity, while activation of macrophages through the "alternative pathway" (via IL-4 and/or IL-13) leads to promotion of a Th2 and antibody-mediated immune response (Mair *et al.*, 2014). At the end of infectious process, anti-inflammatory cytokines are produced, and the situation is resolved. Therefore, porcine macrophages are an important innate immune cells involved in the first lines of defence against various viral, bacterial, and parasitic pathogens.

9.2.5 Neutrophils

Pig neutrophils are highly phagocytic and capable of killing a variety of microorganisms after phagocytosis. In fact, pig neutrophils being the predominant population in pig blood, are the first cell type recruited to the site of an infection and play an important role in innate immunity and the subsequent inflammatory response (Holgersson *et al.*, 2002; Mair *et al.*, 2014; Wu and Kaiser, 2011). Pig neutrophils are shown to display the same morphology as those of humans, but are a smaller size, with lower granularity and higher activation threshold (Crisci *et al.*, 2019). For many years, phagocytosis and subsequent oxidative burst with production of various free-radical related products, an essential weapon to kill pathogens as well as degranulation with protease-dependent killing of pathogens inside of phagolysosomes were thought to be the major neutrophil effector mechanisms. Inflammatory stimuli such as LPS, or various infectious agents cause a dramatic influx of neutrophils and the activation of neutrophils by pathogens or by cytokines induces the expression of various pro-inflammatory cytokines such as IL-1, IL-6 and IL-8 (Juul-Madsen *et al.*, 2008; Mair *et al.*, 2014). Therefore, the first line of defence against various pathogens, is the innate immune system, which is primed even in the absence of infection. It is characterised by a range of various molecules and

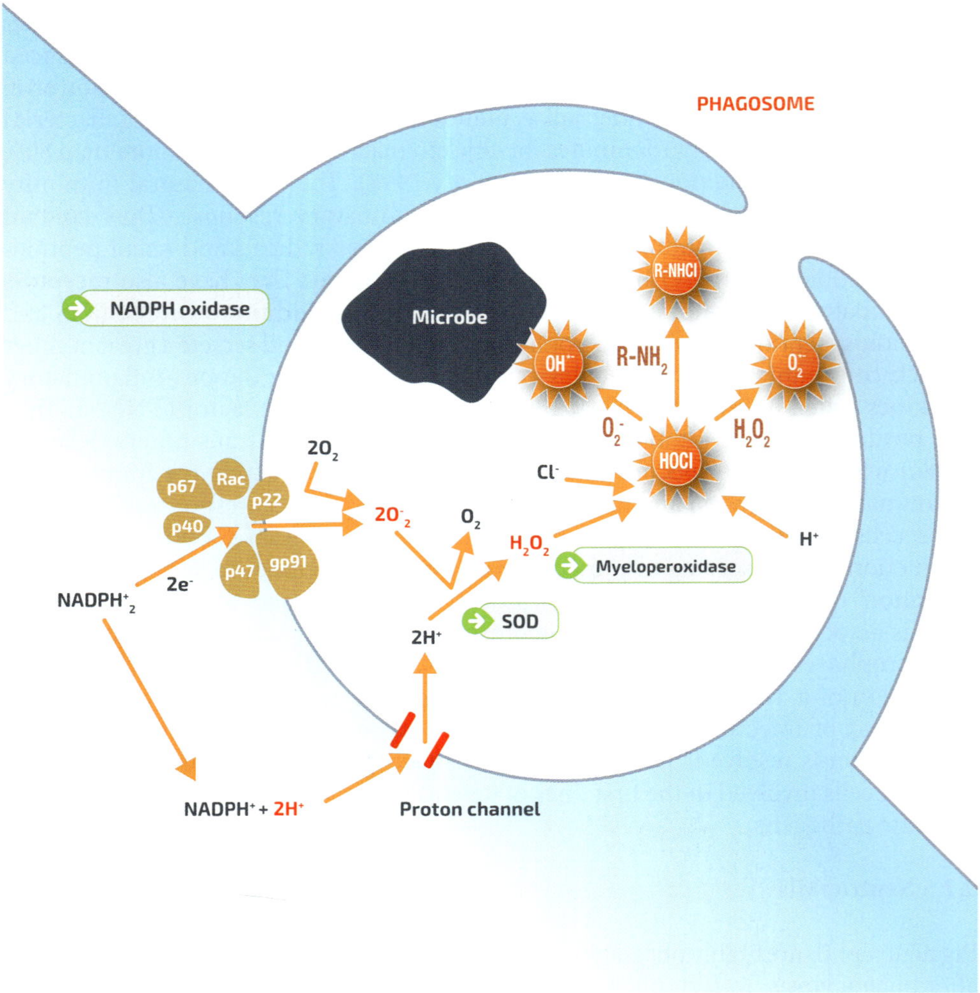

Figure 9.3. Redox balance, the key to macrophage function.

cell-surface receptors – known as pattern recognition receptors (PRRs) that recognise generic molecular structures associated with different groups of pathogens (Maizels, 2009). These receptors perform many different functions, including mobilisation of macrophages and granulocytes, unleashing antimicrobial proteins and reactive metabolites. They are also involved in mobilisation of dendritic cells, which activate the lymphocytes of the adaptive immune system, inducing proliferation of T cells and antibody-producing B cells with variable receptors that specifically recognise the parasite (Maizels, 2009). It should be mentioned that evidence has been accumulated to show that innate immune cells can show memory-like behaviour. One of the key mechanisms that mediates these effects is epigenetic reprogramming. For example, Yoshida *et al.* (2015) showed that the stress response transcription factor ATF7 mediates lipopolysaccharide (LPS)-induced epigenetic changes in macrophages and these changes lead to enhanced protection against pathogens.

Neutrophil activation and phagocytosis of foreign particles are regularly accompanied by a so-called 'respiratory burst', an increase in the production of reactive oxygen species (ROS), exerted by the enzyme complex NADPH oxidase (Figure 9.4).

Therefore neutrophils as well as other phagocyte leukocytes (e.g. macrophages, monocytes and eosinophils) can synthesise toxic oxygen metabolites such as superoxide anion (O_2^-), hydroxyl radical (OH*), singlet oxygen ($^1O_2$), hydrogen peroxide (H_2O_2), nitric oxide (NO), peroxinitrite (ONOO-) hypochlorous acid (HOCl) and chloramines during the respiratory burst (Zhao *et al.*, 1998). For example, a bacterium coming into contact with the plasma membrane is enclosed in a plasma membrane vesicle containing NADPH oxidase and exposed to an intensive flow of superoxide radicals (Gille and Sigler, 1995). Superoxide radical can disproportionate to H_2O_2 which penetrates into the bacterium with a production of hydroxyl radical, which is ultimately a deadly weapon able to damage any biological molecules. In general, the production of ROS and reactive nitrogen species is a characteristic for both mammalian and avian macrophages (Qureshi *et al.*, 1998). It is interesting that,

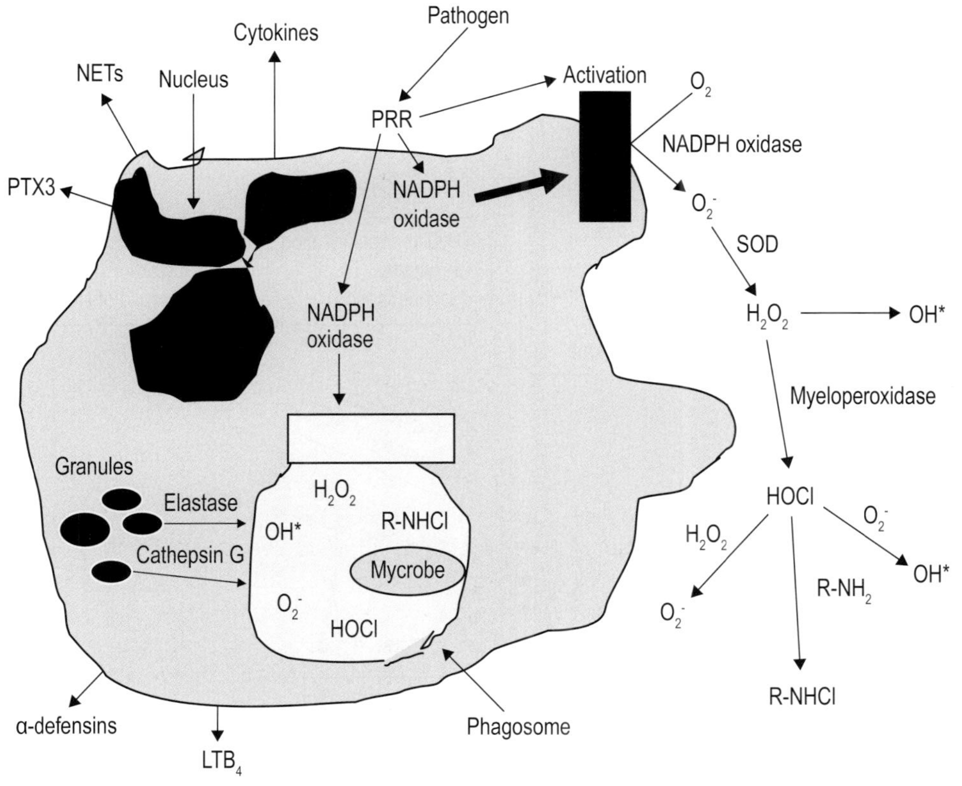

Figure 9.4. Respiratory burst in neutrophils (adapted from Surai, 2006, 2018).

people whose phagocytes possess no functional NADPH oxidase, are shown to suffer from chronic infections of the skin, lung, liver and bones leading to a premature death (Halliwell and Gutteridge, 1999).

However, a traditional view on the neutrophil action has been recently widened by finding that activated neutrophils can release weblike structures of DNA coated with histones, elastase, myeloperoxidase and cathepsin G (Figure 9.5). These structures are called neutrophil extracellular traps (NETs) responsible for extracellular immobilisation/trap and killing bacteria primarily through a cell death process termed NETosis. In general, NETosis was suggested to be a cell death program distinct from apoptosis and necrosis and dependent on the generation of ROS by NADPH oxidase. In fact, NETosis is shown to be dependent on plasma membrane lysis, which followed chromatin decondensation and chromatin mixing with cytoplasmic granule proteins including myeloperoxidase (MPO), neutrophil elastase, calprotectin, cathelicidins, defensins and actin (Castanheira and Kubes, 2019; Papayannopoulos, 2018). NETosis is associated with slow cell death: NETs expand into the extracellular space 3-8 hours after neutrophil activation. There is another alternative mechanism called non-lytic NETosis leading to the rapid release of NETs within minutes of exposure to a pathogen via the secretion of chromatin and granule contents without cell death (Papayannopoulos, 2018).

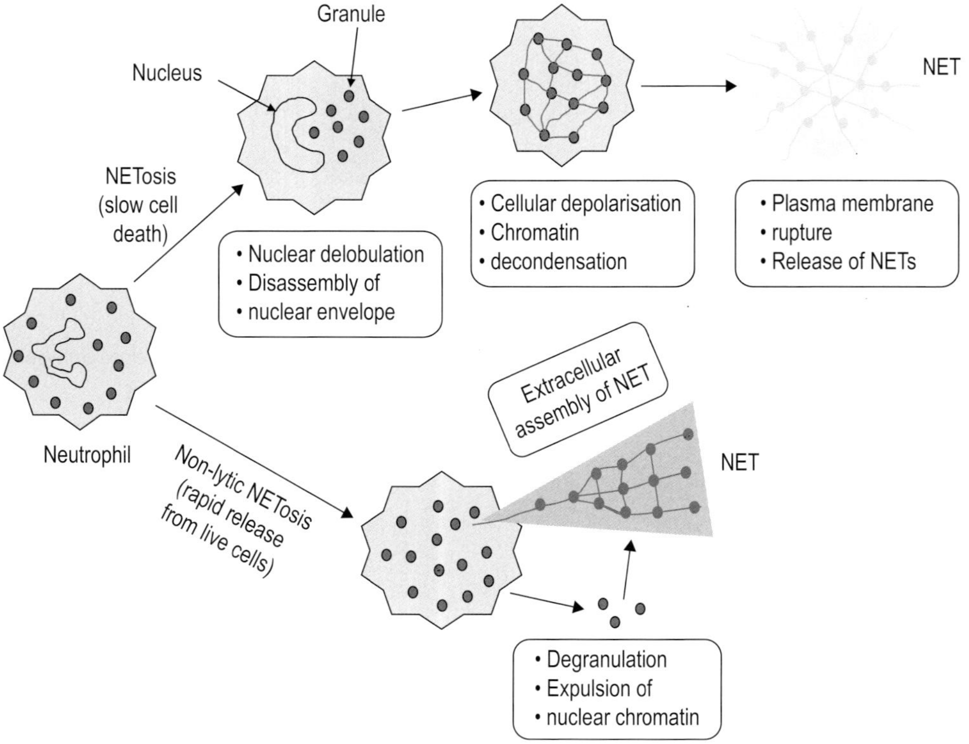

Figure 9.5. Neutrophil extracellular traps (NET) formation pathways (adapted from Papayannopoulos, 2018).

It seems likely that these two mechanisms are working in concert providing an effective mechanism of adaptation to the situation. At first stage of fighting pathogens the non-lytic NETosis is involved providing a rapid response. If the quick response is not able to get rid of pathogen, at the later stage of the protective action of neutrophils more powerful NETosis with slow cell death is involved (Figure 9.5). NETs seem to have different immunogenic effects depending on their composition. Integration of multiple signals, such as environmental triggers, metabolic state and type and amount of a pathogen determine the scenario of neutrophil action including phagocytosis and pathogen killing via oxidative burst, production and release of toxic free radical-related products, degranulation with releasing various enzymes and microbicidal peptides including cathelicidins and defensins, etc. or generate NETs (Wirestam *et al.*, 2019).

The purpose of immune cell products is to destroy invading organisms. However, excessive or inappropriate production of these substances is associated with mortality and morbidity after infection and trauma, and inflammatory diseases. Oxidants enhance IL-1, IL-8, and TNF production in response to inflammatory stimuli by activating the nuclear transcription factor, NF-κB. Therefore, they also can cause a series of harmful effects, including killing host cells, injuring cells and tissue directly via oxidative degradation of essential cellular components and damaging cells indirectly by altering the protease/antiprotease balance that normally exists within the tissue interstitium (Conner and Grisham, 1996). They also can cause cell mutations and sister chromatid exchanges (Weitzman and Stossel, 1981). Therefore, overproduction of ROS or impaired antioxidant defence can result in oxidative damage to host macromolecules (McBride *et al.*, 1991) and extensive pathology and disease (Dietert and Golemboski, 1998). For example, in the case of disease challenge, the overproduction of ROS may occur, and they can leak from fagolysosomes into the surrounding cytoplasm and intracellular space. A number of antioxidant enzymes is expressed at the same time to protect the cells from the cytotoxic effects of ROS directed against engulfed microorganisms. However, low efficiency of these antioxidant mechanisms, for example due to selenium deficiency, could damage cell's microbicidal and metabolic functions (Larsen, 1993). Indeed, macrophage function is disturbed, if there is a lack in antioxidant enzyme activity (Ebert-Dumig *et al.*, 1999). It was shown that TrxR activity as well as GPx activity in these cells can be upregulated by the addition of selenium *in vitro* and *ex vivo*. Recent results revealed that TR1 is both a regulator and a regulated target in the macrophage gene expression network and suggest a link between selenium metabolism and immune signalling (Carlson *et al.*, 2011).

9.2.6 Natural killer cells

In pigs lymphoid cells are the most studied and proper identification and characterisation of B cells, CD4, and CD8T cells, NK, NKT, γδ T cells have been conducted (Blanc *et al.*, 2020). NK cells were originally described as a population of granular lymphocytes with the ability to lyse tumour and virally-infected target cells. They differ from classical lymphocytes being larger in size, containing more cytoplasm and having electron dense granules. However, in contrast to other species, porcine NK cells morphologically resemble small to medium sized cells that lack granules (Mair *et al.*, 2014). The life span of a mature NK cell (in the absence of activation) was shown to be about 7-10 days (Mak *et al.*, 2013). The mechanism of killing is mediated through release of its granule contents (perforins and granzymes) onto the surface of the infected cell (Lydyard *et al.*, 2000). In human NK cells comprise about 10-15% of all circulating lymphocytes. They produce cytokines and chemokines and can lyse target cells without prior sensitisation being crucial components of the innate immune system. They also exert cell cytotoxicity by recognising and inducing lysis of antibody-coated target cells and can be activated non-specifically by several cytokines via various receptors (Middleton *et al.*, 2002). The NK cells were shown to be extremely important in mucosal immunity where they are mainly located (Rogers *et al.*, 2008). Therefore, NK cells recognise and kill infected or transformed cells. First, activated NK cells make a tight contact with the target cell. After that, they polarise lytic granules toward the immunological synapse and release cytolytic proteins toward the target cell (Claus *et al.*, 2016). Indeed, NK cells are important in innate immunity, first described as guardians for the detection and clearance of transformed or virus-infected cells and recently this cell type has been revealed to be also able to recognise and respond to bacteria-infected cells (Adib-Conquy *et al.*, 2014). In fact, NK cells possess receptors allowing them to sense and respond to viral and bacterial patterns, including TLRs. Upon TLR activation, NK cells are an important source of IFN-γ and granulocyte macrophage colony-stimulating factor, cytokines necessary to fight infection (Adib-Conquy *et al.*, 2014). In fact, the consequences of TLR activation are diverse and include activation of cytokine, IFN and chemokine production, cell maturation, induction of degranulation and the direct production of AMP, etc. (Juul-Madsen *et al.*, 2008). On the one hand, porcine NK cells they can produce IFNγ upon activation, thereby steering the adaptive immune response toward a Th1 response. On the other hand, they also show cell surface expression of MHC class II and the co-stimulatory molecules CD80/86, can take up debris of killed target cells, and trigger T cell proliferation, can perform activities that resemble those of professional antigen presenting cells (De Pelsmaeker *et al.*, 2019). In general, data on NK cells in pigs are limited due to the lack of specific markers recognising only NK cells in pigs.

9.2.7 Natural killer T cells

Natural killer T (NKT) cells are described as T lineage cells that recognise glycolipid rather than peptide antigens by their T cell receptors and sharing morphological and functional characteristics with both T cells and NK cells (Mak *et al.*, 2013). NKT cells are known to develop in the thymus from the same NK/T precursors that is converted to αβ T and γδ T cells. Low numbers of NKT cells were shown to be found in peripheral blood, lymph nodes and various tissues including spleen, liver, thymus, bone marrow. Following their activation, NKT cells are characterised by quick response to challenge without a requirement to differentiate into effector cells and this makes the NKT cells important players in innate defence against bacterial and viral pathogens. In fact, activated NKT cells can rapidly secrete chemokines and cytokines simultaneously activating an array of immune cell types, including NK cells, dendritic cells (DCs), B cells and conventional T cells (Yang *et al.*, 2018) that allow these cells to regulate various immune responses. In addition, many of the cytokines secreted by NKT cells were found to have powerful effects on αβ T cell differentiation and functions, linking NKT cells to adaptive defence (Mak *et al.*, 2013). Current understanding about porcine NKT cell biology is rudimentary but porcine NKT cells were shown to display striking phenotypic and functional similarities to human but less to murine NKT cells (Schäfer *et al.*, 2019). Major histocompatibility complex class II (MHC II) and CD25, CD5, ICOS were shown to be activation markers on porcine NKT cells. Furthermore, activated porcine NKT cells were indicated to express IFN-γ, upregulated perforin expression, and displayed degranulation. Interestingly, in steady state, porcine NKT cell frequency was highest in newborn piglets and decreased with age (Schäfer *et al.*, 2019).

9.2.8 γδ T cells

Various T lymphocytes can be distinguished based on the polypeptide chains making up their T cell receptors (TCRs). It is known that in both humans and mice, the majority of T cells in the body are αβ T cells having TCRs built with the TCRα and TCRβ chains. Furthermore, mature classical αβ T cells are characterised by either monomeric CD4 coreceptors or heterodimeric CD8αβ coreceptors. However, in mammals, including pigs, there is another T cell population called γδ T cells expressing TCRs consisted of the TCRγ and TCRδ chains and carrying either homodimeric CD8αα coreceptors or no coreceptor at all (Mak *et al.*, 2013). Interestingly, pigs, together with birds, belongs to the group of species in which γδ T cells may account for 25-85%, depending on the age of all circulating peripheral blood lymphocytes (Crisci *et al.*, 2019). Therefore, in young pigs T cells expressing γδ T cell receptors were found to account for 20-50% of circulating T cells (Toka and Golde, 2013).

Furthermore, γδ TCRs are known to bind to non-peptide antigens derived from pathogen proteins and host stress molecules that are either recognised directly, or presented on CD1c or non-classical MHC class Ib molecules (Mak *et al.*, 2013). In general, γδ T cells are shown to share main features with classical αβ T cells, including high cytotoxic activity and regulatory functions related to maturation of dendritic cell, production of a variety of cytokines, and generation and retaining immunologic memory (Crisci *et al.*, 2019). In pigs, 12 subsets of γδ T cell have been described that belong to two major groups, (1) cluster of differentiation 4 (CD4)– γδ thymocytes that are further subdivided based on CD2/CD8αα expression, and (2) CD4+ γδ thymocytes that stably express CD1, CD2 and CD8αα, but are not exported to the periphery (Toka and Golde, 2013). In addition to their important role in adaptive immune response γδ T cells also expressing innate immune functions such as production of proinflammatory cytokines and non-MHC restricted cytolytic activity, i.e. NK-like activity (Toka and Golde, 2013). Indeed, antigen-presenting and cytolytic functions of porcine γδ T cells (Sipos, 2019) deserve more research.

In fact, NK, NKT and γδ T cells are suggested to connect the innate and adaptive immunity because these cells:

- are closely related to the T lymphocyte lineage;
- recognise antigens in a more general way than do αβ T cells;
- have direct and rapid effects on pathogens;
- affect B lymphocytes and αβ T lymphocytes via cytokine secretion (Mak *et al.*, 2013).

9.2.9 Dendritic cells

It was shown that the peripheral blood mononuclear cell (PBMC) populations include monocytes, plasmacytoid dendritic cells (pDCs), CD21+ B cells, T helper cells (naïve and activated/memory Th cells), cytotoxic T cells (CTLs), regulatory T cells (Tregs), γδ T cells (CD4- subset), and natural killer (NK) cells (Sipos, 2019). Dendritic cells (DC) are the professional antigen-presenting cells (APCs) in pigs effectively linking the innate and adaptive branches of the immune system and orchestrating the immune response. The main DC function can be summarised as follows (Franzoni *et al.*, 2019):

- effective in sensing viruses through a wide range of surface, cytosolic and endosomal receptors, and TLRs;
- priming naïve T lymphocytes and controlling T lymphocyte activation and regulation;
- B-cell activation and the regulation of antibody synthesis;
- directly interacting with B cells and transferring antigens to naïve B cells to initiate specific antibody synthesis;
- NK cell activation and improving their survival, expansion, and function;
- γδ T cell activation, through the release of pro-inflammatory cytokines and expression of ligands for γδ T cell receptors and NK receptors.

Interestingly, the crosstalk between DCs and NK cells is shown to be bidirectional: activated NK cells can affect DC maturation leading to increased stimulation of T cell responses (Franzoni *et al.*, 2019). However, study of these important cells in pigs has been complicated by the low frequency of DC populations in blood and tissue (Kim *et al.*, 2019). In general, DCs can be subdivided into pDCs which have a main function in sensing nucleic acid, and classical (monocyte-derived) DCs which are specialised in antigen presentation (Summerfield and McCullough, 2009). Porcine pDCs, most important immunocytes in defence against viral infections, are shown to be the significant producers of IFN-α, and the proinflammatory cytokines TNF-α, IL-6 and IL-12. They are also known to express TLR7, TLR9 and TLR3 and characterised by the high expression of complement-related genes (C2, C3, C5, and CD93; Sipos, 2019).

In general, some important features of pig innate immunity summarised by Mair *et al.* (2014) are as follows:

- ten different TLRs (TLR1-10) like in humans but less than in mice;
- TLR7 and TLR9 restricted to pDCs and not expressed on classical DCs as it is the case in mice;
- great diversity for interferon type I (39 genes versus 19 in humans);

- no induction of NOS2 in response to LPS like in humans;
- more cathelicidin genes than in humans (10 versus 1);
- no alpha-defensins in pigs;
- duplication of IL-1β gene (IL-1β and IL1-βL);
- duplication of 2',5'-oligoadenylate synthetase 1 (OAS1) gene (OAS1 and OAS1L);
- duplication of CD163 gene (CD163 and CD163L);
- substantial proportion of NKp46$^-$ NK cells;
- very high frequency of γδ T cells in the blood of adolescent animals.

Furthermore, an example of cellular interactions during the innate response to virus infection is shown in Figure 9.6.

Once a pathogen enters to the host, the initial nonspecific response is an inflammatory reaction. In fact, during an inflammatory response, the innate immune cells release signals responsible for recruitment of distant immune cells to the site of infection, and for changing the metabolism of the animal. Indeed, an inhospitable environment for the invading pathogen is created in the body by increasing metabolic rate (fever), reducing feed intake, and initiating the preferential breakdown of skeletal muscle to support gluconeogenesis and synthesis of acute phase proteins in the liver (Korver, 2012). Therefore, inflammatory response results in a series of behavioural, immunologic,

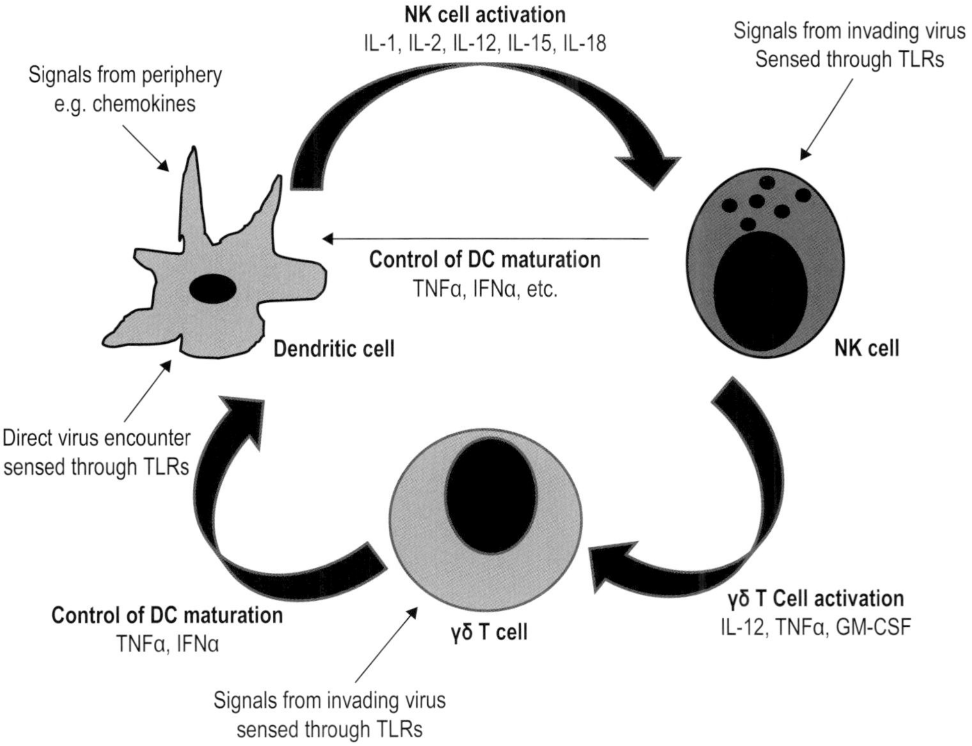

Figure 9.6. Cellular interactions during the innate response to virus infection (adapted from Toka and Golde, 2013).

vascular and metabolic responses (Korver and Klasing, 2001). As a result lethargy and anorexia are observed, growth rate could be slowed, muscle protein degradation increased with possible increased morbidity and mortality and decreased productivity. At the same time acute phase protein (APP) synthesis is increased. These proteins are involved in various aspects of host protection and help homeostasis restoration. In fact in poultry, a number of acute phase proteins have been identified, including metallothionein, ovotransferrin, α1-acid glycoprotein, ceruloplasmin, fibronectin, serum amyloid A, mannan-binding protein and C-reactive protein, etc. (Korver, 2012). In pigs, C-reactive protein, haptoglobin, serum amyloid A and pig major acute phase protein (Pig-MAP) are shown to be the main positive APP (Wierzchosławski *et al.*, 2018). In particular, pig-MAP is considered to be one of the main APPs in pigs and its concentration is shown to increase in pigs with different acute phase conditions, including viral, bacterial and parasitic infections (Piñeiro *et al.*, 2013). In general, a price for immunological defence against pathogens could be quite high.

A scheme presented in the Figure 9.7 shows how various antigens could potentially affect T cells.

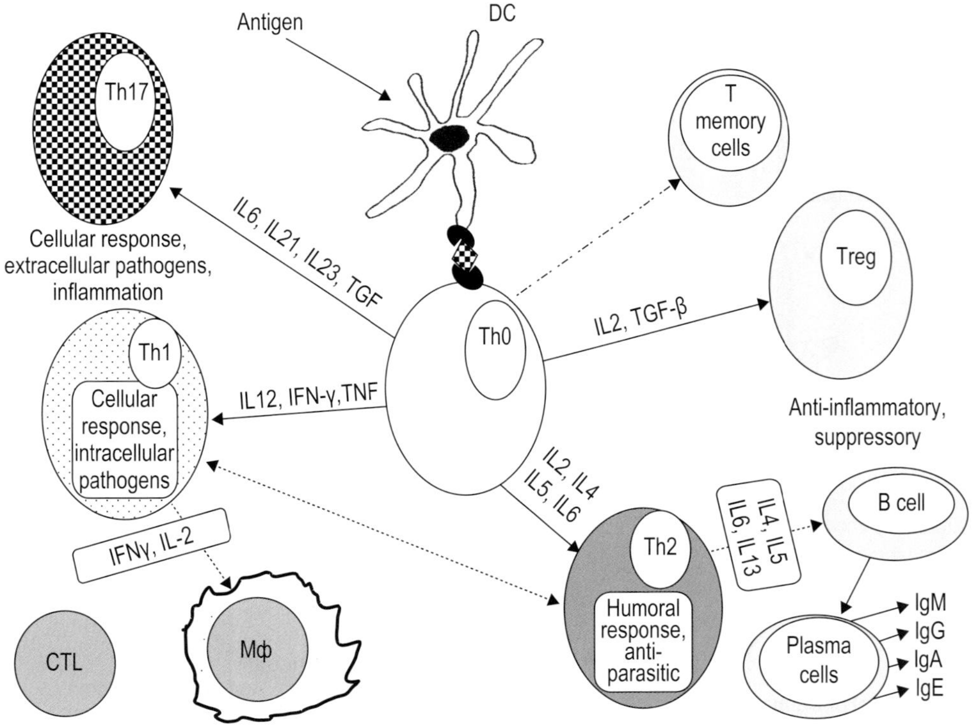

Figure 9.7. Antigen-dependent T cell differentiation (adapted from Elmadfa and Meyer, 2019; Gostner *et al.*, 2013).

Indeed, stimulation of naïve T cells (Th0) by dendritic cells (DC) is shown to be responsible for the differentiation with following effects on various parts of the immune response. It is important to mention that redox balance is a major player in this process (Gostner *et al.*, 2013). On the one hand, a Th1-type immune response is characterised by the secretion of pro-inflammatory cytokines, such as interferon gamma (IFNγ) and IL-2 leading to the activation of macrophages (Mφ) and cytotoxic T cells (CTL). On the other hand, a Th2-type immune response is related to production of IL-4, -5, -6, and -13 by these cells to stimulate B cells and to produce antibodies. In fact, Th1and Th2-type immune responses are shown to be cross-regulated: e.g. suppression of Th1 reactions favours Th2 responses and vice versa. Whereas oxidative conditions are believed to favour Th1 reactions, increased AO defences preferentially activates Th2-type responses (Gostner *et al.*, 2013). Indeed, the dendritic cells can affect T cell differentiation into various subsets of T cells, including Th1, Th2, Th17 and T-regulatory cells and, depending on antigen, they provide an important link between natural and adaptive immunity.

9.2.10 Acquired immunity

Acquired or specific immunity includes humoral immunity and cell-mediated immunity (Figure 9.8). There are two major types of lymphocytes, B cells and T cells. Humoral immunity is mediated by antibodies that are released by B-lymphocytes into the bloodstream. This immunity is based on the production of immunoglobulins (Table 9.5). They are responsible for specific recognition and elimination of various antigens: they bind and remove from the host invading organisms/substances.

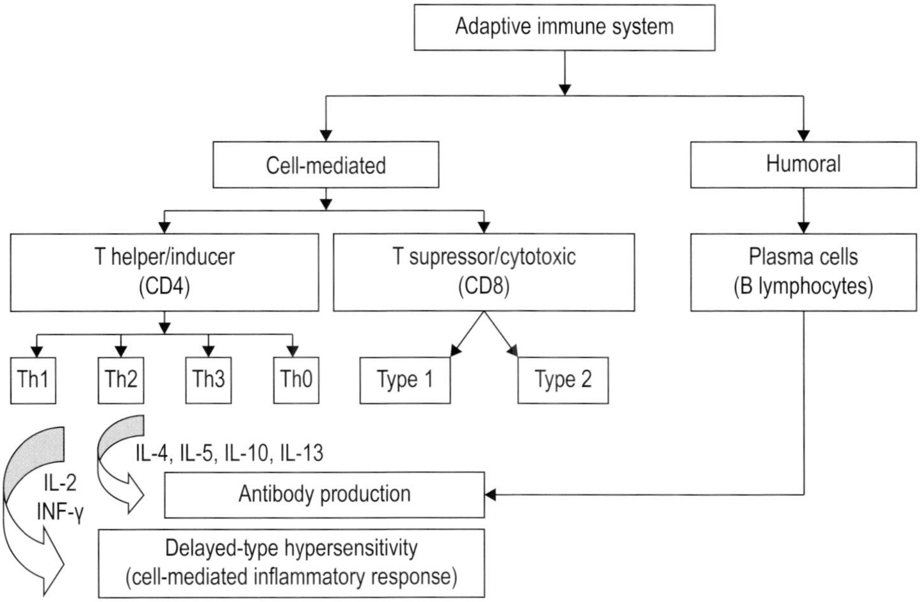

Figure 9.8. General scheme of adaptive immune system (adapted from Surai, 2006, 2018).

Table 9.5. Main immunoglobulin (Ig) classes (adapted from Surai, 2006, 2018).

Immunoglobulin	Characteristics
IgG1; IgG2; IgG3; IgG4	Largest quantity; provide the bulk of immunity to the most blood borne infectious agents; the only antibody class to cross the placenta to provide humoral immunity to the infant; vaccination asks the immune system to produce IgG specific for a particular antigen
IgA	A first line of defence against microbes entering through mucosal surfaces directly communicating with the environment; synthesised by plasma cells; prevents colonisation of mucosal surfaces by pathogens
IgM	The first antibody produced in an immune response in large quantity
IgD	Present in humans, not documented in animals; functions as an antigen receptor on B cells
IgE	Involved in allergy development and in immediate hypersensitivity syndromes such as hay fever and asthma

Cell-mediated immunity is based on specific antigen recognition by thymus-derived T lymphocytes. Due to this immunity, cells infected with a foreign agent, for example virus, are destroyed via a direct contact between an activated T cell and target (infected) cell (Qureshi *et al.*, 1998; Rahe and Murtaugh, 2017). Cell-mediated immunity is responsible for delayed-type hypersensitivity reactions, foreign graft rejections, resistance to many pathogenic microorganisms and tumour immunosurveillance (Kaser *et al.*, 2011; Wu and Maydani, 1998). Therefore, adaptive immunity is based on a tightly regulated interplay between antigen-presenting cells and T and B lymphocytes. They facilitate pathogen-specific immunologic effector pathways, formation of immunologic memory, and regulation of host immune homeostasis. Interactions between T and B-cells, as well as antigen presenting cells, are responsible for the development of specific immunity. These defence mechanisms of specific immunity are induced or stimulated by exposure to foreign substances, are specific for distinct macromolecules and increase in magnitude with each successive exposure to a particular macromolecule (Miles and Calder, 1998).

In comparison to natural immunity, specific immunity takes longer to develop, but is highly specific for antigens and has memory (Table 9.5). These two parts of the immune system work together through direct cell contact and through interactions involving such chemical mediators as cytokines and chemokines (Figure 9.9). The pig immune system requires the co-operation of macrophages, bursa-derived B-lymphocytes and thymus-derived T lymphocytes with various other types of cells. The immune response involves cellular proliferation (T lymphocytes), enhanced protein synthesis (including immunoglobulin synthesis by B-lymphocytes and acute phase protein synthesis by liver) and inflammatory mediator production. Physiological changes resulting from stimulation of the immune system include fever, anorexia and loss of tissue components (Grimble, 1997).

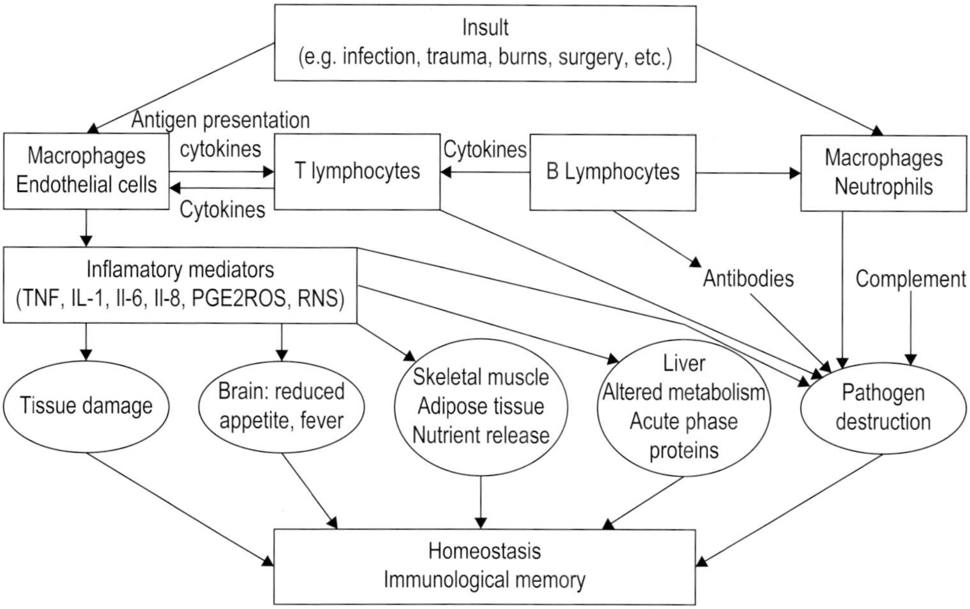

Figure 9.9. The natural and adaptive immunity interactions (adapted from Surai, 2006, 2018).

Therefore, the two arms of the immune system (innate and acquired) are complementary and they operate in an integrated fashion to provide an effective protection of the host. As mentioned above, from one hand, antigen is recognised and presented to the cells of the acquired immune system by innate immune cells. From the other hand, adaptive immune system cells (T helpers) regulate the bactericidal activity of phagocytes (e.g. macrophages and neutrophils). Indeed, the innate immune system provides a rapid response to novel pathogens, but such a protective action can be metabolically and physiologically quite costly to the host. The acquired immune system can respond in a much more pathogen-specific fashion with minimal impact on the host but depends on the interactions with the innate immune system (Korver, 2012; Rahe and Murtaugh, 2017; Table 9.6).

9.2.11 Immune system evaluation

When analysing data related to immunomodulating properties of various nutrients it is necessary to pay special attention to methods used to assess immunological functions. For example, *in vivo* methods of immune function assessment are based on two main approaches: antibody response to vaccine or delayed-type hypersensitivity (DTH) reaction. In the first method immunisation with appropriate antigens (viral or bacterial) can elicit serum antibodies. So called haemagglutination assay measures serum antibody concentration (titer) against antigens. Sheep red blood cells (SRBC) are often used as antigens. This assay provides information about humoral immunity (B-cell responsiveness) and its association with cell-mediated immunity (T cell co-operation). The second (DTH) method is used to assess cell-mediated immune function.

Table 9.6. Key features of innate and adaptive immunity (adapted from Surai, 2006, 2018).

	Innate	Adaptive
Appearance in evolution	Primitive organisms	Vertebrates
Induction time	Fast (hours to days)	Slow (days to decades)
Recognises	Common pathogen-associated microbial patterns (PAMPs)	Unique epitopes on each pathogen/ antigen
Cellular components	Phagocytes (macrophages and neutrophils); natural killer (NK) cells; mast cells; dendritic cells	T and B cells
Generation of specificity	Encoded in germline; has some specificity, no memory	Somatic rearrangement; highly specific and has memory
Effector mechanisms	Complement (Alternative pathway); cytokines; chemokines; cell-mediated cytotoxicity	Antibodies; cytotoxic T cells (CTL); classical complement activation; antibody-dependent cell-mediated cytotoxicity; cytokines; chemokines
Soluble mediators	Macrophage-derived cytokines	Lymphocyte-derived cytokines
Characteristic transcription factors	NF-κB (+JNK/AP1)	Jak/STAT, NF-κB, etc.
Physiological barriers	Skin	Cutaneous and mucosal immune systems
	Mucosal membranes	Antibodies in mucosal secretions
	Lysozyme	
	Stomach acid	
	Commensal bacteria	

In vitro indices of immune function include (Wu and Meydani, 1998):

- **Lymphocyte proliferation assay**. Lymphocyte proliferation assay provides information about cell-mediated immune response and includes measuring the number of cells in culture with or without addition of stimulatory agent (mitogen). In this assay, isolated lymphocytes are incubated with mitogens which activate division of either T-or B-lymphocytes. Decreased proliferation may indicate impaired cell-mediated immunity. Various mitogens are used in such assays, but most often they include (Hayek *et al.*, 1996):
  - concanavalin A (a type of protein found in the jack bean, T cell mitogen);
  - phytohemagglutin (PHA, a protein obtained from plant seeds, T cell mitogen);
  - lipopolysaccharide (LPS, obtained from the membrane of Gram-negative bacteria, B-cell mitogen); and
  - pokeweed mitogen (T- and B cell mitogen).
- **Measuring cytokine production**. T cells produce a range of protein mediators called cytokines which regulate cell activation, growth, differentiation, inflammation and immunity.

- **Cytotoxicity assay**. This assay assesses activity of cytotoxic T lymphocytes (a group of T cells that kill other cells by recognising their cell-surface antigens) and Natural killer (NK) cells (a group of non-T and non-B lymphocytes that kill virus-infected and tumour cells).
- **Flow cytometric analysis**. The assay deals with identifying the cells with different surface markers. The results can be used for understanding the cellular basis of immune response.
- **Plaque-forming cell test.** Which shows the number of antibody-producing cells.

Effect and role of animal immune system in modern animal production systems is difficult to overestimate. Banning feed grade antibiotics in Europe has made immune system competence the major factor determining efficiency of animal production. Molecular immunology is developing very quickly and mechanisms of immunocompetence have received substantial attention (McCorkle, 1998; Saif and Swayne, 1998) and nutritional modulation of resistance to infectious diseases (Dalgaard, *et al.*, 2018; Colitti *et al.*, 2019; Klasing, 1998b; Venter *et al.*, 2020) is a frontline for future research.

Among different nutrients playing an important role in modulating immune response (Klasing, 1998b; Venter *et al.*, 2020) natural antioxidants, including Se (Surai, 2006, 2018) are shown to be very promising. However, antioxidant roles in animal and human immune system modulation received only limited attention. Mechanisms of nutritional modulation of resistance to infectious diseases were divided by Klasing (1998b) into seven categories. However, selenium was not mentioned there. It is well known that several indicators of immune responsiveness are depressed when animals are selenium and/or vitamin E deficient. In particular, decreases in both cellular and humoral immune function in man, laboratory and farm animals and chickens are also observed (Dalgaard, *et al.*, 2018; Lauridsen, 2019; Surai, 2006; 2018). Since these nutrients serve as antioxidants, membrane integrity may be affected by a deficiency. However, cellular integrity is very important for receiving, and responding to the messages needed to co-ordinate an immune response (Latshaw, 1991). Therefore, the antioxidant status of the host is a critical consideration in the optimal functioning of the immune system. Furthermore, modulating immune responses of pigs by nutritional means in many cases targets specific immunity (Dalgaard, *et al.*, 2018). In the next sections effect of Se on phagocyte and lymphocytes *in vitro* and *in vivo* will be considered. Unfortunately, the data in this area obtained with pigs are quite limited and therefore relevant experimental results obtained with other species (laboratory animals, poultry and farm animals) will be presented with a special emphasis to pigs. The results will help understand biological features of immunomodulating actions of selenium.

9.3 Selenium and phagocyte functions

Phagocytosis is the process by which leukocytes and other cells ingest particulate ligands whose size exceeds about 1 μm. By ingesting microbial pathogens, phagocytes accomplish two important immune functions: initiate a microbial death pathway and build a bridge (send specific signals) between the natural and adaptive immune responses (Greenberg and Grinstein, 2002). As mentioned above phagocytes play an important role in natural immunity. Among 6 major specific rationales for modulating macrophage function analysed by Klasing (1998a), mitigation of immunosuppression arising from infectious diseases, dietary toxins or stress could be affected by dietary antioxidants. In fact, the availability of certain substrates and enzymatic cofactors can greatly influence the capacity for metabolite production by macrophages. Therefore, Se deficiency compromises the microbiocidal activity of phagocytes (Serfass and Ganther, 1975) (Figure 9.10). In particular, dietary deficiencies of Se and/or vitamin E have been shown to impair neutrophil and macrophage activities in chickens including decrease in peritoneal macrophages and decreased phagocytosis of red blood cells (Dietert *et al.*, 1990). The intracellular killing of yeast and bacteria by neutrophils and macrophages from chicken with selenium deficiency is reduced (Larsen, 1993). Similar effects were observed in rat, goat, cattle and fish (for review see Larsen, 1993; Surai, 2006, 2018).

Thus, selenoproteins appear to be important in protecting those aspects of phagocytic function that are sensitive to the destructive properties of exogenous H_2O_2. In fact, Se-deficient granulocytes did not metabolise H_2O_2 as well as replete granulocytes and H_2O_2 caused damage to the O_2-generating system. For example, prolonged incubation with stimulants and prior incubations with an H_2O_2-generating system caused loss of activity of the membrane-bound, NADPH-dependent, O^-_2-generating system in Se-deficient granulocytes with low GPx activity (Baker and Cohen, 1984). It seems likely that macrophage function is insufficient in the case of altering the generation of a respiratory burst (e.g. hereditary chronic granulomatous disease) or when there is a lack in antioxidant enzyme activity. Interestingly, thioredoxin has been identified as a lymphocyte growth factor and might therefore be involved in the crosstalk between macrophages and lymphocytes (Ebert-Dumig *et al.*, 1999).

Antioxidant-related action of selenium on phagocytes is confirmed by similarity in Se and vitamin E action. For example, regulatory effects of Se and vitamin E on phagocyte functions have been shown in experiments with farm and laboratory animals as well as in various *in vitro* systems. For example, both vitamin E and selenium enhanced the chemotactic and random migration of PMN and increased the production of superoxide following stimulation with phorbol myristate acetate (Ndiweni and Finch, 1996). *In vitro* Se supplementation of the J774.1 macrophage cell line enhanced phagocytosis, degranulation by the release of beta-glucuronidase after N-formyl-methionyl-leucyl-phenylalanine or cytochalasin B, and the production of superoxide anion after phorbol myristate acetate stimulation of these cells (Safir *et al.*, 2003). Selenium supplementation also substantially increased the release of tumour necrosis factor, IL-1 and IL-6 after lipopolysaccharide stimulation compared to Se-deficient cells.

Figure 9.10. Selenium deficiency impacts immunity.

The *in vitro* effects of an inorganic selenium salt on phagocytic functions of human neutrophilic granulocytes from donors with Se deficiency (as manifested by a low activity of GPx) have been investigated (Urban and Jarstrand, 1986). The phagocytic and bactericidal activities were significantly increased in granulocytes exposed to Se in physiological concentrations (100 and 200 ng Se/ml). However, at 2,000 ng Se/ml these activities were found to be equal to or lower than control levels probably reflecting Se toxicity. It is necessary to bear in mind that effect of supplemental dietary Se on phagocyte function depends on many factors, including background Se level, dose and form of Se used, age and sex of animals tested, etc. For example, ingestion of sodium selenite 400 μg/day (182.8 μg pure Se) by volunteers resulted in a significant increase in plasma Se levels (Greenman *et al.*, 1988). However, the phagocytic function of PMNs measured by ingestion of Oil Red O paraffin droplets and chemiluminescence tests was not significantly affected. Therefore, it was concluded that inorganic selenium was not an efficient stimulating agent of phagocytosis in humans. In fact, high Se doses could be detrimental for immune cells. For example, data reported by Shilo and Tirosh (2003) showed that exposing Jurkat T cells or J774.2 macrophages to >5 ~μM sodium selenite induced cell death. They suggested that selenite induces changes in the balance between mitochondrial superoxide and hydrogen peroxide production, which can facilitate cell death in immune system cells. The authors considered these changes as a mechanism by which Se down-regulates the immune system's inflammatory response and protects against overproduction of peroxides. It seems likely that the level of NO and HSP expression levels in neutrophils can be influenced by Se status. In particular, it was shown that Se deficiency increased the mRNA levels of HSP and iNOS and induced higher level of NO in chicken neutrophils (Chen *et al.*, 2014b). Selenoproteins are important for the balanced biosynthesis of pro- and anti-inflammatory oxylipids during inflammation. Indeed, decreased selenoprotein activity resulted in the accumulation of ROS, increased cyclooxygenase and lipoxygenase expression and decreased oxylipids with known anti-inflammatory properties such as arachidonic acid-derived lipoxin A_4 (LXA_4) and linoleic acid-derived 9-oxo-octadecadienoic acid (9-oxoODE). Furthermore, treating macrophages with LXA_4 or 9-oxoODE diminished oxidant-induced macrophage inflammatory response as indicated by decreased production of TNFα (Mattmiller *et al.*, 2014). Selenoproteins in macrophages enhance 15-PGDH-dependent oxidation of PGE2 to alleviate inflammation (Kaushal *et al.*, 2014).

Interestingly, Se accelerates the differentiation and maturation of chicken dendritic cells (DCs) and selenoproteins are involved in regulation of the differentiation and maturation of DCs and are closely correlated to surface markers of chicken DCs (Sun *et al.*, 2017). Furthermore, Se inclusion (10^{-7} mol/l) into induced and cultured chicken DCs was shown to accelerates the differentiation and maturation of chicken DCs. Se accelerates the differentiation and maturation of chicken DCs associated with increasing the expression of cell surface markers including CD11c, CD40, CD86, and MHC II (Sun *et al.*, 2017). In peritoneal macrophages cultured with *Staphylococcus aureus*, Se was shown to act in a dose-dependent manner on macrophage activation, phagocytosis and bacterial killing (Aribi *et al.*, 2015). These data suggest that inadequate Se status may cause a loss of macrophage bactericidal activities and therefore, in such cases selenium supplementation could enhance the *in vivo* macrophage activity. The role of Se in DC function deserves more attention. Indeed, recent results indicated that Se deficiency inhibited the expression of selenoproteins and changed the secretion of IL-10, IL-12p40 and IFN-γ in chickens. In fact, Se deficiency was shown to suppress the differentiation and immune function of chicken dendritic cells by down-regulating the expression of CD11c, CD40, CD86, MHC II associated with Th1/Th2 imbalance as evidenced by enhancing Th1-type cytokine IL-12p40 and IFN-29γ secretion and reducing Th2-type cytokine IL-10 (Sun *et al.*, 2018). Selenium treatment was found to enhance clearance of *Salmonella* in chicken macrophages and inhibit salmonella-induced macrophage apoptosis via an extrinsic apoptotic signalling pathway involving SIVA1 (Liu *et al.*, 2019). Selenium was found to have antagonistic effect on lead-induced neutrophil apoptosis in chickens via miR-16-5p targeting of PiK3R1 and IGF1R (Yin *et al.*, 2020). In chicken neutrophils, Pb treatment, was shown to increase the levels of IL-(1β, 1R, 4, 8, 10, and 12β), TGF-β4 and levels of IL-2 and IFN-γ mRNA were decreased compared with those in the control group due to Pb exposure. Co-administration of Se (1 mg/kg/day) and Pb resulted in a reversal of the Pb-induced cytokine changes in neutrophils (Xing *et al.*, 2018). Se was reported to play an anti-inflammation role in RAW264.7 macrophages infected with *S. aureus* by suppressing the activation of the NF-κB and MAPK signalling pathways (Bi *et al.*, 2016). Treatment of macrophage U937 cells (provisionally transformed to the macrophage form) with H_2O_2 was shown to result in significant production of the proinflammatory cytokines, including TNF-α, IL-1 and IL-6, and NF-κB activation over control, while supplementation of the cells with Se (2.5 μM) before incubation with H_2O_2 was indicated to significantly reduce the production of such cytokines (Ndoni and Okoko, 2017; Okoko, 2018). Se-supplementation was indicated to suppress the production of MPO, mmu-miR-155, TNF-α, IL-1β, and TLR2 and significantly inhibited the phosphorylation of NF-κB and MAPKs *in vivo* and *in vitro* leading to inhibition of inflammation in *S. aureus*-induced mastitis in mice (Zhang *et al.*, 2019). Recently it has been shown that Se is involved in promotion of the autophagy in macrophages infected by *S. aureus,* it is also alleviating the blockade of autophagic flow, depressing the transcription of MAPK and NF-κB signalling pathways, and finally inhibiting the proliferation of *S. aureus* in RAW264.7 macrophages (Zang *et al.*, 2020). In a mouse model of endometriosis induced by LPS, Se was shown to significantly reverse the LPS-induced uterine histopathological changes, MPO activity and inflammatory cytokine levels *in vivo* by inhibiting TLR4-mediated NF-κB and IRF3 signal transduction pathways (Chen *et al.*, 2020). Se-deficient macrophages were shown to undergo severe inflammation through the NF-κB pathway due to the accumulation of ROS. In fact, Se deficiency was indicated to induce an increased expression of inflammation factors including iNOS, IL-1β, IL-12, IL-10, PTGe, and NF-κB and restrained macrophage production of TNF-α leading to compromised their phagocytic capacity (Xu *et al.*, 2020).

Since various mycotoxins are shown to reduce macrophage viability and effector functions (Bondy and Pestka, 2000; Surai and Dvorska, 2005) a protective immunomodulating effect of antioxidants, including Se, could be especially important in commercial conditions associated with mycotoxicoses (Hegazy and Adachi, 2000). Furthermore, optimised management of macrophage metabolite production by nutritional means is a crucial factor for bacterial resistance of poultry (Dietert and Golemboski, 1998).

9.4 Selenium and antibody production

Research data accumulated for the last 50 years clearly indicate that antioxidant deficiency is associated with impaired immune system in human and laboratory and farm animals. In fact, effects of Se on immunity have been studied since 1972 when Bernstein showed that Se enhanced humoral immunity with following a series of publications by Spallholz who tested immunomodulating properties of Se in mice. His results showed that adding 2.8 mg/kg Se to the diet stimulated antibody synthesis to SRBC while 1.25 mg/kg Se increased number of plaque-forming and concentration of IgM (Spallholz *et al.*, 1973b). It is interesting that concentration of IgG was also affected by Se supplementation of mice at 1-3 mg/kg (Spallholz *et al.*, 1973a). Sodium selenite administered to mice i.p. (5 μg Se) enhanced the primary immune response to the sheep red blood cell antigen (Spallholz *et al.*, 1975). Enhancement of the primary immune response was greatest when Se was administered prior to or simultaneously with the sheep red blood cell antigen.

At the next stage of the research related to immunomodulating properties of Se a chicken model was widely used. For example, when two-week-old chickens were maintained on Se- or Se-vitamin E deficient diets from hatching, they had reduced antibody titres to SRBC (Marsh *et al.*, 1981). In the same experiment three-week old chicks had decreased antibody production only in combined Se-vitamin E deficiency. Therefore it was suggested that Se and vitamin E may be important components of immune function, and their effects depend on antigen concentration, sex, and the age (Larsen, 1993). Se-deficient diet was shown to lead to the inhibition of growth of chicken spleen, bursa of Fabricius, and thymus and decreased the SelT expression in the tissues, imposed oxidative stress and inhibited a pleiotropic mechanism of the immune response leading to immunosuppression (Pan *et al.*, 2018).

In contrast to selenium and/or vitamin E deficiency, antioxidant supplementation of the diet is shown to have immunomodulating effect and the level of vitamin E or Se required for optimal immune function may be higher than that required for growth and other physiological functions. The simplest tests with antibody production against SRBC clearly showed a beneficial effect of vitamin E and selenium. Newcastle disease challenge is often used to assess immunostimulating properties of various antioxidants. For example, the immune response of chicks vaccinated with live Newcastle disease vaccines was significantly improved by Se and vitamin E. Diets were supplemented 14 days before vaccination, at a ratio of 0.25 mg/kg and

300 mg/kg respectively and improvement in hemagglutination inhibition (HI) antibody titres and protection rate against challenge with velogenic Newcastle disease virus were observed (Bassiouni *et al.*, 1990). When Se was added to the feed of White Leghorn chickens (between 0.1 and 0.8 mg/kg) prior to challenge with either *Escherichia coli* or sheep erythrocyte antigen, an antibody titre against sheep erythrocytes increased by 77% (Larsen *et al.*, 1997). Following chilling, antibody titre response was substantially reduced and the titre reduction was prevented with dietary additions of Se between 0.1 and 1.2 mg/kg (Larsen *et al.*, 1997). Furthermore, significantly higher antibody titres at 10 d post-immunisation for Newcastle Disease Virus were attributed to 0.06 mg/kg and 150 IU/kg Se and vitamin E, respectively (Swain *et al.*, 2000). The antibody immune response in White Leghorn chicks was also significantly improved for the Se enrichment diet in the *Salmonella* or *Salmonella* + aflatoxin inoculated groups (Hegazy and Adachi, 2000). Selenium can also be used with drinking water. For example, Deng *et al.* (1999) studied effects of sodium selenite solution on the humoral immunity and erythrocyte immunity function against Newcastle disease in chicks. Their results showed that the HI titre of the control group was much lower than that of the 4 Se-supplemented groups. At the same time, the erythrocyte immune function of the control group was lower than that of the tested groups. Similarly, when chicken were vaccinated with attenuated strain C30-86 of Newcastle disease virus via drinking water at 3 and 21 days of age supplementation with Se through water increased HI antibody titre by stimulating antibody production (Yang *et al.*, 2000). Furthermore, Se supplementation helps increasing post vaccination humoral immune response against IBD in broiler chicks (Arshad *et al.*, 2005). It was shown that organic Se has more pronounced effect on humoral immunity of the growing chickens in comparison to sodium selenite. In particular it was effective in improving antibody production against Newcastle disease (Kovalenko *et al.*, 2008). In fact, at day 39 IgG concentration in chicken plasma increased from 7.42 to 9.66 mg/ml (P<0.05) and IgM from 0.80 to 1.06 mg/ml (P<0.05) due to organic Se supplementation. Feeding Se-biofortified alfalfa hay was shown to increase Se and IgG1 levels in colostrum but does not affect short-term serum antibody concentrations in calves (Wallace *et al.*, 2017). When finishing pigs fed on BD containing 0.1 mg Se/kg was supplemented with organic Se (SeMet or MeSeCys at 0.25 mg/kg), a significant increase in the immunoglobulin A (IgA), IgG, and IgM serum levels compared with BD and sodium selenite groups (Zhang *et al.*, 2020) was observed. Sow supplementation with OH-SeMet was shown to enhance mainly passive immunity through IgG maternal transfer and can influence piglet innate and acquired immunity (Li *et al.*, 2020).

Immunomodulating properties of antioxidants were shown with a range of animal species, including various farm and laboratory animals. Therefore selenium and/or vitamin E deficiencies are associated with a compromised humoral immune response not only in chicken but also in mouse, horse, pig, cattle, calves, lambs and sheep (Larsen, 1993) and inclusion of increased doses of these nutrients into the diet is proven to improve the immune response (Surai, 2006, 2018).

9.5 Selenium and lymphocyte functions

It has been suggested that Se and vitamin E deficiencies affected T lymphocytes to a greater extent than B lymphocytes (Larsen, 1993). This was explained to be a result of higher level of PUFAs in T lymphocytes and higher membrane fluidity. In fact, vitamin E and Se deficiencies may affect both the maturation of specific lymphocyte subpopulations and the functional and proliferative capabilities of the peripheral lymphocytes. A combine Se- and vitamin E deficiency in chicks showed a more severe depression of T cell response than just Se-deficient chicks. For example, in an experiment of Chang *et al.* (1994) the dietary deficiencies in vitamin E and selenium (basal diet without vitamin E and Se supplementation starting from hatching) resulted in a significant inhibition of T lymphocyte proliferation. In particular a decreased proportion of peripheral T cells and more specifically a decreased the number of CD4(+) peripheral blood leukocytes were observed. It is interesting to note that so-called regulatory $CD4^{+}$ T cells are capable of controlling the activity of other lymphocytes and they protect the integrity of tissues and organs *in vivo* and also play a major role in the systemic homeostatic mechanisms that control total lymphocyte numbers (Annacker *et al.*, 2001). The proliferative response to both ConA and PHA was impaired by the vitamin E and Se dietary deficiencies. However, the proliferative response could be fully reconstituted after vitamin E and Se supplementation for periods longer than 1 week.

Selenium deficiency in growing chickens was associated with impaired bursal growth and Se-vitamin E deficiency caused inhibition in thymus growth (Marsh *et al.*, 1986). In this experiment reduced number of lymphocytes were seen in both the thymus and bursa in combined Se-vitamin E deficiency. It has been shown that Se deficiency alone or in combination with vitamin E is associated with depressed splenocyte ability to proliferate in culture (Marsh *et al.*, 1987). Such a depression was not due to reduced lymphocyte viability in culture. Marsh *et al.* (1986) suggested that the primary lymphoid organs are major targets of Se and vitamin E dietary deficiencies and provide a possible mechanism by which immune function may be impaired. In their experiment with chickens specific deficiencies of Se or vitamin E significantly impaired bursal growth. Thymic growth was impaired only by the combined vitamin E-selenium deficient diet. Severe histopathological changes in the bursa resulted from the combined deficiency and these were detectable by 10-14 days after hatching (Marsh *et al.*, 1986). These changes were characterised by a gradual degeneration of the epithelium and an accompanying depletion of lymphocytes. On the other hand, a significant increase in relative weight of the bursa of Fabricius was observed in broiler chicks at a level of 0.10 mg of selenium and 150 IU of vitamin E (Swain *et al.*, 2000). Chicks receiving Se (1 mg/kg) and vitamin E (300 IU/kg) had significantly higher cellular immune responses in terms of percent leukocyte migration inhibition (Swain *et al.*, 2000). It is interesting to note that bursectomy in chickens caused a small but significant fall in spleen Se concentration, but not in that of other tissues (Abdel-Ati *et al.*, 1984).

In a series of recent publications, a relationship between Se deficiency and immune organ growth, development and structure have been described in detail. In particular, it has been shown that chickens fed Se-deficient diets exhibited oxidative stress associated with lesions in immune organs, decreased serum IL-1β, IL-2 content, and serum TNF content and impaired the immune function of chickens (Zhang *et al.*, 2012). In fact, Se concentrations in thymus, bursa and spleen of the low-Se (0.032 mg/kg Se) chicken group were significantly lower than in the control group. Furthermore, low-Se diet compromised antioxidant defences indicative by a decrease in the activities of T-AOC, SOD, GPx, and an increase in XOD activity and MDA content in the immune organs. In addition, pathological lesions and DNA damage of immune tissues were observed in low-Se group, while the serum IL-1β and IL-2 contents decreased, and TNF content increased (Zhang *et al.*, 2012). Similarly, it was shown that low Se diet intake (0.0342 mg/kg Se) caused oxidative stress, evidenced by decrease in GPx and catalase activities and increase in MDA contents, associated with increased apoptosis, arrested cell cycle, and injured structure and immune function of the spleen in chickens (Peng *et al.*, 2012).

Indeed, the weight and relative weight of the spleen were significantly decreased in the low Se group when compared with those of the control group (0.2 mg/kg Se) and splenic lesions were characterised by lymphocyte depletion and congestion of red pulp. In addition, the percentage of apoptotic splenocytes was greatly increased in the low Se group when compared with that of control group (Peng *et al.*, 2012). It seems likely that low-Se diet inhibits the development of thymus by arresting the cell cycle and decreasing the IL-2 content. For example, low-Se diet caused a decrease in S-phase cells in thymus and mitochondria injury with increased apoptotic cells were observed. Furthermore, low-Se diet decreased the serum IL-2 contents and mitogenesis of peripheral blood lymphocytes to concanavalin A in comparison with those of control group (Peng *et al.*, 2011e) and decreased T cell subsets (Peng *et al.*, 2011b) and restrained the development of bursa of Fabricius and thymus by cell cycle arrest, apoptosis and decreasing the IL-2 content (Peng *et al.*, 2011d). It seems likely that the oxidative stress and NO overproduction play a causative role in the apoptosis of immune tissues induced by Se deficiency. For example, free radical production, NO and inducible NO synthases (iNOS) activity in blood and the immune organs (bursa, thymus and spleen) of the chicken in low-Se group (0.032 mg/kg Se) were significantly increased compared to the control group. In addition, the expression of Fas and caspase-3 mRNA increased and apoptosis was observed in chicken immune organ in the low-Se group. The degree and the number of apoptotic cells rose in a time-dependent manner (Zhang *et al.*, 2013).

In fact, selenoproteins, including SelW, Sel15 and SelT play important roles in the protection of immune organs of birds from inflammatory injury by the regulations of inflammation-related genes. In fact, SelW is widely expressed in immune organs of birds and Se-supplementation of the feed increases SelW expression in the thymus and the bursa of Fabricius (Yu *et al.*, 2011). In addition, the histopathological analysis showed that immune tissues were injured in the low-Se groups. Furthermore, chicken Sep15 preserves the typical characteristic of selenoprotein and may play some roles in the redox regulation and chicken Sep15 was shown to be decreased by Se deficiency in immune organs (Sun *et al.*, 2014). Furthermore, Sep15 deficiency induced the occurrence of higher oxidative stress and enhanced the sensitivity of cells to H_2O_2. Se deficiency in chickens also induced lower levels of SelT and induced oxidative stress in immune organs, including reduced catalase (CAT) activity, and increased levels of hydrogen peroxide (H_2O_2) and hydroxyl radical (-OH; You *et al.*, 2014). It has been also shown that Se-deficient diets significantly decreased the mRNA expression of SelW, and induced a significantly up-regulation of pro-inflammatory signalling: cyclooxygenase-2 (COX-2), iNOS, NF-κB, PTGEs and TNF-α mRNA levels in chicken immune organs including spleen, thymus and bursa of Fabricius (Yu *et al.*, 2015). Furthermore, *in vitro*, H_2O_2 induced a significantly up-regulation of the mRNA levels of inflammation-related genes (iNOS, COX-2, NF-κB, PTGEs, and TNF-α) in cultured splenic lymphocyte. However, when lymphocytes were pretreated with Se before treatment with H_2O_2, the inflammation-related genes were significantly decreased. On the other hand, silencing of SelW significantly up-regulated the inflammation-related genes in cultured splenic lymphocyte. The results suggested that the expression levels of inflammatory factors and SelW can be influenced by Se in birds (Yu *et al.*, 2015).

It is well established that Se deficiency causes defects in the chicken bursa of Fabricius and Se-related immunosuppression is usually accompanied by a downregulation of mRNA expression levels of selenoproteins and an upregulation of the HSP mRNA expression levels. In fact, It was shown that in chicken bursa of Fabricius mRNA expression levels of selenoproteins, such as TrxR1, TrxR2, TrxR3, Dio1, Dio2, Dio3, GPx1, GPx2, GPx3 GPx4, SelP1, SelO, Sel15, SelX1, SelS, SelI, SelU, SelH, and SPS2, were significantly decreased due to Se deficiency (0.033 mg/kg for 15-55 days; Khoso *et al.*, 2015). Furthermore, the mRNA and protein expression levels of HSPs (HSP27, HSP40, HSP60, HSP70, and HSP90) were also significantly increased in comparison to Se adequate chickens (0.15 mg/kg). The expression levels of IL-2, IL-6, IL-8, IL-10, IL-17, IL-1β, IFN-α, IFN-β, and IFN-γ were significantly decreased and TNF-α was significantly increased due to Se deficiency compared with the control group (Khoso *et al.*, 2015).

Recently it has been shown that in chicken erythrocytes the expression of 24 selenoproteins and 7 cytokines (IL-2, IL-4, IL-8, IL-10, IL-12β, TGF-β4, and IFN-γ) decreased, and the expression of 3 cytokines (IL-1γ, IL-6 and IL-7) increased due to Se-deficiency (Luan *et al.*, 2016). The increase of pro-inflammatory cytokines (IL-1γ, IL-6, and IL-7) suggested that the immune system of chickens was damaged by the Se deficiency. There was a close correlation between selenoprotein and cytokine expression levels and a Se diet revealing an important relationship between Se and the chicken immune system (Luan *et al.*, 2016). Se deficiency in chickens caused defects in immune organ morphology and function, as evidenced by abnormal histological structures: red pulp broadening and lymphocytes in the cortex and medulla of the thymic lobule decreased distinctly and distributed loosely (Yang *et al.*, 2016a). In Se deficiency in broiler chicks with signs of exudative diathesis, the antioxidant function declined remarkably, and most of the HSP expression levels increased significantly in the spleen, thymus, and bursa of Fabricius (Yang *et al.*, 2016a). At the same time mRNA levels of HSP27, HSP40, and HSP70 decreased in thymus tissues from the Se-deficient chickens. It was shown that Se deficiency decreased the mRNA levels of 23 selenoproteins in the thymus, spleen, and bursa of the Fabricius tissues of broiler chickens (Yang *et al.*, 2016b). Among those 23 selenoproteins, the mRNA levels of Dio1 in the thymus, Txnrd2 in the spleen, and Txnrd3 in the bursa of Fabricius decreased significantly (90.9, 83.3 and 96.8%, respectively). Indeed, Se deficiency mainly influenced the expression of antioxidative selenoproteins, especially GPx, TRs and iodothyronine deiodinases in chicken immune organs (Yang *et al.*, 2016b). Se deficiency was shown to decrease the expression of all major AO enzymes namely TRs, GPxs, as well as Dios in the chicken spleen (Khoso *et al.*, 2019). Furthermore, Se deficiency decreased the level of IL-2 and IL-1β and increased the level of TNF-α in the chicken spleen, reflecting tissue damage in chicken spleens due to Se deficiency. In fact, the expression levels of IL-2, IL-1β, IL-6, IFN-α, and IL-17 were significantly decreased while the levels of IL-8, IL-10, IFN-γ, IFN-β, and TNF-α were significantly increased in the Se-deficient group (Khoso *et al.*, 2019). Significantly decreased Se concentrations compared with controls was reported the thymus samples of Se deficient pigs (Oropeza-Moe *et al.*, 2019).

Inclusion of selenium in the chicken diet (0.3-0.6 mg/kg) for 2 months significantly increased the response of peripheral blood lymphocytes to phytohemagglutinin and NK cell activities, accelerating the development of cell-mediated immunity in chickens (Huang and Chen, 1999). Therefore, Se could significantly increase the response of peripheral blood lymphocytes to phytohemagglutinin and NK cell activity, accelerating the development of cell-mediated immunity in chickens (Huang and Chen, 1999). In general, splenic Natural Killer cells activity is enhanced in selenium-supplemented, healthy animals (Dhur *et al.*, 1990). The results of Colnago *et al.* (1984b) indicate that dietary Se supplementation modifies the number of peripheral blood leukocytes in chickens infected with coccidia. In particular selenium significantly increased blood leukocyte number 8 hours after challenge in one experiment and produced numerically higher leukocyte number in three other experiments. Immunisation of chickens against coccidiosis can be enhanced by Se supplementation (Colnago *et al.*, 1984a).

It seems likely that Se alters the kinetics of expression of high affinity (p55/p75) interleukin 2 receptors (IL-2R). For example, dietary (2 mg/kg for 8 weeks) or *in vitro* (1×10^{-7} M) supplementation with Se results in a significant upregulation of the expression of both the p55 and p70/75 IL-2 binding sites on the surface of Con A-stimulated lymphocytes from mice (Roy *et al.*, 1993). This resulted in the formation of significantly higher numbers of high affinity IL-2R/cell with preservation of the normal ratio of high affinity to total IL-2 binding sites/cell. The authors suggested that Se could accelerate clonal expansion of activated lymphocytes.

As mentioned above, Se is proven to have an important function in maintaining proliferative capacity of T- and B-cells (Surai, 2006, 2018). It has been suggested that Se may modulate the expression of IL-2 receptors on the cell surface, which could lead to the altered ability of lymphocytes from Se-deficient animals to respond to mitogen and antigens (Larsen, 1993). Again this phenomenon is a characteristic for various farmed and laboratory animals (Surai, 2006, 2018).

9.6 *In vitro* effects of selenium on lymphocytes

Natural antioxidants are essential nutritional factors that affect the development and expression of cell-mediated immune responses. Various *in vitro* models were used to assess effects of Se on lymphocyte functions with rodent and human lymphocytes been used most frequently. It has been shown that dietary (2 mg/kg for 8 weeks) or in *in vitro* (1×10^{-7} M) supplementation with Se (as sodium selenite) results in a significant enhancement of the proliferative responses of spleen lymphocytes from C57B1/6J in response to mitogen stimulation. In contrast, Se deficiency (0.02 mg/kg for 8 weeks) had an opposite effect. The alterations in the ability of cells to proliferate were apparently related to the ability of Se to alter the kinetics of expression of high-affinity IL-2 receptors on the surface of activated lymphocytes. This was associated with enhanced or delayed clonal expansion of the cells. The changes in tumour cytotoxicity were paralleled by changes in the amounts of lymphotoxin produced by the activated cells. The results also suggested that Se exerts its effect 8-24 h after stimulation, and that it most likely affects processes in the cytoplasmic and/or nuclear compartments of activated lymphocytes (Kiremidjian-Schumacher *et al.*, 1992).

The impairment in lymphocyte proliferative response to antigens in Se and/or vitamin E deficiency could be a result of lipid peroxidation and damages to membrane structures and more importantly to membrane receptors (Larsen, 1993). As a result, cell-to-cell communication could be compromised. In this case a protective effect of antioxidants could be a crucial factor in immune system competence. For example, it was found that lipid peroxidation in lymphocytes before Con A stimulation was lower than that after stimulation and that SOD promoted lymphocyte proliferation dose dependently. The addition of Na_2SeO_3 to lymphocytes culture or supplementation in drinking water to mice decreased the lipid peroxidation in lymphocytes stimulated by Con A. In the presence of Se, there was an inverse correlation between the levels of LPO in lymphocyte and the stimulated proliferation (Sun *et al.*, 1995). *In vitro* vitamin E and selenium supplementation was able to enhance significantly the depressed PMN-mediated chemotaxis and phagocytosis or monocyte chemoattractant protein-1 production in elderly subjects (Ventura *et al.*, 1994).

Thus, selenium appears to be capable of selectively regulating the generation of functional lymphocyte subsets *in vitro*. For example, Se at physiologic concentrations can inhibit human lymphocyte proliferation in response to irradiated tumour cells in mixed lymphocyte/tumour cell cultures (MLTC; Petrie *et al.*, 1989). Furthermore, the various lymphocyte functional activities generated in these cultures exhibited different levels of sensitivity to the effects of selenium. In particular, the generation of suppressor-cell activity in MLTC was strongly inhibited by the presence of physiologic levels of Se, while the development of cytotoxic T lymphocyte activity in identical cultures was not affected by Se. Production of interleukin-2 in these cultures showed an intermediate sensitivity to the effects of Se (Petrie *et al.*, 1989).

The effect of Se on NK and lymphokine-activated killer cell activities and proliferative responses of human lymphocytes was studied *in vitro* (Nair and Schwartz, 1990).

Direct addition of Se at 1.0 μg/ml to the mixture of target and effector cells significantly suppressed the NK activity of normal lymphocytes. Similarly, when lymphocytes were preincubated with Se at concentrations as low as 0.2 μg/ml for a period of 48 h, a significant inhibitory effect on NK activity was observed. Lymphocyte proliferative responses to T cell mitogens such as PHA and Con A were also significantly suppressed by direct addition of Se at 0.5-1.0 μg/ml. The inhibitory effect of Se was not due to non-specific toxicity of effector cells as demonstrated by viability nor was the effect directed against target cells. Therefore, although Se is an essential micronutrient for various immune mechanisms, an excess of Se may have a deleterious effect on certain immunological functions.

As mentioned above there are principal differences in metabolism of organic and inorganic selenium. This could be a reason of differences in stimulating activity of different forms of selenium on immune system. For example, sodium selenite and SeMet were tested in parallel, and their capability to inhibit or to increase the antibody production by human lymphocytes *in vitro* was investigated (Borella *et al.*, 1996). Low doses of Se (0.5-2.0 μM) added as sodium selenite or SeMet did not alter the secretion of antibodies. When Se was added at higher levels, instead, an inhibitory effect was found using selenite, whereas a progressive increase in immunoglobulin production was observed after exposure to SeMet. Therefore there is an advantage in using organic selenium for immune system stimulation in comparison to selenite.

In general, it is well known that selenium, especially in the form of selenite, is clastogenic for cultured lymphocytes. For example, human lymphocyte cultures were treated with increasing concentrations (from 8.0×10^{-8} M to 8.0×10^{-5} M) of sodium selenite and SeMet 24 h after stimulation with PHA and were assessed for chromosomal aberrations at 48 hours (Khalil, 1989). The yield of abnormal metaphases was dependent on the dose and the form of selenium used. At 8.0×10^{-5} M the proportion of aberrant cells reached 53.5 and 43.0% for selenite and SeMet, respectively. The selenium-induced chromosomal aberrations were primarily of the chromatid type and included breaks and fragments. *In vitro* exposure of human peripheral blood lymphocytes to high concentration of two inorganic salts of selenium sodium selenite (2.9×10^{-5} M) and sodium selenate (2.65×10^{-5} M) was found to be lethal. Lower concentrations of sodium selenite (from 2.32×10^{-7} M to 2.9×10^{-6} M) and sodium selenate (from 1.06×10^{-6} to 5.3×10^{-6} M) induced chromosomal aberrations and reduced cell division in proportions directly related to the dose (Biswas *et al.*, 2000). Clearly, sodium selenite was considerably more clastogenic than sodium selenate.

Selenium is suggested to play a critical role in immune response, enhancing lymphocyte proliferation and T cell function in chickens (Boscolo *et al.*, 2005). It seems likely that only low levels of selenium are essential for T cell mitogenesis even in selenium-insufficient splenic cells. For example, T cell proliferation of the selenium-insufficient splenic cells induced by concanavalin A was increased by addition of Na_2SeO_3, Na_2SeO_4, Na_2Se, seleno-DL-cystine, seleno-L-methionine and selenocystamine (Ueno *et al.*, 2008). Their promoting action was observed at levels lower than 0.1 μmol/l and was completely suppressed at the highest concentration

(1 μmol/l). However, recovery of cGPx activity in the selenium-insufficient cells by supplementary Na_2SeO_3 was only partial, while TrxR activity was readily recovered from selenium deficiency. The stimulation of T cell mitogenic response by the physiological levels of selenite is predominantly caused by increased TrxR activity, which may lead to reduction of Trx-1 dependent on the intracellular expression level and promotion of DNA binding of NF-κB (Ueno *et al.*, 2007).

Se supplementation increases the production of 15d-PGJ2 as an adaptive response to protect cells against oxidative stress-induced pro-inflammatory gene expression (Vunta *et al.*, 2007). It seems likely that Se supplementation of macrophages negatively regulates the LPS-dependent production of inducible NO synthase (iNOS), a proinflammatory gene. Se plays an important role as an anti-inflammatory agent by tightly regulating the expression of pro-inflammatory genes in immune cells (Vunta *et al.*, 2008). It seems likely that Se down-regulates iNOS gene expression and NO production in the LPS-stimulated macrophages through inhibition of the NF-κB activation pathway (Yun *et al.*, 2007). Indeed, selenium is important in the shunting of arachidonic acid metabolism toward the production of PGD(2) metabolites. For example, treatment of selenium-deficient macrophages with rosiglitazone, a peroxisome proliferator-activated receptor γ ligand, up-regulated hematopoietic-PGD(2) synthase, while the expression of nuclear factor-κB-dependent thromboxane synthase and microsomal PGE(2) synthase was down-regulated by selenium (Gandhi *et al.*, 2011).

Therefore, the anti-inflammatory activity of Se using LPS and IL-4-treated macrophages from mice fed Se-deficient and Se-adequate diets were investigated (Nelson *et al.*, 2011). Supplementation with Se (100 nmol/l) of IL-4-treated macrophages significantly increased the expression of alternatively activated macrophage (M2) markers, Arg-I, Fizz1, and Mrc-1. At the same time, expression of classically activated macrophage (M1) markers, TNFα, and IL-1β, was significantly decreased in LPS-treated macrophages that were cultured in Se and IL-4. Furthermore, macrophages treated with inhibitors of PPARγ and STAT6, pivotal transcription factors that mediate the activity of Se and IL-4, respectively, showed complete ablation of Se-dependent expression of M2 markers (Nelson *et al.*, 2011). It is interesting to note that subjects with a decreased serum Se concentration may be exposed to a greater chronic oxidative stress due to neutrophil ROS production (Lee *et al.*, 2011). Recently, the effects of Se on T cell proliferation and IL-2 production were studied in primary porcine splenocytes (Ren *et al.*, 2012). It seems likely that Se can improve the redox status of splenocytes and enhance mitogen-induced T cell activation. For example, splenocytes were treated *in vitro* with different mitogens in the presence of 0.5-4 μmol/l sodium selenite. It was shown that Se significantly promoted T cell receptor (TCR) or concanavalin A (ConA)-induced T cell proliferation and IL-2 production but failed to regulate T cell response to phytohemagglutinin (PHA). In addition, Se significantly increased the levels of cytosolic GPx1 and TrxR1 mRNA, the activity of GPx1 and the concentration of reduced glutathione (GSH) in the unstimulated or activated splenocytes. In the same experiment, N-acetylcysteine (NAC), a pharmacological antioxidant, increased T cell proliferation and IL-2 production by TCR and ConA stimulated splenocytes confirming that T cell proliferation is sensitive to the redox status (Ren *et al.*, 2012).

A new role for selenoproteins in the epigenetic modulation of proinflammatory genes has been suggested (Narayan *et al.*, 2015). In particular the authors showed that selenium supplementation decreased acetylation of histone H4 at K12 and K16 in COX-2 and TNFα promoters, and of the p65 subunit of the redox sensitive transcription factor NF-κB in primary and immortalised macrophages. It seems likely that the ability of selenoproteins to skew the metabolism of arachidonic acid contributes, in part, to their ability to inhibit histone acetylation (Narayan *et al.*, 2015). In bovine Sertoli cell system *in vitro*, medium supplemented with selenium was found to significantly promote cell viability, to upregulate TLR4 gene as well as anti-inflammatory cytokines (IL-4, IL-10, TGFβ1), and to modulate expressions of pro-inflammatory cytokines (TNF-α, IL-1β, IFN-γ) (Adegoke *et al.*, 2018). Transcriptome analysis of goose T cells showed that the selenoproteins and lysosome pathway related genes were promoted by Se stimulation, while ILs and interferons were mainly downregulated. In fact, there were 2,228 differentially expressed genes primarily involved in immune and environmental response and infectious disease (Cao *et al.*, 2017).

9.7 Selenium and disease resistance

The final goal of the improvement of the immune system is to increase resistance to various diseases. Indeed, this option was extensively studied with chickens in 1980-1990. For example, a combination of vitamin E with Se resulted in reduced mortality and increased body weight gain in chickens infected with *Eimeria tenella* (Colnago *et al.*, 1984a,b). The same authors showed that dietary supplementation with selenium or vitamin E reduced mortality and increased body weight gain of non-immunised chickens infected with *E. tenella* in three of four experiments. When chicks were inoculated with virulent Marek's disease (MD) virus at 10 days of age, selenium (0.6 mg/kg) decreased the morbidity and mortality from MD. In particular, Se increased the ability of cells to remove ROS and lipid peroxides, and decreased the degree of tissue damage caused by ROS (Huang and Chen, 1996). In another experiment, from one day of age chicks were fed on a basal diet containing Se at 0.086 mg/kg (group I) or the basal diet supplemented with Se at 0.3 mg/kg (group II) or 0.6 mg/kg (group III). The chickens were infected with infectious bursal disease virus at 39 days of age. Ten days later the mortality rates in groups I, II and III were 33.3, 12.4 and 10.6%, respectively, and the infection induced inhibition of T lymphocyte transformation was less in the Se supplemented birds (Bu *et al.*, 1996). When Se was added to the feed of White Leghorn chickens prior to challenge with either *E. coli* or sheep erythrocyte antigen, the incidence of death or lesions was reduced from 86 to 21% at the optimal dose of Se (0.4 mg/kg feed; Larsen *et al.*, 1997). Lower Se values were measured in infected with *Ascaridia galli* chickens compared with controls, and this was related to a lower degree of Se absorption and the regeneration of the intestinal mucosa in infected birds (Damyanova *et al.*, 1995). Immunostimulating and disease-preventing effects of natural antioxidants are not restricted to avian species, but obvious with other farm animals (Teige *et al.*, 1982).

Therefore, the results presented above clearly showed that various components of the immune system as well as general animal health have been improved when Se and/or vitamin E have been added to deficient diets or supplemented at levels far above those required for growth. The benefit of Se supplementation would be greatest in situations when animals are infected with a particular pathogen. In this case clinical signs of infection could be reduced. Furthermore, for optimising the animal's resistance to disease, Se requirement is higher than that for adequate growth, feed efficiency, egg production or even reproduction (Nockels, 1988). The optimal doses of Se for maximum disease protection depends on many factors and need further elucidation (Tengerdy *et al.*, 1990).

It is well known that Se and other antioxidants are able to protect cells from free radical damage, reduce the detrimental effects of certain eicosanoids, and enhance humoral and cellular immune responses in disease (Nockels, 1988). Improved immune system could ultimately lead to higher resistance of animals to various diseases. It is important to remember that during disease challenge nutrient assimilation from the diet could be further compromised as a result of absorption impairment or decreased feed consumption. This could lead to decreased efficiency of antioxidant system leading to immunocompetence declining. For-example Hao *et al.* (1999) studied the activities of Se-GPx and lipid peroxidation in central and peripheral immune organs and main visceral organs of broilers experimentally infected at one day of age with virulent MD virus. They showed that in infected birds the Se-GPx activity was significantly decreased and lipid peroxidation was enhanced. Therefore, preventing these changes in antioxidant defence system is believed to be an effective means to maintain immune system efficiency and this could be associated with better survival of infected chicks.

9.8 Immunoprotective effects of selenium in stress conditions

Since more than half of known selenoproteins are involved in cell signalling, maintenance of redox balance of the cell and antioxidant defences, protective role of Se on immunity would be most pronounced in various stress conditions. This includes immunosuppressive action of mycotoxins, oxidative stress-related changes in heat stress conditions, heavy metal contamination, and usage of various drugs, including antimicrobials, gut challenge with pathogens or vaccinations.

9.8.1 Mycotoxins

Our previous research established that oxidative stress is the major mechanism of the mycotoxin toxicity and their immunosuppressive action (Surai, 2002, 2006; Surai and Dvorska, 2005). Among various mycotoxins, aflatoxins are considered to be most toxic and immunosuppressive. It was shown that dietary Se, through a mechanism of apoptosis regulation, may ameliorate aflatoxin B1 (AFB1)-induced lesions of thymus and accordingly improve the impaired cellular immune function. In fact, in the chicken fed AFB1-contaminated diet with 0.2-0.6 mg/kg Se the percentages of apoptotic thymocytes and the expression of Caspase-3 and Bax, were decreased, while the expression of Bcl-2 was increased compared to AFB1 group (Chen *et al.*, 2013b). In addition, 0.6 and 0.8 mg/kg Se could also restore the decreased percentages of peripheral blood T cell subsets and the contents of serum IL-2 and IFN-γ induced by 0.3 mg/kg AFB1 in the diets improving cellular immune function in chickens (Chen *et al.*, 2013a). Indeed, dietary Se (0.2-0.6 mg/kg), increased the relative weight of bursa of Fabricius and contents of serum immunoglobulin, and ameliorate histopathological lesions caused by AFB1 (Chen *et al.*, 2014a,c). The percentages of apoptotic bursal cells, in Se-supplemented groups were lower than those in the AFB1 group and compared with the AFB1 group, moreover, the mRNA expressions of Bax and Caspase-3 in the Se-supplemented groups were decreased, while the expression of Bcl-2 was increased.

Therefore, dietary Se can protect chicken from AFB1-induced impairment of humoral immune function by reducing bursal histopathological lesions and percentages of apoptotic bursal cells (Chen *et al.*, 2014c). In fact, 0.3 mg/kg AFB1 could reduce the humoral immune function of the ileum mucosa of chickens, but 0.4 mg/kg supplemented dietary selenium could protect the mucosal humoral immune function from AFB1-induced impairment. In particular, compared with those in the AFB1 group, the numbers of IgA(+) cells as well as the IgA, IgG, and IgM contents were increased in the AFB1 + Se group, and these data had no difference between AFB1 + Se group and control group (He *et al.*, 2014b). Similarly, compared with those in control group, the percentages of CD3+, CD3+CD4+, CD3+CD8+ intraepithelial lymphocytes (IELs) and LPLs, the CD4+/CD8+ ratio of IELs, and the mRNA contents of IL-2, IL-6, and TNF-α were decreased in AFB1 group (He *et al.*, 2014a). However, compared with those in AFB1 group, these parameters of AFB1+Se group (0.4 mg/kg) were increased to be close to those in control group showing immune-protective effects in mycotoxicosis. Furthermore, dietary Se (0.2-0.6 mg/kg) could also improve the cellular immune function impaired by AFB1 through increasing the relative weight of spleen and percentages of splenic T cell subsets, and alleviating histopathological spleen damage (Chen *et al.*, 2014a).

Mitochondrial swelling assays showed that opening of the mitochondrial permeability transition pores was increased in ducklings that had received AFB1 for 14 and 21 days and Se significantly attenuated these adverse effects of AFB1(Shi *et al.*, 2015). By preventing oxidative liver damage, suppressing pro-apoptotic proteins and improving immune status in AFB1 affected animals, Se shows specific protection against AFB1 toxicity (Limaye *et al.*, 2018). SeMet was shown to protect AFB1 challenged porcine splenocytes via enhancing mRNA and protein expression of GPx1, SelS and thioredoxin reductase 1 and restricting GSH depletion (Hao *et al.*, 2016). AFB1 contaminated feed in pigs creates a lot of health-related problems associated with reduced feed intake and body weight gain, impaired liver and immune functions. In fact, dietary AFB1 exposure in pigs was shown to impair lymphocyte activation, negatively affect vaccine efficiency and delayed cell-mediated immune response leading to the heightened expression of pro-inflammatory cytokines IL-6, IFN-, and IL-10 (Limaye *et al.*, 2018).

Dietary supplementation of 0.4 mg/kg Se was shown to diminish AFB1-induced immune toxicity in chicken's Bursa of Fabricius by decreasing oxidative damage and cell cycle arrest through an ATM-Chk2-cdc25 route and the ATM-Chk2-p21 pathway (Hu *et al.*, 2018). AFB1 was shown to significantly decrease the percentages of $CD3^+$, $CD3^+CD4^+$, $CD3^+CD8^+$ T cells, and the $CD4^+/CD8^+$ ratio, and suppressed the expressions of IL-2, IL-4, TNF-α, and IFN-γ mRNA in the caecal tonsil of chickens. Se supplementation (0.4 mg/kg) was found to restore the percentages of T cell subsets, the $CD4^+/CD8^+$ ratio, and mRNA expressions of cytokines in the AFB1 group was found to return almost to the control group levels (Wang *et al.*, 2019).

Selenium could assert an important effect against the immunotoxic effects of T-2 toxin against T lymphocytes. For example, after a sublethal dose of T-2 toxin alone, the number of CD8(+) T lymphocytes was significantly decreased at 12 h and normalised at 48 h, while level of CD3(+) and CD4(+) T lymphocytes were significantly increased at 24 h and returned to normal after 48 h. When Se was injected into the mice 24 h before or concurrent with the T-2 toxin, the effects on CD8(+) cells were mitigated. It is interesting to note that only when Se was given with the toxin could the effects on the CD3(+) and CD4(+) cells be altered (Salimian *et al.*, 2014). Furthermore, in mice after injection of a sublethal dose of T-2 toxin, the number of B cells (CD19+) significantly decreased at 12 h and became normal at 72 h. When selenium was injected both 24 h before and simultaneously with T-2 toxin, it was able to inhibit B lymphocyte (CD19+) reduction (Ahmadi *et al.*, 2015). In piglet splenic lymphocytes, Se can alleviate deoxynivalenol (DON)-induced damage to antioxidant enzymes by improving glutathione peroxidase activity (Wang *et al.*, 2016b).

It seems likely that Se could have a protective effect during DON contamination of the poultry feed. For example, chicken blood granulocyte phagocytic activity was not reduced by dietary DON treatment but numbers of heterophils were increased. In the DON plus Se yeast group phagocytic activity was the same as in the CON group. The Se-yeast and DON plus Se-yeast groups had increased numbers of CD3(+), CD4(+), and CD8(+) T cells as well as IgM(+) B-cells in their blood compared to both CON and DON-groups (Levkut *et al.*, 2009). Furthermore, Se-yeast supplemented to the DON contaminated diet prevented suppression of blood phagocytic activity of broilers (Placha *et al.*, 2009). DON exposure was shown to significantly and dose dependently decrease the expression levels of IL-2, IL-4, IL-6, IL-10, IFN-γ, IgG, and IgM mRNA and protein in piglet splenic lymphocytes. At the same time, co-treatment with DON + Se significantly increased the mRNA and protein levels of all factors examined, except IL-4 and IL-6, compared with DON treatment alone (Wang *et al.*, 2018).

9.8.2 Heat stress

Scientific evidence is accumulating to show compromised immunocompetence under heat-stress (HS) condition in pigs. It has been shown HS to compromise intestinal barrier integrity with local and systemic inflammatory responses. The resulting immune activation is believed to be a cause of many negative consequences HS has on reproduction, growth, and lactation (Mayorga *et al.*, 2020). It seems likely that compromised gut microbiota due to HS also contribute to negative effects of HS of pig production and immunity (Xiong *et al.*, 2020). Furthermore, pigs exposed to in utero heat stress were reported to develop a variety of postnatal phenotypes associated with decreased profitability and compromised health and welfare in commercial production systems (Johnson *et al.*, 2020). Since HS impose oxidative stress at the level of gut and compromise immunity (Surai, 2018) it would be logical to use antioxidant nutritional supplements, including Se, to decrease negative consequences of HS on pigs immunity. However, there are no experimental pig data available to prove this point and we would consider some proof of the concept from other species. Indeed, HS was shown to compromise immunity in avian species, including altered the heterophil to lymphocyte (H/L) ratio, reduced antibody production and decreased phagocytic potential (Habibian *et al.*, 2015). In particular, it was shown that HS severely reduced growth performance and immunocompetence of broilers, whereas organic selenium supplementation (0.2-0.4 mg/kg) was able to alleviate the deleterious effects of heat stress. (Niu *et al.*, 2009). In fact, Se supplementation was associated with improved immunity indicative by increased numbers of abdominal exudate cells (AEC), percentage of macrophages in AEC, phagocytic macrophages, internalised opsonised and unopsonised SRBC. Furthermore, both primary and secondary antibody responses were characterised by increasing titres of antibody to SRBC by dietary Se when birds were exposed to HS (Niu *et al.*, 2009). Similarly, the production of antibody against SRBC and against IBD was significantly increased due to organic selenium supplementation (0.3 or 0.5 mg/kg) in heat-stressed broiler chicks (Da Silva *et al.*, 2010). The liver and lymphoid organ weights as well as IgM and IgG, antibody titers for primary and secondary antibody responses to SRBC were reduced significantly under HS and Se showed a significant protective effect (Habibian *et al.*, 2014). It seems likely that under heat stress conditions Se supplementation can increase antioxidant status and cell-mediated immunity. For example, an experiment was conducted to determine the effect of supplementing various concentrations (0, 0.1, 0.2, 0.3, or 0.4 mg/kg diet) of Se on growth performance, carcass traits, oxidative stress, and immune responses in commercial broiler chickens reared in open-sided poultry house under tropical climatic conditions. It was shown that lipid peroxidation in plasma decreased, while activities of GPx and GR in plasma increased linearly with Se concentration in diet. Furthermore, the cell-mediated immunity (lymphocyte proliferation ratio) increased linearly with increasing dietary Se concentration (Rao *et al.*, 2013). Similarly, heat-stressed chicks fed Se-supplemented diets had higher average daily feed intake, Se concentrations in liver and breast muscle, liver GPx activity, serum antibody titers against H5N1(Re-4 strain), H5N1(Re-5 strain) and lower mortality compared with the control. It is interesting to note that 0.30 mg/kg of Se was more effective than 0.15 mg/kg of Se (Liao *et al.*, 2012). More recent results indicate that HS (the chickens were exposed to 8 h of 23.9 °C, 4 h of 23.9 to 37 °C, 8 h of 37 °C, and 4 h of 37 to 23.9 °C) induced higher levels of TNF-α, IL-4, HSP27, HSP70, and MDA levels but lower level of IFN-γ, IL-2, GPx and SOD in spleen and the responses were ameliorated by the Se treatment (0.3 mg/kg). Indeed, selenium supplementation decreased the elevated levels of TNF-α and IL-4 and increased the diminished levels of TNF-γ and IL-2 induced by heat stress (Xu *et al.*, 2014). The immune dysfunction induced by Heat Shock (HS) was alleviated by treatment with

Se (0.3 mg/kg) and polysaccharide of *Atractylodes macrocephala* Koidz in different immune organs. Indeed, Se specifically regulated the TNF-α and IFN-γ pathways in immune organs (Xu and Tian, 2015).

9.8.3 Antimicrobials

Enrofloxacin (EFX) is an important antimicrobial used in veterinary practice but it is known to exert immune suppression and oxidative stress. In particular, it was shown that the activity of cellular, humoral immune response and enzymatic and non-enzymatic antioxidants, has significantly been decreased as a result of EFX treatment of broiler chickens. In such conditions nano-Se supplementation (0.6 mg/kg feed) greatly restores these values towards the control (Shirsat *et al.*, 2016).

9.8.4 Cadmium and other toxicants

Cadmium (Cd) is a toxic heavy metal of increasing environmental concern due to its wide variety of adverse effects. In fact, Cd consumption causes oxidative stress and has immunosuppressive action (Pappas *et al.*, 2010). *In vitro*, a statistically significant increase in the mRNA expression of HSPs in chicken splenic lymphocytes was observed due to Se treatment. Furthermore, treatment of chicken splenic lymphocytes with Se in combination with Cd enhanced the mRNA expression of HSPs which were reduced by Cd treatment (Chen *et al.*, 2012b). It was shown that cadmium induced nitric oxide-mediated apoptosis of immune organs, and selenium played protective effects against cadmium-induced apoptosis in the immune organs of chickens. Indeed, Se supplementation (10 mg/kg) during dietary cadmium (150 mg/kg CdCl2) reduced the production of nitric oxide, the mRNA level, and activity of inducible nitric oxide synthase, ultrastructural damage, and apoptosis in the immune organs of chicken (Liu *et al.*, 2014a). A high level of MDA and ROS productions were observed in chicken splenic lymphocytes of Cd treatment group; the activities of catalase (CAT), GPx, SOD, and the mitochondrial inner transmembrane potential (ΔΨm) were significantly lower in Cd treatment group than those in controls. Se supplementation significantly improved the activities of antioxidant enzymes and reduced MDA and ROS levels compared to Cd treatment alone group, although not restored to the levels of control group (Liu *et al.*, 2014b). Furthermore, the protective effects of Se against Cd toxicity in chicken splenic lymphocytes were confirmed by the increase in select pro-inflammatory indexes (NF-κB, iNOS, COX-2, TNF-α, and PGE2), NO content and iNOS activity (Liu *et al.*, 2015b). Indeed, compared with the control group and the Se-alone-treated group, the mRNA expression levels of IL-2, IL-4, IL-10, IL-17, and IFN-γ in chicken splenic lymphocytes decreased significantly in the Cd-treated group. In such conditions Se showed a significant protective effect by slowing down a decrease in cytokine production due to Cd contamination (Xu *et al.*, 2015a). Selenium significantly increased the expressions of selenoprotein K, selenoprotein N, selenoprotein S, and selenoprotein T, which were reduced by Cadmium in chicken splenic lymphocytes (Zhao *et al.*, 2014). Therefore, Se partly attenuate immune toxicity induced by Cd in chicken splenic lymphocytes. It was also shown that Pb poisoning induced the mRNA expression of HSPs and inflammatory cytokines in the peripheral blood lymphocytes of chickens, while Se alleviated the Pb-induced increase in HSP and inflammatory cytokine (Sun *et al.*, 2016). In beluga whales lymphocyte proliferative responses were reduced following exposure to 1 μM of HgCl2 and 0.33 μM of MeHgCl. Concurrent exposure of Se provided a degree of protection against the highest concentrations of inorganic Hg or organic Hg for T lymphocytes (Frouin *et al.*, 2012). It was also shown that, non-adherent splenic cells treated *in vitro* with an extract of *Pteridium aquilinum* showed diminished NK cell activity that was not only prevented by selenium co-treatment but also fully reversed by selenium post-treatment (Latorre *et al.*, 2011). Cd treatment was shown to promote a mitochondrial dynamic imbalance and reduced energy metabolism, leading to apoptosis and immune dysfunction in chicken spleens, and Se had an antagonistic effect on Cd-induced apoptosis (Xu *et al.*, 2017). Expression of a range of selenoproteins including Gpx1, Dio2, Dio3, Txnrd1, Txnrd2, Txnrd3, SPS2, Sels, Selm, Selh, Seln, Seli, Selt, Selo was reduced in chicken peripheral blood lymphocytes due to Pb toxicity and this lead to immune system dysregulation. Additional Se supplementation (1 mg/kg vs 0.49 mg/kg as sodium selenite) was shown to partly restore affected selenoprotein expression and improve Th1/Th2 in chicken lymphocytes (Fu *et al.*, 2019).

9.8.5 Gut challenge

Immunomodulating effect of antioxidants is shown to be maximal at their supplementation usually above the requirement for maximal growth and maintenance of reproduction. It could well be that the Se and vitamin E doses that are adequate for maximal productivity in healthy, unchallenged animals are not optimal for maximal immunocompetence and disease resistance.

There is only limited information on the effect of stress (Broom and Kogut, 2018) and antioxidants (Fisinin and Surai, 2013; Hachimura *et al.*, 2018) on gut immunity, which could be crucial barriers to external pathogens (Qureshi *et al.*, 1998). Since free radicals can be damaging to intestinal structures (Hoerr, 1998) antioxidant functions of selenium are responsible for prevention damages to intestinal lymphoid structure as well as damages to intestinal enterocyte membranes. This could be especially important in relation to gut immunosuppression caused by toxins/mycotoxins, nutritional deficiencies and infectious agents (Hoerr, 1998). Dietary supplementation of newly hatched broiler chicks with 0.25 Se mg/kg (as sodium selenite) from hatch significantly reduced necrotic enteritis-induced gut lesions compared with infected birds given a non-supplemented diet. This was associated with increased levels of serum antibody against the NetB toxin in the chicks fed with 0.25 and 0.50 mg/kg Se in comparison to the non-supplemented control group. Furthermore, the transcripts for IL-1β, IL-6, IL-8, iNOS, LITAF, and GPx7, as well as AvBD6, 8, and 13 were shown to be significantly increased in the intestine and spleen of Se-supplemented chickens (Xu *et al.*, 2015b).

Results of various experiments have shown that dietary Se supplementation can regulate the WNT-, TNF-α-, NF-κB-, and PPARγ-dependent pathways in animals, thereby inhibiting inflammatory reactions and maintaining epithelial integrity in the intestines in experimental colitis (Kaushal *et al.*, 2014) and dextran sodium sulphate-induced epithelial injury (Barett *et al.*, 2013). It has been reported that Se plays an important role in regulating cellular and paracellular permeability, cellular redox balance and inflammatory cell responses (Cao *et al.*, 2017; Jaspers *et al.*, 2007). One-week supplementation with SeMet was found to be relatively more effective in alleviating colitis than that with sodium selenite (Hiller *et al.*, 2015), whereas sodium selenite was more effective than SeMet in regulating inflammatory gene expression in macrophages (Narayan *et al.*, 2015).

Indeed, Se is proven to have an important role in resolution of inflammation in various tissues, including gut, via affecting redox balance and expression of various transcription factors including NF-κB and PPARγ (Nettleford and Prabhu, 2018). It seems likely that Se supplementation can affect intestinal barrier and immune responses by modulating gut microbiota (Zhai *et al.*, 2018). Dietary sodium selenite and SeMet were shown to have different effects on the overall gene expression patterns in the small intestine of mice (Barger *et al.*, 2012). Se-deficiency in mouse was shown to adversely affect the gut barrier and intestinal immune responses and to increase their susceptibility to experimental colitis and pathogen infection. Compared with adequate Se (0.15 mg/kg Se) supplementation, Se-deficiency was found to adversely affect the gut barrier and intestinal immune responses of the untreated mice and increased their susceptibility to experimental colitis and pathogen infection. Indeed, compared with the Se-deficient diet, dietary Se supplementations was found to significantly decrease the intestinal permeability, up-regulated the tight junction protein expression, and modulated the inflammation-related cytokine profile in mice (Zhai *et al.*, 2019). An increased Se supplementation (0.40 mg/kg) was indicated to improve mouse intestinal physiology. Interestingly sodium selenite and selenomethionine were shown to have similar effects on the gut barrier and intestinal immune homeostasis but differentially affected fecal metabolites and induced significant proteomic variations in various pathways, including the xenobiotic detoxification pathway and glutathione metabolism (Zhai *et al.*, 2019).

In comparison to adequate Se supplementation (0.15 mg/kg) increased Se (0.40 mg/kg) supplementation was shown to affect the expressions of proteins involved in mouse innate and adaptive immune responses (e.g. Reg3 family, Ighv1 family, and 478 Igkv family proteins) in the gut (Zhai *et al.*, 2019). Se deficiency was shown to cause injury of the mucosal immune barrier and physical barrier of the small intestine, and reduced growth and compromised antioxidant defences in chickens (He *et al.*, 2020). Interestingly, higher (0.4 mg/kg) Se supplementation was found to be more effective than lower (0.15 mg/kg) dose in improving gut barrier function and intestinal immune homeostasis.

9.8.6 Selenium deficiency and viruses

It is generally accepted that the nutritional status of the host is associated with both severity and susceptibility to infectious disease. In particular, oxidative stress via RNA virus infections can contribute to several aspects of viral disease pathogenesis including apoptosis, loss of immune function, viral replication, inflammatory response, and loss of body weight (Reshi *et al.*, 2014). Indeed, the excess production of superoxide's in the influenza A virus infection is detrimental and the downregulation of the superoxide markedly alleviates lung injuries caused by the influenza virus and viral replication, irrespective of the infected viral strain (Suliman *et al.*, 2001). Influenza infection also leads to the thymus-specific elevation of the mitochondrial superoxide, which interferes with the normal functioning of T cell lymphocytes in influenza A virus infections (Allen and Balin, 1989).

Furthermore, inadequate nutrition impairs the functioning of the immune system, thus resulting in increased susceptibility to infection. It has been suggested (Beck and Levander, 2000) that the nutritional status of the host affect not only the immune response, but it can also substantially affect the viral pathogen. In particular, in a mouse model, a benign strain of coxsackievirus B3 became virulent and caused myocarditis in Se and vitamin E-deficient mice (Beck and Levander, 2000). The change in pathogenicity was a result of mutations in the viral genome, which changed an avirulent virus into a virulent one. It is interesting that six nucleotide changes were found in the virus that replicated in the deficient mice, and once these mutations occurred, even mice with normal nutrition became susceptible to disease (Beck, 1999). Therefore mice deficient in Se were more susceptible to infection with coxsackievirus, as well as with influenza virus. It was also suggested that the immune system was altered in the Se-deficient animals, as was the viral pathogen itself. Mice fed a diet deficient in Se suffer much more severe lung pathology after infection with the influenza virus than Se-adequate controls (Beck *et al.*, 2001).

Sequencing of viral isolates recovered from Se-deficient mice demonstrated mutations in the viral genome of both coxsackievirus and influenza virus (Beck, 2001). These changes in the viral genome are associated with the increased pathogenesis of the virus. The effects of 10 week of Se supplementation (5 mg/kg) in drinking water on immune responses and resistance to a myocarditic Coxsackie virus B3 (CBV3) infection were studied in female Balb/c mice (Ilback *et al.*, 1998). Se supplementation reduced CBV3-induced mortality: at day 14 post-inoculation, survival was 58% in the Se-treated group as compared to 25% in the untreated group. Anti-viral protective effect of Se has been shown in the recall responses to polio virus vaccination of healthy volunteers with marginal selenium status (Broome *et al.*, 2004). Experimental mice were kept on a Se-replete or Se-deficient diet for 28 days, infected them intraperitoneally with CVB3 and maintaining previous diets, they were examined for next 90 days for several parameters indicative of the infection or disease (Jun *et al.*, 2011). It was shown that the mice on the Se-deficient diet exhibited a higher mortality, lower serum GPx activity, evident histopathological changes indicative of myocarditis, and a higher level of viral RNA in the heart. Therefore, Se-deficiency creates a chronic myocarditis-prone condition by fostering the active virus replication. In general, in some RNA viruses, including coxsackievirus B3 (CVB3/0) (cause of Keshan disease), human immunodeficiency virus (HIV), influenza A virus, SARS coronavirus, and Ebola virus, Se deficiency may cause accumulation of mutations in their genome, leading to changes in the virulence-associated genetic structures (Harthill, 2011). In such conditions antioxidant protection is a key for overcoming virus-related changes in the cell/body. For example, sodium selenite was shown to suppresses hepatitis B virus protein expression, transcription, and genome replication in hepatoma cell models in a dose- and time-dependent manner (Cheng *et al.*, 2016).

Se-supplementation in healthy human adults with marginal Se status resulted in both beneficial and detrimental effects on cellular immunity to flu that was affected by the form of Se, supplemental dose and delivery matrix (Ivory *et al.*, 2017). It seems likely that Se can attenuated the porcine circovirus type 2 (PCV2) infection in animals through altering the systemic inflammation and maintaining the normal organ morphology. For example, dietary Se supplementation had little effect on the PCV2 virus load in the liver, spleen and lung. However, mice in the Se-supplemented group showed a significant decrease in microscopic lesion scores in the lung and spleen compared with those in the control group (Liu *et al.*, 2015a). In another study PCV2 replication was inhibited by SeMet at a high concentration (6 μM) and the increase in PCV2 replication by oxidative stress was blocked by SeMet at physiological concentrations (2 or 4 μM). Furthermore, PCV2 infection caused a decrease in GPx1 activity. It is interesting to note that SeMet did not significantly block the promotion of PCV2 replication in GPx-knockdown cells. Indeed, GPx plays a key role in blocking the promotion of PCV2 replication (Chen *et al.*, 2012a). Se supplementation can be considered as widely available adjuvant therapy in viral and bacterial infections to support the therapy and/or to improve immunocompetence of farm animals (Steinbrenner *et al.*, 2015). A general scheme of Se-virus interaction is shown in Figure 9.11.

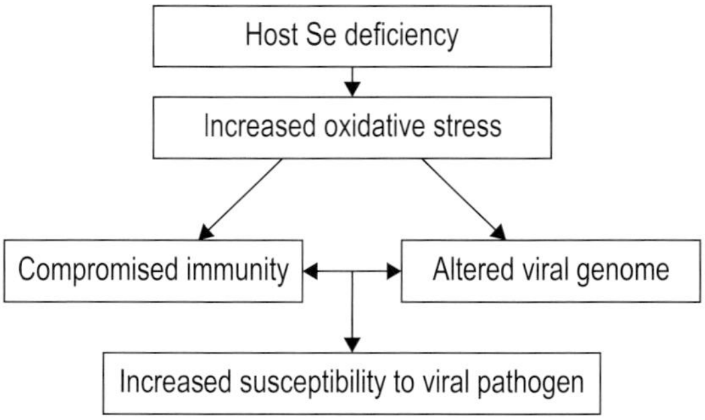

Figure 9.11. Effect of selenium (Se) on viral diseases (adapted from Surai, 2006).

Dietary selenium supplementation was shown to enhance antiviral immunity in chickens challenged with low pathogenic avian influenza virus subtype H9N2 (Shojadoost *et al.,* 2019) The authors showed that supplementation of chicken feed with the organic form of selenium (Se-Yeast) was more effective at reducing virus shedding compared to the inorganic selenium (SS) supplemented det.

9.8.7 Selenium overdose and immunity

Selenium is toxic at high doses, so it is necessary to be careful choosing the correct dose of this trace element for an experiment. Therefore, over-supplementation with Se, in the presence or absence of vitamin E and in amounts insufficient to cause clinical toxicity, may impair vaccination responses. Se supplementation with sodium selenite should be carefully monitored, since Se excess (more than 5 mg/kg) could cause profound immunologic inhibition (Peng *et al.,* 2009). Therefore, Se excessive (more than 5 mg/kg as sodium selenite) intake could impair cellular immunity in broilers causing lesions of thymus and decrease of T cell subsets (Peng *et al.,* 2011a) causing growth retardation of spleen by cell cycle blockage in young chickens (Peng *et al.,* 2011c).

The effects of supranutritional dietary Se (3 mg/kg) on selenoproteins expression in three immune organs of chickens were studied. The results showed that high dietary selenite depressed growth performance of chicken and down-regulated nine and three selenoprotein genes in thymus and spleen, respectively, and only SepP1 was up-regulated in the bursa of Fabricius. Also three, three and seven inflammation-related genes were up-regulated in three organs, respectively. Supranutritional Se elevated activities of SOD, total antioxidant capacity and GPx, mainly in early stages (Tang *et al.,* 2017). Similarly, excessive Se supplementation (5-15 mg/kg) decreased the activities of GPx and SOD but increased the levels of MDA in chicken spleen in a dose- and time-dependent manner. In addition, the ER stress genes GRP78 and ATF6 were highly expressed, and the apoptosis genes caspase 3 and caspase 12 were increased, but Bcl 2 was decreased by Se treatment (Wang *et al.,* 2016a). Immunomodulating properties of selenium are summarised in Figure 9.12.

9.9 Molecular mechanisms of immunomodulating properties of selenium

To elucidate the role of Se-containing proteins in the immune function, the expression of this protein class in T cells or macrophages of mice were knocked out (Carlson *et al.,* 2010). The results of deficiency in Se-proteins can be summarised as follows:

- Reduced pools of mature and functional T cells in lymphoid tissues and an impairment in T cell-dependent antibody responses.
- An inability of T cells to suppress reactive oxygen species production, which in turn affected their ability to proliferate in response to T cell receptor stimulation.
- An altered regulation in extracellular matrix-related gene expression and a diminished migration of macrophages in a protein gel matrix.

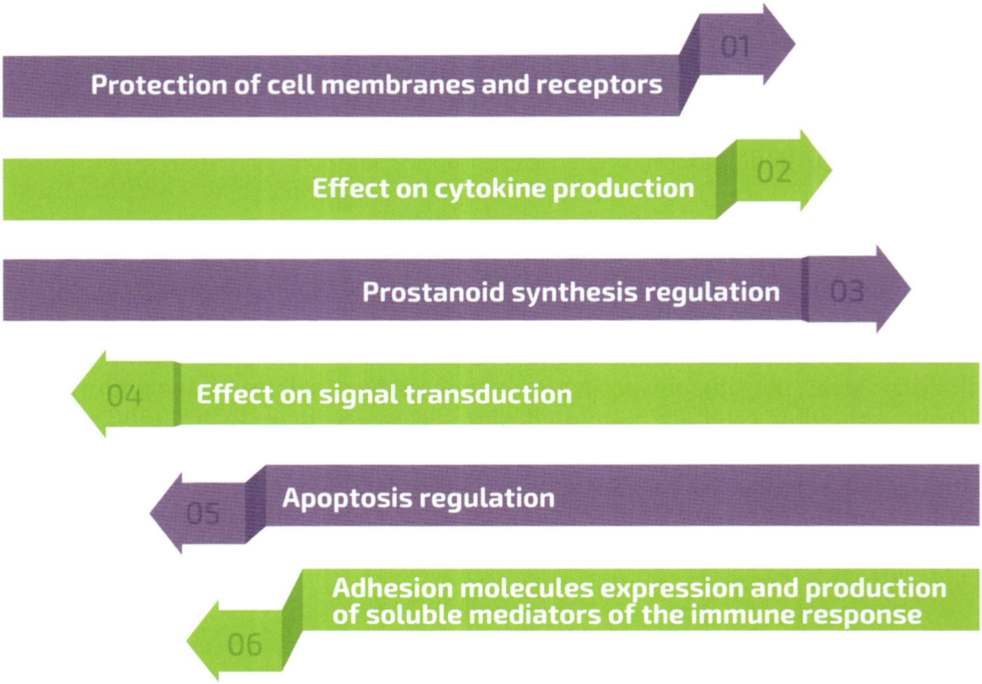

Figure 9.12. Immunomodulating properties of selenium.

It is believed that several mechanisms are involved in Se-dependent modulation of immune system (Wu and Meydani, 1998; Surai, 2002, 2006, 2018):

- **Protection of cell membranes and receptors.** It is well recognised that sophisticated antioxidant defences directly and indirectly protect the host against the damaging effects of cytokines and oxidants. In particular, indirect protection is afforded by antioxidants, which reduce activation of NF-κB, thereby preventing up-regulation of cytokine production by oxidants. On the other hand, cytokines increase both oxidant production and antioxidant defences, thus minimising damage to the host. Antioxidants prevent oxidative stress-induced damage to immune cells. It is necessary to take into account that cellular integrity is very important for receiving and responding to the messages needed to coordinate an immune response. As mentioned above, phagocytosis is the major mechanism for microbe removal from the body. The immune system generates ROS as part of its defence function and these ROS are an important weapon to kill pathogens. However, chronic overproduction of ROS can cause damage to immune cells and compromise their function (Wu and Meydani, 1998). In fact, immune cells are rich in PUFAs which are very susceptible to free radical attack. It is well recognised that many immunological functions are membrane-dependent. These are antigen recognition, receptor expression, secretion of antibodies and cytokines, lymphocyte transformation, and contact cell lysis (Wu and Meydani, 1998). In particular, the receptors are important for antigen recognition and the secretion of various chemical mediators such as interferon, tumour necrosis factor, prostaglandins and interleukins. Therefore, lipid peroxidation can change membrane structure and properties (e.g. fluidity, permeability, flexibility, etc.) which would affect immune cell functions. In contrast antioxidants are able to prevent those damaging effects of ROS and maintain immune function. For example, H_2O_2 depressed lymphocyte proliferation (Metzger *et al.*, 1980), while vitamin E decreased H_2O_2 formation by PMN (Baehner *et al.*, 1977). It was found that H_2O_2 injured lymphocytes immunocompetence deeply and administration of Se counteracts this damage (Sun *et al.*, 1995). It is well known that macrophage activation and phagocytosis of foreign particles are regularly accompanied by a so-called 'respiratory burst', an increase in the production of ROS, exerted by the enzyme complex NADPH oxidase. A number of selenoproteins is expressed at the same time to protect the cells from the cytotoxic effects of ROS directed against engulfed microorganisms. Selenoproteins participating in antioxidant defences (GPx, thioredoxin reductases, methionine sulfoxide reductase B, etc.) are able to protect neutrophils from oxygen-derived radicals that are produced to kill ingested foreign organisms (Ebert-Dumig *et al.*, 1999). Therefore, as a constituent of selenoproteins, Se is needed for the proper functioning of neutrophils, macrophages, NK cells, T lymphocytes and some other immune mechanisms (Ferencik and Ebringer, 2003; Surai, 2006, 2018).

- **Effect on immunomodulator production.** There is a range of regulatory molecules produced by immune cells. For example, IL-2, a lymphocyte growth factor, is recognised as an important immunomodulatory molecule. Oxidative stress suppresses IL-2 production and antioxidants can help to overcome this suppression. Therefore, Se up-regulates the expression of the T cell high affinity IL-2 receptor and provides a vehicle for enhanced T cell responses. Binding of IL-2 by the IL-2 receptor induces proliferation of T lymphocytes. For example, it has been suggested that Se can increase the inducibility of IL-2 receptor whereas vitamin E may counteract the down-regulatory effect of cAMP on IL-2 activity (McCarthy, 1997). The mRNA for IFN-γ was much less abundant in Se-deficient vs Se-adequate mice and mRNA for IL-2 was also lower in the Se-deficient mice (Beck *et al.*, 2001).
- **Prostanoid synthesis regulation.** Antioxidants alter the production of immunomodulatory molecules such as prostaglandins and leukotrienes altering a ratio between immunosuppressive and immunostimulating eicosanoids. The relationship between antioxidants and inflammatory reactions are shown in Figure 9.13. From the figure it is clear that poor antioxidant defence is associated with enhanced inflammation, overproduction PGE_2 resulting in suppression of lymphocyte activity (Grimble, 1997). For example, at low concentrations, PGE_2 is essential for cellular immunity; however, increased PGE_2 concentration is associated with a suppression of cellular and humoral immunity, including antibody formation, DTH, lymphocyte proliferation, and cytokine production (Surai, 2018). Lymphocytes from Se-deficient rats stimulated by an ionophore produced significantly less prostaglandins in comparison to Se-supplemented animals and characterised by decreased activation of phospholipase D (Cao *et al.*, 2002). It was postulated that dietary Se status, which in turn determines tissue Se concentration, plays an important role in the regulation of arachidonate metabolism affecting the 5-lipoxygenase

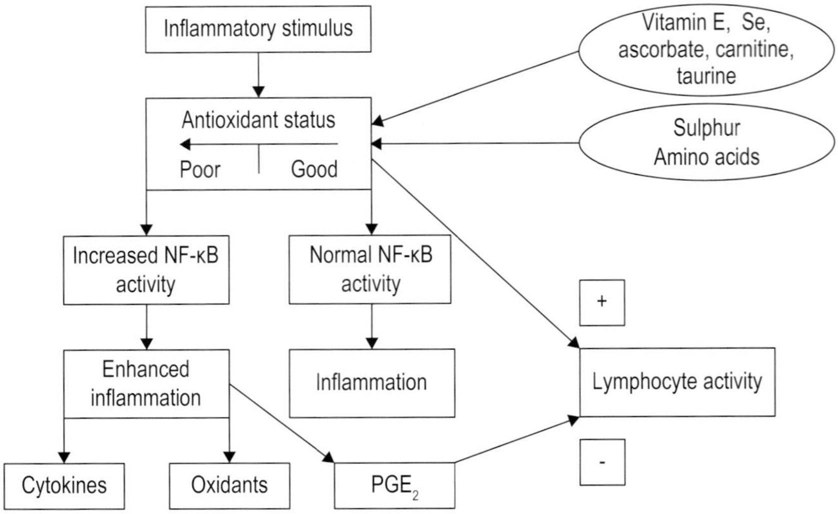

Figure 9.13. Antioxidants and inflammation (adapted from Surai, 2006).

pathway. This may be one of the biochemical mechanisms underlying the inhibition of lymphocyte proliferation and the decrease in resistance to infectious diseases observed in Se-deficient animals (Cao *et al.,* 1992). 5-Lipoxigenase (5-LO), a key enzyme in the biosynthesis of proinflammatory leukotrienes, is regulated by the cellular redox status with a requirement in hydroperoxides (Werz *et al.,* 2000). In fact, granulocyte-derived ROS can activate B-lymphocyte 5-LO. Since leukotrienes are involved in regulation of cell proliferation, activation and maturation, they regulate immune responses.

- **Effect on signal transduction**. Natural antioxidants such as Se and vitamin E may protect against oxidant-mediated inflammation and tissue damage by virtue of their ability to scavenge free radicals and by their ability to inhibit the activation of NF-κB (and possibly other oxidant-sensitive transcription factors). *In vitro* Se supplementation at low levels increased production of interferon by human peripheral lymphocytes and high Se doses were detrimental for interferon production (Watson *et al.,* 1986). In fact, NF-κB is required for maximal transcription of many inflammatory cytokines and adhesion molecules (Hughes, 1999). Furthermore, Se at physiological levels mediates inhibition of the activation of the transcription factor NF-κB which regulates genes that encode inflammatory cytokines (Maehira *et al.,* 2003). Thus, maintaining adequate antioxidant status may provide a useful approach in attenuating the cellular injury and dysfunction observed in some inflammatory disorders. (Conner and Grisham, 1996). In fact, there is an inverse relationship between cellular Se status and inducible form of nitric oxide synthase expression in LPS-stimulated cells (Prabhu *et al.,* 2002). Following LPS stimulation, the nuclear localisation of NF-κB was significantly increased in Se-deficient macrophages, thereby leading to increased expression of pro-inflammatory enzyme cyclo-oxygenase-2 (Zamamiri-Davis *et al.,* 2002). It is necessary to underline that non-toxic concentrations of reactive metabolites of oxygen and nitrogen play an important role in regulating the expression of genes involved in the inflammatory response and in modulating apoptosis (Jourd'heuil *et al.*, 1997). At the same time an immune response requires extensive communication between a wide range of cell types (Klasing, 1998a,b) and special cell receptors are of great importance in this communication. Therefore, protective effect of antioxidants in prevention of membrane and receptor damages due to peroxidation could provide an important way of enhancing the immune system.

- **Apoptosis regulation.** Antioxidants are considered to prevent apoptosis caused by oxidative stress. This could have a great effect on immune cell apoptosis preventing immunosuppression. It has been shown that *in vitro* selenium deficiency in a subset of hepatocellular carcinoma-derived cell lines causes oxidative stress and cytochrome c release with subsequent cell death by apoptosis (Irmak *et al.*, 2003). New techniques are available and they can be used to re-evaluate old data on Se-deficiency. For example, Se and vitamin E deficiency in chickens significantly increased caspase-like activity suggesting that cell death associated with exudative diathesis can follow the apoptotic pathway (Nunes *et al.*, 2003). Immunomodulating properties of Se could be mediated via prevention of apoptosis of immune cells in the case of mycotoxicoses. Indeed, many mycotoxins are immunosupressive and they can cause apoptosis (Da Silva *et al.*, 2018; Surai, 2002, 2006, 2018), therefore Se supplementation of the mycotoxin-contaminated feed could potentially have a beneficial effect.
- **Adhesion molecules expression and production of soluble mediators of the immune response.** It has been shown that Se is able to affect the adhesion molecules expressions that are crucial in the inflammatory process (Jahnova *et al.*, 2002). In fact, the inhibitory effect of Se on the adhesion molecule expression has been studied in cultured endothelial cells after interferon-gamma stimulation (Horvathova *et al.*, 1999). The data suggest that Se affects the expression of P-selectin, ICAM-1, VCAM-1, and ELAM-1 in a dose-dependent manner.

9.10 Immunocommunication, free radicals, redox balance and selenium

As mentioned above the immune system of the animal is based on natural and adaptive immunity. Natural immunity is dependent on the efficient function of phagocytic cells, namely neutrophils and macrophages. These cells are equipped with an array of microbicidal weapons, such as proteases, enzymes that hydrolyse proteins disrupting membranes. This weaponry is stored in granules in the cytoplasm. Furthermore, these cells have a powerful system for generating large amounts of ROS and they use them as an effective chemical weapon to kill pathogen. However, on escape from the phagosome the same free radicals become dangerous and can damage biological molecules, which compromises phagocyte function and reduces adaptive immunity. Phagocytes also produce communication molecules (eicosanoids, cytokines, etc.), that are used for effective communications between various immune cells.

Invading pathogens are controlled by the natural and adaptive branches of the immune system. It seems likely that in pigs natural immunity is responsible for recognition of invading pathogens by specific receptors. Binding of pathogen to those receptors induces the production of ROS and RNS, pro-inflammatory cytokines as well as communicating molecules which are responsible for sending regulatory signals to the adaptive immunity (Werling and Jungi, 2003). As mentioned above, adaptive immunity is based on activity of B- and T lymphocytes, which produce antibodies to specific non-self substances (B-lymphocytes) or directly attach to them (T lymphocytes) and remove them from the cell. In general, in chicken an overall pool includes about 100 million T cells (Juul-Madsen *et al.,* 2008). Specific adaptive immune responses rely on the major histocompatibility complex's recognition of peptide antigens derived from pathogen to activate a variety of effector T cells (helper and cytotoxic cells) that interact with B cells that produce effector proteins called antibodies (Castle, 2000). The adaptive immunity is characterised by high plasticity to recognise up to 10^{11} distinct structures and is tightly regulated to turn on or off a response aiming in eradication of pathogens but not destruction of self (Surai, 2018).

In the healthy animal or human resistance to infection relies on a balance between the natural and adaptive immunity. Regulation of the immune system is extremely complex. We are only starting to understand how the immune system co-ordinates the body's response to a disease or invading pathogen. It seems likely, that communication between immune cells is a crucial factor of immunocompetence (Surai, 2002, 2006, 2018). Interaction between the different immune cell types that make up these components of host defence is carried out by the relative mix of cytokines, hormone-like proteins, as well as other communicating molecules (Castle, 2000). The innate response, including its inflammatory component, reacts initially to the stimulus, acting directly to eliminate it by the activities of complement or phagocytosis.

Cytokines produced by monocytes and macrophages regulate this response and also act on the liver, skeletal muscle, adipose and brain changing their metabolism and stimulating various responses. The cytokines also interact with T lymphocytes (Calder, 2001). If we imagine that immune system is an army fighting against invaders (microorganisms, viruses, etc.) than we would expect them to have something like mobile phones to receive and send signals to each other. Remarkably enough, major immune cells (macrophages, neutrophils, dendrites, NK cells, T- and B-lymphocytes) have on their surface something like 'mobile phones' called receptors. Those receptors are extremely sensitive to communicating molecules, but they are also sensitive to redox balance and free radicals and can be easily compromised/damaged. In such a situation without proper communication all those huge armies of immune cells would become useless.

They also can start fighting each other as well as eventually destroying immunocompetence and causing autoimmune reactions. If we imagine that immune cells are soldiers using chemical weapon to kill enemy, then special ammunition protecting them from their own weapon would be crucial for effective battle. In the case of immune cells such ammunition is represented by natural antioxidants with Se-GPx, thioredoxin reductase and other selenoproteins being major defences. Indeed, if not properly protected, macrophage functions could be compromised including initial overproduction of free radicals with consecutive damages to specific enzymatic systems resulting in decreasing efficiency of oxidative burst and apoptosis. Based on the presented model it is clear that antioxidant defence is a crucial factor of immune defence in the body and Se plays a central role in such defence.

Information is quickly accumulating to confirm this idea. For example, Roy *et al.* (1993) indicated that supplementation of C57BL/6 mice with Se resulted in increased expression of IL-2R on con A-stimulated lymphocytes resulting in a greater number of high-affinity IL-2R per cell and enhanced proliferation and differentiation of lymphocytes. Similar effect of Se was seen on IL-2R in response to allergens: increases in CD25 were found on T helper (Th) cells with increases in dietary Se (Hoffmann *et al.*, 2007). A clear correlation between Se supplementation and lymphocyte proliferation, which was preceded by enhanced expression of high-affinity IL-2R was shown in human study (Roy *et al.*, 1994). It should be mentioned that Th cells are involved in providing effective immune responses to a wide variety of antigens and they depend on IL-2R signalling for activation, proliferation, and differentiation (Zhu *et al.*, 2006).

The PBMC membrane fluidity, IL-2 production and IL-2R expression in patients with chronic hepatitis were significantly lower than those in healthy blood donators, while MDA concentration was significantly increased. Both *in vitro* and *in vivo* administration of selenium could reverse the above parameters (He *et al.*, 2004). It was also shown that Se can suppress the activation of transcription factor NF-κB and IRF3 induced by TLR3 or TLR4 agonists confirming an ability of Se to modulate signalling pathways of TLRs leading to decreased inflammatory gene expression (Youn *et al.*, 2008). Furthermore, in comparison to a group of men supplemented with 300 μg Se/day in the form of Se-Yeast, the low-Se yeast group of healthy free living men showed increased counts of NK cells and T lymphocytes expressing both subunits of the high affinity IL-2 receptor (Hawkes *et al.*, 2009).

The receptor integrity and expression were not studied in the aforementioned study, but it could well be that they were affected. Clearly, selenoproteins mediate T cell immunity through an antioxidant mechanism. It seems likely that dietary Se levels could modulate free thiol levels, receptor expression/integrity and specific signalling events during CD4(+) T cell activation, which influence their proliferation and differentiation. For example, mice fed low (0.08 mg/kg), medium (0.25 mg/kg), or high (1.0 mg/kg) Se diets for 8 wk were challenged with peptide/adjuvant (Hoffmann *et al.*, 2010). It was shown that antigen-specific CD4(+) T cell responses were increased in the high Se group compared with the low and medium Se groups. T cell receptor signalling in *ex vivo* CD4(+) T cells increased with increasing dietary Se, in terms of calcium mobilisation, oxidative burst, translocation of nuclear factor of activated T cells, and proliferation. Interestingly, Se was shown to maintain Ca^{2+} homeostasis in sheep lymphocytes challenged by oxidative stress (Proietti *et al.*, 2018).

Furthermore, the high Se diet increased expression of IL-2 and the high affinity chain of the IL-2 receptor compared with the low and medium Se diets. In the low Se diet group there was a decrease in free thiols compared with the medium and high Se diets. Addition of exogenous free thiols eliminated differences in CD4(+) T cell activation among the dietary groups. (Hoffmann *et al.*, 2010). Similarly, selenoprotein deficiency in T cells led to an inability of these cells to suppress reactive oxygen species (ROS) production, which in turn affected their ability to proliferate in response to T cell receptor stimulation (Carlson *et al.*, 2010; Shrimali *et al.*, 2008). It has been also confirmed that Se can influence immune response by changing the expression of cytokines and their receptors or by making immune cells more resistant to oxidative stress (Naziroglu *et al.*, 2012). Furthermore, supranutritional Se-yeast supplementation to sheep increased whole blood neutrophil expression of genes involved in innate immunity including L-selectin, interleukin-8 receptor and toll-like receptor 4 (Hugejiletu *et al.*, 2013).

It was shown that spleen oxidative stress induced by a high-fat diet in mice induced the decreased expression of genes associated with antioxidant defence, as well as Fc receptor and an antioxidant (lipoic acid) supplementation attenuated the aforementioned changes (Cui *et al.*, 2012). It is interesting that receptor-mediated Ca(2+) flux was decreased in T cells, neutrophils, and macrophages from Selenoprotein K-deficient ($SelK^{-/-}$) mice compared with controls. Furthermore, Ca(2+)-dependent functions including T cell proliferation, T cell and neutrophil migration, and Fcγ receptor-mediated oxidative burst in macrophages were decreased in cells from $SelK^{-/-}$ mice compared with that in cells from controls (Verma *et al.*, 2011). Recently it has been shown that organic selenium supplementation in sheep was associated with the enrichment of B and T cell receptors signalling pathways (Elgendy *et al.*, 2016). Furthermore, effective activation of macrophages through phagocytic Fcγ receptors has been shown to require selenoprotein K (SelK; Norton *et al.*, 2017). In fact, SelK deficiency through genetic deletion or low selenium in culture media leads to low expression of the IP3 receptor (an important element for the activation and function of immune cells) due to a defect in the receptor palmitoylation (Fredericks *et al.*, 2014).

It has been learned a lot recently about the innate receptors, such as Toll-like receptors, that activate APCs in response to microbial products and initiate an immune response. Indeed, it has been shown that Se status was associated with IL-2 receptor levels. It is well established that the immune response is regulated by a very complex interactions between immune cells via cytokine system as well as via other signalling molecules. In fact, there are many various cytokines responsible for immune cell communications. The most difficult question remains to answer is 'how do cells interpret these signals to modulate the immune response?'. A key aspect of the answer lies at the receptor and the multiple modifications and regulatory proteins that collectively shape a cytokine response (Delgoffe *et al.*, 2011). For example, for an optimal and appropriate immune response, T cells require activation through the T cell receptor (TCR), which recognises specific antigen presented in the context of MHC (Williams and Kwon, 2004). This recognition also confers the ability of T cell responses to distinguish between 'self' and 'nonself'. Engagement of the TCR, throughout the lifetime of the T cell, controls the survival, proliferation, and/or differentiation of T cells. Thus, signalling through the TCR has important consequences for proper immune response and the effectiveness of that response. In fact, receptor expression can be dynamically modulated by the cell, based on cell type, stimulation, and cytokine activity. Some receptors are constitutively expressed, while others require additional signals in order to be upregulated and properly expressed. In addition, a lot of cytokine receptors are formed from multiple chains that have distinct modes of regulation. Furthermore, cells can temporally modulate the expression of certain receptors thereby altering the cytokines to which a particular cell is responsive (Delgoffe *et al.*, 2011).

As a result of cytokine stimulation, cells can become activated and differentiate and most of the physiologic outcomes observed in the immune system are mediated by cytokines. It seems likely that T cell activation requires two distinct signals: the first signal (recognition) relies on the interaction between the T cell receptor (TCR) and a MHC-peptide complex, while the second signal is provided by the cross-linking between co-stimulatory molecules on activated T cells and compromise or loss of one or more of such regulatory and/or co-stimulatory molecules can be detrimental for immunocompetence (Pizzi *et al.*, 2016).

The immune response to pathogens is characterised by a biphasic response: the initial non-specific innate immune response and the subsequent pathogen-specific adaptive immune response. It is well established that the innate immune response is mediated mainly by macrophages, dendritic cells and neutrophils, and occurs rapidly after these cells encounter a pathogen, while the adaptive immune response is controlled by T and B cells, which takes place several days after pathogen invasion (Kingeter and Lin, 2012). It is necessary to underline that the innate immune response is crucial for the development of the adaptive immune response. First, phagocytic cells such as dendritic cells and/or macrophages engulf extracellular pathogens, or infected host cells, and present pathogen-associated molecular patterns (PAMPs) to T cells. In this way, the innate immune system indicates to the cells of the adaptive immune system presence of pathogen and the appropriate pathogen-specific adaptive lymphocytes are activated (Kingeter and Lin, 2012).

Therefore, signal transduction pathways are triggered resulting in the activation of various transcription factors, such as NF-κB, responsible for the production of pro-inflammatory cytokines and chemokines, which are involved in further regulation of the immune response. Indeed, immune cell receptor expression and integrity are major determinants of porcine immunocompetence. In fact, we can consider immunocompetence as an effective communication between all major cells of the immune system. It should be mentioned that both the innate and adaptive immunity are closely related and there is a complex interplay between innate and adaptive responses. In fact, the innate immune system can be considered as an effector arm of the adaptive response. For example, interactions between T cells and macrophages are of great importance. In fact, T cell responses are associated with the production of IFNγ a cytokine possessing the ability to activate macrophages, increasing their capacity to phagocytose and kill invading microorganisms. Similarly, macrophages express Fc receptors that bind immunoglobulin (produced by B cells) and enhance recognition and engulfment of foreign pathogen-derived material. There is also a close interaction between T cells and goblet cells resulting in increased secretion of defensins and altered mucus composition (Juul-Madsen *et al.,* 2008).

Indeed, initial encounter of pathogens with the innate system leads not only to the destruction of the pathogens but also initiates a cascade of important immunological events, including recruitment of various immune components, as well as induction and modulation of the adaptive immune system. In such interactions immune receptors are of great importance and their protection in stress conditions is a key for effective immunocompetence. For example, in the cellular innate immune system up to 100 different receptors are expressed at a relatively high frequency (Juul-Madsen *et al.,* 2008) and each receptor is usually expressed on millions of innate and adaptive immune cells. Based on common structural features, cytokine receptors can be grouped into several families including the class I (haematopoietin) cytokine receptor family, the class II (IFN/IL-10) cytokine receptor family, the TNFRSF, the IL-1 receptor family, the TGF-β receptor family and the chemokine receptors (Kaiser and Staheli, 2008).

In fact, Se increases lymphocyte proliferation, expression of the high-affinity IL-2R, cytolytic T lymphocyte efficacy/activity, and NK-cell function, enhancing resistance to infections through modulation of interleukin production and subsequently the Th1/Th2 response. In fact, it seems likely that Se can upregulate the expression or protect from disruptions/damages by ROS of receptors for IL-2 and other cytokines on the surface of activated lymphocytes and NK cells. This event favours the interaction of this cytokine with its respective receptors (Puertillano *et al.,* 2011; Surai, 2002, 2006, 2018).

Over the last decade crucial roles of ROS and redox balance in regulating T- and B-cell functions was well recognised. In particular, the redox balance is believed to be regulated by dithiol/disulphide equilibrium and additional components of the antioxidant defence system including thioredoxin, thioredoxin reductase, NADPH, SOD, catalase, peroxiredoxins, glutaredoxins, etc. (Patwardhan *et al.,* in press). Indeed, the redox balance in general was shown to be able to affect/determine the phenotype and effector functions of T cells (Checker *et al.,* 2010; Gostner *et al.,* 2013; Mak *et al.,* 2017; Sena *et al.,* 2013). In fact, T cell activation is believed to depend on the redox balance at the site of their action and it is associated with ROS production in mitochondria. Furthermore, defects in mitochondria ETC are associated with impaired TCR-dependent ROS production and defects in antigen specific proliferation (Patwardhan *et al.,* in press). In particular, ROS are considered to be essential for T lymphocyte activation, expansion and effector function (Gambhir *et al.,* 2019; Moro-García *et al.,* 2018).

In general, T cell activation is associated with a range of important changes. For example, during T cell activation mitochondria are characterised by elevated membrane potential (Wherry, 2011), metabolic re-programming (Wang *et al.*, 2011) and located to the immunological synapse (Quintana *et al.*, 2007). On the one hand, macrophage derived ROS are shown to induce Tregs in an NADPH oxidase dependent fashion (Kraaij *et al.*, 2010). On the other hand, there is also an activation of NADPH oxidase in T cells (Jackson *et al.*, 2004) with Duox1 (NOX family member) being responsible for TCR-mediated H_2O_2 production and involved in TCR signalling (Kwon *et al.*, 2010) with simultaneous changes in antioxidant defence mechanisms (Gambhir *et al.*, 2019; Sukumar *et al.*, 2016). It seems likely that GSH plays a special regulatory role in this process (Mak *et al.*, 2017), since GSH depletion by different means was shown to impair inflammatory responses (Checker *et al.*, 2010; Gambhir *et al.*, 2014; Ulrich and Jacob, 2019). In fact, GSH is considered to be a central metabolic integrator in immune responses mediated by T cells (Patwardhan *et al.*, in press) with the redox balance in T cells to control metabolic reprogramming (Almeida *et al.*, 2016). Indeed, both excess and insufficient levels of ROS was suggested to impair T cell metabolic reprogramming and functional activity (Patwardhan *et al.*, in press).

It should be mentioned that beyond GSH/GSSG involvement in the redox balance maintenance in immune cells a special role of thioredoxins, low molecular weight (12-kDa), ubiquitously expressed, molecules deserve more attention. Indeed, by reducing proteins as a result of cysteine thiol-disulphide exchange, thioredoxins are considered to be major players in redox balance maintenance and immune cell activation. It seems likely that thioredoxins are responsible for integrity and functional activity of TCRs. For example, trx1 is shown to play an essential role in maintaining surface thiol density/activity of Tregs thereby offering them enhanced tolerance to oxidative stress and regulating their proliferation (Mougiakakos *et al.*, 2011). Indeed, Trx1 was found to exert antioxidant effect controlling ROS accumulation and redox status leading to anti-inflammatory effects and downregulating T cell alloresponses in various model systems (Sofi *et al.*, 2019). The Trx1 system is believed to be essential for controlling DNA synthesis during T cell metabolic reprogramming and proliferation (Muri *et al.*, 2018). Furthermore, Trxs are involved in NF-κB (Lorenzen *et al.*, 2017) and AP1 signalling (Patwardhan *et al.*, in press) in immune cells. Interestingly, the Trx-1 and GSH/Grx1 systems are shown to redundantly fuel murine B-cell development and responses (Muri *et al.*, 2019).

The interaction of cell surface receptors on immune cells, including lymphocytes and phagocytes, with their cognate ligands on interacting cells or such soluble mediators as cytokines, chemokines and growth factors is a vital process responsible for generating and controlling an appropriate immune response including specific cell activation, proliferation and/or migration to sites of infection (Stegmann *et al.*, 2018). It is well appreciated that interacting proteins in intact three-dimensional structures are necessary to present their complementary binding interfaces for mediating receptor-ligand interactions. In fact, post-translation protein modifications are considered to be important mechanisms controlling receptor-ligand interactions and downstream events of immune response (Stegmann *et al.*, 2018).

Disulphide bonds are known to be important determinants of protein structure and function. In particular, they can sense and control changes in the redox environment. They also regulate protein folding, structural stabilisation, activity, localisation and interactions and therefore disturbances of protein disulphide bonds lead to various detrimental consequences to cellular/body homeostasis, including immune system dysregulation. In particular, oxidative stress-induced disulphide formation and thiol-disulphide exchange-mediated reduction of allosteric disulphides could affect immune cell structure and functions (Bechtel and Weerapana, 2017). Protein disulphide bonds serve not only structural role, but they also control the function of the mature protein in which they reside. On the one hand, structural disulphides are shown to be formed during protein folding and they are involved in the folding process and responsible for stability of the resultant protein architecture. On the other hand, functional disulphides can be divided into two groups, namely catalytic disulphides present in specific enzymes (e.g. protein-disulphide isomerase) or antioxidant compounds (e.g. thioredoxin), and responsible for manipulation of disulphides in other proteins and allosteric disulphides inducing changes in the tertiary structure and controlling the function of the protein in which they are located (Chiu and Hogg, 2019). Indeed, allosteric disulphide bond formation or cleavage can switch the structure and function of a protein (Kuroi *et al.*, 2020). In particular, these so called 'allosteric disulphides' allowing post-translation modification of proteins, have been shown to control various physiological processes, including the immune response (Chiu and Hogg, 2019). Indeed, modification of these disulphide bonds is believed to cause conformational changes of the protein and altering its activity and function a process observed as a result of the entry of several enveloped viruses into their host cells, e.g. HIV, hepatitis C virus and Newcastle disease virus (Stegmann *et al.*, 2013). These labile disulphide bonds are common, with several classes of proteins being identified including integrins, receptors, transporters and cell-cell recognition proteins (Metcalfe *et al.*, 2011). The importance of disulphide bonds and central role for redox chemistry at the cell surface in creating an adequate immune response have been reviewed and analysed by Metcalfe *et al.* (2012) summarising that:

- Disulphide bonds in integrins maintain their low affinity state and functional activity and their activities can be modulated by reducing agents.
- Anti-microbial activity of defensins can be increased by reduction of disulphide bonds.
- IL-4 contains a disulphide bond susceptible to mild reduction.
- There are some redox-sensitive ligands for cell surface receptors.
- Vaccination is associated with increase in the number of free thiols in lymph nodes.
- Activation of T cells by dendritic cells is associated with secretion of thioredoxin (Trx) into the extracellular space, where it can catalyse reduction of disulphide bonds on the surface of immune cells, including T cells.
- Trx 1 is shown to reduce disulphide bonds in the TNF receptor family member CD30 leading to altered cytokine binding and signalling.

- The Cys183-Cys232 disulphide bond in mouse CD132 (an important part of the receptor for several cytokines, including IL-4, IL-21, IL-7, IL-9, IL-15) is shown to be susceptible to reduction by enzymes such as gamma interferon-inducible lysosomal thiolreductase, protein disulphide isomerase and antioxidant Trx, which are commonly secreted during immune activation.
- The Cys183-Cys232 disulphide bond is indicated to be reduced in an *in vivo* LPS-induced acute model of inflammation.
- Reduction of the Cys183-Cys232 disulphide in CD132 was reported to inhibit IL-2 binding to the receptor complex (Metcalfe *et al.*, 2012).

Therefore, the activity of a range of cytokine receptors is shown to be affected by redox sensitive disulphide bonds and this could be an important mechanism controlling the levels of responses of cells to exogenous cytokines (Metcalfe *et al.*, 2012), associated with redox balance in the cell reflecting efficacy of antioxidant defence and adaptive vitagene-related signalling (Surai *et al.*, 2019). In fact, disulphide bond redox changes could be considered as a universal mechanism of immune receptor regulation. For example, the CD4, an integral membrane glycoprotein expressed on the surface of specific leukocytes, including T cells, is shown to be a co-receptor stabilising T cell receptor interactions with MHC class II molecules on antigen presenting cells (Owen *et al.*, 2018). Therefore, CD4 is suggested to play a vital role in the immune response and redox exchange of the CD4 disulphides was shown to regulate its functions. Indeed, each human CD4 monomer was indicated to comprise four immunoglobulin like ecto-domains (D1-D4), a transmembrane segment, and a cytoplasmic tail and 3 domains are characterised by presence of a single disulphide bond located in their hydrophobic cores (Owen *et al.*, 2018). In fact, ablation of the allosteric disulphide bond in domain 2 was shown to lead to both a favourable structural collapse and an increase in the stability of CD4 (Owen *et al.*, 2016).

Interestingly, blocking cell surface redox exchanges with both a membrane-impermeable sulfhydryl blocker (DTNB) and specific antibody inhibitors of Trx1 was shown to induce translocation of CD4 into detergent-resistant membrane domains, while Trx1 inactivation did not affect the localisation of the chemokine receptor CCR5 (Moolla *et al.*, 2016). Furthermore, the αβ TCR is known to be a multimeric transmembrane complex consisting of a disulphide-linked antigen binding heterodimer associated with three signal-transducing CD3 subunits (Kim *et al.*, 2012). The TCR is known to mediate recognition of antigenic peptides bound to MHC molecules (pMHC), while the CD3 molecules transduce activation signals to the T cell (Mariuzza *et al.*, 2020). The CD3 delta chain was identified as having a labile disulphide bond and all three of the CD3 delta chains of the heterodimeric ectodomains of the TCR was shown to contain a conserved CysXXCys motif in their membrane proximal stalk. Those cysteines are believed to be essential for TCR-mediated T cell activation and form a disulphide bond which needs to be oxidised for signalling to occur (Brazin *et al.*, 2014). CD44, the primary leukocyte cell surface receptor for hyaluronic acid was also shown to have labile disulphide bond formed by Cys77 and Cys97 being susceptible to chemical and enzymatic reducing agents (Kellett-Clarke *et al.*, 2015).

Immune cell function can be controlled by modulating the structure of either the receptor or the ligand (Stegmann *et al.*, 2018). In fact, the authors identified, quantified and monitored a reduction of labile disulphide bonds in primary cells during immune activation. Indeed, a reduction of labile disulphide bonds was shown to be thiol oxidoreductase-dependent and to affect activatory (e.g. CD132, SLAMF1) and adhesion (CD44, ICAM1) molecules (Stegmann *et al.*, 2018). Increased levels of oxidative stress associated with several diseases are counteracted by the integrated antioxidant defence system which includes activities of various oxidoreductase enzymes, such as thioredoxin (Trx), which are effective reductases of allosteric disulphide bonds in proteins (Gurjar *et al.*, 2019).The importance of disulphide bonds management in stress conditions was demonstrated in experiments with monoclonal antibodies (mAbs). In particular, it was shown that Trx can effectively reduce the interchain disulphide bonds of the mAbs altering their function: increasing antigen-binding capacity and decreasing the Fc receptor binding with significant loss in both complement-dependent cytotoxicity and antibody-dependent cellular cytotoxicity activity (Gurjar *et al.*, 2019).

The immune cells possess an important capacity to produce ROS and various inflammatory compounds as a part of their functional activities. However, as a result of oxidative stress leading to redox disbalance cell signalling/communications can be compromised leading to compromised immune response and over-inflammation, a situation observed in adult mice with premature aging (Garrido *et al.*, 2019). In carp head kidney lymphocyte models, TCR γ knockdown was shown to significantly increase the mRNA expression of IFN-γ, IL-1β, IL-8, IL-10, Nrf2 and NF-κB. It simultaneously decreased the mRNA expression of SOD and CAT and reduced the activities of GSH, GPx, T-AOC, CAT and SOD and increased the content of MDA and H_2O_2 (Yang *et al.*, 2020).

Importantly, ROS are also involved in BCR signalling. Indeed, BCR stimulation was shown to result in extracellular ROS generation by NADPH oxidase enzymatic complex with following participation in a range of cellular events leading to the B cell activation and signalling (Bertolotti *et al.*, 2012). Furthermore, it was calculated that a single plasma cell secreting $\geq 10^3$ IgM per second, each with $\geq 10^2$ disulphides, has to form $\geq 10^5$ bonds per second during Ig production (Bertolotti *et al.*, 2010; Anelli *et al.*, 2015). In fact, disulphide bond formation is associated with formation of H_2O_2, an important signalling molecules which can be converted into a range of various ROS if not under antioxidant system control. It was suggested that BCR, TCR and other Ig family members evolved the H_2O_2-generating activity to facilitate and amplify their signalling output (Reth, 2002).The authors suggested that the second messenger H_2O_2 is the molecule that communicates between BCR and TLR receptor systems, while BCR coreceptors could also be involved by increasing H_2O_2 production around the activated BCR. There is also a possibility that activated macrophages can function as APCs for T cell priming as a result of H_2O_2 diffusing from the macrophage to the T cell during their close interaction phase (Reth, 2002).

It is important to mention that Se can affect both GSH concentration (Surai and Kochish, 2019) and thioredoxin metabolism and efficacy via regulation of TR activity (Surai, 2018) and therefore, optimal Se dietary supplementation is considered as an effective way to maintain optimal redox status of immune cells and intact their receptors responsible for effective immunocommunication. Indeed, as shown above reversible disulphide bond formation can become irreversible in stress conditions when the antioxidant defence network is not able to maintain optimal redox status. This could change immune receptor (including TCR and BCR) structure, expression and activity affecting immune response.

It is proven that Se- and vitamin E deficiencies are associated with compromised functions of natural and adaptive immunity. In particular phagocytic functions, lymphocyte proliferation and antibody production are compromised (Surai, 2002, 2006, 2018). On the other hand, Se supplementation is shown to improve immunocompetence and increase resistance to various diseases. This is true for variety of farm and companion animals including poultry, cows, sheep, horses, pigs, fish, cats and dogs. A summary of the effect of compromised antioxidant system on immune system is shown in Figure 9.14.

The importance of a delicate turning of the immune system is reflected by various observations showing that over-reacting immune system has as detrimental consequences as immunosuppression. For example, in some individuals, the immune system recognises host antigens as 'non-self', attacking them and producing tissue damage leading to chronic inflammatory or autoimmune diseases. The immune system can also become sensitised to usually benign antigens from the environment causing allergies (Calder, 2001).

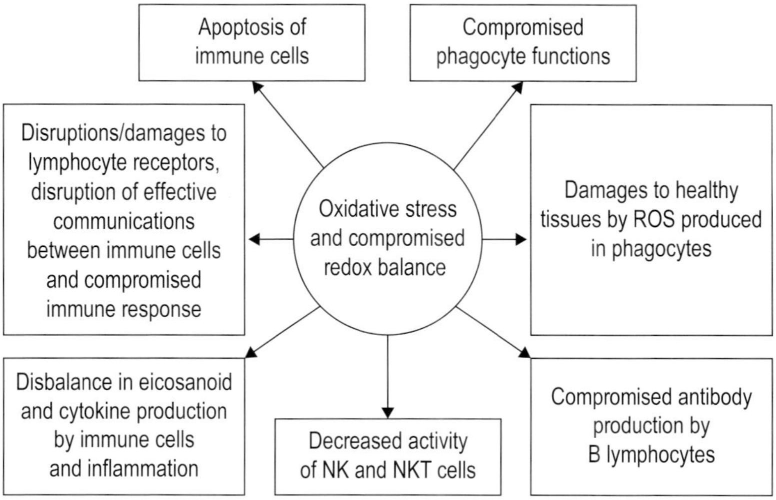

Figure 9.14. Oxidative stress and the immune system (adapted from Surai 2002, 2006, 2018).

It seems likely, that in these immune system miscommunication between immune cells plays a crucial role. The porcine immune system is functionally immature at birth. Therefore, postnatal development of the immune system is associated with accumulation of polyunsaturated fatty acids and a desperate need for antioxidant protection. Therefore, expression of selenoproteins in immune cells in early development would be a crucial factor in a regulation of the immunocompetence development. However, selenium reserves in the newly born piglets are very limited when inorganic selenium is used in the maternal diet. In great contrast, organic selenium is shown in a number of studies to be able to significantly increase Se reserves in newly born piglet. This selenium is absolutely essential for the formation of effective antioxidant defences resulting in effective immune system maturation and immunocompetence building. The expression of IFN-γ and COX-2, was shown to be lower in Se supplemented pigs versus control (Falk *et al.,* 2018). After LPS exposure GPx1 and SelenoN were shown to be more reduced in pigs fed NaSe compared with pigs fed organic Se. This could be a result of Se reserves in form of SeMet usage in stress conditions.

9.11 Conclusions

Selenium affects all components of the immune system, including the development and expression of nonspecific, humoral, and cell-mediated responses (Figure 9.15). In general, a deficiency in Se appears to result in immunosuppression, whereas optimal Se supplementation appears to result in augmentation and/or restoration of immunologic functions. On the one hand, a deficiency of Se has been shown (Kiremidjian-Schumacher and Stotzky, 1987) to inhibit:

- resistance to microbial and viral infections;
- neutrophil function;
- antibody production;
- proliferation of T and B lymphocytes in response to mitogens; and
- cytodestruction by T lymphocytes and NK cells.

On the other hand, Se supplementation has been shown (Kiremidjian-Schumacher and Stotzky, 1987) to stimulate:

- the function of neutrophils;
- production of antibodies;
- proliferation of T and B lymphocytes in response to mitogens;
- production of lymphokines;
- NK cell-mediated cytodestruction;
- delayed-type hypersensitivity reactions;
- allograft rejection;
- the ability of a host to reject transplanted malignant tumours.

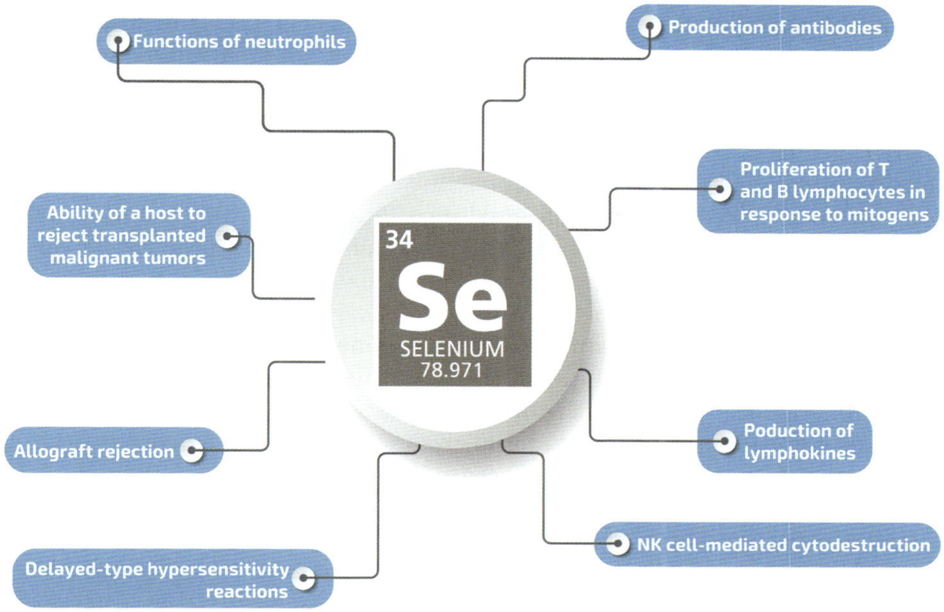

Figure 9.15. Effects of selenium on immunity.

When considering immuno-facilitating properties of selenium, it is necessary to take into account several points:

- Individual antioxidants in the body interacts with each other (Se, vitamin E, ascorbic acid, etc.) and prooxidants (iron, high level of PUFAs, mycotoxins) which themselves have immunostimulating or immunosuppressive effects. There is a range of environmental, technological, nutritional and internal stresses which affect redox balance and immune response. Animal health status and age are also important determinants of immunocompetence. Therefore, in every experiment the obtained results would reflect a sum of all these interactions and if, for example, background dietary concentrations of nutritional antioxidants or stress factors differ, results could be completely different. This could explain inconsistency of some results published for the last 30 years. Furthermore, antioxidants can suppress respiratory burst, however, it is not clear at present if there is a limit of this suppression after which the phagocyte anti-microbial activity would be compromised.
- Immunomodulating effect of antioxidants is shown to be maximal at their supplementation usually above the requirement for maximal growth and maintenance of reproduction. It could well be that the Se and vitamin E doses that are adequate for maximal productivity in healthy, unchallenged animals are not optimal for maximal immunocompetence and disease resistance.
- There is only limited information on the effect of stress (Broom and Kogut, 2018) and antioxidants (Fisinin and Surai, 2013; Hachimura *et al.*, 2018) on gut immunity, which could be crucial barriers to external pathogens (Qureshi *et al.*, 1998). Since free radicals can be damaging to intestinal structures (Hoerr, 1998) antioxidant functions of selenium are responsible for prevention damages to intestinal

lymphoid structure as well as damages to intestinal enterocyte membranes. This could be especially important in relation to gut immunosuppression caused by toxins/mycotoxins, nutritional deficiencies and infectious agents (Hoerr, 1998). Se-deficiency in mouse was shown to adversely affect the gut barrier and intestinal immune responses and to increase their susceptibility to experimental colitis and pathogen infection. In contrast, Se supplementation (0.40 mg/kg) was indicated to improve mouse intestinal physiology. Interestingly sodium selenite and selenomethionine were shown to have similar effects on the gut barrier and intestinal immune homeostasis but differentially affected faecal metabolites and induced significant proteomic variations in various pathways, including the xenobiotic detoxification pathway and glutathione metabolism (Zhai *et al.,* 2019). Dietary supplementation of newly hatched broiler chicks with 0.25 mg/kg Se (as sodium selenite) from hatch significantly reduced necrotic enteritis-induced gut lesions compared with infected birds given a non-supplemented diet. This was associated with increased levels of serum antibody against the NetB toxin in the chicks fed with 0.25 and 0.50 mg/kg Se in comparison to the non-supplemented control group. Furthermore, the transcripts for IL-1β, IL-6, IL-8, iNOS, LITAF, and GPx7, as well as AvBD6, 8, and 13 were shown to be significantly increased in the intestine and spleen of Se-supplemented chickens (Xu *et al.,* 2015b). Indeed, Se is proven to have an important role in resolution of inflammation in various tissues, including gut, via affecting redox balance and expression of various transcription factors including NF-κB and PPARγ (Nettleford and Prabhu, 2018).

- Selenium source (organic vs inorganic) seems to be an important element of its immunomodulating properties. Organic selenium appears to be at an advantage because it is better assimilated from feed and better accumulated in tissues. Indeed, with the same dose of supplementation organic selenium can deliver more element to the target tissues and because of toxicity of high selenium levels this could be a solution to avoid adverse effect of selenium overdose. For example, results presented by Johnson *et al.* (2000) indicated that splenic macrophages and lymphocytes are sensitive to Se intoxication and there is a disparity in the immune system toxicity of inorganic and organic forms of Se administered via the drinking water, inorganic Se being more toxic. In fact, sodium selenite at 9 mg/kg in the drinking water for 14 days increased the production of proinflammatory cytokines, tumour necrosis factor alpha and interleukin-1 beta, in LPS-stimulated splenic macrophages. However, mice exposed to Se as SeMet in the drinking water did not display any effects on the parameters examined at the dose range in the study. Furthermore, Se supplementation in organic form could have a beneficial effect during acute phase response (APR) in many infectious diseases. The APR is characterised by the synthesis of acute phase proteins, fever, accelerated whole-body protein turnover, high rates of hepatic gluconeogenesis (Klasing, 1998a) and decreased Se concentration in plasma (Sattar *et al.,* 1997). It is interesting that acute infections decrease serum Se levels regardless of the infective agent (Sammalkorpi *et al.,* 1988). Therefore if the body has selenium reserves in the form of SeMet in muscles, during APR it would be liberated as a result of skeletal muscle protein catabolism by proteasome action and would be used for re-synthesis of new selenoproteins which could decrease oxidative stress.

- Because of complexity of regulation of the immune response and lack of understanding of molecular mechanisms involved in such a regulation there is a reasonable suggestion (Klasing, 1998b) to direct immunomodulation firstly toward the correction of dysfunctional situation created by immune system immaturity, stress, immunosuppressive disease, or genetics. In this respect natural antioxidants, especially Se, could have a prominent role.
- As a result of antioxidant (selenium) deficiency increased oxidative stress of a host can lead to increased virus mutation rate and change in a viral pathogen (Beck, 1999) resulting in emerging viral pathogens with new pathogenic properties. Therefore, Se deficiency was associated with a change to the viral genotype, converting the virus from a benign to a highly virulent strain (Beck, 1998). This possibility was not fully exploited in animal production, but seems important to study more extensively.
- There is a need to study antioxidant composition and fatty acid profile of immunocompetent tissues depending on animal age and nutritional supplementation of antioxidants.

Acknowledgements

Peter F. Surai is supported by a grant of the Government of Russian Federation (Contract No. 14.W03.31.0013).

References

Abdel-Ati, K.A., Latshaw, J.D. and Donahoe, J.P., 1984. Distribution of selenium in chicken tissues as affected by bursectomy. Poultry Science 63: 518-523.

Adegoke, E.O., Wang, X., Wang, H., Wang, C., Zhang, H. and Zhang, G., 2018. Selenium ($Na(_2)SeO(_3)$) upregulates expression of immune genes and blood-testis barrier constituent proteins of bovine sertoli cell *in vitro*. Biological Trace Element Research 185: 332-343.

Adib-Conquy, M., Scott-Algara, D., Cavaillon, J.M. and Souza-Fonseca-Guimaraes, F., 2014. TLR-mediated activation of NK cells and their role in bacterial/viral immune responses in mammals. Immunology and Cell Biology 92: 256-262.

Ahmadi, A., Poursasan, N., Amani, J. and Salimian, J., 2015. Adverse effect of T-2 toxin and the protective role of selenium and vitamin E on peripheral blood B lymphocytes. Iranian Journal of Immunology 12: 64-69.

Allen, R.G. and Balin, A.K., 1989. Oxidative influence on development and differentiation: an overview of a free radical theory of development. Free Radical Biology and Medicine 6: 631-661.

Almeida, L., Lochner, M., Berod, L. and Sparwasser, T., 2016. Metabolic pathways in T cell activation and lineage differentiation. Seminars in Immunology 28: 514-524.

Anelli, T., Sannino, S. and Sitia, R., 2015. Proteostasis and "redoxtasis" in the secretory pathway: tales of tails from ERp44 and immunoglobulins. Free Radical Biology and Medicine 83: 323-330.

Annacker, O., Pimenta-Araujo, R., Burlen-Defranoux, O. and Bandeira, A., 2001. On the ontogeny and physiology of regulatory T cells. Immunological Reviews 182: 5-17.

Aribi, M., Meziane, W., Habi, S., Boulatika, Y., Marchandin, H. and Aymeric, J.L, 2015. Macrophage bactericidal activities against *Staphylococcus aureus* are enhanced *in vivo* by selenium supplementation in a dose-dependent manner. PLoS One 10: e0135515.

Arshad, M., Siddique, M., Ashraf, M. and Khan, H.A., 2005. Effect of selenium supplementation on antibody titres against infectious bursal disease vaccine in broiler chicks. Pakistan Veterinary Journal 25: 203-204.

Baehner, R.L., Boxer, L.A., Allen, J.M. and Davis, J., 1977. Autooxidation as a basis for altered function by polymorphonuclear leukocytes. Blood 50: 327-335.

Baker, S.S. and Cohen, H.J., 1984. Increased sensitivity to H_2O_2 in glutathione peroxidase-deficient rat granulocytes. Journal of Nutrition 114: 2003-2009.

Barger, J.L., Kayo, T., Pugh, T.D., Vann, J.A., Power, R., Dawson, K., Weindruch, R. and Prolla, T.A., 2012. Gene expression profiling reveals differential effects of sodium selenite, selenomethionine, and yeast-derived selenium in the mouse. Genes and Nutrition 7: 155-165.

Barrett, C.W., Singh, K., Motley, A.K., Lintel, M.K., Matafonova, E., Bradley, A.M., Ning, W., Poindexter, S.V., Parang, B., Reddy, V.K., Chaturvedi, R., Fingleton, B.M., Washington, M.K., Wilson, K.T., Davies, S.S., Hill, K.E., Burk, R.F. and Williams, C.S., 2013. Dietary selenium deficiency exacerbates DSS-induced epithelial injury and AOM/DSS-induced tumorigenesis. PloS One 8(7): e67845.

Bassiouni, A.A., Zaki, M.M. and Hady, M.M., 1990. Effect of vitamin E and selenium on the immune response of chickens against living Newcastle disease vaccine. Veterinary Medical Journal Giza 38: 145-155.

Bechtel, T.J. and Weerapana, E., 2017. From structure to redox: the diverse functional roles of disulfides and implications in disease. Proteomics 17: 6.

Beck, M.A., 1998. The influence of antioxidant nutrients on viral infection. Nutrition Reviews 56: S140-S146.

Beck, M.A., 1999. Selenium and host defence towards viruses. Proceedings of the Nutrition Society 58: 707-711.

Beck, M.A., 2001. Antioxidants and viral infections: host immune response and viral pathogenicity. Journal of the American College of Nutrition 20, Suppl. 5: 384S-388S.

Beck, M.A. and Levander, O.A., 2000. Host nutritional status and its effect on a viral pathogen. Journal of Infectious Disease 182, Suppl. 1: S93-S96.

Beck, M.A., Nelson, H.K., Shi, Q., Van Dael, P., Schiffrin, E.J., Blum, S., Barclay, D. and Levander, O.A., 2001. Selenium deficiency increases the pathology of an influenza virus infection. FASEB Journal 15: 1481-1483.

Bertolotti, M., Sitia, R. and Rubartelli, A., 2012. On the redox control of B lymphocyte differentiation and function. Antioxidants & Redox Signaling 16: 1139-1149.

Bertolotti, M., Yim, S.H., Garcia-Manteiga, J.M., Masciarelli, S., Kim, Y.J., Kang, M.H., Iuchi, Y., Fujii, J., Vené, R., Rubartelli, A. and Rhee, S.G., 2010. B-to plasma-cell terminal differentiation entails oxidative stress and profound reshaping of the antioxidant responses. Antioxidants & Redox Signaling 13: 1133-1144.

Bi, C.L., Wang, H., Wang, Y.J., Sun, J., Dong, J.S., Meng, X. and Li, J.J., 2016. Selenium inhibits *Staphylococcus aureus*-induced inflammation by suppressing the activation of the NF-κB and MAPK signalling pathways in RAW264. 7 macrophages. European Journal of Pharmacology 780: 159-165.

Biswas, S., Talukder, G. and Sharma, A., 2000. Chromosome damage induced by selenium salts in human peripheral lymphocytes. Toxicology in Vitro 14: 405-408.

Blanc, F., Prévost-Blondel, A., Piton, G., Bouguyon, E., Leplat, J.J., Andréoletti, F., Egidy, G., Bourneuf, E., Bertho, N. and Vincent-Naulleau, S., 2020. The composition of circulating leukocytes varies with age and melanoma onset in the MeLiM pig biomedical model. Frontiers in Immunology 11: 291.

Blom, A.M., 2017. The role of complement inhibitors beyond controlling inflammation. Journal of Internal Medicine 282: 116-128.

Bondy, G.S. and Pestka, J.J., 2000. Immunomodulation by fungal toxins. Journal of Toxicology and Environmental Health, Part B 3: 109-143.

Borceux, P., Morelle, J. and Goffin, E., 2020. Complement system activation and peritoneal membrane alterations: Culprit or innocent bystander? Peritoneal Dialysis International 40: 115-123.

Borella, P., Bargellini, A. and Medici, C.I., 1996. Chemical form of selenium greatly affects metal uptake and responses by cultured human lymphocytes. Biological Trace Element Research 51: 43-54.

Boscolo, P., Di Giampaolo, L., Qiao, N., Reale, M., Castellani, M.L., Lucci, I., Travaglini, P., Kouri, M., Verna, N., Volpe, A.R. and Carmignani, M., 2005. Inhibitory effects of cadmium on peripheral blood mononuclear cell proliferation and cytokine release are reversed by zinc and selenium salts. Annals of Clinical & Laboratory Science 35: 115-120.

Brazin, K.N., Mallis, R.J., Li, C., Keskin, D.B., Arthanari, H., Gao, Y., Wu, S.L., Karger, B.L., Wagner, G. and Reinherz, E.L., 2014. Constitutively oxidized CXXC motifs within the CD3 heterodimeric ectodomains of the T cell receptor complex enforce the conformation of juxtaposed segments. The Journal of Biological Chemistry 289(27): 18880-18892. https://doi.org/10.1074/jbc.M114.574996

Broom, L.J. and Kogut, M.H., 2018. Gut immunity: its development and reasons and opportunities for modulation in monogastric production animals. Animal Health Research Reviews 19: 46-52.

Broome, C.S., McArdle, F., Kyle, J.A., Andrews, F., Lowe, N.M., Hart, C.A., Arthur, J.R. and Jackson, M.J., 2004. An increase in selenium intake improves immune function and poliovirus handling in adults with marginal selenium status. American Journal of Clinical Nutrition 80: 154-162.

Bu, Z.G., Huang, K.H. and Chen, W.F., 1996. Study on the cell mediated immunity mechanism in the selenium-enhanced resistance of chicks to infectious bursal disease. Chinese Journal of Veterinary Science 16: 273-276.

Calder, P.C., 2001. Polyunsaturated fatty acids, inflammation, and immunity. Lipids 36: 1007-1024.

Cao, N., Li, W., Li, B., Tian, Y. and Xu, D., 2017. Transcriptome profiling reveals the immune response of goose T cells under selenium stimuli. Animal Science Journal 88: 2001-2009.

Cao, Y.Z., Maddox, J.F., Mastro, A.M., Scholz, R.W., Hildenbrandt, G. and Reddy, C.C., 1992. Selenium deficiency alters the lipoxygenase pathway and mitogenic response in bovine lymphocytes. Journal of Nutrition 122: 2121-2127.

Cao, Y.Z., Weaver, J.A., Reddy, C.C. and Sordillo, L.M., 2002. Selenium deficiency alters the formation of eicosanoids and signal transduction in rat lymphocytes. Prostaglandins and Other Lipid Mediators 70: 131-143.

Carlson, B.A., Yoo, M.H., Conrad, M., Gladyshev, V.N., Hatfield, D.L. and Park, J.M., 2011. Protein kinase-regulated expression and immune function of thioredoxin reductase 1 in mouse macrophages. Molecular Immunology 49: 311-316.

Carlson, B.A., Yoo, M.H., Shrimali, R.K., Irons, R., Gladyshev, V.N., Hatfield, D.L. and Park, J.M., 2010. Role of selenium-containing proteins in T cell and macrophage function. Proceedings of the Nutrition Society 69: 300-310.

Castanheira, F.V.S. and Kubes, P., 2019. Neutrophils and NETs in modulating acute and chronic inflammation. Blood 133: 2178-2185.

Castle, S.C., 2000. Clinical relevance of age-related immune dysfunction. Clinical Infectious Diseases 31: 578-585.

Cerutti, A., Puga, I. and Cols, M., 2011. Innate control of B cell responses. Trends in Immunology 32: 202-211.

Chang, W.P., Hom, J.S., Dietert, R.R., Combs Jr., G.F. and Marsh, J.A., 1994. Effect of dietary vitamin E and selenium deficiency on chicken splenocyte proliferation and cell surface marker expression. Immunopharmacology and Immunotoxicology 16: 203-223.

Checker, R., Sharma, D., Sandur, S.K., Subrahmanyam, G., Krishnan, S., Poduval, T.B. and Sainis, K.B., 2010. Plumbagin inhibits proliferative and inflammatory responses of T cells independent of ROS generation but by modulating intracellular thiols. Journal of Cellular Biochemistry 110: 1082-1093.

Chen, K., Fang, J., Peng, X., Cui, H., Chen, J., Wang, F., Chen, Z., Zuo, Z., Deng, J., Lai, W. and Zhou, 2014c. Effect of selenium supplementation on aflatoxin B1-induced histopathological lesions and apoptosis in bursa of Fabricius in broilers. Food and Chemical Toxicology 74: 91-97.

Chen, K., Peng, X., Fang, J., Cui, H., Zuo, Z., Deng, J., Chen, Z., Geng, Y., Lai, W., Tang, L. and Yang, Q., 2014a. Effects of dietary selenium on histopathological changes and T cells of spleen in broilers exposed to aflatoxin B1. International Journal of Environmental Research and Public Health 11: 1904-1913.

Chen, K., Shu, G., Peng, X., Fang, J., Cui, H., Chen, J., Wang, F., Chen, Z., Zuo, Z., Deng, J., Geng, Y. and Lai, W., 2013b. Protective role of sodium selenite on histopathological lesions, decreased T cell subsets and increased apoptosis of thymus in broilers intoxicated with aflatoxin B1. Food and Chemical Toxicology 59: 446-454.

Chen, K., Yuan, S., Chen, J., Peng, X., Wang, F., Cui, H. and Fang, J., 2013a. Effects of sodium selenite on the decreased percentage of T cell subsets, contents of serum IL-2 and IFN-γ induced by aflatoxin B1 in broilers. Research in Veterinary Science 95: 143-145.

Chen, X., Ren, F., Hesketh, J., Shi, X., Li, J., Gan, F. and Huang, K., 2012a. Selenium blocks porcine circovirus type 2 replication promotion induced by oxidative stress by improving GPx1 expression. Free Radical Biology and Medicine 53: 395-405.

Chen, X., Yao, H., Yao, L., Zhao, J., Luan, Y., Zhang, Z. and Xu, S., 2014b. Selenium deficiency influences the gene expressions of heat shock proteins and nitric oxide levels in neutrophils of broilers. Biological Trace Element Research 161: 334-340.

Chen, X., Zhu, Y.H., Cheng, X.Y., Zhang, Z.W. and Xu, S.W., 2012b. The protection of selenium against cadmium-induced cytotoxicity via the heat shock protein pathway in chicken splenic lymphocytes. Molecules 17: 14565-14572.

Chen, Y., Zhao, Y.F., Yang, J., Jing, H.Y., Liang, W., Chen, M.Y., Yang, M., Wang, Y. and Guo, M.Y., 2020. Selenium alleviates lipopolysaccharide-induced endometritis via regulating the recruitment of TLR4 into lipid rafts in mice. Food & Function 11: 200-210.

Cheng, Z., Zhi, X., Sun, G., Guo, W., Huang, Y., Sun, W., Tian, X., Zhao, F. and Hu, K., 2016. Sodium selenite suppresses hepatitis B virus transcription and replication in human hepatoma cell lines. Journal of Medical Virology 88: 653-663.

Chiu, J. and Hogg, P.J., 2019. Allosteric disulfides: Sophisticated molecular structures enabling flexible protein regulation. Journal of Biological Chemistry 294: 2949-2960.

Claus, M., Dychus, N., Ebel, M., Damaschke, J., Maydych, V., Wolf, O.T., Kleinsorge, T. and Watzl, C., 2016. Measuring the immune system: a comprehensive approach for the analysis of immune functions in humans. Archives of Toxicology 90: 2481-2495.

Coffey, T.J. and Werling, D., 2011. Therapeutic targeting of the innate immune system in domestic animals. Cell and Tissue Research 343: 251-261.

Colitti, M., Stefanon, B., Gabai, G., Gelain, M.E. and Bonsembiante F., 2019. Oxidative stress and nutraceuticals in the modulation of the immune function: current knowledge in animals of veterinary interest. Antioxidants 8(1): 28. https://doi.org/10.3390/antiox8010028

Colnago, G.L., Jensen, L.S. and Long, P.L., 1984a. Effect of selenium on peripheral blood leukocytes of chickens infected with *Eimeria*. Poultry Science 63: 896-903.

Colnago, G.L., Jensen, L.S. and Long, P.L., 1984b. Effect of natural feedstuffs added to a semi-purified diet on *Eimeria tenella* infection. Poultry Science 63: 2145-2152.

Conner, E.M. and Grisham, M.B., 1996. Inflammation, free radicals, and antioxidants. Nutrition 12: 274-277.

Crisci, E., Fraile, L. and Montoya, M., 2019. Cellular innate immunity against PRRSV and swine influenza viruses. Veterinary Sciences 6:1, p.26.

Cui, J., Xiao, Y., Shi, Y.H., Wang, B. and Le, G.W., 2012. Lipoic acid attenuates high-fat-diet-induced oxidative stress and B-cell-related immune depression. Nutrition 28: 275-280.

Da Silva, E.O., Bracarense, A.P.F.L. and Oswald, I.P., 2018. Mycotoxins and oxidative stress: where are we? World Mycotoxin Journal 11: 113-133.

Da Silva, I.C.M., Ribeiro, A.M.L., Canal, C.W., Trevizan, L., Macagnan, M., Gonçalves, T.A. and Pereira, R.A., 2010. The impact of organic and inorganic selenium on the immune system of growing broilers submitted to immune stimulation and heat stress. Revista Brasileira de Ciência Avícola 12: 247-254.

Dalgaard, T.S., Briens, M., Engberg, R.M. and Lauridsen, C., 2018. The influence of selenium and selenoproteins on immune responses of poultry and pigs. Animal Feed Science and Technology 238: 73-83.

Damyanova, A., Teodorova, S. and Gabrashanska, M., 1995. Content of some microelements in chickens with ascaridiasis under combined drug treatment. Parasitology Research 81: 549-552.

De Pelsmaeker, S., Devriendt, B., De Regge, N. and Favoreel, H.W., 2019. Porcine NK cells stimulate proliferation of pseudorabies virus-experienced CD8(+) and CD4(+)CD8(+) T cells. Frontiers in Immunology 9: 3188.

Delgoffe, G.M., Murray, P.J. and Vignali, D.A., 2011. Interpreting mixed signals: the cell's cytokine conundrum. Current Opinion in Immunology 23: 632-638.

Deng, H., Yang, H., Liu, Y.Q. and He, Y.M., 1999. Influence of feeding sodium selenite solution on chick's humoral immunity and erythrocyte immunity function against Newcastle disease. Chinese Journal of Veterinary Science and Technology 29: 21-22.

Dhur, A., Galan, P. and Hercberg, S., 1990. Relationship between selenium, immunity and resistance against infection. Comparative Biochemistry and Physiology 96C: 271-280.

Dietert, R.R. and Golemboski, K.A., 1998. Avian macrophage metabolism. Poultry Science 77: 990-997.

Dietert, R.R., Combs, G.F., Lin, H.K. and Puzzi, J.V., 1990. Impact of combined vitamin E and selenium deficiency on chicken macrophage function. Annals of the New York Academy of Sciences 587: 281-282.

Ebert-Dumig, R., Seufert, J., Schneider, D., Kohrle, J., Schutze, N. and Jakob, F., 1999. Expression of selenoproteins in monocytes and macrophages – implications for the immune system. Medizinische Klinik 9494: 29-34.

Elgendy, R., Giantin, M., Castellani, F., Grotta, L., Palazzo, F., Dacasto, M. and Martino, G., 2016. Transcriptomic signature of high dietary organic selenium supplementation in sheep: a nutrigenomic insight using a custom microarray platform and gene set enrichment analysis. Journal of Animal Science 94: 3169-3184.

Elmadfa, I. and Meyer, A.L., 2019. The role of the status of selected micronutrients in shaping the immune function. Endocrine, Metabolic & Immune Disorders-Drug Targets 19(8): 1100-1115.

Falk, M., Bernhoft, A., Framstad, T., Salbu, B., Wisløff, H., Kortner, T.M., Kristoffersen, A.B. and Oropeza-Moe, M., 2018. Effects of dietary sodium selenite and organic selenium sources on immune and inflammatory responses and selenium deposition in growing pigs. Journal of Trace Elements in Medicine and Biology 50: 527-536.

Ferencik, M. and Ebringer, L., 2003. Modulatory effects of selenium and zinc on the immune system. Folia Microbiologica 48: 417-426.

Fisinin, V.I. and Surai, P.F., 2013. Gut immunity in birds: facts and thoughts. Selskokhozaistvennaya Biologia 4: 1-25.

Franzoni, G., Graham, S.P., Dei Giudici, S. and Oggiano, A., 2019. Porcine dendritic cells and viruses: an update. Viruses 11(5): E445.

Fredericks, G.J., Hoffmann, F.W., Rose, A.H., Osterheld, H.J., Hess, F.M., Mercier, F. and Hoffmann, P.R., 2014. Stable expression and function of the inositol 1,4,5-triphosphate receptor requires palmitoylation by a DHHC6/selenoprotein K complex. Proceedings of the National Academy of Sciences of the USA 111: 16478-16483.

Frouin, H., Loseto, L.L., Stern, G.A., Haulena, M. and Ross, P.S., 2012. Mercury toxicity in beluga whale lymphocytes: limited effects of selenium protection. Aquatic Toxicology 109: 185-193.

Fu, J., Yang, T., Wang, W. and Xu, S., 2019. Effect of selenium antagonist lead-induced damage on Th1/Th2 imbalance in the peripheral blood lymphocytes of chickens. Ecotoxicology and Environmental Safety 175: 74-82.

Gambhir, L., Checker, R., Thoh, M., Patwardhan, R.S., Sharma, D., Kumar, M. and Sandur, S.K., 2014. 1, 4-Naphthoquinone, a pro-oxidant, suppresses immune responses via KEAP-1 glutathionylation. Biochemical Pharmacology 88: 95-105.

Gambhir, L., Sharma, V., Kandwal, P. and Saxena, S., 2019. Perturbation in cellular redox homeostasis: Decisive regulator of T cell mediated immune responses. International Immunopharmacology 67: 449-457.

Gandhi, U.H., Kaushal, N., Ravindra, K.C., Hegde, S., Nelson, S.M., Narayan, V., Vunta, H., Paulson, R.F. and Prabhu, K.S., 2011. Selenoprotein-dependent up-regulation of hematopoietic prostaglandin D2 synthase in macrophages is mediated through the activation of peroxisome proliferator-activated receptor (PPAR) gamma. Journal of Biological Chemistry 286: 27471-27482.

Garcia, M.M., Goicoechea, C., Molina-Álvarez, M. and Pascual, D., 2020. Toll-like receptor 4: a promising crossroads in the diagnosis and treatment of several pathologies. European Journal of Pharmacology 874: 172975.

Garrido, A., Cruces, J., Ceprián, N., Vara, E. and De la Fuente, M., 2019. Oxidative-inflammatory stress in immune cells from adult mice with premature aging. International Journal of Molecular Sciences 20(3): 769.

Gille, G. and Sigler, K., 1995. Oxidative stress and living cells. Folia Microbiologica 40: 131-152.

Gordon, S. and Martinez, F.O., 2010. Alternative activation of macrophages: mechanism and functions. Immunity 32: 593-604.

Gostner, J.M., Becker, K., Fuchs, D. and Sucher, R., 2013. Redox regulation of the immune response. Redox Report 18: 88-94.

Greenberg, S. and Grinstein, S., 2002. Phagocytosis and innate immunity. Current Opinion in Immunology 14: 136-145.

Greenman, E., Phillipich, M.J., Meyer, C.J., Charamella, L.J. and Dimitrov, N.V., 1988. The effect of selenium on phagocytosis in humans. Anticancer Research 8: 825-828.

Grimble, R.F., 1997. Effect of antioxidative vitamins on immune function with clinical applications. International Journal for Vitamin and Nutrition Research 67: 312-320.

Gurjar, S.A., Wheeler, J.X., Wadhwa, M., Thorpe, R., Kimber, I., Derrick, J.P., Dearman, R.J. and Metcalfe, C., 2019. The impact of thioredoxin reduction of allosteric disulfide bonds on the therapeutic potential of monoclonal antibodies. Journal of Biological Chemistry 294: 19616-19634.

Habibian, M., Ghazi, S., Moeini, M.M. and Abdolmohammadi, A., 2014. Effects of dietary selenium and vitamin E on immune response and biological blood parameters of broilers reared under thermoneutral or heat stress conditions. International Journal of Biometeorology 58: 741-752.

Habibian, M., Sadeghi, G., Ghazi, S. and Moeini, M.M., 2015. Selenium as a feed supplement for heatstressed poultry: a review. Biological Trace Element Research 165: 183-193.

Hachimura, S., Totsuka, M. and Hosono, A., 2018. Immunomodulation by food: impact on gut immunity and immune cell function. Bioscience, Biotechnology, and Biochemistry 82: 584-599.

Halliwell, B. and Gutteridge, J.M.C., 1999. Free radicals in biology and medicine, 3rd edition. Oxford University Press, New York, NY, USA.

Hao, S., Hu, J., Song, S., Huang, D., Xu, H., Qian, G., Gan, F. and Huang, K., 2016. Selenium alleviates aflatoxin B1-induced immune toxicity through improving glutathione peroxidase 1 and selenoprotein S expression in primary porcine splenocytes. Journal of Agricultural and Food Chemistry 64: 1385-1393.

Hao, Y.H., Li, Q.Z., Qu, Q.H., Xu, B.R. and Wei, P., 1999. Changes of activity of Se-GSH-PX and the content of LPO in chickens infected with vMDV. Chinese Journal of Veterinary Science 19: 218-220.

Harthill, M., 2011. Review: micronutrient selenium deficiency influences evolution of some viral infectious diseases. Biological Trace Element Research 143: 1325-1336.

Hawkes, W.C., Hwang, A. and Alkan, Z., 2009. The effect of selenium supplementation on DTH skin responses in healthy North American men. Journal of Trace Elements in Medicine and Biology 23: 272-280.

Hayek, M.G., Boissonneault, G.A. and Mitchell Jr., G.E., 1996. Influence of vitamin E on immune response of livestock and poultry. BASF Manual. Selenium and immunity Selenium in poultry nutrition and health, pp. 185-190.

He, S.X., Wu, B., Chang, X.M., Li, H.X. and Qiao, W., 2004. Effects of selenium on peripheral blood mononuclear cell membrane fluidity, interleukin-2 production and interleukin-2 receptor expression in patients with chronic hepatitis. World Journal of Gastroenterology 10: 3531-3533.

He, X., Lin, Y., Lian, S., Sun, D., Guo, D., Wang, J. and Wu, R., 2020. Selenium deficiency in chickens induces intestinal mucosal injury by affecting the mucosa morphology, SIgA secretion, and GSH-Px activity. Biological Trace Element Research 197: 660-666. https://doi.org/10.1007/s12011-019-02017-6

He, Y., Fang, J., Peng, X., Cui, H., Zuo, Z., Deng, J., Chen, Z., Geng, Y., Lai, W., Shu, G. and Tang, L., 2014b. Effects of sodium selenite on aflatoxin B1-induced decrease of ileal IgA+ cell numbers and immunoglobulin contents in broilers. Biological Trace Element Research 160: 49-55.

He, Y., Fang, J., Peng, X., Cui, H., Zuo, Z., Deng, J., Chen, Z., Lai, W., Shu, G. and Tang, L., 2014a. Effects of sodium selenite on aflatoxin B1-induced decrease of ileac T cell and the mRNA contents of IL-2,IL-6, and TNF-α in broilers. Biological Trace Element Research 159: 167-173.

Hegazy, S.M. and Adachi, Y., 2000. Comparison of the effects of dietary selenium, zinc, and selenium and zinc supplementation on growth and immune response between chick groups that were inoculated with *Salmonella* and aflatoxin or *Salmonella*. Poultry Science 79: 331-335.

Hiller, F., Oldorff, L., Besselt, K. and Kipp, A.P., 2015. Differential acute effects of selenomethionine and sodium selenite on the severity of colitis. Nutrients 7: 2687-2706.

Hoerr, F.J., 1998. Pathogenesis of enteric diseases. Poultry Science 77: 1150-1155.

Hoffmann, F.W., Hashimoto, A.C., Shafer, L.A., Dow, S., Berry, M.J. and Hoffmann, P.R., 2010. Dietary selenium modulates activation and differentiation of CD4+ T cells in mice through a mechanism involving cellular free thiols. Journal of Nutrition 140: 1155-1161.

Hoffmann, P.R., Jourdan-Le Saux, C., Hoffmann, F.W., Chang, P.S., Bollt, O., He, Q., Tam, E.K. and Berry, M.J., 2007. A role for dietary selenium and selenoproteins in allergic airway inflammation. Journal of Immunology 179: 3258-3267.

Holgersson, J., Ehrnfelt, C., Hauzenberger, E. and Serrander, L., 2002. Leukocyte endothelial cell interactions in pig to human organ xenograft rejection. Veterinary Immunology and Immunopathology 87: 407-415.

Horvathova, M., Jahnova, E. and Gazdik, F., 1999. Effect of selenium supplementation in asthmatic subjects on the expression of endothelial cell adhesion molecules in culture. Biological Trace Element Research 69: 15-26.

Hu, P., Zuo, Z., Wang, F., Peng, X., Guan, K., Li, H., Fang, J., Cui, H., Su, G., Ouyang, P. and Zhou, Y., 2018. The protective role of selenium in AFB(1)-induced tissue damage and cell cycle arrest in chicken's bursa of Fabricius. Biological Trace Element Research 185(2): 486-496.

Huang, K.H. and Chen, W.F., 1996. Effect of selenium on the resistance of chickens to Marek's disease and its mode of action. Acta Veterinaria et Zootechnica Sinica 27: 448-455.

Huang, K.H. and Chen, W.F., 1999. Effect of selenium on T lymphocyte transformation rate and natural killer cell activities in chickens. Journal of Nanjing Agricultural University 22: 76-79.

Hugejiletu, H., Bobe, G., Vorachek, W.R., Gorman, M.E., Mosher, W.D., Pirelli, G.J. and Hall, J.A., 2013. Selenium supplementation alters gene expression profiles associated with innate immunity in whole blood neutrophils of sheep. Biological Trace Element Research 154: 28-44.

Hughes, D.A., 1999. Effects of dietary antioxidants on the immune function of middle-aged adults. Proceedings of the Nutrition Society 58: 79-84.

Ilback, N.G., Fohlman, J. and Friman, G., 1998. Effects of selenium supplementation on virus-induced inflammatory heart disease. Biological Trace Element Research 63: 51-66.

Irmak, M.B., Ince, G., Ozturk, M. and Cetin-Atalay, R., 2003. Acquired tolerance of hepatocellular carcinoma cells to selenium deficiency: a selective survival mechanism? Cancer Research 63: 6707-6715.

Ivory, K., Prieto, E., Spinks, C., Armah, C.N., Goldson, A.J., Dainty, J.R. and Nicoletti, C., 2017. Selenium supplementation has beneficial and detrimental effects on immunity to influenza vaccine in older adults. Clinical Nutrition 36: 407-415.

Jackson, S.H., Devadas, S., Kwon, J., Pinto, L.A. and Williams, M.S., 2004. T cells express a phagocyte-type NADPH oxidase that is activated after T cell receptor stimulation. Nature immunology 5: 818-827.

Jahnova, E., Horvathova, M., Gazdik, F. and Weissova, S., 2002. Effects of selenium supplementation on expression of adhesion molecules in corticoid-dependent asthmatics. Bratislavske Lekarske Listy 103: 12-16.

Jaspers, I., Zhang, W., Brighton, L.E., Carson, J.L., Styblo, M., Beck, M.A., 2007. Selenium deficiency alters epithelial cell morphology and responses to influenza. Free Radical Biology and Medicine 42(12): 1826-1837.

Jin, C., Henao-Mejia, J. and Flavell, R.A., 2013. Innate immune receptors: key regulators of metabolic disease progression. Cell Metabolism 17: 873-882.

Johnson, J.S., Stewart, K.R., Safranski, T.J., Ross, J.W. and Baumgard, L.H., 2020. *In utero* heat stress alters postnatal phenotypes in swine. Theriogenology 154: 110-119.

Johnson, V.J., Tsunoda, M. and Sharma, R.P., 2000. Increased production of proinflammatory cytokines by murine macrophages following oral exposure to sodium selenite but not to seleno-L-methionine. Archives of Environmental Contamination and Toxicology 39: 243-250.

Jourd'heuil, D., Morise, Z., Conner, E.M. and Grisham, M.B., 1997. Oxidants, transcription factors, and intestinal inflammation. Journal of Clinical Gastroenterology 25, Suppl. 1: S61-S72.

Jun, E.J., Ye, J.S., Hwang, I.S., Kim, Y.K. and Lee, H., 2011. Selenium deficiency contributes to the chronic myocarditis in coxsackievirus-infected mice. Acta Virologica 55: 23-29.

Juul-Madsen, H.R., Viertlboeck, B., Smith, A.L. and Gobel, T.W.F., 2008. Avian innate immune responses. In: Davison, F., Kaspers, B. and Schat, K.A. (eds) Avian immunology. Elsevier, Amsterdam, the Netherlands, pp. 107-128.

Kaiser, P. and Staheli, P., 2008. Avian cytokines and chemokines. In: Davison, F., Kaspers, B. and Schat, K.A. (eds) Avian immunology. Elsevier, Amsterdam, the Netherlands, pp. 203-222.

Käser, T., Gerner, W. and Saalmüller, A., 2011. Porcine regulatory T cells: mechanisms and T cell targets of suppression. Developmental and Comparative Immunology 35(11): 1166-1172.

Kaspers, B., Kothlow, S. and Butter, C., 2008. Avian antigen presenting cells, In: Davison, F., Kaspers, B. and Schat, K.A. (eds) Avian immunology. Elsevier, Amsterdam, the Netherlands, pp. 183-202.

Kaushal, N., Kudva, A.K., Patterson, A.D., Chiaro, C., Kennett, M.J., Desai, D., Amin, S., Carlson, B.A., Cantorna, M.T. and Prabhu, K.S., 2014. Crucial role of macrophage selenoproteins in experimental colitis. Journal of Immunology 193: 3683-3692.

Kellett-Clarke, H., Stegmann, M., Barclay, A.N. and Metcalfe, C., 2015. CD44 binding to hyaluronic acid is redox regulated by a labile disulfide bond in the hyaluronic acid binding site. PloS One 10(9): e0138137.

Khalil, A.M., 1989. The induction of chromosome aberrations in human purified peripheral blood lymphocytes following *in vitro* exposure to selenium. Mutation Research 224: 503-506.

Khoso, P.A., Yang, Z., Liu, C. and Li, S., 2015. Selenoproteins and heat shock proteins play important roles in immunosuppression in the bursa of Fabricius of chickens with selenium deficiency. Cell Stress and Chaperones 20: 967-978.

Khoso, P.A., Zhang, Y., Yin, H., Teng, X. and Li, S., 2019. Selenium deficiency affects immune function by influencing selenoprotein and cytokine expression in chicken spleen. Biological Trace Element Research 187: 506-516.

Kim, S.E., Hwang, J.H., Kim, Y.K. and Lee, H.T., 2019. Heterogeneity of porcine bone marrow-derived dendritic cells induced by GM-CSF. PloS One 14: 11.

Kim, S.T., Shin, Y., Brazin, K., Mallis, R.J., Sun, Z.Y., Wagner, G., Lang, M.J. and Reinherz, E.L., 2012. TCR mechanobiology: torques and tunable structures linked to early T cell signaling. Frontiers in Immunology 18(3): 76.

Kingeter, L.M. and Lin, X., 2012. C-type lectin receptor-induced NF-κB activation in innate immune and inflammatory responses. Cell Molecular Immunology 9: 105-112.

Kiremidjian-Schumacher, L. and Stotzky, G., 1987. Selenium and immune responses. Environmental Research 42: 277-303.

Kiremidjian-Schumacher, L., Roy, M., Wishe, H.I., Cohen, M.W. and Stotzky, G., 1992. Regulation of cellular immune responses by selenium. Biological Trace Element Research 33: 23-35.

Klasing, K.C., 1998a. Avian macrophages: regulators of local and systemic immune responses. Poultry Science 77: 983-989.

Klasing, K.C., 1998b. Nutritional modulation of resistance to infectious diseases. Poultry Science 77: 1119-1125.

Korver, D.R., 2012. Implications of changing immune function through nutrition in poultry. Animal Feed Science and Technology 173: 54-64.

Korver, D. and Klasing, K., 2001. Influence of nutrition on immune status of the bird. Proceedings of 24th Technical Turkey Conference. April 25-27, 2001. Shrigley Hall Hotel, Macclesfield, UK, pp. 43-46.

Kovalenko, M.V., Stepchenko, L.M., Shevtsova, A.I., Brazaluk, O.Z. and Surai, P.F., 2008. Effect of selenium-containing supplements on the indices of specific immunity and nonspecific resistance in chicken. Fiziology Zhurmal 54: 69-73.

Kraaij, M.D., Savage, N.D., Van der Kooij, S.W., Koekkoek, K., Wang, J., Van den Berg, J.M., Ottenhoff, T.H., Kuijpers, T.W., Holmdahl, R., Van Kooten, C. and Gelderman, K.A., 2010. Induction of regulatory T cells by macrophages is dependent on production of reactive oxygen species. Proceedings of the National Academy of Sciences 107: 17686-17691.

Kumar, H., Kawai, T., Akira, S., 2011. Pathogen recognition by the innate immune system. International Reviews of Immunology 30: 16-34.

Kuroi, K., Kamijo, M., Ueki, M., Niwa, Y., Hiramatsu, H. and Nakabayashi, T., 2020. Time-resolved FTIR study on the structural switching of human galectin-1 by light-induced disulfide bond formation. Physical Chemistry Chemical Physics 22(3): 1137-1144.

Kwon, J., Shatynski, K.E., Chen, H., Morand, S., De Deken, X., Miot, F., Leto, T.L. and Williams, M.S., 2010. The nonphagocytic NADPH oxidase Duox1 mediates a positive feedback loop during T cell receptor signaling. Science Signaling 3:133, ra59.

Larsen, C.T., Pierson, F.W. and Gross, W.B., 1997. Effect of dietary selenium on the response of stressed and unstressed chickens to *Escherichia coli* challenge and antigen. Biological Trace Element Research 58: 169-176.

Larsen, H.J., 1993. Relations between selenium and immunity. Norwegian Journal of Agricultural Sciences 11: 105-119.

Latorre, A.O., Caniceiro, B.D., Wysocki Jr., H.L., Haraguchi, M., Gardner, D.R. and Górniak, S.L., 2011. Selenium reverses *Pteridium aquilinum*-induced immunotoxic effects. Food and Chemical Toxicology 49: 464-470.

Latshaw, J.D., 1991. Nutrition – mechanisms of immunosuppression. Veterinary Immunology and Immunopathology 30: 111-120.

Lauridsen, C., 2019. From oxidative stress to inflammation: redox balance and immune system. Poultry Science 98(10): 4240-4246.

Lee, S., Takahashi, I., Matsuzaka, M., Yamai, K., Danjo, K., Kumagai, T., Umeda, T., Itai, K. and Nakaji, S., 2011. The relationship between serum selenium concentration and neutrophil function in peripheral blood. Biological Trace Element Research 144: 396-406.

Levkut, M., Revajová, V., Levkutova, M., Sevcíková, Z., Herich, R., Borutová, R. and Leng, L., 2009. Leukocytic responses of broilers following dietary contamination with deoxynivalenol and/or treatment by dietary selenium supplementation. British Poultry Science 50: 181-187.

Li, N.Y., Sun, Z.J., Ansari, A.R., Cui, L., Hu, Y.F., Li, Z.W., Briens, M., Kai, L., Sun, L.H., Karrow, N.A. and Liu, H.Z., 2020. Impact of maternal selenium supplementation from late gestation and lactation on piglet immune function. Biological Trace Element Research 194: 159-167.

Liao, X., Lu, L., Li, S., Liu, S., Zhang, L., Wang, G., Li, A. and Luo, X., 2012. Effects of selenium source and level on growth performance, tissue selenium concentrations, antioxidation, and immune functions of heat-stressed broilers. Biological Trace Element Research 150: 158-165.

Limaye, A., Yu, R.C., Chou, C.C., Liu, J.R. and Cheng, K.C., 2018. Protective and detoxifying effects conferred by dietary selenium and curcumin against AFB1-mediated toxicity in livestock: a review. Toxins 10(1): 25. https://doi.org/10.3390/toxins10010025

Liu, G., Yang, G., Guan, G., Zhang, Y., Ren, W., Yin, J., Aguilar, Y.M., Luo, W., Fang, J., Yu, X., Li, T. and Yin, Y., 2015a. Effect of dietary selenium yeast supplementation on porcine circovirus type 2 (PCV2) infections in mice. PLoS ONE 10: e0115833.

Liu, L.L., Zhang, J.L., Zhang, Z.W., Yao, H.D., Sun, G. and Xu, S.W., 2014a. Protective roles of selenium on nitric oxide-mediated apoptosis of immune organs induced by cadmium in chickens. Biological Trace Element Research 159: 199-209.

Liu, S., Xu, F., Fu, J. and Li, S., 2015b. Protective roles of selenium on nitric oxide and the gene expression of inflammatory cytokines induced by cadmium in chicken splenic lymphocytes. Biological Trace Element Research 168: 252-260.

Liu, S., Xu, F.P., Yang, Z.J., Li, M., Min, Y.H. and Li, S., 2014b. Cadmium-induced injury and the ameliorative effects of selenium on chicken splenic lymphocytes: mechanisms of oxidative stress and apoptosis. Biological Trace Element Research 160: 340-351.

Liu, Z., Huang, J., Nie, Y., Qazi, I.H., Cao, Y., Wang, L., Ai, Y., Zhou, G., Wu, K. and Han, H., 2019. Selenium treatment enhanced clearance of *Salmonella* in chicken macrophages (HD11). Antioxidants 8(11): 532.

Lorenzen, I., Mullen, L., Bekeschus, S. and Hanschmann, E.M., 2017. Redox regulation of inflammatory processes is enzymatically controlled. Oxidative Medicine and Cellular Longevity 2017: 8459402. https://doi.org/10.1155/2017/8459402

Luan, Y., Zhao, J., Yao, H., Zhao, X., Fan, R., Zhao, W., Zhang, Z. and Xu, S., 2016. Selenium deficiency influences the mRNA expression of selenoproteins and cytokines in chicken erythrocytes. Biological Trace Element Research 171: 427-436.

Lydyard, P.M., Whelan, A. and Fanger, M.W., 2000. Instant notes in immunology. BIOS Scientific Publishers Ltd., Oxford, UK.

Maehira, F., Miyagi, I. and Eguchi, Y., 2003. Selenium regulates transcription factor NF-kappaB activation during the acute phase reaction. Clinica Chimica Acta 334: 163-171.

Mair, K.H., Sedlak, C., Käser, T., Pasternak, A., Levast, B., Gerner, W., Saalmüller, A., Summerfield, A., Gerdts, V., Wilson, H.L. and Meurens, F., 2014. The porcine innate immune system: an update. Developmental & Comparative Immunology 45: 321-343.

Maizels, R.M., 2009. Parasite immunomodulation and polymorphisms of the immune system. Journal of Biology 8: 62.

Mak, T.W., Grusdat, M., Duncan, G.S., Dostert, C., Nonnenmacher, Y., Cox, M., Binsfeld, C., Hao, Z., Brüstle, A., Itsumi, M. and Jäger, C., 2017. Glutathione primes T cell metabolism for inflammation. Immunity 46: 675-689.

Mak, T.W., Saunders, M.E. and Jett, B.D., 2013. Primer to the immune response. Elsevier, London, UK.

Mariuzza, R.A., Agnihotri, P. and Orban, J., 2020. The structural basis of T cell receptor (TCR) activation: an enduring enigma. Journal of Biological Chemistry 295: 914-925.

Marsh, J.A., Combs Jr., G.F., Whitacre, M.E. and Dietert, R.R., 1986. Effect of selenium and vitamin E dietary deficiencies on chick lymphoid organ development. Proceedings of the Society for Experimental Biology and Medicine 182: 425-436.

Marsh, J.A., Dietert, R.R. and Combs Jr., G.F., 1981. Influence of dietary selenium and vitamin E on the humoral immune response of the chick. Proceedings of the Society for Experimental Biology and Medicine 166: 228-236.

Marsh, J.A., Dietert, R.R. and Combs Jr., G.F., 1987. Effect of dietary selenium and vitamin E deficiencies in the chicken on Con A induced splenocyte proliferation. Progress in Clinical and Biological Research 238: 333-345.

Mattmiller, S.A., Carlson, B.A., Gandy, J.C. and Sordillo, L.M., 2014. Reduced macrophage selenoprotein expression alters oxidized lipid metabolite biosynthesis from arachidonic and linoleic acid. Journal of Nutritional Biochemistry 25: 647-654.

Mayorga, E.J., Ross, J.W., Keating, A.F., Rhoads, R.P. and Baumgard, L.H., 2020. Biology of heat stress; the nexus between intestinal hyperpermeability and swine reproduction. Theriogenology 154: 73-83.

McBride, T.J., Preston, B.D. and Loeb, L.A., 1991. Mutagenic spectrum resulting from DNA damage by oxygen radicals. Biochemistry 30: 207-213.

McCarty, M.F., 1997. Promotion of interleukin-2 activity as a strategy for 'rejuvenating' geriatric immune function. Medical Hypotheses 48: 47-54.

McCorkle, F.M., 1998. Symposium: nonlymphoid cells and their factors in immune response. Introduction to the symposium. Poultry Science 77: 963.

McCormack, R., De Armas, L., Shiratsuchi, M. and Podack, E.R., 2013. Killing machines: three pore-forming proteins of the immune system. Immunologic Research 57: 268-278.

Metcalfe, C., Cresswell, P. and Barclay, A.N., 2012. Interleukin-2 signalling is modulated by a labile disulfide bond in the CD132 chain of its receptor. Open Biology 2:1, 110036.

Metcalfe, C., Cresswell, P., Ciaccia, L., Thomas, B. and Barclay, A.N., 2011. Labile disulfide bonds are common at the leucocyte cell surface. Open Biology 1(3): 110010.

Metzger, Z., Hoffeld, J.T. and Oppenheim, J.J., 1980. Macrophage-mediated suppression. I. Evidence for participation of both hydrogen peroxide and prostaglandins in suppression of murine lymphocyte proliferation. Journal of Immunology 124: 983-988.

Middleton, D., Curran, M. and Maxwell, L., 2002. Natural killer cells and their receptors. Transplant Immunology 10: 147-164.

Miles, E.A. and Calder, P.C., 1998. Modulation of immune function by dietary fatty acids. Proceedings of the Nutrition Society 57: 277-292.

Moolla, N., Killick, M., Papathanasopoulos, M. and Capovilla, A., 2016. Thioredoxin (Trx1) regulates CD4 membrane domain localization and is required for efficient CD4-dependent HIV-1 entry. Biochimica et Biophysica Acta (BBA)-General Subjects, 1860: 1854-1863.

Moro-García, M.A., Mayo, J.C., Sainz, R.M. and Alonso-Arias, R., 2018. Influence of inflammation in the process of T lymphocyte differentiation: proliferative, metabolic, and oxidative changes. Frontiers in Immunology 9: 339.

Mougiakakos, D., Johansson, C.C., Jitschin, R., Böttcher, M. and Kiessling, R., 2011. Increased thioredoxin-1 production in human naturally occurring regulatory T cells confers enhanced tolerance to oxidative stress. Blood 117: 857-861.

Muri, J., Heer, S., Matsushita, M., Pohlmeier, L., Tortola, L., Fuhrer, T., Conrad, M., Zamboni, N., Kisielow, J. and Kopf, M., 2018. The thioredoxin-1 system is essential for fueling DNA synthesis during T cell metabolic reprogramming and proliferation. Nature Communications 9: 1, 1851.

Muri, J., Thut, H., Heer, S., Krueger, C.C., Bornkamm, G.W., Bachmann, M.F. and Kopf, M., 2019. The thioredoxin-1 and glutathione/glutaredoxin-1 systems redundantly fuel murine B-cell development and responses. European Journal of Immunology 49: 709-723.

Nair, M.P. and Schwartz, S.A., 1990. Immunoregulation of natural and lymphokine-activated killer cells by selenium. Immunopharmacology 19: 177-183.

Narayan, V., Ravindra, K.C., Liao, C., Kaushal, N., Carlson, B.A. and Prabhu, K.S., 2015. Epigenetic regulation of inflammatory gene expression in macrophages by selenium. Journal of Nutritional Biochemistry 26: 138-145.

Nazıroğlu, M., Yıldız, K., Tamtürk, B., Erturan, İ. and Flores-Arce, M., 2012. Selenium and psoriasis. Biological Trace Element Research 150: 3-9.

Ndiweni, N. and Finch, J.M., 1996. Effects of *in vitro* supplementation with alpha-tocopherol and selenium on bovine neutrophil functions: implications for resistance to mastitis. Veterinary Immunology and Immunopathology 51: 67-78.

Ndoni, S.A. and Okoko, T., 2017. Comparative effect of selenium and glycine on hydrogen peroxide-induced cell death and activation of macrophage U937 cells. Journal of Genetic Engineering and Biotechnology 15(2): 521-526.

Nelson, S.M., Lei, X. and Prabhu, K.S., 2011. Selenium levels affect the IL-4-induced expression of alternative activation markers in murine macrophages. Journal of Nutrition 141: 1754-1761.

Nettleford, S.K. and Prabhu, K.S., 2018. Selenium and selenoproteins in gut inflammation – a review. Antioxidants 7(3): 36.

Niu, Z.Y., Liu, F.Z., Yan, Q.L. and Li, L., 2009. Effects of different levels of selenium on growth performance and immunocompetence of broilers under heat stress. Archives of Animal Nutrition 63: 56-65.

Nockels, C.F., 1988. The role of vitamins in modulating disease resistance. Veterinary Clinics of North America: Food Animal Practice 4: 531-542.

Norton, R.L., Fredericks, G.J., Huang, Z., Fay, J.D., Hoffmann, F.W. and Hoffmann, P.R., 2017. Selenoprotein K regulation of palmitoylation and calpain cleavage of ASAP2 is required for efficient FcγR-mediated phagocytosis. Journal of Leukocyte Biology 101: 439-448.

Nunes, V.A., Gozzo, A.J., Juliano, M.A., Cesar, M.C., Sampaio, M.U., Sampaio, C.A. and Araujo, M.S., 2003. Antioxidant dietary deficiency induces caspase activation in chick skeletal muscle cells. Brazilian Journal of Medical Biological Research 36: 1047-1053.

O'Neill, L.A. and Bowie, A.G., 2007. The family of five: TIR-domain-containing adaptors in toll-like receptor signalling. Nature Reviews Immunology 7: 353-364.

Okoko, T., 2018. Kolaviron and selenium reduce hydrogen peroxide-induced alterations of the inflammatory response. Journal of Genetic Engineering and Biotechnology 16: 485-490.

Oropeza-Moe, M., Falk, M., Vollset, M., Wisløff, H., Bernhoft, A., Framstad, T. and Salbu, B., 2019. A descriptive report of the selenium distribution in tissues from pigs with mulberry heart disease (MHD). Porcine Health Management 5(1): 17.

Owen, G.R., Channell, J.A., Forsyth, V.T., Haertlein, M., Mitchell, E.P., Capovilla, A., Papathanasopoulos, M. and Cerutti, N.M., 2016. Human CD4 metastability is a function of the allosteric disulfide bond in domain 2. Biochemistry 55: 2227-2237.

Owen, G.R., Le, D., Stoychev, S., Cerutti, N.M. and Papathanasopoulos, M., 2018. Redox exchange of the disulfides of human two-domain CD4 regulates the conformational dynamics of each domain, providing insight into its mechanisms of control. Biochemical and Biophysical Research Communications 497: 811-817.

Pan, T., Liu, T., Tan, S., Wan, N., Zhang, Y. and Li, S., 2018. Lower selenoprotein T expression and immune response in the immune organs of broilers with exudative diathesis due to selenium deficiency. Biological Trace Element Research 182: 364-372.

Papayannopoulos, V., 2018. Neutrophil extracellular traps in immunity and disease. Nature Reviews Immunology 18: 134-147.

Pappas, A.C., Zoidis, E., Fegeros, K., Surai, P.F. and Zervas, G., 2010. Cadmium toxicity and the antioxidant system. Nova Science Publishers Inc., Hauppauge, NY, USA.

Patwardhan, R.S., Singh, B., Pal, D., Checker, R., Bandekar, M., Sharma, D. and Sandur, S.K., in press. Redox regulation of regulatory T cell differentiation and functions. Free Radical Research https://doi.org/10.1080/10715762.2020.1745202

Peng, X., Cui, H., Cui, Y., Deng, J., Zuo, Z. and Fang, J., 2011a. Lesions of thymus and decreased percentages of the peripheral blood T cell subsets in chickens fed on diets excess in selenium. Human and Experimental Toxicology 30: 1972-1978.
Peng, X., Cui, H., Deng, J., Zuo, Z. and Lai, W., 2011c. Histological lesion of spleen and inhibition of splenocyte proliferation in broilers fed on diets excess in selenium. Biological Trace Element Research 140: 66-72.
Peng, X., Cui, H., Fang, J., Zuo, Z., Deng, J., Pan, K., Lai, W. and Zhou, Y., 2012. Low selenium diet alters cell cycle phase, apoptotic population and modifies oxidative stress markers of spleens in broilers. Biological Trace Element Research 148: 182-186.
Peng, X., Cui, H., Yuan, J., Cui, W., Fang, J., Zuo, Z., Deng, J., Pan, K., Zhou, Y. and Lai, W., 2011e. Low selenium diet induces cell cycle arrest of thymocytes and alters serum IL-2 content in chickens. Biological Trace Element Research 144: 688-694.
Peng, X., Cui, H.M, Deng, J., Zuo, Z. and Cui, W., 2011b. Low dietary selenium induce increased apoptotic thymic cells and alter peripheral blood T cell subsets in chicken. Biological Trace Element Research 142: 167-173.
Peng, X., Cui, Y., Cui, W., Deng, J. and Cui, H., 2009. The decrease of relative weight, lesions, and apoptosis of bursa of Fabricius induced by excess dietary selenium in chickens. Biological Trace Element Research 131: 33-42.
Peng, X., Cui, Y., Cui, W., Deng, J., Cui, H. and Yang, F., 2011d. The cell cycle arrest and apoptosis of bursa of Fabricius induced by low selenium in chickens. Biological Trace Element Research 139: 32-40.
Petrie, H.T., Klassen, L.W. and Kay, H.D., 1989. Selenium and the immune response: 1. Modulation of alloreactive human lymphocyte functions *in vitro*. Journal of Leukocyte Biology 45: 207-214.
Piñeiro, M., Morales, J., Vizcaíno, E., Murillo, J.A., Klauke, T., Petersen, B. and Piñeiro, C., 2013. The use of acute phase proteins for monitoring animal health and welfare in the pig production chain: the validation of an immunochromatographic method for the detection of elevated levels of pig-MAP. Meat science 95: 712-718.
Pizzi, M., Boi, M., Bertoni, F. and Inghirami, G., 2016. Emerging therapies provide new opportunities to reshape the multifaceted interactions between the immune system and lymphoma cells. Leukemia 30: 1805-1815.
Placha, I., Borutova, R., Gresakova, L., Petrovic, V., Faix, S. and Leng, L., 2009. Effects of excessive selenium supplementation to diet contaminated with deoxynivalenol on blood phagocytic activity and antioxidative status of broilers. Journal of Animal Physiology and Animal Nutrition 93: 695-702.
Prabhu, K.S., Zamamiri-Davis, F., Stewart, J.B., Thompson, J.T., Sordillo, L.M. and Reddy, C.C., 2002. Selenium deficiency increases the expression of inducible nitric oxide synthase in RAW 264.7 macrophages: role of nuclear factor-kappaB in up-regulation. Biochemistry Journal 366: 203-209.
Proietti, P., Marinucci, M.T., Del Pino, A.M., D'Amato, R., Regni, L., Acuti, G., Chiaradia, E. and Palmerini, C.A., 2018. Selenium maintains Ca2+ homeostasis in sheep lymphocytes challenged by oxidative stress. PloS One 13(7): e0201523.
Puertollano, M.A., Puertollano, E., De Cienfuegos, G.Á. and De Pablo, M.A., 2011. Dietary antioxidants: immunity and host defense. Current Topics in Medicinal Chemistry 11: 1752-1766.
Quintana, A., Schwindling, C., Wenning, A.S., Becherer, U., Rettig, J., Schwarz, E.C. and Hoth, M., 2007. T cell activation requires mitochondrial translocation to the immunological synapse. Proceedings of the National Academy of Sciences, 104: 14418-14423.
Qureshi, M.A., 1998. Role of macrophages in avian health and disease. Poultry Science 77: 978-982.
Qureshi, M.A., Hussain, I. and Heggen, C.L., 1998. Understanding immunology in disease development and control. Poultry Science 77: 1126-1129.

Rahe, M.C. and Murtaugh, M.P., 2017. Mechanisms of adaptive immunity to porcine reproductive and respiratory syndrome virus. Viruses 9(6): 148.

Rao, S.V., Prakash, B., Raju, M.V., Panda, A.K., Poonam, S. and Murthy, O.K., 2013. Effect of supplementing organic selenium on performance, carcass traits, oxidative parameters and immune responses in commercial broiler chickens. Asian-Australian Journal of Animal Science 26: 247-252.

Ren, F., Chen, X., Hesketh, J., Gan, F. and Huang, K., 2012. Selenium promotes T cell response to TCR stimulation and ConA, but not PHA in primary porcine splenocytes. PLoS ONE 7: e35375.

Reshi, M.L., Su, Y.C. and Hong, J.R., 2014. RNA viruses: ROS-mediated cell death. International Journal of Cell Biology 2014: 467452.

Reth, M., 2002. Hydrogen peroxide as second messenger in lymphocyte activation. Nature Immunology 3: 1129-1134.

Rogers, S.L., Viertlboeck, B.C., Göbel, T.W. and Kaufman, J., 2008. Avian NK activities, cells and receptors. Seminars in Immunology 20: 353-360.

Roy, M., Kiremidjian-Schumacher, L., Wishe, H.I., Cohen, M.W. and Stotzky, G., 1993. Selenium supplementation enhances the expression of interleukin 2 receptor subunits and internalization of interleukin 2. Proceedings of the Society for Experimental Biology and Medicine 202: 295-301.

Roy, M., Kiremidjian-Schumacher, L., Wishe, H.I., Cohen, M.W. and Stotzky, G., 1994. Supplementation with selenium and human immune cell functions. I. Effect on lymphocyte proliferation and interleukin 2 receptor expression. Biological Trace Element Research 41: 103-114.

Safir, N., Wendel, A., Saile, R. and Chabraoui, L., 2003. The effect of selenium on immune functions of J774.1 cells. Clinical Chemistry and Laboratory Medicine 41: 1005-1011.

Saif, Y.M. and Swayne, D.E., 1998. Symposium: infectious poultry diseases. Introduction. Poultry Science 77: 8.

Salimian, J., Arefpour, M.A., Riazipour, M. and Poursasan, N., 2014. Immunomodulatory effects of selenium and vitamin E on alterations in T lymphocyte subsets induced by T-2 toxin. Immunopharmacology and Immunotoxicology 36: 275-281.

Sammalkorpi, K., Valtonen, V., Alfthan, G., Aro, A. and Huttunen, J., 1988. Serum selenium in acute infections. Infection 16: 222-224.

Sattar, N., Eatock, F., Fell, G.S. and O'Reilly, D., 1997. Selenium: an acute-phase reactant? Annals of Clinical Biochemistry 34: 437-439.

Schäfer, A., Hühr, J., Schwaiger, T., Dorhoi, A., Mettenleiter, T.C., Blome, S., Schröder, C. and Blohm U., 2019. Porcine invariant natural killer T cells: functional profiling and dynamics in steady state and viral infections. Frontiers in Immunology 10: 1380.

Sena, L.A., Li, S., Jairaman, A., Prakriya, M., Ezponda, T., Hildeman, D.A., Wang, C.R., Schumacker, P.T., Licht, J.D., Perlman, H. and Bryce, P.J., 2013. Mitochondria are required for antigen-specific T cell activation through reactive oxygen species signaling. Immunity 38: 225-236.

Serfass, R.E. and Ganther, H.E., 1975. Defective microbicidal activity in glutathione peroxidase-deficient neutrophils of selenium-deficient rats. Nature 255: 640-641.

Shi, D., Liao, S., Guo, S., Li, H., Yang, M. and Tang, Z., 2015. Protective effects of selenium on aflatoxin B1-induced mitochondrial permeability transition, DNA damage, and histological alterations in duckling liver. Biological Trace Element Research 163: 162-168.

Shilo, S. and Tirosh, O., 2003. Selenite activates caspase-independent necrotic cell death in Jurkat T cells and J774.2 macrophages by affecting mitochondrial oxidant generation. Antioxidants and Redox Signaling 5: 273-279.

Shirsat, S., Kadam, A., Mane, R.S., Jadhav, V.V., Zate, M.K., Naushad, M. and Kim, K.H., 2016. Protective role of biogenic selenium nanoparticles in immunological and oxidative stress generated by enrofloxacin in broiler chicken. Dalton Transactions 45: 8845-8853.

Shojadoost, B., Kulkarni, R.R., Yitbarek, A., Laursen, A., Taha-Abdelaziz, K., Alkie, T.N., Barjesteh, N., Quinteiro-Filho, W.M., Smith, T.K. and Sharif, S., 2019. Dietary selenium supplementation enhances antiviral immunity in chickens challenged with low pathogenic avian influenza virus subtype H9N2. Veterinary Immunology and Immunopathology 207: 62-68.

Shrimali, R.K., Irons, R.D., Carlson, B.A., Sano, Y., Gladyshev, V.N., Park, J.M. and Hatfield, D.L., 2008. Selenoproteins mediate T cell immunity through an antioxidant mechanism. Journal of Biological Chemistry 283: 20181-20185.

Sipos, W., 2019. Shifts in porcine PBMC populations from adolescence to adulthood. Veterinary Immunology and Immunopathology 211: 35-37.

Sofi, M.H., Wu, Y., Schutt, S.D., Dai, M., Daenthanasanmak, A., Voss, J.H., Nguyen, H., Bastian, D., Iamsawat, S., Selvam, S.P. and Liu, C., 2019. Thioredoxin-1 confines T cell alloresponse and pathogenicity in graft-versus-host disease. The Journal of Clinical Investigation 129: 2760-2774.

Spallholz, J.E., Martin, J.L., Gerlach, M.L. and Heinzerling, R.H., 1973a. Enhanced immunoglobulin M and immunoglobulin G antibody titers in mice fed selenium. Infection and Immunity 8: 841-842.

Spallholz, J.E., Martin, J.L., Gerlach, M.L. and Heinzerling, R.H., 1973b. Immunologic responses of mice fed diets supplemented with selenite selenium. Proceedings of the Society for Experimental Biology and Medicine 143: 685-689.

Spallholz, J.E., Martin, J.L., Gerlach, M.L. and Heinzerling, R.H., 1975. Injectable selenium: effect on the primary response of mice. Proceedings of the Society for Experimental Biology and Medicine 148: 37-40.

Stegmann, M., Barclay, A.N. and Metcalfe, C., 2018. Reduction of leucocyte cell surface disulfide bonds during immune activation is dynamic as revealed by a quantitative proteomics workflow (SH-IQ). Open Biology 8(9): 180079.

Stegmann, M., Metcalfe, C. and Barclay, A.N., 2013. Immunoregulation through membrane proteins modified by reducing conditions induced by immune reactions. European Journal of Immunology 43: 15-21.

Steinbrenner, H., Al-Quraishy, S., Dkhil, M.A., Wunderlich, F. and Sies, H., 2015. Dietary selenium in adjuvant therapy of viral and bacterial infections. Advances in Nutrition 6: 73-82.

Sukumar, M., Liu, J., Mehta, G.U., Patel, S.J., Roychoudhuri, R., Crompton, J.G., Klebanoff, C.A., Ji, Y., Li, P., Yu, Z. and Whitehill, G.D., 2016. Mitochondrial membrane potential identifies cells with enhanced stemness for cellular therapy. Cell Metabolism 23: 63-76.

Suliman, H.B., Ryan, L.K., Bishop, L. and Folz, R.J., 2001. Prevention of influenza-induced lung injury in mice overexpressing extracellular superoxide dismutase. American Journal of Physiology – Lung Cellular and Molecular Physiology 280: L69-L78.

Summerfield, A. and McCullough, K.C., 2009. The porcine dendritic cell family. Developmental & Comparative Immunology 33: 299-309.

Sun, E., Xu, H., Liu, Q., Zhou, J., Zuo, P. and Wang, J., 1995. The mechanism for the effect of selenium supplementation on immunity. Biological Trace Element Research 48: 231-238.

Sun, G.X., Chen, Y., Liu, C.P., Li, S. and Fu, J., 2016. Effect of selenium against lead-induced damage on the gene expression of heat shock proteins and inflammatory cytokines in peripheral blood lymphocytes of chickens. Biological Trace Element Research 172: 474-480.

Sun, H., Deng, T. and Fu, J., 2014. Chicken 15-kDa selenoprotein plays important antioxidative function in splenocytes. Biological Trace Element Research 161: 288-296.

Sun, Z., Liu, C., Pan, T., Yao, H. and Li, S., 2017. Selenium accelerates chicken dendritic cells ifferentiation and affects selenoproteins expression. Developmental and Comparative Immunology 77: 30-37.

Sun, Z., Xu, Z., Wang, D., Yao, H. and Li, S., 2018. Selenium deficiency inhibits differentiation and immune function and imbalances the Th1/Th2 of dendritic cells. Metallomics 10: 759-767.

Surai, P.F., 2002. Natural antioxidants in avian nutrition and reproduction. Nottingham University Press, Nottingham, UK.

Surai, P.F., 2006. Selenium in nutrition and health. Nottingham University Press, Nottingham, UK.

Surai, P.F., 2018. Selenium in poultry nutrition and health. Wageningen Academic Publishers, Wageningen, the Netherland, 430 pp.

Surai, P.F. and Dvorska, Y.E., 2005. Effects of mycotoxins on antioxidant status and immunity. In: Diaz, D.E. (ed.) The mycotoxin blue book. Nottingham University Press, Nottingham, UK, pp. 93-137.

Surai, P.F. and Kochish, I.I., 2019. Nutritional modulation of the antioxidant capacities in poultry: the case of selenium. Poultry Science 98: 4231-4239.

Surai, P.F., Kochish, I.I., Fisinin, V.I. and Kidd, M.T., 2019. Antioxidant defence systems and oxidative stress in poultry biology: an update. Antioxidants 8(7): 235.

Swain, B.K., Johri, T.S. and Majumdar, S., 2000. Effect of supplementation of vitamin E, selenium and their different combinations on the performance and immune response of broilers. British Poultry Science 41: 287-292.

Tang, J., Huang, X., Wang, L., Li, Q., Xu, J., Jia, G., Liu, G., Chen, X., Shang, H. and Zhao, H., 2017. Supranutritional dietary selenium depressed expression of selenoprotein genes in three immune organs of broilers. Animal Science Journal 88: 331-338.

Teige, J., Tollersrud, S., Lund, A. and Larsen, H.J., 1982. Swine dysentery: the influence of dietary vitamin E and selenium on the clinical and pathological effects of *Treponema hyodysenteriae* infection in pigs. Research in Veterinary Science 32: 95-100.

Tengerdy, R.P., Lacetera, N.G. and Nockels, C.F., 1990. Effect of beta-carotene on disease protection and humoral immunity in chickens. Avian Diseases 34: 848-854.

Toka, F.N. and Golde, W.T., 2013. Cell mediated innate responses of cattle and swine are diverse during foot-and-mouth disease virus (FMDV) infection: a unique landscape of innate immunity. Immunology Letters 152: 135-143.

Ueno, H., Hasegawa, G., Ido, R., Okuno, T. and Nakamuro, K., 2008. Effects of selenium status and supplementary seleno-chemical sources on mouse T cell mitogenesis. Journal of Trace Elements in Medicine and Biology 22: 9-16.

Ueno, H., Kajihara, H., Nakamura, H., Yodoi, J. and Nakamuro, K., 2007. Contribution of thioredoxin reductase to T cell mitogenesis and NF-kappaB DNA-binding promoted by selenite. Antioxidants and Redox Signaling 9: 115-121.

Ulrich, K. and Jakob, U., 2019. The role of thiols in antioxidant systems. Free Radical Biology and Medicine 20(140): 14-27.

Urban, T. and Jarstrand, C., 1986. Selenium effects on human neutrophilic granulocyte function *in vitro*. Immunopharmacology 12: 167-172.

Venter, C., Eyerich, S., Sarin, T. and Klatt, K.C., 2020. Nutrition and the immune system: a complicated tango. Nutrients 12: 3.

Ventura, M.T., Serlenga, E., Tortorella, C. and Antonaci, S., 1994. *In vitro* vitamin E and selenium supplementation improves neutrophil-mediated functions and monocyte chemoattractant protein-1 production in the elderly. Cytobios 77: 225-232.

Verma, S., Hoffmann, F.W., Kumar, M., Huang, Z., Roe, K., Nguyen-Wu, E., Hashimoto, A.S. and Hoffmann, P.R., 2011. Selenoprotein K knockout mice exhibit deficient calcium flux in immune cells and impaired immune responses. Journal of Immunology 186: 2127-2137.

Vorup-Jensen, T. and Boesen, T., 2011. Protein ultrastructure and the nanoscience of complement activation. Advanced Drug Delivery Reviews 63: 1008-1019.

Vunta, H., Belda, B.J., Arner, R.J., Channa Reddy, C., Vanden Heuvel, J.P. and Sandeep Prabhu, K., 2008. Selenium attenuates pro-inflammatory gene expression in macrophages. Molecular Nutrition & Food Research 52: 1316-1323.

Vunta, H., Davis, F., Palempalli, U.D., Bhat, D., Arner, R.J., Thompson, J.T., Peterson, D.G., Reddy, C.C, and Prabhu, K.S., 2007. The anti-inflammatory effects of selenium are mediated through 15-deoxy-delta12,14-prostaglandin J2 in macrophages. Journal of Biological Chemistry 282: 17964-17973.

Wallace, L.G., Bobe, G., Vorachek, W.R., Dolan, B.P., Estill, C.T., Pirelli, G.J. and Hall, J.A., 2017. Effects of feeding pregnant beef cows selenium-enriched alfalfa hay on selenium status and antibody titers in their newborn calves. Journal of Animal Science 95: 2408-2420.

Wang, F., Zuo, Z., Chen, K., Peng, X., Fang, J., Cui, H., Shu, G., He, M. and Tang, L., 2019. Selenium rescues aflatoxin B(1)-inhibited T cell subsets and cytokine levels in cecal tonsil of chickens. Biological Trace Element Research 188(2): 461-467.

Wang, R., Dillon, C.P., Shi, L.Z., Milasta, S., Carter, R., Finkelstein, D., McCormick, L.L., Fitzgerald, P., Chi, H., Munger, J. and Green, D.R., 2011. The transcription factor Myc controls metabolic reprogramming upon T lymphocyte activation. Immunity 35(6): 871-882. https://doi.org/10.1016/j.immuni.2011.09.021

Wang, X., Zuo, Z., Deng, J., Zhang, Z., Chen, C., Fan, Y., Peng, G., Cao, S., Hu, Y., Yu, S. and Chen, C., 2018. Protective role of selenium in immune-relevant cytokine and immunoglobulin production by piglet splenic lymphocytes exposed to Deoxynivalenol. Biological Trace Element Research 184: 83-91.

Wang, X., Zuo, Z., Zhao, C, Zhang, Z., Peng, G., Cao, S., Hu, Y., Yu, S., Zhong, Z., Deng, J. and Ren, Z., 2016b. Protective role of selenium in the activities of antioxidant enzymes in piglet splenic lymphocytes exposed to deoxynivalenol. Environmental Toxicology and Pharmacology 47: 53-61.

Wang, Y., Jiang, L., Li, Y., Luo, X. and He, J., 2016a. Excessive selenium supplementation induced oxidative stress and endoplasmic reticulum stress in chicken spleen. Biological Trace Element Research 172: 481-487.

Watson, R.R., Moriguchi, S., McRae, B., Tobin, L., Mayberry, J.C. and Lucas, D., 1986. Effects of selenium *in vitro* on human T lymphocyte functions and K-562 tumor cell growth. Journal of Leukocyte Biology 39: 447-456.

Weitzman, S.A. and Stossel, T.P., 1981. Mutation caused by human phagocytes. Science 212: 546-547.

Werling, D. and Jungi, T.W., 2003. Toll-like receptors linking innate and adaptive immune response. Veterinary Immunology and Immunopathology 91: 1-12.

Werz, O., Szellas, D. and Steinhilber, D., 2000. Reactive oxygen species released from granulocytes stimulate 5-lipoxygenase activity in a B-lymphocytic cell line. European Journal of Biochemistry 267: 1263-1269.

Wherry, E.J., 2011. T cell exhaustion. Nature Immunology 12: 492-499.

Wierzchosławski, K., Kwit, K., Pejsak, Z. and Pomorska-Mól, M., 2018. Selected serum acute-phase proteins in peripartum sows and evaluation of their diagnostic usefulness. Animal Reproduction Science 191: 44-55.

Williams, M.S. and Kwon, J., 2004. T cell receptor stimulation, reactive oxygen species, and cell signaling. Free Radical Biology and Medicine 37: 1144-1151.

Wirestam, L., Arve, S., Linge, P. and Bengtsson, A.A., 2019. Neutrophils-important communicators in systemic lupus erythematosus and antiphospholipid syndrome. Frontiers in Immunology 10: 2734.

Wu, D.O. and Meydani, S.N., 1998. Antioxidants and immune function. In: Papas, A.M. (ed.) Antioxidant status, diet, nutrition, and health. CRC Press, Boca Raton, FL, USA, pp. 371-400.

Wu, Z. and Kaiser, P., 2011. Antigen presenting cells in a non-mammalian model system, the chicken. Immunobiology 216: 1177-1183.

Xing, M., Jin, X., Wang, J., Shi, Q., Cai, J. and Xu, S., 2018. The antagonistic effect of selenium on lead-induced immune dysfunction via recovery of cytokine and heat shock protein expression in chicken neutrophils. Biological Trace Element Research 185(1): 162-169.

Xiong, Y., Yi, H., Wu, Q., Jiang, Z. and Wang, L., 2020. Effects of acute heat stress on intestinal microbiota in grow-finishing pigs, and associations with feed intake and serum profile. Journal of Applied Microbiology 128: 840-852.

Xu, D. and Tian, Y., 2015. Selenium and polysaccharides of *Atractylodes macrocephala* Koidz play different roles in improving the immune response induced by heat stress in chickens. Biological Trace Element Research 168: 235-241.

Xu, D., Li, W., Huang, Y., He, J. and Tian, Y., 2014. The effect of selenium and polysaccharide of *Atractylodes macrocephala* Koidz. (PAMK) on immune response in chicken spleen under heat stress. Biological Trace Element Research 160: 232-237.

Xu, F., Liu, S. and Li, S., 2015a. Effects of selenium and cadmium on changes in the gene expression of immune cytokines in chicken splenic lymphocytes. Biological Trace Element Research 165: 214-221.

Xu, J., Gong, Y., Sun, Y., Cai, J., Liu, Q., Bao, J., Yang, J. and Zhang, Z., 2020. Impact of selenium deficiency on inflammation, oxidative stress, and phagocytosis in mouse macrophages. Biological Trace Element Research 194: 237-243.

Xu, S.Z., Lee, S.H., Lillehoj, H.S. and Bravo, D., 2015b. Dietary sodium selenite affects host intestinal and systemic immune response and disease susceptibility to necrotic enteritis in commercial broilers. British Poultry Science 56: 103-112.

Xu, Z., Jin, X., Pan, T., Liu, T., Wan, N. and Li, S., 2017. Antagonistic effects of selenium on cadmium-induced apoptosis by restoring the mitochondrial dynamic equilibrium and energy metabolism in chicken spleens. Oncotarget 8: 52629-52641.

Yang, G., Richt, J.A. and Driver, J.P., 2018. Harnessing invariant NKT cells to improve influenza vaccines: a pig perspective. International Journal of Molecular Sciences 19(1): 68.

Yang, H., Deng, H., Liu, Y.Q., He, Y.M. and Zhang, T.Y., 2000. Effect of supplementation of selenium through water on immunoefficacy against Newcastle disease. Chinese Journal of Veterinary Medicine 26: 17-18.

Yang, J., Gong, Y., Cai, J., Zheng, Y., Liu, H. and Zhang, Z., 2020. Chlorpyrifos induces redox imbalance-dependent inflammation in common carp lymphocyte through dysfunction of T-cell receptor γ. Journal of Fish Diseases 43: 423-430.

Yang, Z., Liu, C., Liu, C., Teng, X. and Li, S., 2016b. Selenium deficiency mainly influences antioxidant selenoproteins expression in broiler immune organs. Biological Trace Element Research 172: 209-221.

Yang, Z., Liu, C., Zheng, W., Teng, X. and Li, S., 2016a. The functions of antioxidants and heat shock proteins are altered in the immune organs of selenium-deficient broiler chickens. Biological Trace Element Research 169: 341-351.

Yin, K., Cui, Y., Sun, T., Qi, X., Zhang, Y. and Lin, H., 2020. Antagonistic effect of selenium on lead-induced neutrophil apoptosis in chickens via miR-16-5p targeting of PiK3R1 and IGF1R. Chemosphere 246: 125794.

Yoshida, K., Maekawa, T., Zhu, Y., Renard-Guillet, C., Chatton, B., Inoue, K., Uchiyama, T., Ishibashi, K., Yamada, T., Ohno, N., Shirahige, K., Okada-Hatakeyama, M. and Ishii, S., 2015. The transcription factor ATF7 mediates lipopolysaccharide-induced epigenetic changes in macrophages involved in innate immunological memory. Nature Immunology 16: 1034-1043.

You, L., Liu, C., Yang, Z.J., Li, M. and Li, S., 2014. Prediction of selenoprotein T structure and its response to selenium deficiency in chicken immune organs. Biological Trace Element Research 160: 222-231.

Youn, H.S., Lim, H.J., Choi, Y.J., Lee, J.Y., Lee, M.Y. and Ryu, J.H., 2008. Selenium suppresses the activation of transcription factor NF-kappa B and IRF3 induced by TLR3 or TLR4 agonists. International Immunopharmacolog 8: 495-501.

Yu, D., Li, J.L., Zhang, J.L., Gao, X.J. and Xu, S., 2011. Effects of dietary selenium on selenoprotein W gene expression in the chicken immune organs. Biological Trace Element Research 144: 678-687.

Yu, D., Zhang, Z., Yao, H., Li, S. and Xu, S.W., 2015. The role of selenoprotein W in inflammatory injury in chicken immune tissues and cultured splenic lymphocyte. Biometals 28: 75-87.

Yun, C.H., Yang, J.S., Kang, S.S., Yang, Y., Cho, J.H., Son, C.G. and Han, S.H., 2007. NF-kappaB signaling pathway, not IFN-beta/STAT1, is responsible for the selenium suppression of LPS-induced nitric oxide production. International Immunopharmacology 7: 1192-1198.

Zamamiri-Davis, F., Lu, Y., Thompson, J.T., Prabhu, K.S., Reddy, P.V., Sordillo, L.M. and Reddy, C.C., 2002. Nuclear factor-kappaB mediates over-expression of cyclooxygenase-2 during activation of RAW 264.7 macrophages in selenium deficiency. Free Radical Biology and Medicine 32: 890-897.

Zang, H., Qian, S., Li, J., Zhou, Y., Zhu, Q., Cui, L., Meng, X., Zhu, G. and Wang, H., 2020. The effect of selenium on the autophagy of macrophage infected by *Staphylococcus aureus*. International Immunopharmacology 83: 106406.

Zhai, Q., Cen, S., Li, P., Tian, F., Zhao, J., Zhang, H. and Chen, W., 2018. Effects of dietary selenium supplementation on intestinal barrier and immune responses associated with its modulation of gut microbiota. Environmental Science & Technology Letters 5(12): 724-730.

Zhai, Q., Xiao, Y., Li, P., Tian, F., Zhao, J., Zhang, H. and Chen, W., 2019. Varied doses and chemical forms of selenium supplementation differentially affect mouse intestinal physiology. Food & Function 10: 5398-5412.

Zhang, K., Zhao, Q., Zhan, T., Han, Y., Tang, C. and Zhang, J., 2020. Effect of different selenium sources on growth performance, tissue selenium content, meat quality, and selenoprotein gene expression in finishing pigs. Biological Trace Element Research 196(2): 463-471.

Zhang, Z.B., Guo, Y.F., Li, C.Y., Qiu, C.W. and Guo, M.Y., 2019 Selenium influences mmu-miR-155 to inhibit inflammation in *Staphylococcus aureus*-induced mastitis in mice. Food & Function 16(10): 6543-6555.

Zhang, Z.W., Wang, Q.H., Zhang, J.L., Li, S., Wang, X.L. and Xu, S.W., 2012. Effects of oxidative stress on immunosuppression induced by selenium deficiency in chickens. Biological Trace Element Research 149: 352-361.

Zhang, Z.W., Zhang, J.L., Gao, Y.H., Wang, Q.H., Li, S., Wang, X.L. and Xu, S.W., 2013. Effect of oxygen free radicals and nitric oxide on apoptosis of immune organ induced by selenium deficiency in chickens. Biometals 26: 355-365.

Zhao, W., Han, Y., Zhao, B., Hirota, S., Hou, J. and Xin, W., 1998. Effect of carotenoids on the respiratory burst of rat peritoneal macrophages. Biochimica et Biophysica Acta 1381: 77-88.

Zhao, W., Liu, W., Chen, X., Zhu, Y., Zhang, Z., Yao, H. and Xu, S., 2014. Four endoplasmic reticulum resident selenoproteins may be related to the protection of selenium against cadmium toxicity in chicken lymphocytes. Biological Trace Element Research 161: 328-333.

Zhu, J., Yamane, H., Cote-Sierra, J., Guo, L. and Paul, W.E., 2006. GATA-3 promotes Th2 responses through three different mechanisms: induction of Th2 cytokine production, selective growth of Th2 cells and inhibition of Th1 cell-specific factors. Cell Research 16: 3-10.

Chapter 10
Looking ahead: from the past to the future

P.F. Surai[1-7]
[1]*Vitagene and Health Research Centre, Bristol, BS4 2RS, United Kingdom;* [2]*Department of Hygiene and Poultry Sciences, Moscow State Academy of Veterinary Medicine and Biotechnology Named After K.I. Skryabin, 109472 Moscow, Russia;* [3]*Department of Microbiology and Biochemistry, Faculty of Veterinary Medicine, Trakia University, 6000 Stara Zagora, Bulgaria;* [4]*Department of Animal Nutrition, Faculty of Agricultural and Environmental Sciences, Szent Istvan University, 2103 Gödöllo, Hungary;* [5]*Saint-Petersburg State Academy of Veterinary Medicine, 196084 St. Petersburg, Russia;* [6]*Sumy National Agrarian University, 40000 Sumy, Ukraine;* [7]*Odessa National Academy of Food Technologies, 65039 Odessa, Ukraine; psurai@feedfood.co.uk*

Abstract

Our analysis of the literature reinforces the importance of Se in pig nutrition and health. In fact, from 25 known porcine selenoproteins more than half are involved in antioxidant defences and redox homeostasis maintenance. An important consideration of optimal Se nutrition is related to maintenance of redox homeostasis in the gut under various stress condition and effects of Se nutrition/status on Nrf2 and NF-κB expressions as well as anti-inflammatory properties of Se deserves more attention and additional research. Furthermore, several selenoproteins, including thioredoxin reductase and glutathione peroxidases are regulated at the vitagene level and their involvement in stress adaptation is an important topic of current research. In particular within precise nutrition concept there is a move from preventing Se deficiencies to meeting Se requirement and organic Se sources are shown to be an effective tool to achieve this goal. It seems likely that OH-SeMet is the most effective organic Se source for farm animals and poultry to date showing advantages in comparison to sodium selenite, Se-Yeast or pure SeMet/Zn-SeMet and it is just the matter of time before sodium selenite would be replaced by organic Se sources in commercial pig nutrition.

Keywords: selenium, gut, inflammation, oxidative stress, pigs

10.1 Introduction: a historical view

Selenium (Se) is an essential element for pig nutrition that was discovered by the Swedish chemist Berzelius more than 200 years ago. Interest in this enigmatic element is growing since 1957, when its essentiality was described by Schwarz and Foltz (1957, 1958; Figure 10.1).

Peter F. Surai (ed.) **Selenium in pig nutrition and health**
DOI 10.3920/978-90-8686-912-1_10, © Wageningen Academic Publishers 2021

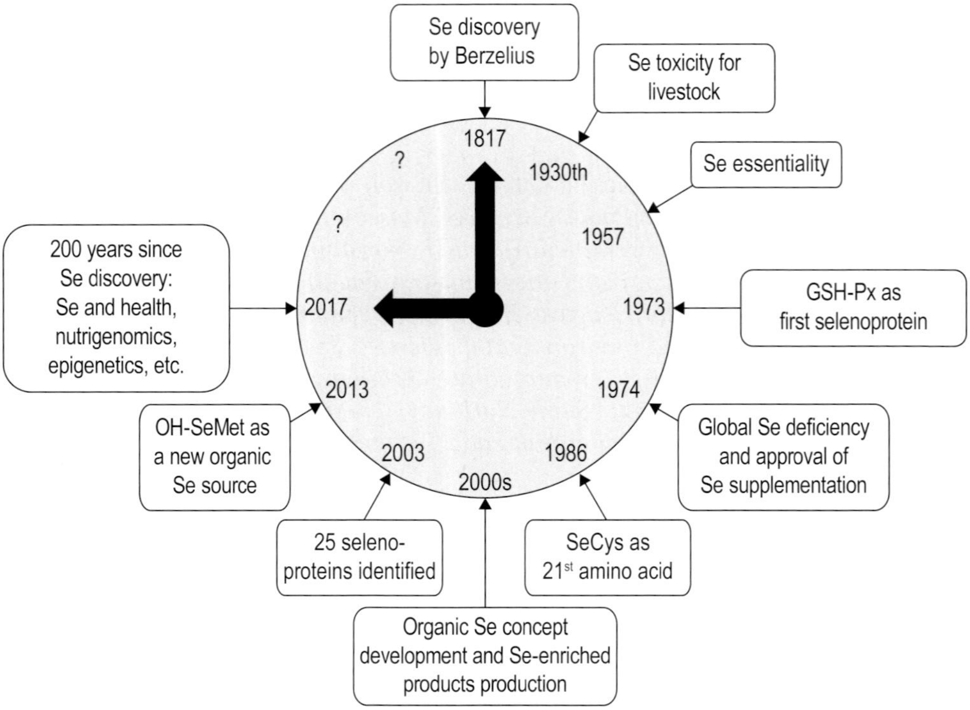

Figure 10.1. History of selenium (Se) research and usage (adapted from Surai *et al.*, 2017).

Selenium was discovered by the Swedish chemist Berzelius in 1817. In 1930, Se toxicity for livestock was described (Surai, 2006). In 1957, Se essentiality was discovered by Schwarz and Foltz (1957, 1958). The first selenoprotein, GPx was described by Rotruck *et al.* (1973). In 1970, a global Se deficiency in livestock was admitted and the FDA approved Se supplements for poultry and pigs in 1974 in the form of selenite or selenate. In 1986, SeCys was identified as the 21st amino acid encoded by the stop codon UGA (Chambers *et al.*, 1986). The organic Se concept was developed and Se-Yeast found its way on the market as an organic source of Se in 2000's and a range of Se-enriched products appeared on the market (Surai, 2006). In 2003, mammalian proteomes were characterised and 25 selenoproteins were identified (Kryukov *et al.*, 2003). Later, 25 genes encoding different selenoproteins in pigs were identified (Chen *et al.*, 2018). Organic selenium sources found their way into animal nutrition, and a new effective source of organic Se (OH-SeMet) combining major advantages of Se-yeast and pure SeMet was successfully tested and found its way to pig/poultry industry (Jlali *et al.*, 2014; Briens *et al.*, 2014). Two hundred years have passed since the discovery of Se, and interest in the chemistry, biochemistry, and practical application of this element in animal industry increases steadily.

10.2 From preventing selenium deficiencies to meeting Se requirement: current knowledge

Severe Se deficiency has been shown to be associated with the development of various disorders that no longer occur in modern commercial pig industry. However, decreased productive and reproductive performance due to suboptimal levels of dietary Se and inadequate antioxidant defences in stress conditions can still be observed in farms. The introduction of modern genetics to the pig industry has substantially improved the growth rates of piglets. However, a major downside of such improvements in performance is that the pigs are often highly sensitive to various stresses. Therefore, in modern pig production, there is an important movement from prevention of Se deficiency to meeting the exact Se requirement of pigs for optimising their performance. In particular, the discovery and characterisation of major selenoproteins, as well as a better understanding of the relationships between different antioxidants of the antioxidant system, provide new insights in this area. It is generally accepted that in biological systems, Se participates in various biochemical pathways and physiological functions as an integral part of a range of important selenoproteins. In pigs, 25 genes encoding different selenocysteine (SeCys)-containing proteins have been identified (for details see Chapter 2; Zhao *et al.*, 2021). Interestingly, more than half of the known selenoproteins are directly or indirectly involved in antioxidant defences and maintaining redox balance in the cell (Figure 10.2).

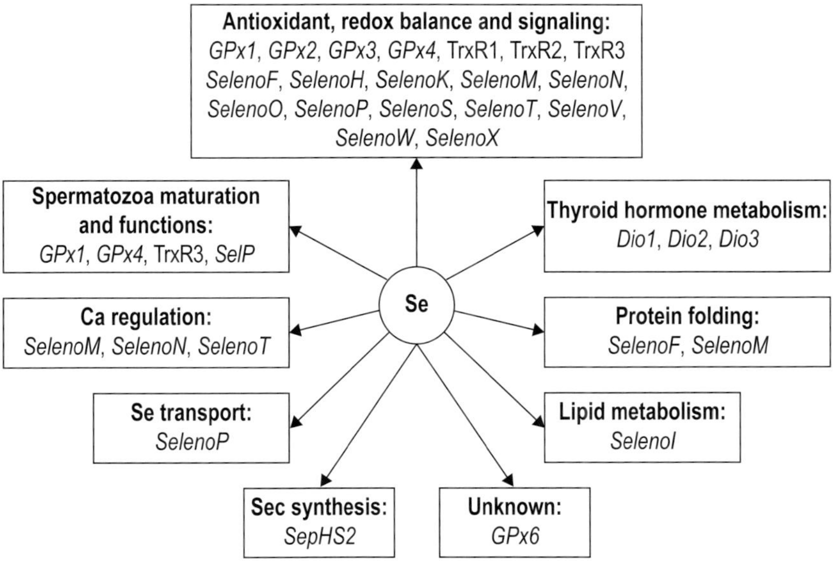

Figure 10.2. Established and suggested selenoprotein functions in pigs (adapted from Surai, 2018).

For example, selenoproteins are involved in GSH-dependent hydroperoxide removal, reduction of thioredoxins, selenophosphate synthesis, activation and inactivation of thyroid hormones, thioredoxin-dependent repair of oxidised methionine residues, endoplasmic reticulum-associated protein degradation, and other important biochemical processes. This explains the role of Se in animal health, including gut health and immune system regulation. Selenoprotein expression is characterised by high tissue specificity, depends on Se availability, can be regulated by hormones, and contributes to various pathological conditions if compromised (Surai, 2006).

During the last 40 years, a mandatory approach to regulate Se supplementation in commercial feed has been used in the pig industry worldwide. To meet Se requirements, the pig industry has relied completely on supplemental Se delivered with premixes (usually at 0.2-0.3 mg/kg diet), while the Se levels in the feed ingredients have not been taken into account (Surai, 2006, 2018; Surai and Fisinin, 2016). Selenium is considered to be one of the most controversial trace elements. From the one hand, it is toxic at high doses and there is a great body of information related to environmental issues of Se contamination. On the other hand, Se deficiency is a global problem and Se supplementation with premix has become a common procedure for commercial feed producers.

The importance of Se is related to the quickly developed understanding of the role of free radicals in biology and medicine, including redox balance maintenance and signalling. Indeed, detrimental effects of free radicals are well established and their involvement in the development of various diseases is proven. It is also generally accepted that mitochondria are the major source of free radicals in biological systems. Therefore, even in physiological conditions free radicals are produced and they can damage all types of biological molecules, including DNA, lipids and proteins. In fact, lipid peroxidation is shown to be an important detrimental process in biological systems. Indeed, various calculations of electrons escaping from the electron-transport chain in mitochondria gave an estimation of about 20 billion free radicals in every cell every day to be produced in physiological conditions. Therefore, an old perception that free radicals are produced in stress conditions was corrected to that stress conditions increase free radical production, which is already substantial in physiological conditions.

In recent years it has been established that damage to DNA and proteins are also of great importance from the point of view of understanding the pathology of many different changes in the body caused by free radicals. During evolution an integrated antioxidant system was developed in the body and in particular in every cell to deal with those free radicals. In fact, there are three major levels of antioxidant defence and Se participates in all of them. The first level is based on the activity of antioxidant enzymes with SOD to be the main device to detoxify superoxide radicals, the most important radicals produced in biological systems. Se-dependent GPx, catalase and metal-binding proteins are also integral part of the first level of antioxidant defence. It is important to underline, that GPx also belongs to the second level of antioxidant defence, since it is responsible for the detoxification of lipid hydroperoxides formed as a result of the reaction of lipid peroxides with antioxidants, such as vitamin E. Indeed, an optimal interaction between vitamin E and GPx provides an important mechanism of antioxidant defence. Recently, it has been shown that SeMet could affect activity of DNA repairing enzymes, while selenoproteins methionine sulfoxide reductase B (MsrB) and TrxR are responsible for repairing oxidised proteins. This means that Se also belongs to the third level of antioxidant defence.

The understanding that all antioxidants in the body are working as a team, called the antioxidant system, creates an additional interest in antioxidant interactions. Indeed, vitamin E recycling is shown to be the most important mechanism of the antioxidant defences. This means that if recycling is effective, even low levels of vitamin E could provide a substantial antioxidant protection. For example, in the brain, the vitamin E level is usually comparatively low, while levels of polyunsaturated fatty acids are high, but in physiological conditions it is very difficult to detect any products of peroxidation there. On the other hand, if the recycling process is broken, even high vitamin E supplementation would not provide the adequate antioxidant protection. There are various additional mechanisms of antioxidant protection, including redox signalling and changes in vitagene expression, stress-protein synthesis as well as apoptosis.

Recent understanding of essentiality of free radicals in various physiological processes shifted research to the antioxidant-prooxidant balance rather than studying simply the lipid peroxidation. For example, free radicals are synthesised in phagocyte cells and used as a weapon to kill pathogens. This process is under strict metabolic control, since an excess of the free radical production could escape from the phagosome and damage various biological molecules including damages to phagocytes themselves.

It seems likely that some toxicants, such as mycotoxins could affect this regulation causing an excessive free radical production and creating oxidative stress. In general, ingestion of excessive amounts of antioxidants is presumed to shift the oxidant-antioxidant balance toward the antioxidant side. This is supposed to be beneficial; however, this may also adversely affect key physiological processes that are dependent on free radicals, including prostaglandin production, cell division and differentiation. This is not the case for Se, since its activity depends on selenoprotein synthesis, a metabolically expensive process and therefore selenoproteins are not produced in advance but synthesised as needed in response to stress-conditions and the synthesis requires an adequate Se supply. Indeed, interactions of selenoproteins and various transcription factors (Nrf2, NF-κB, AP-1, etc.) and their participation in redox balance maintenance and signalling are of great importance for future research.

In 1973 GPx was described as a selenoprotein and antioxidant role of Se was established. Since then a family of selenoproteins in animals/poultry was shown to include 25 different selenoproteins (Surai, 2018). In fact, many of new selenoproteins have been described only recently and therefore the knowledge of their functions and regulation is far from being complete (For more details see Chapter 2; Zhao *et al.*, 2021). Traditionally selenoproteins have been considered to function as antioxidants. This is particularly true for GPx, but also for TrxR and Msr. In fact, there are five different forms of Se-dependent GPx in pigs and they are located in different parts of the cell, in different tissues and in some cases are differently regulated. It seems likely that this is an adaptive mechanism to deal more effectively with free radical production. For example, PH-GPx is specifically located in membranes and is able to deal with lipid peroxides directly, without the necessity to release them from membranes by phospholipases. GI-GPx is considered to be the most important defensive mechanism against lipid hydroperoxide absorption. In fact, GI-GPx destroys lipid peroxides in the GI tract preventing them from absorption. The role of GI-GPx in the maintenance of the antioxidant-prooxidant balance in the GI tract needs further investigation. For the question if GPx family is the main selenoprotein in pigs, my answer would be, probably not. A great body of information related to the role and functions of GPx accumulated for the last 40 years put this enzyme at the top of the selenoprotein list. However, TrxR, described as a selenoprotein 23 years later than GPx and MsrB, described as a selenoprotein only in 2003 are going to change that view on the importance of different selenoproteins. TrxR is involved in regulation of processes, such as DNA synthesis, cell proliferation and apoptosis, as well as in vitamin E recycling. At the same time, understanding functions of MsrB is going to change our perception to lipid peroxidation. Indeed, free radicals induce oxidation of not only polyunsaturated fatty acids (PUFAs), but also oxidised proteins and DNA. It is generally accepted now that MsrB is an enzyme responsible for the prevention of protein oxidation. Indeed, methionine molecules located in active centre of various enzymes are considered to be 'body guards' for active cysteine molecules. When free radicals attack proteins, methionine would be oxidised first, thus protecting cysteine from oxidation thereby maintaining enzymatic activity. Msr is responsible for reducing oxidised methionine back to an active form. Therefore, the more we understand the function of Msr, the more we realise the importance of protein oxidation and its prevention in biological systems. Taking into account the possible roles of all 25 known selenoproteins from which about 18 members are involved in antioxidant defences, maintaining redox balance of the cell and cell signalling, it was suggested that Se is 'the chief-executive of the antioxidant system'. This means that Se is responsible for regulation of the major antioxidant protections via direct antioxidant activities of GPx, TrxR, Msr, or the other less studied selenoproteins (SelH, SelK, SelM, SelN, SelO, SelP, SelS, SelT, SelW, etc.) as well as indirectly via vitamin E recycling and other important interactions. Therefore, Se deficiency could compromise many important physiological functions as a result of the detrimental changes in the antioxidant defence system (Surai, 2018).

The plant absorbs Se from the soil in the form of selenite or selenate and synthesises selenoamino acids with SeMet representing more than 50% of the Se in cereal grains and other important feed ingredients. Organic Se, which can be found in grains and forages is primarily in the form of SeMet and is metabolised in the same way as methionine. It is actively transported through intestinal membranes during absorption and actively accumulated in tissues such as muscle. It is well known that methionine is not synthesised by animals and therefore it is an essential amino acid. The same is true for SeMet, which is not synthesised in animals and must be derived from feed sources. In fact, SeMet is considered to be a storage form of Se in the body. Indeed, when organic Se is used in the diet, the Se reserve is built in muscles in the form of SeMet.

These reserves can be used in stress conditions when the Se requirement increases but feed consumption decreased. The skeletal muscles are the major Se-storage organ, accounting for more than 46% of the total Se in the human body, while the kidney contains only 4% of the Se reserves. In fact, in humans, whole body Se depends on the regional location and varies from 3-6 mg up to 13-20 mg (Surai, 2018). In stress conditions protein catabolism by proteasomes can release SeMet, which could serve as a source of Se for newly synthesised selenoproteins, such as GPx, thioredoxin reductase and methionine sulphoxide reductase. Those enzymes can deal with overproduction of free radicals and prevent a decrease in productive and reproductive performance of farm animals/pigs. It was proven that Se from both selenite and SeMet is readily available for synthesis of the selenoenzyme GPx in animal tissues. There are also several lines of evidence confirming the idea that Se accumulated in tissues in the form of SeMet can be available for selenoprotein synthesis. It should be noted that only a small proportion of the methionine pool can be replaced by SeMet, since only a very small part of methionine could be replaced by SeMet in the diet. Furthermore, the protein turnover prevents accumulation of SeMet to toxic levels in the organism.

There is a great discrepancy in results of comparative evaluation of Se bioavailability from various sources. The problem is that it is difficult to choose a proper endpoint for such an evaluation. A range of techniques has been used for this purpose, including prevention of specific Se-responsive diseases in animals, measurement of the uptake of Se in different tissues after ingestion of a test compound, activity of GPx in plasma, red blood cells, whole blood, liver and other tissues and selenium excretion with urine. However, none of those techniques can give an ultimate answer as for Se bioavailability or Se status of animals/poultry.

It seems likely, that a combination of various techniques needs to be used for such an evaluation. In general, absorption of various forms of Se is quite high. However, the main difference between organic and inorganic forms of Se is in its accumulation in muscle tissues and building Se reserves in the body. Since SeMet is the main form of Se in muscles, and animals cannot synthesise it, only organic selenium can substantially increase Se concentrations in those tissues. This can give advantages to the developing animals/pigs in terms of improvement of their Se status, as well an important tool for the production of Se-meat. It is well appreciated that via selenoprotein expression Se is involved in regulation of cell growth, apoptosis and modifying the action of cell signalling systems and transcription factors and therefore its adequate dietary supply is a crucial factor for many physiological processes in the animal/poultry body (for more details see Chapter 3; Surai *et al.*, 2021).

Selenium has important applications in pig nutrition. In particular, commercial pig production uses modern pig genetics characterised by increased muscles percentage in the body and high growth rate. However, the price for such improvements is related to high sensitivity of pigs to various stresses. Therefore, in modern pig production there is an important move from preventing Se deficiency to meeting Se requirement. The data presented in Chapter 5 (Surai, 2021a) indicate that organic selenium is a choice for diets designed to maintain high productive and reproductive performance of sows. In particular, replacement of sodium selenite by organic selenium in the sow's diet is related to improvement of antioxidant defences of sows, developing foetus, newly born and weaned piglets. Indeed, organic selenium (SeMet) is more effectively transferred via placenta from the diet to the developing piglets and further to the newly-born piglet. Furthermore, an improved Se status of colostrum and milk of sows due to organic Se supplementation further improve Se status and antioxidant defences in the weaned piglets. This improves antioxidant defences and helps piglets to overcome the oxidative stress of the birth and weaning. It is well known that at piglet birth, many physiological systems, including digestive and immune systems, are not mature and continue to develop during early postnatal development. Therefore, this is the most vulnerable period of ontogenesis of the piglets.

Advantages of organic selenium for boars (Chapter 6; Surai and Fisinin, 2021) are related to maintenance of the antioxidant defences of the spermatozoa and their fertilising ability in stress condition of semen manipulation (e.g. semen dilution, storage, deep freezing, etc.). Indeed, extremely high proportions of long chain PUFAs in boar spermatozoa require adequate antioxidant defences and selenoproteins are of paramount importance in this process. Organic Se has important roles in growing pig nutrition by supporting antioxidant defences and positively affecting immunocompetence and gut health (Chapter 7; Surai and Kochish, 2021). Additional benefits of organic Se in pig nutrition is shown to come from studies showing possible beneficial effects of organic Se in muscle redox balance maintenance and improvement meat quality (Chapter 8; Mercier *et al.*, 2021). In fact, roles of protein oxidation in meat quality deterioration, including drip loss (Estevez *et al.,* 2020) warrants more investigations in this exciting area.

Pig Se requirement in physiological conditions is quite low and in many cases can be met by the Se found in the main feed ingredients (in Europe it is about 0.1 mg/kg and in the USA it is about 0.2-0.3 mg/kg). However, commercial pig production is associated with a range of stresses which substantially increase the Se requirement. It is impossible to predict the strength, frequency and duration of stresses and therefore dietary Se supplementation with mineral or vitamin-mineral premixes is considered to be an 'insurance policy', ensuring the best Se status for maximum expression of the main selenoproteins and antioxidant protection. Therefore, the commercial level of Se supplementation is about 0.3 mg/kg, and when supplemented in organic form the producer can get highest benefit of it, a 'comprehensive insurance policy'. Unfortunately, recently the European Union implemented a directive restricting organic Se supplementation of poultry and pigs at 0.2 mg/kg. In this scenario, producers are looking for Se supplements with the highest efficacy.

The most important application of the dietary Se supplementation is related to its immunomodulating properties is described in Chapter 9 (Surai, 2021b). In fact, the immune system is the most complex system in the body and until now it is still not known how this system is regulated at the molecular level. It seems likely, that an effective communication between different types of the immune cells (e.g. neutrophils, macrophages, dendrites, NK cells, NKT cells, B-, T lymphocytes, etc.) is a key element in the immunocompetence. It has been suggested that receptors on the surfaces of these cells are working like mobile phones receiving and sending various signals using so called signalling molecules (e.g. cytokines, chemokines, eicosanoids, etc.). These receptors are extremely sensitive to low concentrations of communicating molecules, but they are also sensitive to redox balance and to excess of free radicals, which could have detrimental effects of expression and integrity of those receptors. In fact, under stress conditions free radical production can exceed the ability of the antioxidant system to detoxify them leading to oxidative stress. Those radicals can damage communication between immune cells and the great number of immune cells will be useless without effective communication and immunocompetence would be substantially compromised. In fact, phagocytic cells produce free radicals and use them as a weapon to kill pathogens. This process is redox-sensitive being strictly regulated. It could be comparable to a nuclear power station where the control chain reaction produces energy. However, if this control is broken (e.g. Chernobyl power station) the power station could become an atomic bomb. A similar scenario could happen when the regulation in phagocytic cells is compromised. An example of such a process would likely be the action of mycotoxins on immune cells. The free radicals produced by phagocytes could be considered as a 'chemical weapon' used to kill pathogens. However, the problem is that the chemical weapon is not specific and could kill anything and anywhere. Protection of the immune cells in general and the receptors in particular from their own weapon by antioxidants is of great importance for the immunocompetence. Therefore, selenoproteins expressed in immune cells together with other antioxidants (e.g. thioredoxins and GSH) could be considered as a protective mechanism maintaining the expression and integrity of receptors and preventing declining immunocompetence in stress conditions. It has been shown that Se affects all components of the immune system, including the development and expression of non-specific (innate), humoral, and cell-mediated (adaptive) responses. In general, a deficiency in Se appears to result in immunosuppression, whereas supplementation with low doses of Se appears to result in augmentation and/or restoration of immunologic functions.

In fact, a deficiency of Se has been shown to inhibit resistance to microbial and viral infections, neutrophil function, antibody production, proliferation of T- and B-lymphocytes in response to mitogens, and cytodestruction by T lymphocytes and NK cells. On the other hand, optimal Se supplementation has been shown to maintain the function of neutrophils, production of antibodies, proliferation of T- and B-lymphocytes in response to mitogens, production of lymphokines, NK cell-mediated cytodestruction, delayed-type hypersensitivity reactions and allograft rejection, and the ability of a host to reject transplanted malignant tumours.

When considering immuno-modulating properties of selenium, it is necessary to take into account several important points. Individual antioxidants in the body interact with each other (vitamin E, ascorbic acid, GSH, thioredoxins, AO enzymes, selenoproteins, etc.) and with prooxidants (iron, high level of PUFAs, mycotoxins) which themselves have immunostimulating or immunosuppressive effects. Therefore, in every experiment the results reflect a sum of all these interactions and if background dietary concentrations of those nutrients differ, the results could be completely different. This could explain the inconsistency of some results published during the last 20 years. Furthermore, high doses of nutritional antioxidants can suppress respiratory burst in phagocytes, however, it is not clear at present if there is a limit of this suppression after which the phagocyte antimicrobial activity would be compromised. Immunomodulating effects of antioxidants are usually shown to be maximal when supplementation is above the requirement for maximal growth and maintenance of reproduction. It could well be that the Se and vitamin E doses that are adequate for maximal productivity in healthy, unchallenged animals are not optimal for immunocompetence and disease resistance. There is no data available on the effect of antioxidants on intestinal immune structures, which could be crucial barriers to external pathogens. Since free radicals can be damaging to intestinal structures, antioxidant functions of selenium are responsible for the prevention of damages to the intestinal lymphoid structure as well as to intestinal enterocyte membranes. This could be especially important in relation to digestive immunosuppression caused by toxins/mycotoxins, nutritional deficiencies and infectious agents. Selenium source (organic vs inorganic) seems to be an important element of its immunomodulating properties. Organic selenium appears to be an advantage because it is better assimilated from feed and better accumulated in tissues. Indeed, with the same dose of supplementation, organic selenium can deliver more of the element to the target tissues and because of toxicity of high sodium selenite doses this could be a solution to avoid adverse effects of selenium overdose. If the body has selenium reserves in the form of SeMet in muscles, it would be liberated during the acute phase response as a result of skeletal muscle protein catabolism by proteasome action, and used for re-synthesis of new selenoproteins which could decrease oxidative stress.

As a result of antioxidant (selenium) deficiency increased oxidative stress of a host can lead to increased virus mutation rate, and changes in a viral pathogen can result in emerging viral pathogens with new pathogenic properties. Se deficiency has been associated with a change to the viral genotype, converting the virus from a benign to a virulent strain. This possibility has not been exploited in pig production but seems important to study more extensively. There is a need to study antioxidant composition and fatty acid profile of immunocompetent tissues depending on animal age and nutritional supplementation of antioxidants and n-3 PUFAs.

10.3 Selenium and gut redox balance: future challenges

It is well known that antioxidant-prooxidant balance in the digestive tract is a major determinant of animal health (Surai, 2002, 2006, 2018, 2020b; Surai and Fisinin, 2015; Surai *et al.*, 2003, 2004) (Figure 10.3).

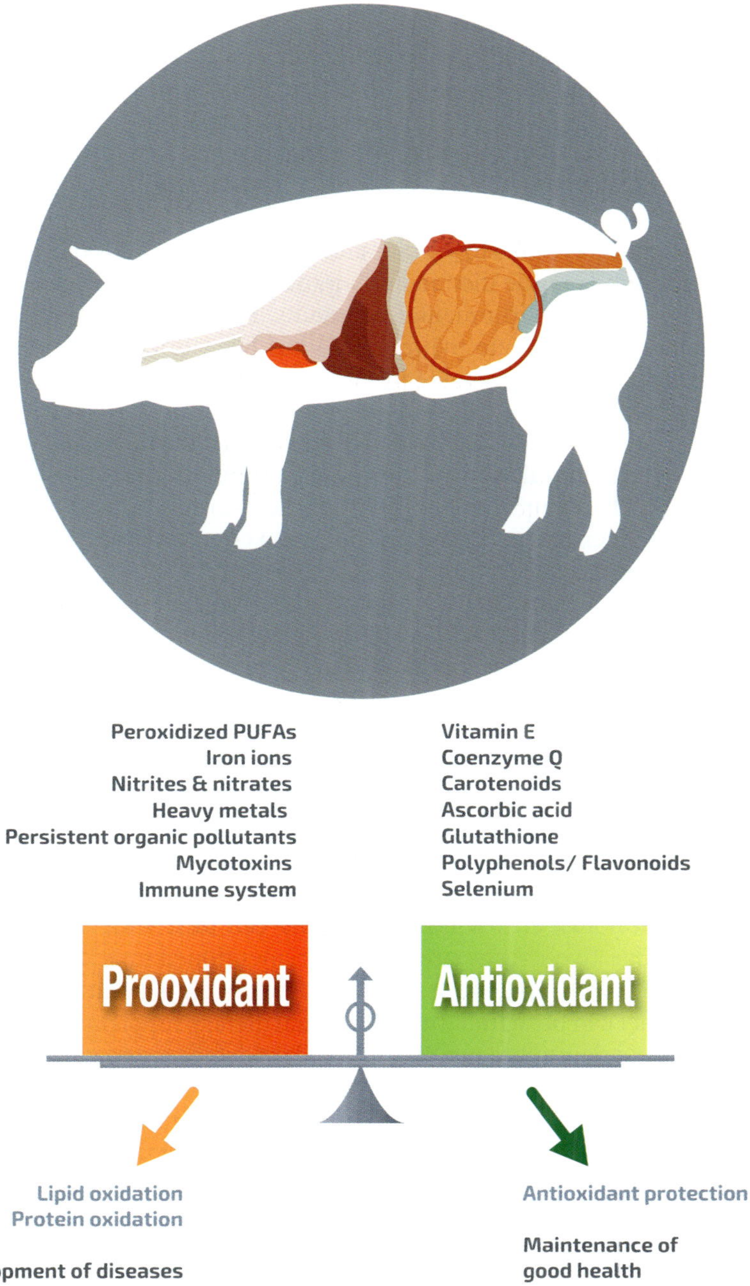

Figure 10.3. Redox balance in the gut.

In fact, redox signalling is proven to play a critical role in the physiology and pathophysiology of gastrointestinal tract (Perez *et al.*, 2017). Progression of gut inflammation is believed to rely on the balance between pro-inflammatory redox-sensitive pathways, such as NLRP3 inflammasome and nuclear factor-kappa B (NF-κB), and the adaptive up-regulation of Nrf2 and vitagenes, including SOD, GPx2 and other protective mechanisms (Perez *et al.*, 2017; Surai, 2020b).

Analysis of the literature related to pig diet showed that on the one hand, there is a range of harmful compounds in the diet, including oxidised PUFAs, traces of mycotoxins, heavy metals and other environmental pollutants. Recently a detrimental role of air pollution, including particulate matter (PM) has been linked to redox disbalance in the gut and detrimental changes in microbiota (Feng *et al.*, 2020). Therefore, emissions of harmful agents and dust in pig production could have detrimental effects on animal gut health by imposing oxidative stress. Furthermore, dietary nitrate can affect gut health via modulation of intestinal microbiota metabolism (Rocha and Laranjinha, 2020). It is important to mention host-immune, host-microbiota and immune-microbiota crosstalk during physiological conditions and disease processes rely on the redox homeostasis in the gut (Campbell and Colgan, 2019). The authors suggested that redox signals derived from immune cells, parenchymal cells and the gut microbiota, are associated with various redox products produced locally (e.g. GSH, thioredoxin, etc.) and it seems likely that dietary antioxidants and pro-oxidants could also have regulatory effects for gut signalling and adaptation.

On the one hand, a combination of prooxidant compounds could potentially stimulate lipid peroxidation, compromise redox homeostasis and cause DNA and protein damages in the intestine leading to decrease absorption efficacy of various nutrients and deteriorating FCR. On the other hand, the diet contains a range of antioxidant compounds, including vitamin E, ascorbic acid, carnitine, taurine, coenzyme Q, flavonoids, etc., which potentially can deal with those free radicals produced in the digestive tract as a result of aforementioned toxic compounds action (Surai and Fisinin, 2015). Therefore, a balance between antioxidants and prooxidants in the digestive tract can be maintained by optimal dietary supplementation of natural antioxidants and nutrients upregulating antioxidant defences and vitagenes in the gut.

In fact, some nutrients which are not well absorbed could have health-promoting properties. For example, it is well known that various flavonoids/polyphenols are not well absorbed in the small intestine (Surai, 2014, 2015). As a result, many of them will be available in the large intestine providing an important antioxidant protection, preventing lipid peroxidation, damages to DNA and proteins. Since depending on condition those compounds can show antioxidant or prooxidant properties it is highly likely that small amounts of dietary flavonoids could alert antioxidant defence systems in the gut and help stress adaptation (Surai, 2020b).

The pig gut is a specific place for dietary selenocompounds to show their biological activities. On the one hand sodium selenite is a reactive compound which in combination with reduced glutathione (which can be found in the digestive tract) could potentially produce free radicals contributing to lipid peroxidation (Surai, 2006). Indeed, prooxidant properties of selenite could have a detrimental effect on the gut. For example, the duodenal and intestinal mucosa of the chicken ileum was negatively affected by inorganic Se at 0.3 mg/kg supplementation (Attia *et al.*, 2010). In fact, in the duodenum of chickens fed on 0.3 mg/kg selenite vacuolar and hydropic degeneration of the epithelial cells lining the intestinal crypts were seen, while in the ileum of chickens from the same group an excess of mononuclear cell infiltration and aggregation between degenerated and necrotic intestinal glands was noticed (Attia *et al.*, 2010). On the other hand, organic selenium in the form of SeMet possesses antioxidant properties per se and could have a completely different protective effect. Furthermore, a beneficial effect of SeMet on DNA repairing enzymes could be relevant to the protective effects of Se in the intestine (Surai, 2018).

Taking into account antioxidant functions of many selenoproteins it is reasonable to expect protective effects of dietary Se against inflammation development in the gut by attenuating oxidative stress. Indeed, numerous studies over the years have shown a relationship between Se status and incidence and severity of intestinal disease (Short *et al.*, 2018). Furthermore, a relationship between Se supplementation, selenoprotein expression and microbiota deserves more attention. In fact, the GI tract contains internally-originated antioxidant enzymes SOD, GPx and CAT and they represent an important mechanism of the enterocyte defence from oxidative damage. A specific gastrointestinal GPx (GI-GPx) has been described in 1993 (Chu *et al.*, 1993). GI-GPx activity was present in both the villus and crypt regions of the rat mucosal epithelium and its activity nearly equalled that of classical GPx throughout the small intestine and colorectal segments (Esworthy *et al.*, 1998). GI-GPx could be considered as a barrier against hydroperoxide resorption (Brigelius-Flohé and Maiorino, 2013). Furthermore, in the gastrointestinal tract there are at least three more selenoproteins, including plasma GPx, selenoprotein P and thioredoxin reductase (Mork *et al.*, 1998).

Glutathione and glutathione-dependent enzymes contribute significantly towards intestinal antioxidant defences. In fact, an important peroxide detoxification pathway in the intestine is based on the GSH redox system (LeGrand and Aw, 1998). In this system GPx reduces peroxides at the expense of GSH oxidation. Oxidised glutathione is reduced back to the active form by glutathione reductase utilising reducing potential of NADPH which is produced in the pentose phosphate pathway. Various studies in mouse models of colon cancer and selenoprotein gene deletion studies indicated that selenoproteins play a pivotal role in the maintenance of gut homeostasis (Irons *et al.*, 2006; Reeves and Hoffman, 2009). In particular, GPx, selenoprotein S (SelS), and selenoprotein P (SepP1) have been extensively studied for their redox regulation, antioxidant and anti-inflammatory roles in preventing chronic intestinal inflammation (Reeves and Hoffman, 2009; Speckmann and Steinbrenner, 2014). Indeed, GPx1 is shown to be expressed in all cell types of the gut, whereas GPx2 is predominantly expressed in the epithelial cells, including the paneth cells, of the gastrointestinal mucosa (Brigelius-Flohe *et al.*, 2001; Esworthy *et al.*, 1998; Florian *et al.*, 2001) and GPx4 is found to be expressed in epithelial cells and the lamina propria of the intestine (Speckmann *et al.*, 2011). Furthermore, SelW is widely expressed in the gastrointestinal tract tissues of birds and the transcription of the SelW gene is very sensitive to dietary Se (Li *et al.*, 2011). In fact, the expression of the SelW mRNA in the gastrointestinal tract tissues of chickens was shown to correlate with the dietary Se concentrations (Gao *et al.*, 2012).

In an *in vitro* model system supplementation with Se was shown to significantly reduce the expression of NF-κB and COX-2 induced by bacterial endotoxin lipopolysaccharides in RAW264.7 macrophages (Zamamiri-Davis *et al.*, 2002). In fact, Se-deficient macrophages were shown to undergo severe inflammation through the NF-κB pathway due to the accumulation of ROS and are hindered in their phagocytic capacity (Xu *et al.*, 2020). The changes in GPx and other selenoprotein expression in the gut due to Se deficiency may be the major cause of injury in the intestinal tract. In fact, recent results indicated that Se deficiency induced oxidative damage in the intestinal tract of chickens and that low levels of GPx and high contents of NO may exert a major role in the injury of the intestinal tract induced by Se deficiency (Yu *et al.*, 2015). It was shown that duodenum and jejunum were the primary organs targeted by Se deficiency, while the rectum has a certain tolerance for the lack of Se (Yu *et al.*, 2015).

In fact, Se deficiency induced high expression levels of PTGE, COX-2, TNF-α, and NF-κB in the gastrointestinal tract tissues (Gao *et al.*, 2016). The effects were more pronounced in the duodenum and small intestine than those in other parts of the gut. Furthermore, Se deficiency induced the production of pro-inflammatory factors, thus aggravating inflammatory lesions in the gastrointestinal tract. In particular, Se deficiency increased the expression of NF-κB which promotes inflammation (Liu *et al.*, 2006). Therefore, one of the most important consequences of Se deficiency in chicken is an induction of inflammation in the gastrointestinal tract (Placha *et al.*, 2009). Furthermore, chickens fed with low-Se diet exhibited histological changes, lower H2S production, and lower mRNA expression of H2S-producing enzymes, as well as higher mRNA expression of intestinal inflammatory factors (TNF-α, NF-B p50, COX-2, and PTGES) compared to controls (Wu *et al.*, 2016). Similarly, a decreased level of GPx induced by Se deficiency has been reported to be the cause of damage to intestinal tissues. In fact, Se deficiency decreased GPx activity in the colon of humans causing oxidative stress (Pawłowicz *et al.*, 1991). GPx activity and mRNA level were significantly depressed in the ileum of Se-deficient mice (Esworthy *et al.*, 2005). Similarly, Se deficiency in rats reduced GPx and other antioxidant enzymes in intestines and led to DNA damage and oxidative stress (Rao *et al.*, 2001) as well as increased lipid peroxidation and decreased GSH levels, thus exacerbating the intestinal mucosal injury in the intestine of rats (Bolkent *et al.*, 2007).

In mouse, supplementation with Se-enriched milk proteins and Se-yeast upregulated the expression of gut antioxidant selenoproteins, enhancing the capacity for cell protection from oxidative damage (Hu *et al.*, 2010). It was also proven that dietary macro-fungal organic Se from Se-enriched *Agaricus bisporus* protected the gastrointestinal tract in rats from the effects of heat induced oxidative stress, by restoring epithelial ion transport and barrier functions, and elevating the expression of GPx-1 and gastrointestinal specific GPx-2, selenoenzymes relevant to mitigating oxidative stress (Maseko *et al.*, 2014a,b). Furthermore, it was shown that dietary supplementation of young broilers with Se (0.25-1.0 mg/kg) is beneficial to reduce the negative consequence of necrotic enteritis (NE; Xu *et al.*, 2015b). In particular, chickens fed with 0.50 mg/kg Se showed significantly increased body weights and antibody levels against NetB, and significantly reduced gut lesions compared with non-supplemented chickens reducing negative consequence of necrotic enteritis induced immunopathology (Xu *et al.*, 2015a). Similarly, in pigs an increase of dietary Se and vitamin E mitigated the impacts of heat stress on intestinal barrier integrity, associated with a reduction in oxidative stress (Liu *et al.*, 2016b). Se deficiency could increase the expression of inflammatory cytokines to induce inflammation in chicken erythrocytes (Luan *et al.*, 2016). The expression of prostaglandin E (PTGEs) synthase, COX-2, NF-κB, and inducible nitric oxide synthase (iNOS) could be increased by Se deficiency in the gastrointestinal tract of chickens, leading to inflammatory lesions of the digestive tract (Gao *et al.*, 2016).

The aforementioned data clearly indicate that Se as a part of various selenoproteins play an important role in protection of gut integrity by regulating antioxidant/prooxidant balance and redox status of the gut. Furthermore, Se deficiency attenuated chicken duodenal mucosal immunity via activation of the NF-κB signalling pathway regulated by redox activity, which suggested that Se is a crucial host factor involved in regulating inflammation (Liu *et al.*, 2016a). Furthermore, anti-inflammatory properties of Se in the gut due to its regulatory effects on transcription factor NF-κB (Narayan *et al.*, 2015; Tyszka-Czochara *et al.*, 2016; Wrobel *et al.*, 2015) are of great importance. Indeed, recent reports demonstrated the ability of selenium to inhibit the acetylation of non-histone and histone proteins by histone acetyltransferase p300 and there was an epigenetic modulation of the expression of proinflammatory genes, including NF-κB member p65, in macrophages (Narayan *et al.*, 2015). On the other hand, direct effect of selenium on the composition of the gut microbiota in animals (Kasaikina *et al.*, 2011) deserve more attention and further investigation.

In recent years several reports indicate that many changes in the chicken and rodent gut related to Se deficiency are also observed in pigs. Indeed, Se deficiency was found to increase the expression of inflammatory cytokines in peripheral blood lymphocyte in pigs and induce inflammation (Liu *et al.*, 2018). In Se deficiency, pig myocardial necroptosis associated with neutrophil dysfunction via Nrf2 pathway suppression (Yang *et al.*, 2017) were observed. A low-Se/high-fat diet was shown to cause relevant selenoprotein expression changes and promote the expression of pro-inflammatory factors and HSPs in pig peripheral blood lymphocytes. Furthermore, low Se can lead to enhancement of the expression of HSPs and inflammation factors induced by high fat (Liu *et al.*, 2018). In a pure line pig model, established by feeding a diet with either 0.07 mg/kg Se or 0.3 mg/kg Se for 16 weeks, a Se deficiency was shown to induce liver inflammation by activating the NF-κB pathway through multiple mechanisms (Tang *et al.*, 2020). Diet with Se deficiency can activate the iNOS/NF-κB pathway to upregulate the expression of inflammatory cytokines, thereby leading to inflammatory lesions in the pig's brain, while HSPs, including HSP70 are activated as the compensatory mechanism of regulation of inflammation (Zhang *et al.,* 2020a). Se deficiency was also shown to significantly change induced porcine small intestinal epithelial cell line (IPEC-J2) morphology and growth status and cause necroptosis and inflammatory responses at the morphological level of pigs ileum (Zhang *et al.,* 2020b). Interestingly, the protective effect of selenium (as SS or SeMet) from heat stress IPEC-J2 injury was shown to be associated with upregulation expression of selenoproteins and downregulation of pro-inflammatory cytokines (Tang *et al.,* 2019), while HSP70 was downregulated. Interestingly, SeMet exhibited a better protective effect than SS. Indeed, a role of relationship between Se supplementation, selenoprotein and vitagene expression and microbiota in the intestine redox homeostasis maintenance and gut health deserves more attention. This is especially important in commercial pig production in relation to feed antibiotic prohibition and removal of high doses of Zn supplements from the pig diets when gut health becomes a hot issue.

10.4 Vitagene-modulating properties of dietary selenium

The term 'vitagene' was introduced in 1988 by Professor Suresh Rattan at Aarhus University (Denmark) and the vitagene concept in medical sciences has been developed by a group of Italian scientists at the University of Catania under the lead of Professor Vittorio Calabrese (for details and references see Surai, 2020b). The vitagene family includes genes encoding a range of protective molecules, including SOD, HSP70, HO-1, Glutathione system, Thioredoxin system and sirtuins (Surai, 2020b). Vitagene activity is regulated by various transcription factors (e.g. HSF for HSP; Nrf2 for SOD, HO-1, Trx, TR GPx, etc.; FOXO for GPx3, SOD; NF-κB for SOD1, SOD2, GPx1, HO-1; HIF1 for SOD2, GPx3, HO-1, etc. as well as by other mechanisms including NAD^+/NADH ratio (SIRTs). It was proven that vitagenes can be regulated by various nutrients, including carnitine, betaine, taurine, vitamins A, E, D and selenium (Surai, 2020b; Surai *et al.,* 2017). Since Se-dependent enzymes namely GPx and TrxR belongs to vitagene family modulating effects of dietary Se on vitagene expression and activity is of great interest for the development of an antistress strategy in commercial pig production (Surai, 2020b) (Figure 10.4). Therefore, modulation of vitagenes by dietary Se could be considered as an important mechanism of protective effects of Se not directly related to expression of selenoproteins. Since data obtained on pigs are quite limited and mainly related to selenoproteins (GPx and TrxR) which were analysed in Chapter 2 (Zhao *et al.*, 2021). We first present data related to other monogastric species namely chicken. In fact, effects of Se on vitagene products (HSP, TrxR and SOD) was shown in various stress conditions, caused by Se deficiency or excess, HS or heavy metal toxicity. For example, it has been observed that mRNA levels and protein levels of HSP70 were significantly increased in spleen and thymus (Khoso *et al.,* 2016) and erythrocytes (Zhao *et al.,* 2016) of chickens in response to stress imposed by the Se-deficiency. However, Se supplementation was indicated to decrease the mRNA levels of HSP70 in the spleen of heat-stressed chickens (Xu *et al.,* 2014a). Of note is the fact that effects of Se on HSPs depend on the Se form used. For example, selenium yeast significantly attenuated HSP70 expression in enteropathogenic *Escherichia coli*-challenged or in heat-stressed male broiler chickens (Mahmoud and Edens, 2005).

Indeed, Se-yeast-fed chickens were less affected by mild HS (indicated by decreased induction of HSP70) than sodium selenite-fed chickens (Mahmoud and Edens, 2003). Of interest, HSP70 concentrations were significantly upregulated in the heat-stress turkey embryo developing in eggs from non-Se-supplemented birds, while in the Se-supplemented groups the HSP70 concentrations in embryos were not affected by HS (Rivera *et al.*, 2005).

Nutritional stress caused by low-Se diet in chickens was associated with a significant decrease in the activities of SOD and GPx, and an increased MDA content in thymus, spleen, bursa of Fabricius and serum (Zhang *et al*, 2012). Interestingly, not only Se deficiency (0.03 mg Se per kg of diet) but also Se excess (3 mg/kg) in chickens significantly lowered SOD and CAT activities in liver and serum (Xu *et al.*, 2014b).

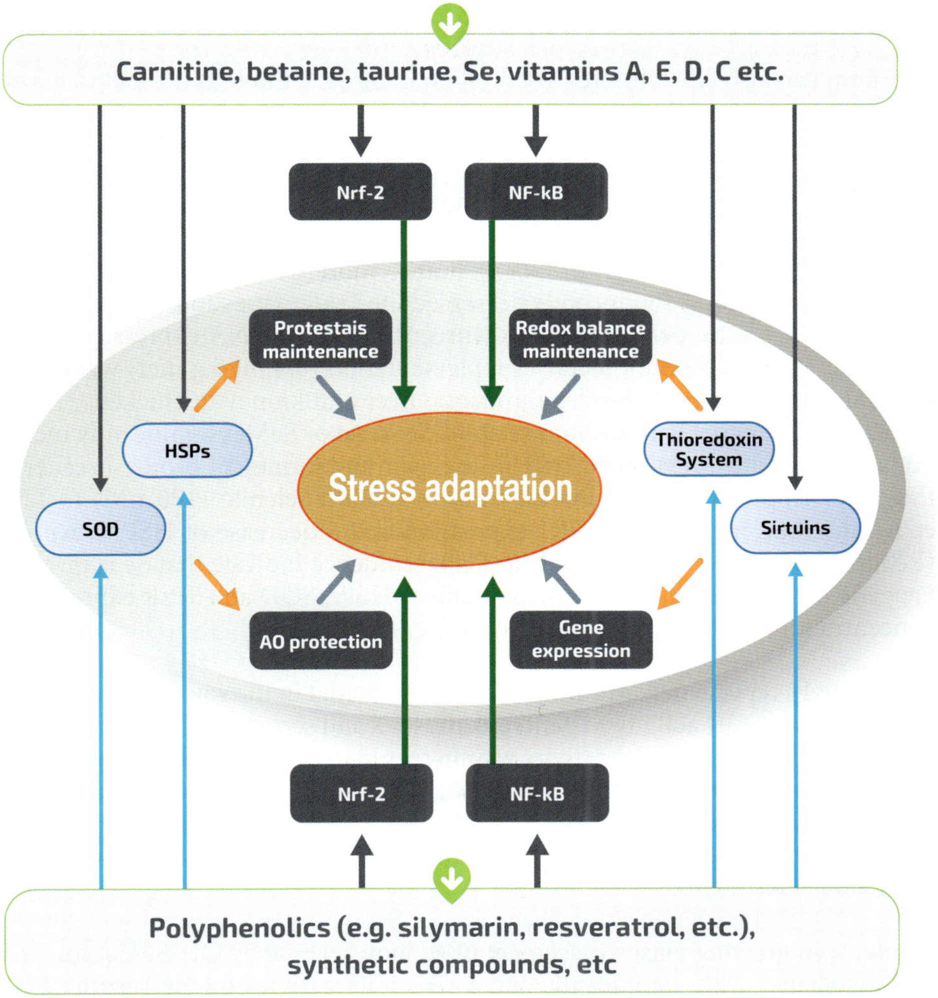

Figure 10.4. Vitagene: a gut defence.

It seems likely that SOD in adult birds is also dependent on Se status. For example, laying hens fed a Se-supplemented diet (0.3 mg/kg diet) were characterised by higher SOD and GPx activity and lower MDA content in plasma compared with those birds fed a control non-supplemented diet containing Se at 0.03 mg/kg (Jing *et al.*, 2015). It has been shown that Se efficacy is dependent on its form in the dietary supplements with organic selenium being more effective than sodium selenite (Surai, 2006, 2018; Surai and Fisinin, 2016). Indeed, GPx and SOD activities in serum were significantly higher in Se-Yeast-fed chickens in comparison to sodium selenite-fed birds (Chen *et al.*, 2014). In line with these findings, dietary Se-Yeast significantly increased the activities of CAT and SOD but reduced the MDA content in chicken breast muscle at 42 days of age (Ahmad *et al.*, 2012). Dietary SeMet (0.15 mg/kg) was observed to increase hepatic SOD activity in broilers in comparison to birds fed on sodium selenite-supplemented diet (Wang *et al.*, 2011). Similarly, SOD activities in chicken breast muscle were significantly elevated due to dietary Se-Met supplementation (0.075-0.225 mg/kg diet; Jiang *et al.*, 2009).

Selenium can also affect thioredoxin system in birds. Of note is the fact that TrxR is a Se-dependent stress-inducible enzyme and in Se deficiency its expression in animal tissues could be increased or decreased depending on experimental conditions (Surai, 2006, 2018). For example, Se deficiency in chickens was associated with reduced TrxR activity in adipose tissue (Liang *et al.*, 2014), decreased TrxR2 expression in thyroids (Lin *et al.*, 2014) and reduced activity of TrxR1 (by 50%), TrxR2 (by 83%) and TrxR3 (by 36%) in pancreas (Zhao *et al.*, 2014). Furthermore, a low Se diet (0.028 mg/kg) or high Se diet (3 mg/kg) significantly decreased the TrxR expression in chicken kidney. In particular, the Se-deficient diet downregulated the mRNA expression of TrxR3 (Xu *et al.*, 2016). Selenium dietary supplementation (0.4 mg/kg diet) was shown to increase TrxR activity in duodenal mucosa, liver and kidney in chickens (Placha *et al.*, 2014). Compared with sodium selenite, Se-Yeast or SeMet significantly increased the activity and expression of Trx1 in the liver and kidney of broiler breeders and their offspring (Yuan *et al.*, 2012). Adding sodium selenite (100 nM) to LNCaP cells caused an increase of TrxR1 expression and a decrease of HSP70 expression (Roveri *et al.*, 2008). Therefore, the available evidence indicate that Se is involved in vitagene network regulation in birds: it affects HSPs, SOD and TrxR expression and concentrations in different tissues.

Interestingly, replacement of sodium selenite by SeMet in the sow diet was associated with increased SOD activity in pancreas, muscles and serum of the progeny piglets at day 28 postpartum and GPx activity was increased in the liver and kidney in addition to the aforementioned tissues (Zhan *et al.*, 2011). Similarly, replacement of sodium selenite with OH-SeMet in the sow diet was associated with increased SOD activity in the sow's serum at day 110 (P110) of pregnancy and at day 0 of lactation (L0; Figure 10.5; Mou *et al.*, 2020).

Similarly, in growing pigs, replacement of sodium selenite by OH-SeMet for 28 days was associated with significantly increased SOD activity in the serum ($P<0.05$) and numerically increased SOD activity in the liver (Figure 10.6). Interestingly, in comparison to the control (unsupplemented) pigs fed on the diet containing 0.13 mg/kg Se, dietary supplementation of OH-SeMet at 0.4 mg/kg significantly increased SOD activity in both serum and liver (Chao *et al.*, 2019).

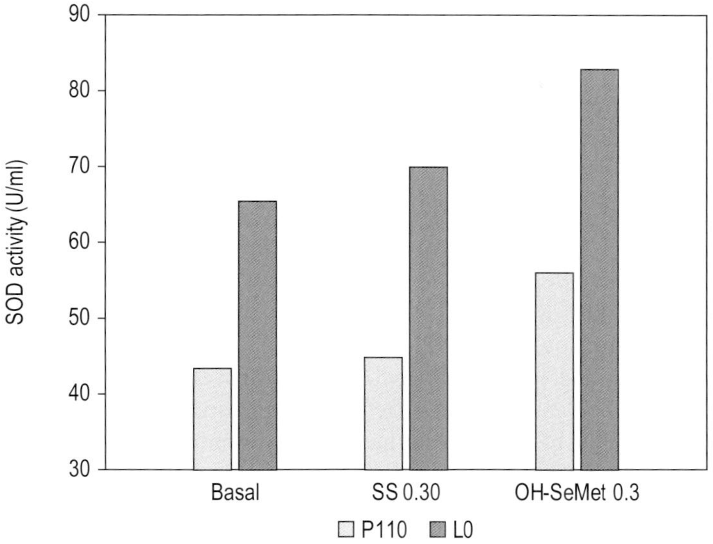

Figure 10.5. Effect of maternal selenium intake during gestation on super oxide dismutase (SOD) activity in sow's serum (adapted from Mou *et al.*, 2020).

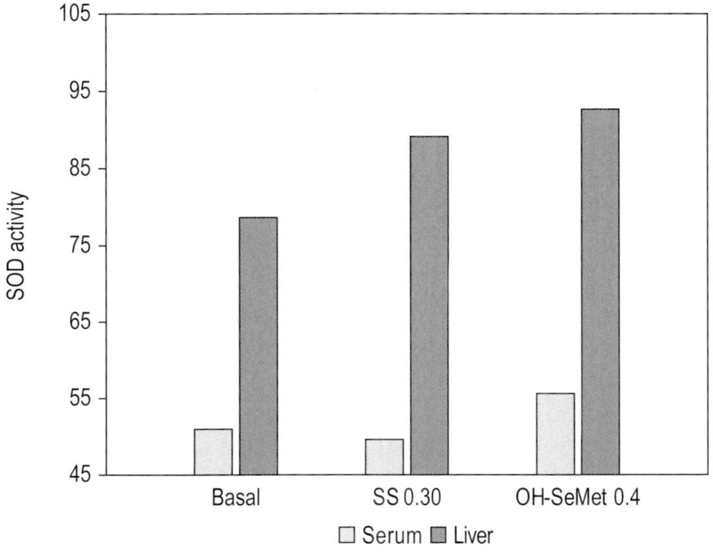

Figure 10.6. Effects of different forms of dietary selenium on super oxide dismutase (SOD) activity in the serum (U/ml) and liver (U/mg protein) of weaned piglets (Chao *et al.*, 2019).

Therefore, dietary selenium in addition to regulate expression and activity of 25 selenoproteins in pigs (see Chapter 2; Zhao *et al.*, 2021) could also affect vitagene-regulated antioxidants in the body including SOD, HO-1, HSP, etc. This could help pigs better adapt to various stress conditions. However, exact mechanisms of interactions between dietary selenium, selenoprotein expression and vitagenes expression/activity await further investigation.

10.5 Selenium and selenoproteins in the integrated antioxidant defence network

Taking into account all previous information provided above, an update on the involvement of selenium and selenoproteins in the integrated antioxidant defence mechanisms and their involvement in stress adaptation can be presented as follows (Figure 10.7; Surai, 2020a,b). Mitochondria and phagocyte cells are major sources of RONS. There are also other sources of RONS production and various stress conditions in pig production increasing RONS production. It is proven that RONS, especially H_2O_2, are signalling molecules; however, when their concentration is above threshold level, they can cause damage to main biological molecules, including proteins, PUFAs and DNA/RNA. In fact RONS excess leads to a stress response program (oxidative stress response) activation.

Since oxidative stress can cause a range of damages to various biological molecules, in addition to oxidative stress response, other stress response programs, including heat shock response, unfolded protein response, hypoxia-induced response, and DNA damage response, are also activated. This leads to activation of various transcription factors, including HSF1, Nrf2, NF-κB, FOXO, HIF, p53, and others. As a result of the upregulation of transcription factors, various genes, including vitagenes, are activated. In fact, activation of HSF increased synthesis of HSP70; Nrf2 activation increased synthesis of SOD, HO-1, important elements of thioredoxin and glutathione systems; and FOXO activation would affect expression of sirtuins. There is a complex system of interplay between vitagenes and transcription factors. In fact, some vitagenes, like SOD, are affected by several transcription factors including Nrf2, NF-κB, p53, etc. Since SOD is responsible for production of H_2O_2, major signalling RONS, control of its concentration in the cell is very strict. Furthermore, products of vitagene activation, e.g. sirtuins, could affect expression and activity of some transcription factors, including Nrf2, NF-κB, FOXO, etc. It seems likely that some vitagenes can be activated directly without transcription factor involvement. This includes transcriptional regulation of SOD in response to RONS as well as activation of sirtuins by changes in NAD+/NADH ratio. In general, redox homeostasis plays a vital role in the regulation of the antioxidant defence network. RONS are also responsible for adaptive production/activation of other antioxidants, which are not included into vitagene family (e.g. various selenoproteins, CoQ, catalase, etc.) and they all responsible for redox balance maintenance, stress resistance and adaptation leading to good health, high immunocompetence, high productive and reproductive performance of pigs. However, when the antioxidant defence system, together

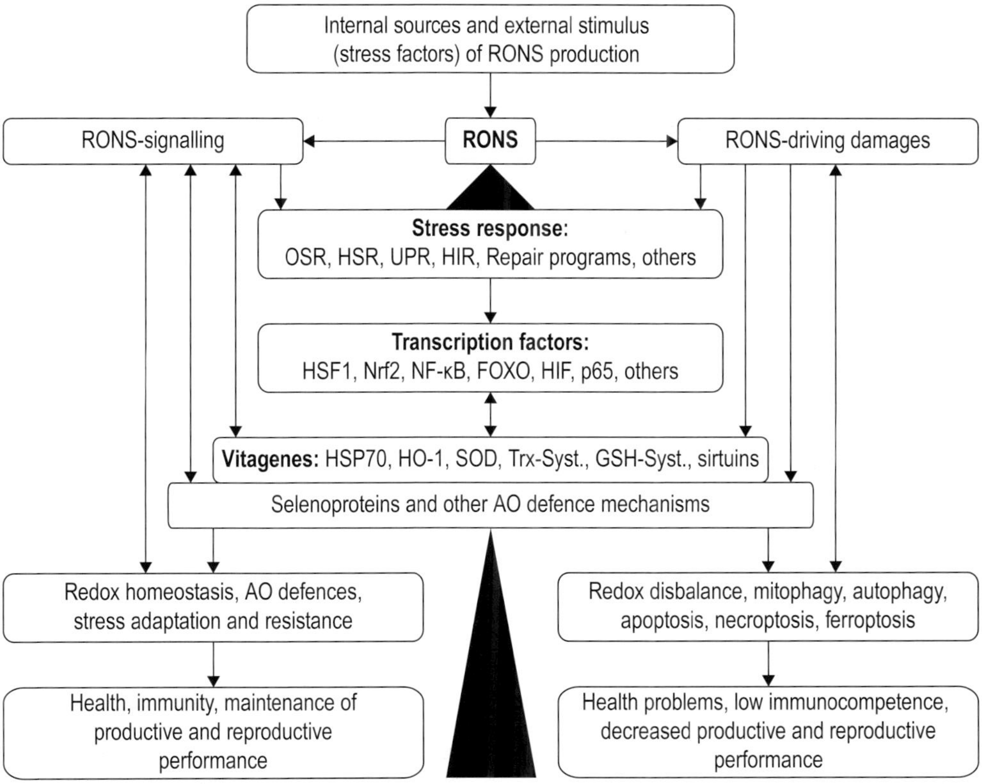

Figure 10.7. Antioxidant defence network in pigs (adapted from Surai, 2020a,b).

with the vitagene network, are not able to prevent or repair damages imposed by RONS to biological molecules, other protective mechanisms including mitophagy, autophagy, apoptosis, necroptosis, and ferroptosis are dealing with terminally damaged molecules, organelles or cells. As a result of disrupted redox balance and accumulation of damages in cells/tissues, health-related problems, including low immunocompetence, appear. In addition, decreased productive and reproductive performance can cause heavy economic losses for pig industry. It should be mentioned that in stress conditions there is an important interplay between various transcription factors. For example, moderate levels of RONS would activate initially Nrf2 leading to an additional synthesis of a range protective molecules, including AO enzymes, which can decrease RONS concentration and re-establish redox homeostasis. In this case selenoprotein expression could have a substantial protective effect. At the same time NF-κB expression would be decreased, synthesis of pro-inflammatory cytokines reduced and this would help to re-establish cell/body homeostasis. In contrary, when RONS concentration is exceeded threshold level and Nrf2-related antioxidants are not able to balance the homeostatic environment, NF-κB would be activated with following synthesis of pro-inflammatory cytokines, inflammation development with a reduction in Nrf2 expression and further development of oxidative stress (Surai,

2020b). In this case additional selenoprotein expression could have a protective effect, however the most likely scenario would be related to cell death via apoptosis/necroptosis/ferroptosis.

10.6 Conclusions

The analysis of the literature presented in this book reinforces the importance of Se in pig nutrition and health. Indeed, optimisation of Se nutrition of pigs and other farm animals will allow to maintain animal health, increase meat production efficiency, and what is also important, to increase meat quality. From the data presented above it is clear that organic selenium is an effective choice to achieve this goal.

Really, diets cure more than lancet

Acknowledgements

Peter F. Surai is supported by a grant of the Government of Russian Federation (Contract No. 14.W03.31.0013).

References

Ahmad, H., Tian, J., Wang, J., Khan, M.A., Wang, Y., Zhang, L. and Wang, T., 2012. Effects of dietary sodium selenite and selenium yeast on antioxidant enzyme activities and oxidative stability of chicken breast meat. Journal of Agricultural and Food Chemistry 60: 7111-7120.

Attia, Y.A., Abdalah, A.A., Zeweil, H.S., Bovera, F., Tag El-Din, A.A. and Araft, M.A., 2010. Effect of inorganic or organic selenium supplementation on productive performance, egg quality and some physiological traits of dual-purpose breeding hens. Czech Journal of Animal Science 55: 505-519.

Bolkent, S., Koyuturk, M., Bulan, O.K., Tunali, S., Yanardag, R. and Tabakoglu, A.O., 2007. The effects of combined alpha-tocopherol, ascorbic acid, and selenium against cadmium toxicity in rat intestine. Journal of Environmental Pathology, Toxicology and Oncology 26: 21-27.

Briens, M., Mercier, Y., Rouffineau, F. and Geraert, P.A., 2014. 2-Hydroxy-4-methylselenobutanoic acid induces additional tissue selenium enrichment in broiler chicken compared to other selenium sources. Poultry Science 93: 85-93.

Brigelius-Flohé, R. and Maiorino, M., 2013. Glutathione peroxidases. Biochimica et Biophysica Acta 1830: 3289-3303.

Brigelius-Flohe, R., Muller, C., Menard, J., Florian, S., Schmehl, K. and Wingler, K., 2001. Functions of GI-GPx: lessons from selenium-dependent expression and intracellular localization. BioFactors 14: 101-106.

Campbell, E.L. and Colgan, S.P., 2019. Control and dysregulation of redox signalling in the gastrointestinal tract. Nature reviews. Gastroenterology & Hepatology 16: 106-120.

Chambers, I., Frampton, J., Goldfarb, P., Affara, N., McBain, W. and Harrison, P.R., 1986. The structure of the mouse glutathione peroxidase gene: the selenocysteine in the active site is encoded by the 'termination' codon, TGA. EMBO Journal 5: 1221-1227.

Chao, Y., Yu, B., He, J., Huang, Z., Mao, X., Luo, J., Luo, Y., Zheng, P., Yu, J. and Chen, D., 2019. Effects of different levels of dietary hydroxy-analogue of selenomethionine on growth performance, selenium deposition and antioxidant status of weaned piglets. Archives of Animal Nutrition 73: 374-383.

Chen, G., Wu, J. and Li, C., 2014. Effect of different selenium sources on production performance and biochemical parameters of broilers. Journal of Animal Physiology and Animal Nutrition 98: 747-754.

Chen, X.D., Zhao, Z.P., Zhou, J.C. and Lei, X.G., 2018. Evolution, regulation, and function of porcine selenogenome. Free Radical Biology & Medicine 127: 116-123.

Chu, F.F., Doroshow, J.H. and Esworthy, R.S., 1993. Expression, characterization, and tissue distribution of a new cellular selenium-dependent glutathione peroxidase, GSHPx-GI. Journal of Biological Chemistry 268: 2571-2576.

Estévez, M., Geraert, P.A., Liu, R., Delgado, J., Mercier, Y. and Zhang, W., 2020. Sulphur amino acids, muscle redox status and meat quality: More than building blocks – invited review. Meat Science 163: 108087.

Esworthy, R.S., Swiderek, K.M., Ho, Y.S. and Chu, F.F., 1998. Selenium-dependent glutathione peroxidase-GI is a major glutathione peroxidase activity in the mucosal epithelium of rodent intestine. Biochimica et Biophysica Acta 1381: 213-22.

Esworthy, R.S., Yang, L., Frankel, P.H. and Chu, F.F., 2005. Epithelium-specific glutathione peroxidase, Gpx2, is involved in the prevention of intestinal inflammation in selenium-deficient mice. Journal of Nutrition 135: 740-745.

Feng, J., Cavallero, S., Hsiai, T. and Li, R., 2020. Impact of air pollution on intestinal redox lipidome and microbiome. Free Radical Biology & Medicine 151: 99-110.

Florian, S., Wingler, K., Schmehl, K., Jacobasch, G., Kreuzer, O.J., Meyerhof, W. and Brigelius-Flohe, R., 2001. Cellular and subcellular localization of gastrointestinal glutathione peroxidase in normal and malignant human intestinal tissue. Free Radical Research 35: 655-663.

Gao, X., Xing, H., Li, S., Li, J., Ying, T. and Xu, S., 2012. Selenium regulates gene expression of selenoprotein W in chicken gastrointestinal tract. Biological Trace Element Research 145: 181-188.

Gao, X., Zhang, Z., Xing, H., Yu, J., Zhang, N. and Xu, S., 2016. Selenium deficiency-induced inflammation and increased expression of regulating inflammatory cytokines in the chicken gastrointestinal tract. Biological Trace Element Research 173: 210-218.

Hu, Y., McIntosh, G.H., Le Leu, R.K. and Young, G.P., 2010. Selenium-enriched milk proteins and selenium yeast affect selenoprotein activity and expression differently in mouse colon. British Journal of Nutrition 104: 17-23.

Irons, R., Carlson, B.A., Hatfield, D.L. and Davis, C.D., 2006. Both selenoproteins and low molecular weight selenocompounds reduce colon cancer risk in mice with genetically impaired selenoprotein expression. Journal of Nutrition 136: 1311-1317.

Jiang, Z., Lin, Y., Zhou, G., Luo, L., Jiang, S. and Chen, F., 2009. Effects of dietary selenomethionine supplementation on growth performance, meat quality and antioxidant property in yellow broilers. Journal of Agricultural and Food Chemistry 57: 9769-9772.

Jing, C.L., Dong, X.F., Wang, Z.M., Liu, S. and Tong, J.M., 2015. Comparative study of DL-selenomethionine vs sodium selenite and seleno-yeast on antioxidant activity and selenium status in laying hens. Poultry Science 94: 965-975.

Jlali, M., Briens, M., Rouffineau, F., Geraert, P.A., Mercier, Y., 2014. Evaluation of the efficacy of 2-hydroxy-4-methylselenobutanoic acid on growth performance and tissue selenium retention in growing pigs. Journal of Animal Science 92, 182-188.

Kasaikina, M.V., Kravtsova, M.A., Lee, B.C., Seravalli, J., Peterson, D.A., Walter, J., Legge, R., Benson, A.K., Hatfield, D.L. and Gladyshev, V.N., 2011. Dietary selenium affects host selenoproteome expression by influencing the gut microbiota. FASEB Journal 25: 2492-2499.

Khoso, P.A., Liu, C., Khoso, M.H. and Li, S., 2016. Selenium deficiency activates heat shock protein expression in chicken spleen and thymus. Biological Trace Element Research 173: 492-500.

Kryukov, G.V., Castellano, S., Novoselov, S.V., Lobanov, A.V., Zehtab, O., Guigó, R. and Gladyshev, V.N., 2003. Characterization of mammalian selenoproteomes. Science 300: 1439-1443.

LeGrand, T.S. and Aw, T.Y., 1998. Chronic hypoxia alters glucose utilization during GSH-dependent detoxication in rat small intestine. American Journal of Physiology 274: G376-G384.

Li, J.L., Li, H.X., Li, S., Jiang, Z.H., Xu, S.W. and Tang, Z.X., 2011. Selenoprotein W gene expression in the gastrointestinal tract of chicken is affected by dietary selenium. Biometals 24: 291-299.

Liang, Y., Lin, S.L., Wang, C.W., Yao, H.D., Zhang, Z.W. and Xu, S.W., 2014. Effect of selenium on selenoprotein expression in the adipose tissue of chickens. Biological Trace Element Research 160: 41-48.

Lin, S.L., Wang, C.W., Tan, S.R., Liang, Y., Yao, H.D., Zhang, Z.W. and Xu, S.W., 2014. Selenium deficiency inhibits the conversion of thyroidal thyroxine (T 4) to triiodothyronine (T 3) in chicken thyroids. Biological Trace Element Research 161: 263-271.

Liu, F., Cottrell, J.J., Furness, J.B., Rivera, L.R., Kelly, F.W., Wijesiriwardana, U., Pustovit, R.V., Fothergill, L.J., Bravo, D.M., Celi, P., Leury, B.J., Gabler, N.K. and Dunshea, F.R., 2016b. Selenium and vitamin E together improve intestinal epithelial barrier function and alleviate oxidative stress in heat-stressed pigs. Experimental Physiology 101: 801-810.

Liu, T., Yang, T., Pan, T., Liu, C. and Li, S., 2018. Effect of low-selenium/high-fat diet on pig peripheral blood lymphocytes: perspectives from selenoproteins, heat shock proteins, and cytokines. Biological Trace Element Research 183: 102-113.

Liu, X., Shen, J., Jin, Y., Duan, M. and Xu, J., 2006. Recombinant humanery thropoietin preconditioning on nuclear factor-kappa B (NF-κB) activation & proinflammatory cytokines inducted by myocardial ischaemia-reperfusion. Indian Journal of Medical Research 124: 343-335.

Liu, Z., Qu, Y., Wang, J. and Wu, R., 2016a. Selenium deficiency attenuates chicken duodenal mucosal immunity via activation of the NF-κb signaling pathway. Biological Trace Element Research 172: 465-473.

Luan, Y., Zhao, J., Yao, H., Zhao, X., Fan, R., Zhao, W., Zhang, Z. and Xu, S., 2016. Selenium deficiency influences the mRNA expression of selenoproteins and cytokines in chicken erythrocytes. Biological Trace Element Research 171: 427-436.

Mahmoud, K.Z. and Edens, F.W., 2003. Influence of selenium sources on age-related and mild heat stress-related changes of blood and liver glutathione redox cycle in broiler chickens (*Gallus domesticus*). Comparative Biochemistry and Physiology Part B: Biochemistry and Molecular Biology 136: 921-934.

Mahmoud, K.Z. and Edens, F.W., 2005. Influence of organic selenium on hsp70 response of heat-stressed and enteropathogenic *Escherichia coli*-challenged broiler chickens (*Gallus gallus*). Comparative Biochemistry and Physiology Part C: Toxicology & Pharmacology 141: 69-75.

Maseko, T., Dunshea, F.R., Howell, K., Cho, H.J., Rivera, L.R., Furness, J.B. and Ng, K., 2014a. Seleniumenriched *Agaricus bisporus* mushroom protects against increase in gut permeability *ex vivo* and up-regulates glutathione peroxidase 1 and 2 in hyperthermally-induced oxidative stress in rats. Nutrients 6: 2478-2492.

Maseko, T., Howell, K., Dunshea, F.R., Ng, K., 2014b. Selenium-enriched *Agaricus bisporus* increases expression and activity of glutathione peroxidase-1 and expression of glutathione peroxidase-2 in rat colon. Food Chemistry 146: 327-333.

Mercier, Y., Batonon-Alavo, D. and Geraert, P.A., 2021. Selenium and pork meat quality: muscle redox status, selenium, sulphur amino acids and beyond. Chapter 8. In: Surai, P.F. (ed.) Selenium in pig nutrition and health. Wageningen Academic Publishers, Wageningen, the Netherlands, pp. 235-260. https://doi.org/10.3920/978-90-8686-912-1_8

Mörk, H., Lex, B., Scheurlen, M., Dreher, I., Schutze, N., Köhrle, J. and Jakob, F., 1998. Expression pattern of gastrointestinal selenoproteins – targets for selenium supplementation. Nutrition and Cancer 32: 64-70.

Mou, D., Ding, D., Yan, H., Qin, B., Dong, Y., Li, Z., Che, L., Fang, Z., Xu, S., Lin, Y., Zhuo, Y., Li, C., Huang, C., Zou, Y., Li, L., Briens, M., Wu, D. and Feng, B., 2020. Maternal supplementation of organic selenium during gestation improves sows and offspring antioxidant capacity, inflammatory status, and promotes embryo survival. Food and Function 11: 7748-7761. https://doi.org/10.1039/D0FO00832J

Narayan, V., Ravindra, K.C., Liao, C., Kaushal, N., Carlson, B.A. and Prabhu, K.S., 2015. Epigenetic regulation of inflammatory gene expression in macrophages by selenium. Journal of Nutritional Biochemistry 26: 138-145.

Pawłowicz, Z., Zachara, B., Trafikowska, U., Maciag, A., Marchaluk, E. and Nowicki, A., 1991. Blood selenium concentrations and glutathione peroxidase activities in patients with breast cancer and with advanced gastrointestinal cancer. Journal of Trace Elements and Electrolytes in Health and Disease 5: 275.

Pérez, S., Taléns-Visconti, R., Rius-Pérez, S., Finamor, I. and Sastre, J., 2017. Redox signaling in the gastrointestinal tract. Free Radical Biology & Medicine 104: 75-103.

Placha, I., Borutova, R., Gresakova, L., Petrovic, V., Faix, S. and Leng, L., 2009. Effects of excessive selenium supplementation to diet contaminated with deoxynivalenol on blood phagocytic activity and antioxidative status of broilers. Journal of Animal Physiology and Animal Nutrition 93: 695-702.

Placha, I., Takacova, J., Ryzner, M., Cobanova, K., Laukova, A., Strompfova, V., Venglovska, K. and Faix, S., 2014. Effect of thyme essential oil and selenium on intestine integrity and antioxidant status of broilers. British Poultry Science 55: 105-114.

Rao, L., Puschner, B. and Prolla, T.A., 2001. Gene expression profiling of low selenium status in the mouse intestine: transcriptional activation of genes linked to DNA damage, cell cycle control and oxidative stress. Journal of Nutrition 131: 3175-3181.

Reeves, M.A. and Hoffmann, P.R., 2009. The human selenoproteome: recent insights into functions and regulation. Cellular and Molecular Life Sciences 66: 2457-2478.

Rivera, R.E., Christensen, V.L., Edens, F.W. and Wineland, M.J., 2005. Influence of selenium on heat shock protein 70 expression in heat stressed turkey embryos (*Meleagris gallopavo*). Comparative Biochemistry and Physiology Part A: Molecular & Integrative Physiology 142: 427-432.

Rocha, B.S. and Laranjinha, J., 2020. Nitrate from diet might fuel gut microbiota metabolism: minding the gap between redox signaling and inter-kingdom communication. Free Radical Biology & Medicine 149: 37-43.

Rotruck, J.T., Pope, A.L., Ganther, H.E., Swanson, A.B., Hafeman, D.G. and Hoekstra, W.G., 1973. Selenium: biochemical role as a component of glutathione peroxidase. Science 179: 588-590.

Roveri, A., Vitale, M.P., Serain, E., Zaccarin, M., Mauri, P., Di Silvestre, D., De Palma, A., Gion, M., Toppo, S., Maiorino, M. and Ursini, F., 2008. Differential liquid phase proteomic analysis of the effect of selenium supplementation in LNCaP cells. Journal of Chromatography B 865: 63-73.

Schwarz, K. and Foltz, C.M., 1957. Selenium as an integral part of factor 3 against dietary liver degeneration. Journal of American Chemical Society 79: 3292-3293.

Schwarz, K. and Foltz, C.M., 1958. Factor 3 activity of selenium compounds. Journal of Biological Chemistry 233: 245-51.

Short, S.P., Pilat, J.M. and Williams, C.S., 2018. Roles for selenium and selenoprotein P in the development, progression, and prevention of intestinal disease. Free Radical Biology & Medicine 127: 26-35.

Speckmann, B. and Steinbrenner, H., 2014. Selenium and selenoproteins in inflammatory bowel diseases and experimental colitis. Inflammatory Bowel Disease 20: 1110-1119.

Speckmann, B., Bidmon, H.J., Pinto, A., Anlauf, M., Sies, H. and Steinbrenner, H., 2011. Induction of glutathione peroxidase 4 expression during enterocytic cell differentiation. Journal of Biological Chemistry 286: 10764-10772.

Surai, K.P., Surai, P.F., Speake, B.K. and Sparks, N.H.C., 2003. Antioxidant prooxidant balance in the intestine: food for thought. 1. Prooxidants. Nutritional Genomics and Functional Foods 1: 51-70.

Surai, K.P., Surai, P.F., Speake, B.K. and Sparks, N.H.C., 2004. Antioxidant prooxidant balance in the intestine: food for thought. 1. Antioxidants. Current Topics in Nutraceutical Research 2: 27-46.

Surai, P.F., 2002. Natural antioxidants in avian nutrition and reproduction. Nottingham University Press, Nottingham, UK.

Surai, P.F., 2006. Selenium in nutrition and health. Nottingham University Press, Nottingham, UK.

Surai, P.F., 2014. Polyphenol compounds in the chicken/animal diet: from the past to the future. Journal of Animal Physiology and Animal Nutrition 98: 19-31.

Surai, P.F., 2015. Silymarin as a natural antioxidant: an overview of the current evidence and perspectives. Antioxidants 4: 204-247.

Surai, P.F., 2018. Selenium in poultry nutrition and health. Wageningen Academic Publishers, Wageningen, the Netherland, 430 pp.

Surai, P.F., 2020a. Antioxidants in poultry nutrition and reproduction: an update. Antioxidants 9(2): 105.

Surai, P.F., 2020b. Vitagene in avian biology and poultry health. Wageningen Academic Publishers, Wageningen, the Netherlands, 544 pp.

Surai, P.F., 2021a. Selenium in sow nutrition. Chapter 5. In: Surai, P.F. (ed.) Selenium in pig nutrition and health. Wageningen Academic Publishers, Wageningen, the Netherlands, pp. 133-172. https://doi.org/10.3920/978-90-8686-912-1_5

Surai, P.F., 2021b. Selenium and immunity. Chapter 9. In: Surai, P.F. (ed.) Selenium in pig nutrition and health. Wageningen Academic Publishers, Wageningen, the Netherlands, pp. 261-347. https://doi.org/10.3920/978-90-8686-912-1_9

Surai, P.F. and Fisinin, V.I., 2015. Antioxidant-prooxidant balance in the intestine: applications in chick placement and pig weaning. Journal of Veterinary Science and Medicine 3: 1-16.

Surai, P.F. and Fisinin, V.I., 2016. Selenium in sow nutrition. Animal Feed Science and Technology 211: 18-30.

Surai, P.F. and Fisinin, V.I., 2021. Selenium in boar nutrition. Chapter 6. In: Surai, P.F. (ed.) Selenium in pig nutrition and health. Wageningen Academic Publishers, Wageningen, the Netherlands, pp. 173-213. https://doi.org/10.3920/978-90-8686-912-1_6

Surai, P.F. and Kochish, 2021. Selenium in piglet nutrition. Chapter 7. In: Surai, P.F. (ed.) Selenium in pig nutrition and health. Wageningen Academic Publishers, Wageningen, the Netherlands, pp. 215-233. https://doi.org/10.3920/978-90-8686-912-1_7

Surai, P.F., Kochish, I.I. and Fisinin, V.I., 2017. Antioxidant systems in poultry biology: Nutritional modulation of vitagenes. European Journal of Poultry Science 81: 1612-9199.

Surai, P.F., Kochish, I.I. and Fisinin, V.I., 2021. Selenium in feed: organic selenium concept development. Chapter 3. In: Surai, P.F. (ed.) Selenium in pig nutrition and health. Wageningen Academic Publishers, Wageningen, the Netherlands, pp. 61-110. https://doi.org/10.3920/978-90-8686-912-1_3

Tang, C., Li, S., Zhang, K., Li, J., Han, Y., Zhan, T., Zhao, Q., Guo, X. and Zhang, J., 2020. Selenium deficiency-induced redox imbalance leads to metabolic reprogramming and inflammation in the liver. Redox Biology 36: 101519.

Tang, J., Cao, L., Jia, G., Liu, G., Chen, X., Tian, G., Cai, J., Shang, H. and Zhao, H., 2019. The protective effect of selenium from heat stress-induced porcine small intestinal epithelial cell line (IPEC-J2) injury is associated with regulation expression of selenoproteins. The British Journal of Nutrition 122: 1081-1090.

Tyszka-Czochara, M., Pasko, P., Zagrodzki, P., Gajdzik, E., Wietecha-Posluszny, R. and Gorinstein, S., 2016. Selenium supplementation of amaranth sprouts influences betacyanin content and improves anti-inflammatory properties via NFκB in murine RAW 264.7 macrophages. Biological Trace Element Research 169: 320-330.

Wang, Y., Zhan, X., Zhang, X., Wu, R. and Yuan, D., 2011. Comparison of different forms of dietary selenium supplementation on growth performance, meat quality, selenium deposition, and antioxidant property in broilers. Biological Trace Element Research 143: 261-273.

Wrobel, J.K., Choi, J.J., Xiao, R., Eum, S.Y., Kwiatkowski, S., Wolff, G., Spangler, L., Power, R.F. and Toborek, M., 2015. Selenoglycoproteins attenuate adhesion of tumor cells to the brain microvascular endothelium via a process involving NF-κB activation. Journal of Nutritional Biochemistry 26: 120-129.

Wu, C., Xu, Z. and Huang, K., 2016. Effects of dietary selenium on inflammation and hydrogen sulfide in the gastrointestinal tract in chickens. Biological Trace Element Research 174: 428-435.

Xu, D., Li, W., Huang, Y., He, J. and Tian, Y., 2014a. The effect of selenium and polysaccharide of *Atractylodes macrocephala* Koidz. (PAMK) on immune response in chicken spleen under heat stress. Biological Trace Element Research 160: 232-237.

Xu, J., Gong, Y., Sun, Y., Cai, J., Liu, Q., Bao, J., Yang, J. and Zhang, Z., 2020. Impact of selenium deficiency on inflammation, oxidative stress, and phagocytosis in mouse macrophages. Biological trace element research 194(1): 237-243.

Xu, J.X., Cao, C.Y., Sun, Y.C., Wang, L.L., Li, N., Xu, S.W. and Li, J.L., 2014b. Effects on liver hydrogen peroxide metabolism induced by dietary selenium deficiency or excess in chickens. Biological Trace Element Research 159: 174-182.

Xu, J.X., Zhang, C., Cao, C.Y., Zhu, S.Y., Li, H., Sun, Y.C. and Li, J.L., 2016. Dietary selenium status regulates the transcriptions of selenoproteome and activities of selenoenzymes in chicken kidney at low or super-nutritional levels. Biological Trace Element Research 170: 438-448.

Xu, S., Lee, S.H., Lillehoj, H.S., Hong, Y.H. and Bravo, D., 2015b. Effects of dietary selenium on host response to necrotic enteritis in young broilers. Research in Veterinary Science 98: 66-73.

Xu, S.Z., Lee, S.H., Lillehoj, H.S. and Bravo, D., 2015a. Dietary sodium selenite affects host intestinal and systemic immune response and disease susceptibility to necrotic enteritis in commercial broilers. British Poultry Science 56: 103-112.

Yang, T., Zhao, Z., Liu, T., Zhang, Z., Wang, P., Xu, S., Lei, X.G. and Shan, A., 2017. Oxidative stress induced by Se-deficient high-energy diet implicates neutrophil dysfunction via Nrf2 pathway suppression in swine. Oncotarget 8: 13428-13439.

Yu, J., Yao, H., Gao, X., Zhang, Z., Wang, J.F. and Xu, S.W., 2015. The role of nitric oxide and oxidative stress in intestinal damage induced by selenium deficiency in chickens. Biological Trace Element Research 163: 144-153.

Yuan, D., Zhan, X.A. and Wang, Y.X., 2012. Effect of selenium sources on the expression of cellular glutathione peroxidase and cytoplasmic thioredoxin reductase in the liver and kidney of broiler breeders and their offspring. Poultry Science 91: 936-942.

Zamamiri-Davis, F., Lu, Y., Thompson, J.T., Prabhu, K.S., Reddy, P.V., Sordillo, L.M. and Reddy, C.C., 2002. Nuclear factor-κB mediates over-expression of cyclooxygenase-2 during activation of RAW 264.7 macrophages in selenium deficiency. Free Radical Biology and Medicine 32: 890-897.

Zhan, X., Qie, Y., Wang, M., Li, X. and Zhao, R.Q., 2011. Selenomethionine: an effective selenium source for sow to improve Se distribution, antioxidant status, and growth performance of pig offspring. Biological Trace Element Research 142: 481-49.

Zhang, Y., Cui, J., Lu, Y., Huang, C., Liu, H. and Xu, S., 2020a. Selenium deficiency induces inflammation via the iNOS/NF-κB pathway in the brain of pigs. Biological Trace Element Research 196: 103-109.

Zhang, Y., Zhang, J., Bao, J., Tang, C. and Zhang, Z., 2020b. Selenium deficiency induced necroptosis, Th1/Th2 imbalance, and inflammatory responses in swine ileum. Journal of Cellular Physiology236(1): 222-234. https://doi.org/10.1002/jcp.29836

Zhang, Z.W., Wang, Q.H., Zhang, J.L., Li, S., Wang, X.L. and Xu, S.W., 2012. Effects of oxidative stress on immunosuppression induced by selenium deficiency in chickens. Biological Trace Element Research 149: 352-361.

Zhao, H., Zhao, Z. and Lei, X.G., 2021. Function and regulation of porcine selenogenome and selenoproteome. Chapter 2. In: Surai, P.F. (ed.) Selenium in pig nutrition and health. Wageningen Academic Publishers, Wageningen, the Netherlands, pp. 39-59. https://doi.org/10.3920/978-90-8686-912-1_2

Zhao, J., Xing, H., Liu, C., Zhang, Z. and Xu, S., 2016. Effect of selenium deficiency on nitric oxide and heat shock proteins in chicken erythrocytes. Biological Trace Element Research 171: 208-213.

Zhao, X., Yao, H., Fan, R., Zhang, Z. and Xu, S., 2014. Selenium deficiency influences nitric oxide and selenoproteins in pancreas of chickens. Biological Trace Element Research 161: 341-349.